OPTICAL FIBER COMMUNICATIONS

McGraw-Hill Series in Electrical and Computer Engineering

SENIOR CONSULTING EDITOR

Stephen W. Director, University of Michigan, Ann Arbor

Circuits and Systems
Communications and Signal Processing
Computer Engineering
Control Theory and Robotics
Electromagnetics
Electronics and VLSI Circuits
Introductory
Power
Antennas, Microwaves, and Radar

PREVIOUS CONSULTING EDITORS

Communications and Signal Processing

SENIOR CONSULTING EDITOR

Stephen W. Director, University of Michigan, Ann Arbor

OPTICAL FIBER COMMUNICATIONS

THIRD EDITION

Gerd Keiser

GTE Systems and Technology Corporation

Boston   Burr Ridge, IL   Dubuque, IA   Madison, WI
New York   San Francisco   St. Louis   Bangkok   Bogotá
Caracas   Lisbon   London   Madrid   Mexico City   Milan
New Delhi   Seoul   Singapore   Sydney   Taipei   Toronto

McGraw-Hill Higher Education

A Division of The **McGraw-Hill** *Companies*

OPTICAL FIBER COMMUNICATIONS

4 5 6 7 8 9 0 DOC/DOC   0 9 8 7 6 5 4

ISBN 0-07-232101-6 (book)
ISBN 0-07-235680-4 (CD)
ISBN 0-07-236076-3 (book bound with CD)

Vice president/Editor in chief: *Kevin T. Kane*
Publisher: *Thomas Casson*
Executive editor: *Elizabeth A. Jones*
Sponsoring editor: *Catherine Fields*
Editorial assistant: *Michelle Flomenhoft*
Senior marketing manager: *John T. Wannemacher*
Project manager: *Christina Thornton-Villagomez*
Senior production supervisor: *Heather D. Burbridge*
Designer: *Sabrina Dupont*
Supplement coordinator: *Sue Lombardi*
New media: *Christopher Styles*
Compositor: *Interactive Composition Corp.*
Typeface: *10/12 Times Roman*
Printer: *R. R. Donnelley & Sons Inc.*

Library of Congress Cataloging-in-Publication Data
Keiser, Gerd.
    Optical fiber communications / Gerd Keiser.—3rd ed.
      p. cm.—(McGraw-Hill series in electrical engineering.
  Communications and signal processing)
    Includes bibliographical references.
    ISBN 0-07-232101-6
    1. Optical communications. 2. Fiber optics. I. Title.
  II. Series.
  TK5103.59.K44 2000
  621.382′75—dc21   99-37108

http://www.mhhe.com

To Ching-yun and Nishla

ABOUT THE AUTHOR

GERD KEISER has been involved with the research, development, and application of optical networking technology and digital switch development for 25 years. The majority of this time was spent at the GTE Systems and Technology Corporation. His current interest is the architectural design and implementation of high-performance telecommunication networks. As a side activity, he has served as an adjunct professor of electrical engineering at Northeastern University and Tufts University, and was an industrial advisor to the Wentworth Institute of Technology. He is a Fellow of the IEEE and is the author of the book *Local Area Networks* (McGraw-Hill). He received his B.A. and M.S. degrees in mathematics and physics from the University of Wisconsin in Milwaukee and a Ph.D. degree in solid-state physics from Northeastern University in Boston.

BRIEF CONTENTS

CONTENTS

Appendixes

PREFACE

TELECOMMUNICATION NETWORKS BASED on optical fiber technology have become a major information-transmission system, with high-capacity optical fiber links encircling the globe in both terrestrial and undersea installations. In the early days of optical fiber communications, the applications involved basically only the optical fiber, a light source, and a photodetector. Now, there are numerous passive and active optical devices within a light-wave link that perform complex networking functions in the optical domain, such as signal restoration, routing, and switching. Along with the need to understand the functions of these devices comes the necessity to measure both component and network performance, and to model and simulate the complex behavior of reliable high-capacity networks.

This book presents the fundamental principles for understanding and applying optical fiber technology to such sophisticated modern telecommunication systems. This text methodically examines the fundamental behavior of the individual optical components, describes their interactions with other devices in an optical fiber link, discusses the behavior of basic analog and digital optical links, and examines the performance characteristics of complex optical links and networks. Key features of the text for accomplishing this are as follows:

- A comprehensive treatment of the theory and behavior of basic constituents, such as optical fibers, light sources, photodetectors, connecting and coupling devices, and optical amplifiers.
- The basic design principles of digital and analog optical fiber transmission links.
- The operating principles of wavelength-division multiplexing (WDM) and the components needed for its realization.

- Descriptions of the architectures and performance characteristics of complex optical networks for connecting users who have a wide range of transmission needs.
- Discussions of advanced optical communication techniques, such as soliton transmission, optical code-division multiple access (optical CDMA), and ultra-fast optical time-division multiplexing (OTDM).
- An entire chapter devoted to measurement standards, basic test equipment, and techniques for verifying the operational characteristics of components in optical fiber communication links.
- A modeling and simulation program on a CD-ROM.

The modeling and simulation program on the CD-ROM is an abbreviated version of the *Photonic Transmission Design Suite*® (PTDS) from Virtual Photonics, Inc. This program is called PTDSlite, and is intended for student use. The software on the CD-ROM will allow students to examine the performance of key components (e.g., laser diodes, optical couplers, optical amplifiers, and photodetectors) and basic links consisting of these components. There are predefined parameter sets for each component, but, using Windows-based input screens, the user can vary any of these parameters (e.g., fiber length) and can turn them on and off to see their effect on link performance. This is a Windows-based program, so it can run on any standard PC that has the appropriate random-access memory and disk size.

This book provides the basic material for an introductory senior-level or graduate course in the theory and application of optical fiber communication technology. It will also serve well as a working reference for practicing engineers dealing with optical fiber communication system designs. The background required to study the book is that of typical senior-level engineering students. This includes introductory electromagnetic theory, calculus and elementary differential equations, basic concepts of optics as presented in a freshman physics course, and the basic concepts of electronics. To refresh readers' memories, concise reviews of several background topics, such as optics concepts, electromagnetic theory, and basic semiconductor physics, are included in the main body of the text. In this edition, various sections dealing with advanced material (e.g., the application of Maxwell's equations to cylindrical waveguides and the mathematical theory of optical receivers) are designated by a star and can be skipped over without loss of continuity. To aid readers in learning the material and applying it to practical designs, numerous examples are given throughout the book. A collection of 266 homework problems is included to help test the reader's comprehension of the material covered, and to extend and elucidate the text. Instructors can obtain the problem solutions from the publisher.

The original concept of this book is attributable to Tri T. Ha, Naval Postgraduate School, who urged me to write it when we were colleagues at GTE. His suggestions for the first two editions were most helpful. For this edition, I am greatly indebted to Ira Jacobs, Virginia Polytechnic Institute, and Don Nicholson, Syracuse University, for critical reviews of the manuscript and suggestions for enhancing and clarifying the material. In addition, I am grateful to Lian-kuan Chen, The Chinese University of Hong Kong; Walter Johnstone, University of Strathclyde, Scotland; and Winston I. Way, National Chiao-Tung University, Taiwan, for critical reviews of material in the newer chapters. Special thanks go to Dirk Seewald, Kay Iverson, Arthur Lowery (also of the University of Melbourne, Australia), and Stefan Georgi of Virtual Photonics for supplying the CD-ROM with the modeling tool and for reviewing the manuscript. Numerous people have helped directly or indirectly with various aspects of this book and its previous editions. Among them are Bert Basch, Joanne LaCourse, and Bill Nelson, GTE Laboratories; Sonia Bélanger, EXFO; C. T. Chang, San Diego State University; Emmanuel Desurvire, Alcatel Alsthom Researche, France; Paul Green, Jr., Tellabs; Katie Hall, MIT Lincoln Laboratory; Mark Jerabek, West Virginia University; Mohsen Kavehrad, University of Pennsylvania; Nishla Keiser, MIT; Arnie Michelson, GTE; Fred Quan, Corning; Don Rice, happily retired from GTE and Tufts University; Paul Schumate, Jr., Telcordia Technologies; Ramakant Srivastava, University of Florida; and Dan Yang, AFC Technologies. Particularly encouraging for doing the third edition were the many positive comments on the previous editions received from users and adapters at the numerous institutions worldwide. This edition especially benefited from the expert guidance of Betsy Jones, Michelle Flomenhoft, and Christina Thornton-Villagomez of McGraw-Hill. As a final personal note, I am grateful to my wife Ching-yun and my daughter Nishla for their patience and encouragement during the time I devoted to writing and revising this book.

Further information on new developments and reference material related to the book can be found on the following McGraw-Hill Web site for this book: http://www.mhhe.com/engcs/electrical/keiser/.

Gerd Keiser

CHAPTER
1

OVERVIEW OF OPTICAL FIBER COMMUNICATIONS

Ever since ancient times, one of the principal needs of people has been to communicate. This need created interest in devising communication systems for sending messages from one distant place to another. Many forms of communication systems have appeared over the years. The basic motivations behind each new form were either to improve the transmission fidelity, to increase the data rate so that more information could be sent, or to increase the transmission distance between relay stations. Before the nineteenth century, all communication systems operated at a very low information rate and involved only optical or acoustical means, such as signal lamps or horns. One of the earliest known optical transmission links, for example, was the use of a fire signal by the Greeks in the eighth century B.C. for sending alarms, calls for help, or announcements of certain events.[1]

The invention of the telegraph by Samuel F. B. Morse in 1838 ushered in a new epoch in communications—the era of electrical communications.[2] In the ensuing years, an increasingly larger portion of the electromagnetic spectrum, shown in Fig. 1-1, was utilized for conveying information from one place to another.[3] The reason for this trend is that, in electrical systems, the data are usually transferred over the communication channel by superimposing the information onto a sinusoidally varying electromagnetic wave, which is known as the *carrier*. At the destination, the information is removed from the carrier wave and processed as desired. Since the amount of information that can be transmitted is directly related to the frequency range over which the carrier operates, increasing

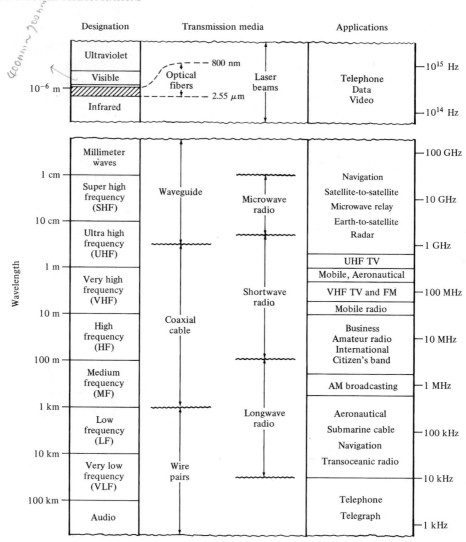

FIGURE 1-1
Examples of communication systems applications in the electromagnetic spectrum. (Used with permission from Carlson,[3] © 1986, McGraw-Hill Book Company.)

the carrier frequency theoretically increases the available transmission bandwidth and, consequently, provides a larger information capacity. Thus, the trend in electrical communication system developments was to employ progressively higher frequencies (shorter wavelengths), which offer corresponding increases in bandwidth or information capacity. This activity led to the birth of radio, television, radar, and microwave links.

Another important portion of the electromagnetic spectrum encompasses the optical region shown in Fig. 1-1. In contrast to electrical communications, transmission of information in an optical format is carried out not by frequency modulation of the carrier, but by varying the intensity of the optical power. Similar to the radio-frequency spectrum, two classes of transmission medium can be used: an atmospheric channel or a guided-wave channel. In optical systems it is customary to specify the band of interest in terms of wavelength, instead of in terms of frequency as used in the radio region. However, with the advent of high-speed multiple-wavelength systems in the mid-1990s, the output of optical sources started to be specified in terms of optical frequency. The reason for this is that in optical sources such as mode-locked semiconductor lasers, it is easier to control the frequency of the output light, rather than the wavelength, in order to tune the device to different emission regions. Of course, the different optical frequencies v are related to the wavelengths λ through the fundamental equation $c = v\lambda$. Thus, for example, a 1552.5-nm wavelength light signal has a frequency of 193.1 THz (193.1×10^{12} Hz). The optical spectrum ranges from about 50 nm (ultraviolet) to about 100 μm (far infrared), the visible region being the 400-to-700-nm band. This book addresses optical fiber communications, which operate in the 800-to-1600-nm wavelength band.

1.1 BASIC NETWORK INFORMATION RATES

The period of the 1990s saw a burgeoning demand on communication-network assets for services such as database queries and updates, home shopping, video-on-demand, remote education, telemedicine, and videoconferencing.[4-7] This demand was fueled by the rapid proliferation of personal computers (PCs), coupled with a phenomenal increase in their storage capacity and processing capabilities, the widespread availability of the Internet, and an extensive choice of remotely accessible programs and information databases. To handle the ever-increasing demand for high-bandwidth services from users ranging from home-based PCs to large businesses and research organizations, telecommunication companies worldwide are using light waves traveling within optical fibers as the dominant transmission system. This optical transmission medium consists of hair-thin glass fibers that guide the light signal over long distances.

Table 1-1 gives examples of information rates for some typical voice, video, and data services. To send these services from one user to another, network providers combine the signals from many different users and send the aggregate signal over a single transmission line. This scheme is known as *time-division multiplexing* (TDM). Here, N independent information streams, each running at a data rate of R b/s, are interleaved electrically into a single information stream operating at a higher rate of $N \times R$ b/s. To get a detailed perspective of this, let us look at the multiplexing schemes used in telecommunications.

Early applications of fiber optic transmission links were largely for trunking of telephone lines. These were digital links consisting of time-division-multiplexed 64-kb/s voice channels. Figure 1-2 shows the digital transmission hierarchy used in the North American telephone network. The fundamental building block is a

TABLE 1-1
Examples of information rates for some typical voice, video, and data services

Type of service	Data rate
Video on demand/interactive TV	1.5–6 Mb/s
Video games	1–2 Mb/s
Remote education	1.5–3 Mb/s
Electronic shopping	1.5–6 Mb/s
Data transfer or telecommuting	1–3 Mb/s
Videoconferencing	0.384–2 Mb/s
Voice (single channel)	64 kb/s

1.544-Mb/s transmission rate known as a T1 rate. It is formed by the time-division multiplexing of 24 voice channels, each digitized at a 64-kb/s rate. Framing bits are added along with these voice channels to yield the 1.544-Mb/s bit stream. At any level, a signal at the designated input rate is multiplexed with other input signals at the same rate.

The system is not restricted to multiplexing of voice signals. For example, at the T1 level, any 64-kb/s digital signal of the appropriate format could be transmitted as one of the 24 input channels shown in Fig. 1-2. As noted in Fig. 1-2 and Table 1-2, the multiplexed rates are designated as T1 (1.544 Mb/s), T2 (6.312 Mb/s), T3 (44.736 Mb/s), and T4 (274.176 Mb/s). Similar hierarchies using different bit-rate levels are employed in Europe and Japan, as Table 1-2 shows. The correct nomenclature used to describe generic digital systems is DS1, DS2, and DS3, where DS stands for *digital system*. In the strict sense, "DSx" refers to the framing format and other interface specifications, whereas "Tx" refers to the transmission medium over which the DSx signal is sent. However, the two designations are often used interchangeably.

With the advent of high-capacity fiber optic transmission lines in the 1980s, service providers established a standard signal format called *synchronous optical*

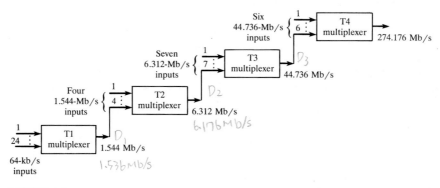

FIGURE 1-2
Digital transmission hierarchy used in the North American telephone network.

TABLE 1-2
Digital multiplexing levels used in North America, Europe, and Japan

Digital multiplexing level	Number of 64-kb/s channels	Bit rate (Mb/s)		
		North America	Europe	Japan
0	1	0.064	0.064	0.064
1	24	1.544		1.544
	30		2.048	
	48	3.152		3.152
2	96	6.312		6.312
	120		8.448	
3	480		34.368	32.064
	672	44.376		
	1344	91.053		
	1440			97.728
4	1920		139.264	
	4032	274.176		
	5760			397.200

network (SONET) in North America and *synchronous digital hierarchy* (SDH) in other parts of the world. These standards define a synchronous frame structure for sending multiplexed digital traffic over optical fiber trunk lines.[8] The basic building block and first level of the SONET signal hierarchy is called the *synchronous transport signal—level 1* (STS-1), which has a bit rate of 51.84 Mb/s. Higher-rate SONET signals are obtained by byte-interleaving N STS-1 frames, which are then scrambled and converted to an *optical carrier—level N* (OC-N) signal. Thus, the OC-N signal will have a line rate exactly N times that of an OC-1 signal. For SDH systems, the fundamental building block is the 155.52-Mb/s *synchronous transport module—level 1* (STM-1). Again, higher-rate information streams are generated by synchronously multiplexing N different STM-1 signals to form the STM-N signal. Table 1-3 shows commonly used SDH and SONET signal levels. Chapter 12 presents more details on this subject.

TABLE 1-3
Commonly used SONET and SDH transmission rates

SONET level	Electrical level	Line rate (Mb/s)	SDH equivalent
OC-1	STS-1	51.84	—
OC-3	STS-3	155.52	STM-1
OC-12	STS-12	622.08	STM-4
OC-24	STS 24	1244.16	STM-8
OC-48	STS-48	2488.32	STM-16
OC-96	STS-96	4976.64	STM-32
OC-192	STS-192	9953.28	STM-64

Beyond the use of fiber optics for telephone trunking lies an enormous world of both analog and digital applications. For example, by putting information in an asynchronous transfer mode (ATM) format, it is possible to transmit simultaneously both narrowband and broadband communication services, such as telephone, videoconferencing, video entertainment, digital imaging, and data on a single subscriber line. In particular, digital imaging is rapidly becoming a major information resource within the Internet.[9] This service can require tremendous bandwidth, which will be a challenge for transmission-link providers. Another key concept is the use of fiber optics for the *integrated services digital network* (ISDN). The ISDN scheme encompasses the ability of a digital communication network to handle voice (telephone), facsimile, data, videotex, telemetry, and broadcast audio and video services.[10,11] Transmission rates for these concepts vary from 155 Mb/s (SONET OC-3) for localized applications to around 10 Gb/s for high-capacity backbone trunks (SONET OC-192).[12,13]

1.2 THE EVOLUTION OF FIBER OPTIC SYSTEMS

The bit-rate–distance product BL, where B is the transmission bit rate and L is the repeater spacing, measures the transmission capacity of optical fiber links. Since the inception of optical fiber communications in 1974, their transmission capacity has experienced a 10-fold increase every 4 years. Several major technology advances spurred this growth. Figure 1-3 shows the operating range of optical fiber systems and the characteristics of the four key components of a link: the optical fiber, light sources, photodetectors, and optical amplifiers. In the figure, the dashed vertical lines indicate the centers of the three main operating windows of optical fiber systems, as detailed in Sec. 1.3. The first-generation links operated at around 850 nm, which was a low-loss transmission window of early silica fibers. These links used existing GaAs-based optical sources, silicon photodetectors, and multimode fibers. Intermodal dispersion and fiber loss limited the capacity of these systems. Some of the initial telephone-system field trials in the USA were carried out in 1977 by GTE in Los Angeles[14] and by AT&T in Chicago.[15] Links similar to these were demonstrated in Europe and Japan. Intercity applications ranged from 45 to 140 Mb/s with repeater spacings of around 10 km.

The development of optical sources and photodetectors capable of operating at 1300 nm allowed a shift in the transmission wavelength from 800 nm to 1300 nm. This resulted in a substantial increase in the repeaterless transmission distance for long-haul telephone trunks, since optical fibers exhibit lower power loss and less signal dispersion at 1300 nm. Intercity applications first used multimode fibers, but in 1984 switched exclusively to single-mode fibers, which have a significantly larger bandwidth. Bit rates for long-haul links typically range between 155 and 622 Mb/s (OC-3 and OC-12), and, in some cases, up to 2.5 Gb/s (OC-48), over repeater spacings of 40 km. Both multimode and single-mode 1300-nm fibers are used in local area networks, where bit rates range from 10 to 100 Mb/s over distances ranging from 500 m to tens of kilometers.[16,17]

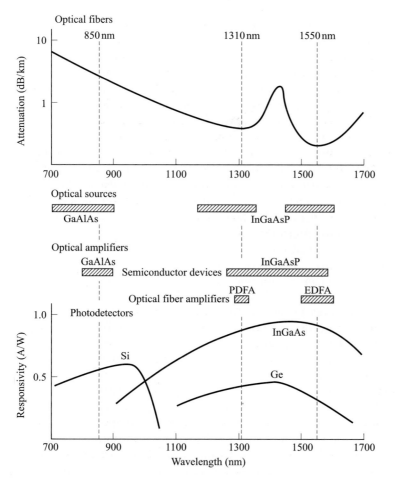

FIGURE 1-3

The operating range of optical fiber systems and the characteristics of the four key link components.

Systems operating at 1550 nm provide the lowest attention, but standard silica fibers have a much larger signal dispersion at 1550 nm than at 1300 nm. Fiber manufacturers overcame this limitation by creating the so-called dispersion-shifted fibers. Thus, 1550-nm systems attracted much attention for high-capacity long-span terrestrial and undersea transmission links.[18,19] These links routinely carry traffic at around 2.5 Gb/s over 90-km repeaterless distances. By 1996, advances in high-quality lasers and receivers allowed single-wavelength transmission rates of around 10 Gb/s (OC-192).

The introduction of optical amplifiers in 1989 gave a major boost to fiber transmission capacity. Although GaAlAs-based solid-state optical amplifiers appeared first,[20] the most successful and widely used devices are erbium-doped fiber amplifiers (commonly called EDFAs) operating at around 1550 nm.[21]

Praseodymium-doped fiber amplifiers (called PDFAs) operating at around 1300 nm have also been developed, but are still undergoing refinement.[22] During the same time period, impressive demonstrations of long-distance high-capacity systems were made using optical soliton signals.[23] A soliton is a nondispersive pulse that makes use of nonlinear dispersion properties in a fiber to cancel out chromatic dispersion effects. As an example, solitons at rates of 10 Gb/s have been sent over a 12,200-km experimental link using optical amplifiers and special modulation techniques.[24]

The use of wavelength-division multiplexing (WDM) offers a further boost in fiber transmission capacity.[25] The basis of WDM is to use multiple sources operating at slightly different wavelengths to transmit several independent information streams over the same fiber. Although researchers started looking at WDM in the 1970s, during the ensuing years it generally turned out to be easier to implement higher-speed electronic and optical devices than to invoke the greater system complexity called for in WDM. However, a dramatic surge in its popularity started in the early 1990s as electronic devices neared their modulation limit and high-speed equipment became increasingly complex. One example of the many worldwide installations of WDM networks is the SEA-ME-WE-3 cable system shown in Fig. 1-4. This undersea network runs from Germany to Singapore, connecting more than a dozen countries in between; hence the name SEA-ME-WE, which refers to Southeast Asia (SEA), the Middle East (ME), and Western Europe (WE). The network has two pairs of undersea fibers with a capacity of eight STM-16 wavelengths per fiber.

Starting in the mid-1990s, a combination of EDFAs and WDM was used to boost fiber capacity to even higher levels and to increase the transmission distance. A major system consideration in these superhigh links is to ensure that there is appropriate link and equipment redundancy, so that alternate paths are available in case of disruptions in communications caused by cable ruptures (e.g., caused by errant digging from a backhoe) or equipment failures at an intermediate node.[26-28] Such disruptions could otherwise have a devastating effect on a large group of users.

1.3 ELEMENTS OF AN OPTICAL FIBER TRANSMISSION LINK

An optical fiber transmission link comprises the elements shown in Fig. 1-5. The key sections are a transmitter consisting of a light source and its associated drive circuitry, a cable offering mechanical and environmental protection to the optical fibers contained inside, and a receiver consisting of a photodetector plus amplification and signal-restoring circuitry. Additional components include optical amplifiers, connectors, splices, couplers, and regenerators (for restoring the signal-shape characteristics). The cabled fiber is one of the most important elements in an optical fiber link, as we shall see in Chaps. 2 and 3. In addition to protecting the glass fibers during installation and service, the cable may contain

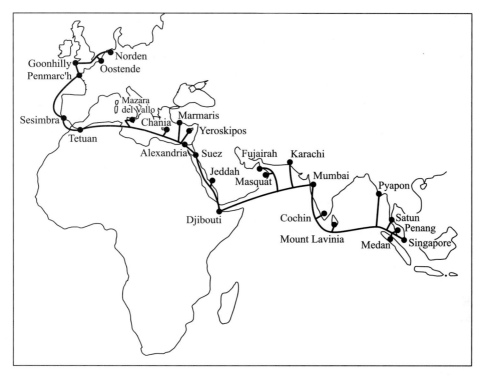

FIGURE 1-4
The SEA-ME-WE-3 undersea WDM cable network connects countries between Germany and Singapore. (Adapted with permission from Trischitta and Marra,[19] © 1998, IEEE.)

copper wires for powering optical amplifiers or signal regenerators, which are needed periodically for amplifying and reshaping the signal in long-distance links.

Analogous to copper cables, optical fiber cables can be installed either aerially, in ducts, undersea, or buried directly in the ground, as Fig. 1-6 illustrates. As a result of installation and/or manufacturing limitations, individual cable lengths will range from several hundred meters to several kilometers. Practical considerations such as reel size and cable weight determine the actual length of a single cable section. Shorter segments tend to be used when the cables are pulled through ducts. Longer lengths are used in aerial, direct-burial, or undersea applications. Splicing together individual cable sections forms continuous transmission lines for these long-distance links. For undersea installations, the splicing and repeater-installation functions are carried out on board a specially designed cable-laying ship.[18]

One of the principal characteristics of an optical fiber is its attenuation as a function of wavelength, as shown in Fig. 1-7. Early technology made exclusive use of the 800-to-900-nm wavelength band, since, in this region, the fibers made at that time exhibited a local minimum in the attenuation curve, and optical sources

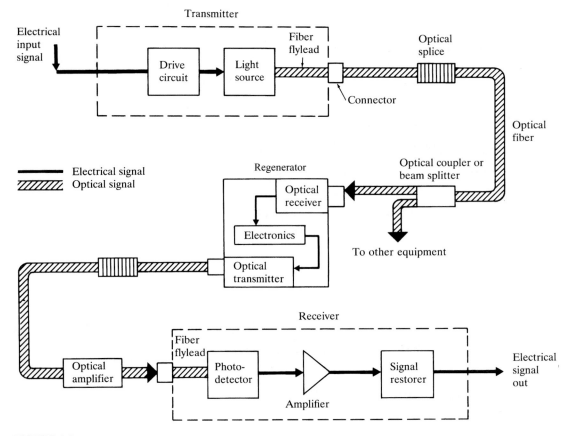

FIGURE 1-5

Major elements of an optical fiber transmission link. The basic components are the light signal transmitter, the optical fiber, and the photodetecting receiver. Additional elements include fiber and cable splices and connectors, regenerators, beam splitters, and optical amplifiers.

and photodetectors operating at these wavelengths were available. This region is referred to as the *first window*. By reducing the concentration of hydroxyl ions and metallic impurities in the fiber material, in the 1980s manufacturers were able to fabricate optical fibers with very low loss in the 1100-to-1600-nm region. This spectral band is referred to as the *long-wavelength region*. Two windows are defined here: the *second window*, centered around 1310 nm, and the *third window*, centered around 1550 nm.

In 1998 a new ultrahigh purifying process patented by Lucent Technologies eliminated virtually all water molecules from the glass fiber material. By dramatically reducing the water-attenuation peak around 1400 nm, this process opens the transmission region between the second and third windows to provide around 100 nm more bandwidth than in conventional single-mode fibers. This particular AllWave™ fiber, which was specifically designed for metropolitan networks, will

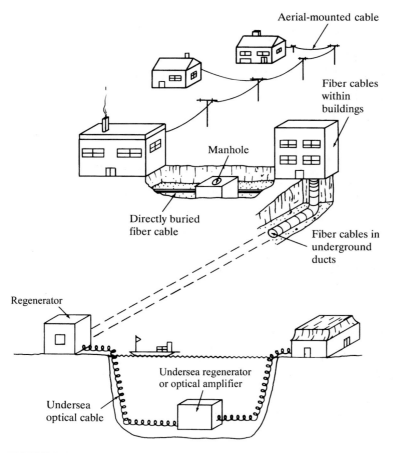

FIGURE 1-6
Optical fiber cables can be installed on poles, in ducts, and undersea, or they can be buried directly in the ground.

give local service providers the ability to deliver cost-effectively up to hundreds of optical wavelengths simultaneously.

Once the cable is installed, a light source that is dimensionally compatible with the fiber core is used to launch optical power into the fiber. Semiconductor light-emitting diodes (LEDs) and laser diodes are suitable for this purpose, since their light output can be modulated rapidly by simply varying the bias current at the desired transmission rate, thereby producing an optical signal. The electric input signals to the transmitter circuitry for the optical source can be of either analog or digital form. For high rate systems (usually greater than 1 Gb/s), direct modulation of the source can lead to unacceptable signal distortion. In this case, an *external modulator* is used to vary the amplitude of a continuous light output from a laser diode source. In the 800-to-900-nm region the light sources are generally alloys of GaAlAs. At longer wavelengths (1100–1600 nm), an

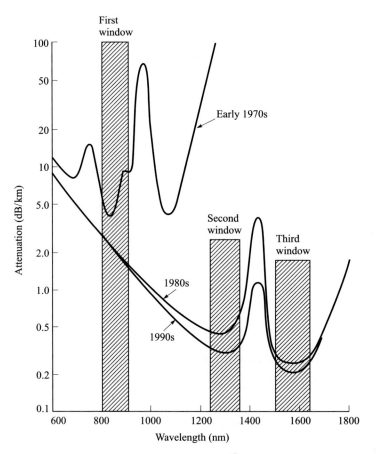

FIGURE 1-7
Optical fiber attenuation as a function of wavelength. Early fiber links exclusively used the 800-to-900-nm range. Achievement of lower attenuation shifted operations to the longer-wavelength windows around 1310 and 1550 nm.

InGaAsP alloy is the principal optical source material. Chapter 4 discusses optical sources in detail.

After an optical signal is launched into a fiber, it will become progressively attenuated and distorted with increasing distance because of scattering, absorption, and dispersion mechanisms in the glass material. At the receiver, a photodiode will detect the weakened optical signal emerging from the fiber end and convert it to an electric current (referred to as a *photocurrent*). Silicon photodiodes are used in the 800-to-900-nm region. The prime material in the 1100-to-1600-nm region is an InGaAs alloy. We address these photodetectors in Chap. 6.

The design of an optical receiver is inherently more complex than that of the transmitter, since it has to interpret the content of the weakened and degraded signal received by the photodetector. The principal figure of merit for a receiver is the minimum optical power necessary at the desired data rate to attain either a given error probability for digital systems or a specified signal-to-noise ratio for an analog system. As we shall see in Chap. 7, the ability of a receiver to achieve a certain performance level depends on the photodetector type, the effects of noise in the system, and the characteristics of the successive amplification stages in the receiver.

Chapter 8 presents the conditions that specify how far one can transmit a signal in a digital fiber optic system before an amplifier or a regenerator is required. In addition, the chapter addresses code formats used on optical data streams for improving the performance of digital links and discusses optically induced noise effects on system performance. Chapter 9 addresses similar considerations for analog systems.

An interesting and powerful aspect of an optical communication link is that many different wavelengths can be sent along a fiber simultaneously in the 1300-to-1600-nm spectrum. The technology of combining a number of wavelengths onto the same fiber is known as *wavelength-division multiplexing* or WDM. Figure 1-8 shows the basic WDM concept. Here, N independent optically formatted information streams, each transmitted at a different wavelength, are combined with an optical multiplexer and sent over the same fiber. Note that each of these streams could be at a different data rate. Each information stream maintains its individual data rate after being multiplexed with the other streams, and still operates at its unique wavelength. Conceptually, the WDM scheme is the same as frequency-division multiplexing (FDM) used in microwave radio and satellite systems. Chapter 10 presents the concepts and implementations of WDM.

Traditionally, when setting up an optical link, one formulates a power budget and adds repeaters when the path loss exceeds the available power margin. To amplify an optical signal with a conventional repeater, one performs photon-to-electron conversion, electrical amplification, retiming, pulse shaping, and then electron-to-photon conversion. Although this process works well for moderate-speed single-wavelength operation, it can be fairly complex and expensive for high-speed multiwavelength systems. Thus, a great deal of effort has been expended to develop all-optical amplifiers for the two long-wavelength transmission windows of optical fibers. Chapter 11 looks at the basic types of optical amplifiers and describes how they are used in optical communication networks.

Whereas Chap. 8 covers the performance features of point-to-point links, in which the optical fiber system serves as a simple connection between two sets of

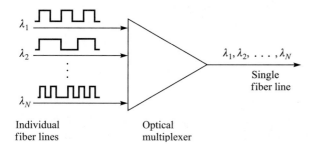

FIGURE 1-8
Optical multiplexing of N independent optical signals onto a single fiber.

electrical signal-processing equipment, Chap. 12 treats more complex networks. The architectures presented therein can be utilized in local-, metropolitan-, or wide-area networks to connect hundreds or thousands of users having as wide range of transmission capacities and speeds. As an example, Fig. 1-9 shows a conceptual SONET or SDH network providing a wealth of services ranging from broadband (high-speed) remote-computing processes to standard 64-kb/s telephone-related services. In this network, a high-speed OC-192 WDM-based optical fiber backbone provides multiple 10-Gb/s channels between cities or countries. The use of multiple wavelengths greatly increases the capacity, configuration flexibility, and growth potential of this backbone. Moderate-speed regional net-

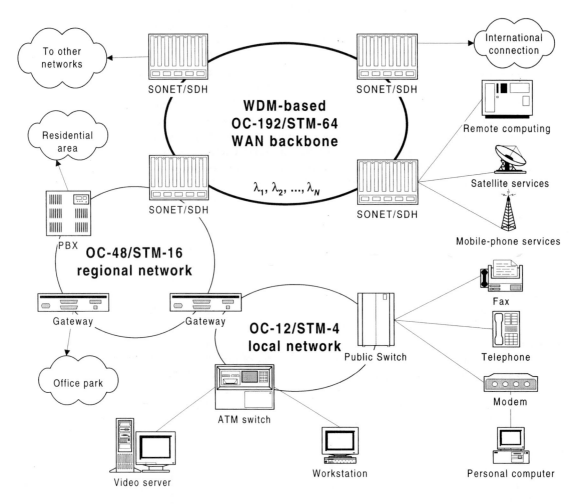

FIGURE 1-9
Conceptual SONET/SDH optical transport network connecting local, metropolitan, and wide-area communications elements.

works attached to this backbone provide applications such as interconnection of telephone switching centers, access to satellite transmission facilities, and access to mobile-phone base stations. More localized, lower-speed networks offer a wide variety of applications, such as telephony services to homes and businesses, distance learning, Internet access, community antenna television (CATV), security surveillance, and electronic mail. A major motivation for developing these sophisticated networks has been the rapid proliferation of information exchange desired by institutions such as commerce, finance, education, health, government, security, and entertainment. The potential for this information exchange arose from the ever-increasing power of computers and data-storage devices.

The design and installation of an optical fiber communication system require measurement techniques for verifying the operational characteristics of the constituent components. In addition to optical fiber parameters, system engineers are interested in knowing the characteristics of passive splitters, connectors, and couplers, and electro-optic components, such as sources, photodetectors, and optical amplifiers. Furthermore, when a link is being installed and tested, the operational parameters of interest include bit-error rate, timing jitter, and signal-to-noise ratio as indicated by the eye pattern. During actual operation, measurements are needed for maintenance and monitoring functions to determine factors such as fault locations in fibers and the status of remotely located optical amplifiers. Chapter 13 first outlines fiber optic component-testing standards and standards for evaluating system performance. This is followed by discussions of various measurement techniques for optical fiber communication systems.

1.4 SIMULATION AND MODELING TOOLS

Numerical methods have been employed for many years to predict and evaluate the behavior of individual fiber optic components,[29-36] links,[37-45] and networks.[46-54] Now with the increased complexity of optical links and networks, computer-based simulation and modeling tools that integrate component, link, and network functions can make the design process more efficient, cheaper, and faster. The rapid proliferation and increase in capabilities of personal computers has led to the development of many sophisticated simulation programs for these machines. These photonic design automation (PDA) tools are based on well-established numerical models. They can simulate factors such as connector losses due to geometric or position mismatches of fibers, efficiencies of coupling optical power from light sources into fibers, large- and small-signal behaviors of passive and active optical components, and the performance of complex optical networks. These software tools can also model devices such as waveguide couplers, optical filters, waveguide grating arrays, optical amplifiers, and optical sources.

This section gives an overview of PDA tools that can run on a personal computer or a desktop workstation. Included in this are discussions of the basic characteristics of the simulation tools and the graphical-programming languages they use. An important part of this section is a description of a powerful simulation and modeling tool called the *Photonic Transmission Design Suite* (PTDS) from Virtual Photonics, Inc.[55] This program is described in Sec. 1.4.3 and a student version is on the CD-ROM that is included in the back cover of this book. The reader can consult Web sites and texts, such as those listed in Refs. 55 and 56, for offerings from other PDA-tool vendors

1.4.1 Characteristics of Simulation and Modeling Tools

Computer-aided PDA tools can offer a powerful method to assist in analyzing the design of an optical component, circuit, or network before costly prototypes are built. Important points to consider, however, are the approximations and modeling assumptions made in the software design. Since most telecommunication systems are designed with several decibels of safety margin, approximations for calculating operating behavior that are reasonably accurate are not only acceptable but also, in general, necessary to allow tractable computation times.

The theoretical models used in computer simulations nominally include the following characteristics:[48]

- Enough detail so that all factors which could influence the performance of the component, circuit, or network can be appropriately evaluated.
- A common set of parameters so that simulated devices can be interconnected with each other to form circuits or networks.
- Interfaces that pass sufficient information between the constituent components so that all possible interactions are identified.
- Computational efficiency that allows a tradeoff between accuracy and speed, so that quick estimates of system performance can be made in the early stages of a design.
- The capability to simulate devices over the desired spectral bandwidth.
- The ability to simulate factors such as nonlinear effects, crosstalk between optical channels, distortion in lasers, and dispersion in optical fibers.

To enable a user to visualize and simulate a system quickly, the simulation programs normally have the following features:

- The ability to create a system schematic based on a library of graphical icons and a graphical user interface (GUI). The icons represent various system components (such as optical fibers, filters, amplifiers) and instrumentation (e.g., data sources, power meters, spectrum analyzers).
- The ability for the user to interact with the program during a simulation. For example, the user may want to modify a parameter or some operating condition in order to evaluate its effect. This is especially important in the early stages of a design when the operating range of interest is being established.
- A wide range of statistical-analysis, signal-processing, and display tools.

- Common display formats, including time waveforms, electrical and optical spectra, eye diagrams, and error-rate curves.

1.4.2 Programming Languages

Commercially available simulation tools are based on graphical-programming languages,[57] such as Ptolemy and LabVIEW®. In these languages, system components (e.g., lasers, modulators, optical amplifiers, optical fibers) are represented by a module library of programmed icons that have bidirectional or unidirectional optical and electrical interfaces. Associated with each icon is a menu window where the user specifies the value of the operational parameters of the component and its interface characteristics. In addition to using preprogrammed modules, users can create their own custom devices with either the underlying software code (such as C or C++) or the graphical-programming language.

The module library can be grouped into four major classes: transmitters, channels, receivers, and visualizers. Table 1-4 lists some of the icons and their characteristics that could be included in these classes. Using such a set of graphical icons, in several minutes one can easily put together a simulation of a more complex component, a simple link, or a sophisticated multichannel transmission path. One simply selects the icons representing the desired components and measurement instruments, and connects them together with a wiring tool to create a model of the optical transmission system. When the design is completed, the diagram compiles rapidly and the simulation can be run using control buttons on a tool bar.

Once the icons have been selected and connected together, the complex part starts for the user. This involves choosing realistic ranges of the parameters of the electrical and optical components and submodules. It is important that the parameter values make sense in an actual application. In some cases, this may entail examining the specification sheets of vendors.

Equally important is to keep in mind the modeling approach used in the simulation software. Two popular methods are the *transmission-line laser model* (TLLM)[58] and *split-step Fourier modeling*.[59,60] These are pseudo-spectral methods, which reduce computational efforts by eliminating the optical carrier frequency from the calculations. The TLLM solves bidirectional propagation problems by using scattering matrices at discrete points along the propagation path. The solution is carried out in the time domain. The inherent advantage is that the information flow in the modeled solution echoes the propagation of optical waves. Thus, propagation delays are realistically represented so that devices in which pulses of energy travel along a cavity are correctly modeled.

Split-step Fourier modeling is a numerical approach in which derivatives are calculated in the frequency domain and multiplications are carried out in the time domain. Its key advantage is that all cross-coupling effects can be properly taken into account. Its main disadvantage is that if certain factors, such has the sampling rate, are not specified properly it can be very time-consuming. This is

TABLE 1-4

Four major classes of simulation-tool icons and some of the functional characteristics that could be included in these classes

Icon group/type	Characteristics of module or function
Transmitter	
Laser	Device may be modeled using rate equations that include emission intensity, phase noise, laser chirp, and polarization. Alternatively, a simplified analytical model might be used.
Encoder	Describes digital coding schemes.
Pulse generator	Pseudo-random bit streams and specification of pulse shapes.
Modulator	Specification of lithium-niobate external modulator parameters.
Channel	
Monomode fiber	Specification of liner and nonlinear effects, such as attenuation, chromatic dispersion, polarization-mode dispersion, self- and cross-phase modulation, four-wave mixing, Raman scattering.
Optical amplifiers	EDFAs and semiconductor amplifiers: unidirectional or bidirectional pumping schemes, reflection levels.
Optical filters	Optical passband and other characteristics of filter types such as Fabry-Perot, Mach-Zehnder, fiber Bragg-grating, or arrayed-waveguide.
Frequency converters	Use of semiconductor optical amplifiers for wavelength conversion at nodal points in a network.
Passive devices	Optical attenuators, splitters, combiners, connectors.
Receiver	
Photodiode	*pin* or avalanche detectors
Preamplifier	Characteristics of EDFA preamplifiers.
Filters	Optical and electrical filters in the receiver.
Samplers	Specification of different ways to calculate the bit-error rate.
Visualizer	
Instruments	Data can be sent to icons representing instruments such as spectrum analyzers, bit-error rate testers, power meters.
Graphs of data	Visualization of the calculated data using time or frequency representations, eye diagrams, histograms, etc.
3-D plots	Optimization problems with several variables can be represented as three-dimensional data plots or contour plots.

particularly critical when the performance of a large number of WDM channels is being evaluated. For example, whereas a single channel can be simulated in a few minutes, simulation of 32 channels with a rigorous computation method can consume hours of central processing unit (CPU) time and 100 channels could require days, unless approximations or semianalytical methods are used. A variety of strategies have been investigated for reducing this computation time and in some cases are incorporated into the simulation software. The popularity of split-step modeling is that it can be used to design fiber and pulse parameters that minimize intersymbol interference and mitigate the appearance of noise factors such as four-wave mixing. In addition, it creates accurate eye diagrams, and tradeoffs between modeled bandwidth and accuracy are well understood. Note that, generally, any cautionary details on how to properly specify the simulation parameters are described in the software user manuals.

1.4.3   The PTDS® Simulation and Modeling Tool

As an example of commercial software, Virtual Photonics, Inc., has a simulation and modeling tool called the *Photonic Transmission Design Suite*® (PTDS). The PTDS can be used for designing either a photonic component to be used in communication systems, a complete optical communication link or system, or an optical network. For component design, PTDS has a comprehensive library of detailed physical models of photonic devices and a graphical user interface to combine them easily. Examples of photonic components that can be modeled include optical amplifiers, transmitters, wavelengths converters and routers, optical TDM components, all-optical switches, wavelength multiplexers, and optical cross-connect devices. In addition, PTDS has an electro-optical co-design feature that includes RF, analog, and digital signal processing (DSP) modeling and simulation capabilities. The system-design tool of PTDS enables the user to simulate a wide range of network architectures: for example, WDM, optical-TDM, and fiber-network-based CATV systems.

When simulating the performance of photonic networks, the PTDS optical network simulation layer (ONSL) controls data exchange between PTDS and other user-defined or third-party simulation tools. When checking the performance of a design, a simulation manager enables intuitive and user-friendly parameter sweeps. For these parameter sweeps, a visualization tool offers interactive displays similar to a digital scope or spectrum analyzer.

The PTDS offers a vertically integrated design process with layered simulation technologies. Thus, it can work independently at the component, transmission-link, and transport-network simulation layers, and can pass the design abstractions and analyses done at a particular layer to adjacent layers. For example, the user can design a laser with sophisticated equations and then pass the characteristic of the laser as a simplified model to the higher-up transmission-link layer. After having incorporated an optical amplifier model into the link, the link layer can then pass the baseband response of the transmission path to the network layer. In this way, network architects can optimize the higher layers efficiently without losing the underlying contributions that the lower layers make to the network. In addition, in a collaborative design effort, users working on one layer can transfer their results to colleagues working on adjacent layers with sufficient detail to be meaningful.

An abbreviated version of this tool for student use is included on the CD-ROM at the back of this book. The CD-ROM contains the working software, instructions for its use, on-line documentation, installation instructions, and licensing conditions for a full-featured version of the software. This is a Windows-based program, so it can run on any standard PC that has the appropriate random-access memory and disk size. More details are in Sec. 1.5.2.

1.5   USE AND EXTENSION OF THE BOOK

The following chapters present an introduction to the field of optical fiber communications. The sequence of topics systematically takes the reader from descriptions of the major building blocks to analyses of complete links and networks, together with methodologies for measuring and simulating their performance. The material presented here will provide a broad and firm basis with which to analyze and design optical fiber communication systems.

As is the case for any active scientific and engineering discipline, optical fiber technology is constantly undergoing changes and improvements. New concepts are being pursued in optical multiplexing, all-optical networking, integrated optics, and device configurations; new materials are being introduced for fibers, sources, and photodetectors; and maintenance and operation methodologies for optical networks are being developed. These changes should not have a major impact on the material presented in this book, since it is based on enduring fundamental concepts. The understanding of these concepts will allow a rapid comprehension of any new technological developments that will undoubtedly arise.

1.5.1   Reference Material

Numerous references are provided at the end of each chapter as a start for delving deeper into any given topic. Since optical fiber communications brings together research and development efforts from many different scientific and engineering disciplines, there are hundreds of articles in the literature relating to the material covered in each chapter. Even though not all of these articles can be cited in the references, the selections represent some of the major contributions to the fiber optics field and can be considered as a good introduction to the literature. Additional references for up-to-date developments can be found in various conference proceedings.[61-64] Further supplementary material can be found in specialized textbooks.[21,23,59,60,65-80]

To help the reader understand and use the material in the book, the table in the inside front cover provides a quick reference for various physical constants and units, and Apps. A through E give an overview of the international system of units, listing of mathematical formulas needed for homework problems, discussions on decibels, and topics from communication theory. Appendix F shows a derivation of pulse broadening or signal dispersion in an optical fiber based on examining the frequency dependence of the propagation constant in the wave equation.

1.5.2   Simulation Program on a CD-ROM

The abbreviated version of the PTDS software on the CD-ROM will allow students to simulate the performance of key components (e.g., laser diodes, optical couplers, optical amplifiers, and photodetectors) and of basic links consisting of these components. There are parameter sets predefined for each component, which generally represent commercially available devices. Using Windows-based input screens, the user can vary any of these parameters (e.g., fiber length) and can turn certain specific parameters on and off to see their effect on link performance. The results can be stored in a file from where the data can be read by any standard spread sheet or plotting routine for graphical display of the results. This interactive process will give the reader a better understanding of the material presented in this book and will indicate the potential of the full simulation and modeling program (available from Virtual Photonics, Inc.).

1.5.3 Photonics Laboratory

In addition to classroom learning, exposure to the operating characteristics of components, links, and instruments in a laboratory course can greatly enhance the understanding of optical fiber technology. Various universities have implemented such courses, and equipment specifically designed for classroom use is available.[81,82]

1.5.4 Web-Based Resources

A Web site for the book is available at http://www.mhhe.com/engcs/electrical/keiser. This Web site includes instructor and student resources, corrections or revisions to the text, contact information, links to other related Web sites, and updates on new technical developments or standards. The site will be updated periodically with material such as suggestions for new homework problems, design projects, or modeling and simulation ideas; descriptions of recent technology or implementation developments; and a listing of web addresses of vendors offering components, transmission equipment, and measurement instruments. The Web site also contains a mechanism for readers to provide feedback and suggestions to the author.

REFERENCES

1. V. Aschoff, "Optische Nachrichtenübertragung im klassischen Altertum," *Nachrichtentechn. Z. (NTZ)*, vol. 30, pp. 23–28, Jan. 1977.
2. H. Busignies, "Communication channels," *Sci. Amer.*, vol. 227, pp. 99–113, Sept. 1972.
3. A. B. Carlson, *Communication Systems*, McGraw-Hill, New York, 3rd ed., 1986.
4. C.-J. L. van Driel, P. A. M. van Grinsven, V. Pronk, and W. A. M. Snijders, "The (r)evolution of access networks for the information superhighway," *IEEE Commun. Mag.*, vol. 35, pp. 104–112, June 1997.
5. G. Hill, L. Fernandez, and R. Cadeddu, "Building the road to optical networks," *Brit. Telecomm. Eng.*, vol. 16, pp. 2–12, Apr. 1997.
6. J. Höller, "Voice and telephony networking over ATM," *Ericsson Rev.*, vol. 75, no. 1, pp. 40–45, 1998.
7. N. Thorsen, *Fiber Optics and the Telecommunication Explosion*, Prentice Hall, New York, 1998.
8. C. A. Siller, Jr. and M. Shafi, eds., *SONET/SDH*, IEEE Press, New York, 1996.

9. H. S. Stone, "Image libraries and the Internet," *IEEE Commun. Mag.*, vol. 37, pp. 99–106, Jan. 1999.

10. J. Griffiths, *ISDN Explained*, Wiley, New York, 2nd ed., 1992.

11. K. Asatani, *Introduction to B-ISDN*, Wiley, New York, 1997.

12. D. J. H. Maclean, *Optical Line Systems*, Wiley, Chichester, UK, 1996. This book gives a detailed discussion of the evolution of optical fiber links and networks.

13. D. H. Rice and G. E. Keiser, "Applications of fiber optics to tactical communication systems," *IEEE Commun. Mag.*, vol. 23, pp. 46–57, May 1985.

14. E. E. Basch, R. A. Beaudette, and H. A. Carnes, "Optical transmission for interoffice trunks," *IEEE Trans. Commun.*, vol. COM-26, pp. 1007–1014, July 1978.

15. M. I. Schwartz, W. A. Reenstra, J. H. Mullins, and J. S. Cook, "Chicago lightwave communications project," *Bell Sys. Tech. J.*, vol. 57, pp. 1881–1888, July–Aug. 1978.

16. G. E. Keiser, *Local Area Networks*, McGraw-Hill, New York, 1989.

17. S. Saunders, *The McGraw-Hill High-Speed LANs Handbook*, McGraw-Hill, New York, 1996.

18. Special Issue on "Undersea Communications Technology," *AT&T Tech. J.*, vol. 74, Jan./Feb. 1995.

19. P. R. Trischitta and W. C. Marra, "Applying WDM technology to undersea cable networks," *IEEE Commun. Mag.*, vol. 36, pp. 62–66, Feb. 1998.

20. Y. Yamamoto and T. Mukai, "Fundamentals of optical amplifiers," *Fiber & Integrated Optics*, vol. 21, Special Issue on "Optical Amplifiers," pp. S1–S14, 1989.

21. E. Desurvire, *Erbium-Doped Fiber Amplifiers*, Wiley, New York, 1994.

22. T. J. Whitley, "A review of recent system demonstrations incorporating 1.3-mm praseodymium-doped fluoride fiber amplifiers," *J. Lightwave Tech.*, vol. 13, pp. 744–760, May 1995.

23. A. Hasegawa and Y. Kodama, *Solitons in Optical Communications*, Clarendon, Oxford, 1995.

24. M. Suzuki, N. Edagawa, H. Taga, S. Yamamoto, and S. Akiba, "10 Gb/s over 12200 km soliton data transmission with alternating-amplitude solitons," *IEEE Photonics Tech. Lett.*, vol. 6, pp. 757–759, 1994.

25. G. E. Keiser, "A review of WDM technology and applications," *Opt. Fiber Technol*, vol. 5, pp. 3–39, Jan. 1999.

26. K. Sato, "Photonic transport network OAM technologies," *IEEE Commun. Mag.*, vol. 34, pp. 86–94, Dec. 1996.

27. C.-S. Li and R. Ramaswami, "Automatic fault detection, isolation, and recovery in transparent all-optical networks." *J. Lightwave Tech.*, vol. 15, pp. 1784–1793, Oct. 1997.

28. Y. Hamazumi and M. Koga, "Transmission capacity of optical path overhead transfer scheme using pilot tone for optical path network," *J. Lightwave Tech.*, vol. 15, pp. 2197–2205, Dec. 1997.

29. A. J. Lowery, P. C. R. Gurney, X.-H. Wang, L. V. T. Nguyen, Y.-C. Chan, and M. Premarante, "Time-domain simulation of photonic devices, circuits, and systems," *SPIE Proc. Physics and Simulation of Optoelectronic Devices*, vol. 2693, pp. 624–635, Feb. 1996.

30. X. J. M. Leijtens, P. Le Louree, and M. K. Smit, "S-matrix oriented CAD tool for simulating complex integrated optical circuits," *IEEE J. Sel. Topics Quantum Electron.*, vol. 2, pp. 257–262, June 1996.

31. B. P. C. Tsou and D. L. Pulfrey, "A versatile SPICE model for quantum-well lasers," *IEEE J. Quantum Electron.*, vol. 33, pp. 246–254, Feb. 1997.

32. D. S. Ellis and J. M. Xu, "Electro-opto-thermal modeling of threshold current dependence on temperature," *IEEE J. Sel. Topics Quantum Electron.*, vol. 3, pp. 640–648, Apr. 1997.

33. W. R. Smith, J. R. King, and B. Tuck, "Mathematical modeling of thermal effects in semiconductor laser operation," *IEEE Proc.—Optoelectron.*, vol. 144, pp. 389–396, Dec. 1997.

34. J. K. Morikuni and S.-M. Kang, *Computer-Aided Design of Optoelectronic Integrated Circuits and Systems*, Prentice Hall, New York, 1997.

35. G. R. Hadley, "Low-truncation error finite difference equations for photonics simulation I: Beam propagation," *J. Lightwave Tech.*, vol. 16, pp. 134–141, Jan. 1998.

36. G. R. Hadley, "Low-truncation error finite difference equations for photonics simulation II: Vertical-cavity surface-emitting lasers," *J. Lightwave Tech.*, vol. 16, pp. 142–151, Jan. 1998.

37. W. H. Tranter and C. R. Ryan, "Simulation of communication systems using personal computers," *IEEE J. Sel. Areas Commun.*, vol. 6, pp. 13–23, Jan. 1988.

38. A. F. Elrefaie, J. K. Townsend, M. B. Romeiser, and K. S. Shanmugan, "Computer simulation of digital lightwave links," *IEEE J. Sel. Areas Commun.*, vol. 6, pp. 94–105, Jan. 1988.

39. M. K. Moaveni and M. Shafi, "A statistical design approach for gigabit-rate fiber optic transmission systems," *J. Lightwave Tech.*, vol. 8, pp. 1064–1072, July 1990.

40. F. R. Shapiro, "The use of a spreadsheet for sinusoidal steady-state transmission line and optics problems," *IEEE Trans. Education*, vol. 36, pp. 269–272, May 1993.

41. D. B. Mortimore, "Optical network simulation software for physical layer design and development," *BT Technol. J.,* vol. 14, pp. 161–167, July 1996.

42. R. A. A. Lima, M. C. R. Carvalho, and L. F. M. Conrado, "On the simulation of digital optical links with EDFAs: An accurate method for estimating BER through Gaussian approximation," *IEEE J. Sel. Topics Quantum Electron.*, vol. 3, pp. 1037–1044, Aug. 1997.

43. M. R. N. Ribeiro, H. Waldman, J. Klein, and L. S. de Souza-Mendes, "Error-rate patterns for the modeling of optically amplified transmission systems," *IEEE J. Sel. Areas Commun.*, vol. 15, pp. 707–716, May 1997.

44. B. K. Whitlock, P. K. Pepeljugoski, D. M. Kuchta, J. D. Crow, and S.-M. Kang, "Computer modeling and simulation of the Optoelectronic Technology Consortium (OETC) optical bus," *IEEE J. Sel. Areas Commun.*, vol. 15, pp. 717–730, May 1997.

45. A. Carena, V. Curri, R. Gaudino, P. Poggiolini, and S. Benedetto, "A time-domain optical transmission system simulation package accounting for nonlinear and polarization-related effects in fiber," *IEEE J. Sel. Areas Commun.*, vol. 15, pp. 751–765, May 1997.

46. J. M. Senior, D. E. Asumu, and T. Finegan, "Software package for designing/planning fiber networks in the subscriber loop," *Fiber Integr. Optics*, vol. 15, no. 1, pp. 45–61, 1996.

47. P. Bell, K. Cobb, and J. Peacock, "Optical power budget calculating tool for fibre in the access network," *BT Technol. J.,* vol. 14, pp. 116–120, Apr. 1996.

48. A. J. Lowery, "Computer-aided photonics design," *IEEE Spectrum*, vol. 34, pp. 26–31, Apr. 1997.

49. N. Antoniades, I. Roudas, R. E. Wagner, and S. F. Habiby, "Simulation of ASE noise accumulation in a wavelength add-drop multiplexer cascade," *IEEE Photonics Tech. Lett.*, vol. 9, pp. 1274–1276, Sept. 1997.

50. A. J. Lowery and P. C. Gurney, "Two simulators for photonic computer-aided design," *Appl. Opt.*, vol. 37, pp. 6066–6077, Sept. 1998.

51. E. Iannone, F. Matera, A. Mecozzi, and M. Settembre, *Nonlinear Optical Communication Networks*, App. A1, Wiley, New York, 1998.

52. M. A. Olabe and J. C. Olabe, "Telecommunications network design using modeling and simulation," *IEEE Trans. Education*, vol. 41, pp. 37–44, Feb. 1998.

53. D. Seewald and A. J. Lowery, "Versatile simulations plot of photonic future," *Fibre Systems*, vol. 2, pp. 31–32, Dec. 1998.

54. H. Hamster and J. Lam, "PDA—challenges for an emerging industry," *Lightwave*, vol. 15 (*see* http://www.light-wave.com), Aug. 1998.

55. The following are examples of organizations that offer software programs for modeling lightwave components and systems:
    (*a*) Virtual Photonics, Inc., Berlin, Germany (http://www.virtualphotonics.com).
    (*b*) IPSIS, Cesson-Sevigne, France (e-mail: telecom@ipsis.com).
    (*c*) Artis Software Corp., Torino, Italy (http://www.artis-software.com).
    (*d*) RSoft, Inc., Ossining, New York (http://www.rsoftinc.com).
    (*e*) BBV Software, Enschede, The Netherlands (http://www.bbv-software.com).
    (*f*) Apollo Photonics, Inc., Ontario, Canada (http://www.ApolloPhoton.com).

56. P. C. Becker, N. A. Olsson, and J. R. Simpson, *Erbium-Doped Fiber Amplifiers*, Academic, New York, 1999.

57. (*a*) LabVIEW® is a trademark of National Instruments Corporation.
    (*b*) Ptolemy was developed at the University of California at Berkeley in 1989, and has become a quasi-standard for complex system simulations.

58. A. J. Lowery, "A new dynamic semiconductor laser model based on the transmission-line modeling method," *IEEE Proc. J.—Optoelectron.*, vol. 134, pp. 281–289, 1987.
59. G. P. Agrawal, *Nonlinear Fiber Optics*, Academic, New York, 2nd ed., 1995.
60. E. Iannone, F. Matera, A. Mecozzi, and M. Settembre, *Nonlinear Optical Communication Networks*, Wiley, New York, 1998.
61. *Optical Fiber Communications (OFC) Conference*, cosponsored annually by the Optical Society of America (OSA), Washington, DC and the Institute of Electrical and Electronic Engineers (IEEE), New York.
62. *International Conference on Communications (ICC)*, sponsored annually by the IEEE.
63. *Optoelectronic and Fiber Optic Devices and Applications Symposium*, sponsored annually by SPIE—The International Society for Optical Engineering, Bellingham, WA.
64. *European Conference on Optical Fibre Communication (ECOC)*, held annually in Europe; sponsored by various European engineering organizations.
65. D. Gloge, *Optical Fiber Technology*, IEEE Press, New York, 1976. This book contains reprints of fundamental research articles that appeared in the literature from 1969 to 1975.
66. C. K. Kao, *Optical Fiber Technology, II*, IEEE Press, New York, 1980. This book contains reprints of basic research articles that appeared in the literature from 1976 to 1979.
67. S. E. Miller and A. G. Chynoweth, eds., *Optical Fiber Telecommunications*, Academic, New York, 1979. This book and the texts listed in Refs. 68 and 69 contain dozens of topics in all areas of optical fiber technology presented by researchers from AT&T Bell Laboratories over a period of more than 20 years.
68. S. E. Miller and I. P. Kaminow, eds., *Optical Fiber Telecommunications—II*, Academic, New York, 1988.
69. I. P. Kaminow and T. L. Koch, eds., *Optical Fiber Telecommunications—III*, vols. A and B, Academic, New York, 1997.
70. J. M. Senior, *Optical Fiber Communications*, Prentice Hall, Englewood Cliffs, NJ, 2nd ed., 1992.
71. P. E. Green, Jr., *Fiber Optic Networks*, Prentice Hall, Englewood Cliffs, NJ, 1993.
72. D. J. G. Mestdagh, *Fundamentals of Multiaccess Fiber Optic Networks*, Artech House, Boston, 1995.
73. G. Einarsson, *Principles of Lightwave Communications*, Wiley, New York, 1995.
74. L. Kazovsky, S. Benedetto, and A. E. Willner, *Optical Fiber Communication Systems*, Artech House, Boston, 1996.
75. B. Mukherjee, *Optical Communication Networks*, McGraw-Hill, New York, 1997.
76. M. Weik, *Fiber Optics Standard Dictionary*, Chapman & Hall, New York, 1997.
77. G. P. Agrawal, *Fiber Optic Communication Systems*, Wiley, New York, 2nd ed., 1997.
78. J. Powers, *An Introduction to Fiber Optic Systems*, Irwin, Chicago, 2nd ed., 1997.
79. J. C. Palais, *Fiber Optics Communications*, Prentice Hall, New York, 4th ed., 1998.
80. R. Ramaswami and K. N. Sivarajan, *Optical Networks*, Morgan Kaufmann, San Francisco, 1998.
81. B. L. Anderson, L. J. Pelz, S. A. Ringel, B. D. Clymer, and S. A. Collins, Jr., "Photonics laboratory with emphasis on technical diversity," *IEEE Trans. Education*, vol. 41, pp. 194–202, Aug. 1998.
82. OptoSci Ltd., 141 St James Road, Glasgow G4 0LT, Scotland (http://ourworld.compuserve.com/homepages/optosci).

CHAPTER 2

OPTICAL FIBERS: STRUCTURES, WAVEGUIDING, AND FABRICATION

One of the most important components in any optical fiber system is the optical fiber itself, since its transmission characteristics play a major role in determining the performance of the entire system. Some of the questions that arise concerning optical fibers are

1. What is the structure of an optical fiber?
2. How does light propagate along a fiber?
3. Of what materials are fibers made?
4. How is the fiber fabricated?
5. How are fibers incorporated into cable structures?
6. What is the signal loss or attenuation mechanism in a fiber?
7. Why and to what degree does a signal get distorted as it travels along a fiber?

The purpose of this chapter is to present some of the fundamental answers to the first five questions in order to attain a good understanding of the physical structure and waveguiding properties of optical fibers. Questions 6 and 7 are answered in Chap. 3.

Since fiber optics technology involves the emission, transmission, and detection of light, we begin our discussion by first considering the nature of light and then we shall review a few basic laws and definitions of optics.[1] Following a description of the structure of optical fibers, two methods are used to describe how an optical fiber guides light. The first approach uses the geometrical or ray optics concept of light reflection and refraction to provide an intuitive picture of the propagation mechanisms. In the second approach, light is treated as an electromagnetic wave which propagates along the optical fiber waveguide. This involves solving Maxwell's equations subject to the cylindrical boundary conditions of the fiber.

2.1 THE NATURE OF LIGHT

The concepts concerning the nature of light have undergone several variations during the history of physics. Until the early seventeenth century, it was generally believed that light consisted of a stream of minute particles that were emitted by luminous sources. These particles were pictured as travelling in straight lines, and it was assumed that they could penetrate transparent materials but were reflected from opaque ones. This theory adequately described certain large-scale optical effects, such as reflection and refraction, but failed to explain finer-scale phenomena, such as interference and diffraction.

The correct explanation of diffraction was given by Fresnel in 1815. Fresnel showed that the approximately rectilinear propagation character of light could be interpreted on the assumption that light is a wave motion, and that the diffraction fringes could thus be accounted for in detail. Later, the work of Maxwell in 1864 theorized that light waves must be electromagnetic in nature. Furthermore, observation of polarization effects indicated that light waves are transverse (i.e., the wave motion is perpendicular to the direction in which the wave travels). In this *wave* or *physical optics viewpoint*, the electromagnetic waves radiated by a small optical source can be represented by a train of spherical wave fronts with the source at the center as shown in Fig. 2-1. A *wave front* is defined as the locus of all

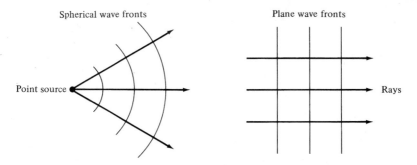

FIGURE 2-1
Representations of spherical and plane wave fronts and their associated rays.

points in the wave train which have the same phase. Generally, one draws wave fronts passing through either the maxima or the minima of the wave, such as the peak or trough of a sine wave, for example. Thus, the wave fronts (also called *phase fronts*) are separated by one wavelength.

When the wavelength of the light is much smaller than the object (or opening) which it encounters, the wave fronts appear as straight lines to this object or opening. In this case, the light wave can be represented as a *plane wave*, and its direction of travel can be indicated by a *light ray* which is drawn perpendicular to the phase front. Thus, large-scale optical effects such as reflection and refraction can be analyzed by the simple geometrical process of *ray tracing*. This view of optics is referred to as *ray* or *geometrical optics*. The concept of light rays is very useful because the rays show the direction of energy flow in the light beam.

2.1.1 Linear Polarization

The electric or magnetic field of a train of *plane linearly polarized waves* traveling in a direction **k** can be represented in the general form

$$\mathbf{A}(\mathbf{x}, t) = \mathbf{e}_i A_0 \exp[j(\omega t - \mathbf{k} \cdot \mathbf{x})] \tag{2-1}$$

with $\mathbf{x} = x\mathbf{e}_x + y\mathbf{e}_y + z\mathbf{e}_z$ representing a general position vector and $\mathbf{k} = k_x\mathbf{e}_x + k_y\mathbf{e}_y + k_z\mathbf{e}_z$ representing the wave propagation vector.

Here, A_0 is the maximum amplitude of the wave, $\omega = 2\pi\nu$, where ν is the frequency of the light; the magnitude of the wavevector **k** is $k = 2\pi/\lambda$, which is known as the *wave propagation constant*, with λ being the wavelength of the light; and $\mathbf{e}_i$ is a unit vector lying parallel to an axis designated by i.

Note that the components of the actual (measurable) electromagnetic field represented by Eq. (2-1) are obtained by taking the real part of this equation. For example, if $\mathbf{k} = k\mathbf{e}_z$, and if **A** denotes the electric field **E** with the coordinate axes chosen such that $\mathbf{e}_i = \mathbf{e}_x$, then the real measurable electric field is given by

$$\mathbf{E}_x(z, t) = \text{Re}(\mathbf{E}) = \mathbf{e}_x E_{0x} \cos(\omega t - kz) \tag{2-2}$$

which represents a plane wave that varies harmonically as it travels in the z direction. The reason for using the exponential form is that it is more easily handled mathematically than equivalent expressions given in terms of sine and cosine. In addition, the rationale for using harmonic functions is that any waveform can be expressed in terms of sinusoidal waves using Fourier techniques.

The electric and magnetic field distributions in a train of plane electromagnetic waves at a given instant in time are shown in Fig. 2-2. The waves are moving in the direction indicated by the vector **k**. Based on Maxwell's equations, it is easily shown[2] that **E** and **H** are both perpendicular to the direction of propagation. Such a wave is called a *transverse wave*. Furthermore, **E** and **H** are mutually perpendicular, so that **E**, **H**, and **k** form a set of orthogonal vectors.

The plane wave example given by Eq. (2-2) has its electric field vector always pointing in the $\mathbf{e}_x$ direction. Such a wave is *linearly polarized* with polarization vector $\mathbf{e}_x$. A general state of polarization is described by considering another

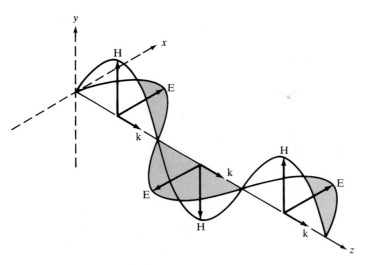

FIGURE 2-2
Electric and magnetic field distributions in a train of plane electromagnetic waves at a given instant in time.

linearly polarized wave which is independent of the first wave and orthogonal to it. Let this wave be

$$\mathbf{E}_y(z, t) = \mathbf{e}_y E_{0y} \cos(\omega t - kz + \delta) \tag{2-3}$$

where δ is the relative phase difference between the waves. The resultant wave is then simply

$$\mathbf{E}(z, t) = \mathbf{E}_x(z, t) + \mathbf{E}_y(z, t) \tag{2-4}$$

If δ is zero or an integer multiple of 2π, the waves are in phase. Equation (2-4) is then also a linearly polarized wave with a polarization vector making an angle

$$\theta = \arctan \frac{E_{0y}}{E_{0x}} \tag{2-5}$$

with respect to $\mathbf{e}_x$ and having a magnitude

$$E = (E_{0x}^2 + E_{0y}^2)^{1/2} \tag{2-6}$$

This case is shown schematically in Fig. 2-3. Conversely, just as any two orthogonal plane waves can be combined into a linearly polarized wave, an arbitrary linearly polarized wave can be resolved into two independent orthogonal plane waves that are in phase.

2.1.2   Ellipitical and Circular Polarization

For general values of δ the wave given by Eq. (2-4) is *elliptically polarized*. The resultant field vector $\mathbf{E}$ will both rotate and change its magnitude as a function of

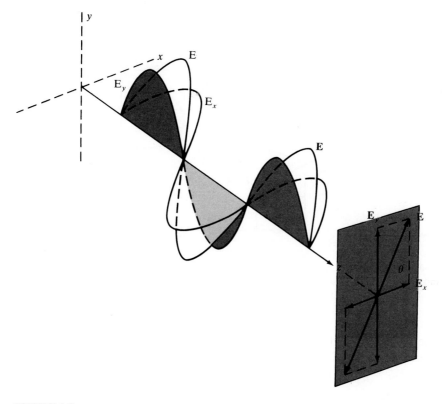

FIGURE 2-3
Addition of two linearly polarized waves having zero relative phase between them.

the angular frequency ω. From Eqs. (2-2) and (2-3) we can show that (see Prob. 2-5) for a general value of δ,

$$\left(\frac{E_x}{E_{0x}}\right)^2 + \left(\frac{E_y}{E_{0y}}\right)^2 - 2\left(\frac{E_x}{E_{0x}}\right)\left(\frac{E_y}{E_{0y}}\right)\cos\delta = \sin^2\delta \qquad (2\text{-}7)$$

which is the general equation of an ellipse. Thus, as Fig. 2-4 shows, the endpoint of **E** will trace out an ellipse at a given point in space. The axis of the ellipse makes an angle α relative to the x axis given by

$$\tan 2\alpha = \frac{2E_{0x}E_{0y}\cos\delta}{E_{0x}^2 - E_{0y}^2} \qquad (2\text{-}8)$$

To get a better picture of Eq. (2-7), let us align the principal axis of the ellipse with the x axis. Then $\alpha = 0$, or, equivalently, $\delta = \pm\pi/2, \pm 3\pi/2, \ldots$, so that Eq. (2-7) becomes

$$\left(\frac{E_x}{E_{0x}}\right)^2 + \left(\frac{E_y}{E_{0y}}\right)^2 = 1 \qquad (2\text{-}9)$$

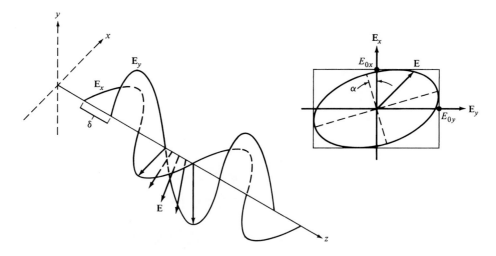

FIGURE 2-4
Elliptically polarized light results from the addition of two linearly polarized waves of unequal amplitude having a nonzero phase difference δ between them.

This is the familiar equation of an ellipse with the origin at the center and semi-axes E_{0x} and E_{0y}.

When $E_{0x} = E_{0y} = E_0$ and the relative phase difference $\delta = \pm\pi/2 + 2m\pi$, where $m = 0, \pm1, \pm2, \ldots$, then we have *circularly polarized* light. In this case, Eq. (2-9) reduces to

$$E_x^2 + E_y^2 = E_0^2 \tag{2-10}$$

which defines a circle. Choosing the positive sign for δ, Eqs. (2-2) and (2-3) become

$$\mathbf{E}_x(z, t) = \mathbf{e}_x E_0 \cos(\omega t - kz) \tag{2-11}$$

$$\mathbf{E}_y(z, t) = -\mathbf{e}_y E_0 \sin(\omega t - kz) \tag{2-12}$$

In this case, the endpoint of **E** will trace out a circle at a given point in space, as Fig. 2-5 illustrates. To see this, consider an observer located at some arbitrary point z_{ref} toward whom the wave is moving. For convenience, we will pick this point at $z = \pi/k$ at $t = 0$. Then, from Eqs. (2-11) and (2-12) we have

$$\mathbf{E}_x(z, t) = -\mathbf{e}_x E_0 \quad \text{and} \quad \mathbf{E}_y(z, t) = 0$$

so that **E** lies along the negative x axis as Fig. 2-5 shows. At a later time, say $t = \pi/2\omega$, the electric field vector has rotated through 90° and now lies along the positive y axis at z_{ref}. Thus, as the wave moves toward the observer with increasing time, the resultant electric field vector **E** rotates *clockwise* at an angular frequency ω. It makes one complete rotation as the wave advances through one wavelength. Such a light wave is *right circularly polarized.*

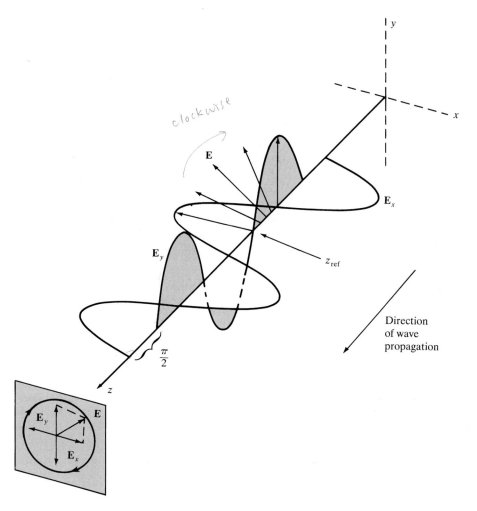

FIGURE 2-5
Addition of two equal-amplitude linearly polarized waves with a relative phase difference $\delta = \pi/2 + 2m\pi$ results in a right circularly polarized wave.

If we choose the negative sign for δ, then the electric field vector is given by

$$\mathbf{E} = E_0[\mathbf{e}_x \cos(\omega t - kz) + \mathbf{e}_y \sin(\omega t - kz)] \qquad (2\text{-}13)$$

Now **E** rotates *counterclockwise* and the wave is *left circularly polarized.*

2.1.3 The Quantum Nature of Light

The wave theory of light adequately accounts for all phenomena involving the transmission of light. However, in dealing with the interaction of light and matter, such as occurs in dispersion and in the emission and absorption of light, neither the particle theory nor the wave theory of light is appropriate. Instead, we must turn to quantum theory, which indicates that optical radiation has particle as well as wave properties. The particle nature arises from the observation that light energy is always emitted or absorbed in discrete units called *quanta* or *photons*. In all experiments used to show the existence of photons, the photon energy is found to depend only on the frequency v. This frequency, in turn, must be measured by observing a wave property of light.

The relationship between the energy E and the frequency v of a photon is given by

$$E = hv \qquad (2\text{-}14)$$

where $h = 6.625 \times 10^{-34}$ J $\cdot$ s is Planck's constant. When light is incident on an atom, a photon can transfer its energy to an electron within this atom, thereby exciting it to a higher energy level. In this process either all or none of the photon energy is imparted to the electron. The energy absorbed by the electron must be exactly equal to that required to excite the electron to a higher energy level. Conversely, an electron in an excited state can drop to a lower state separated from it by an energy hv by emitting a photon of exactly this energy.

2.2 BASIC OPTICAL LAWS AND DEFINITIONS

We shall next review some of the basic optical laws and definitions relevant to optical fibers. A fundamental optical parameter of a material is the *refractive index* (or *index of refraction*). In free space a light wave travels at a speed $c = 3 \times 10^8$ m/s. The speed of light is related to the frequency v and the wavelength λ by $c = v\lambda$. Upon entering a dielectric or nonconducting medium the wave now travels at a speed v, which is characteristic of the material and is less than c. The ratio of the speed of light in a vacuum to that in matter is the index of refraction n of the material and is given by

$$n = \frac{c}{v} \qquad (2\text{-}15)$$

Typical values of n are 1.00 for air, 1.33 for water, 1.50 for glass, and 2.42 for diamond.

The concepts of reflection and refraction can be interpreted most easily by considering the behavior of light rays associated with plane waves traveling in a dielectric material. When a light ray encounters a boundary separating two different media, part of the ray is reflected back into the first medium and the remainder is bent (or refracted) as it enters the second material. This is shown in Fig. 2-6 where $n_2 < n_1$. The bending or refraction of the light ray at the interface is a result of the difference in the speed of light in two materials that have different refractive indices. The relationship at the interface is known as Snell's law and is given by

$$n_1 \sin \phi_1 = n_2 \sin \phi_2 \qquad (2\text{-}16)$$

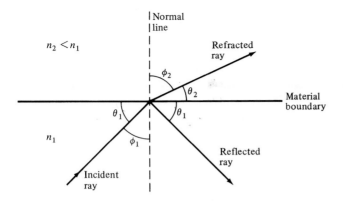

FIGURE 2-6
Refraction and reflection of a light ray at a material boundary.

or, equivalently, as

$$n_1 \cos \theta_1 = n_2 \cos \theta_2 \qquad (2\text{-}17)$$

where the angles are defined in Fig. 2-6. The angle ϕ_1 between the incident ray and the normal to the surface is known as the *angle of incidence*.

According to the law of reflection, the angle θ_1 at which the incident ray strikes the interface is exactly equal to the angle that the reflected ray makes with the same interface. In addition, the incident ray, the normal to the interface, and the reflected ray all lie in the same plane, which is perpendicular to the interface plane between the two materials. This is called the *plane of incidence*. When light traveling in a certain medium is reflected off an optically denser material (one with a higher refractive index), the process is referred to as *external reflection*. Conversely, the reflection of light off of less optically dense material (such as light traveling in glass being reflected at a glass–air interface) is called *internal reflection*.

As the angle of incidence ϕ_1 in an optically denser material becomes larger, the refracted angle ϕ_2 approaches $\pi/2$. Beyond this point no refraction is possible and the light rays become *totally internally reflected*. The conditions required for total internal reflection can be determined by using Snell's law [Eq. (2-16)]. Consider Fig. 2-7, which shows a glass surface in air. A light ray gets bent toward the glass surface as it leaves the glass in accordance with Snell's law. If the angle of incidence ϕ_1 is increased, a point will eventually be reached where the light ray in air is parallel to the glass surface. This point is known as the *critical angle of incidence* ϕ_c. When the incidence angle ϕ_1 is greater than the critical angle, the condition for total internal reflection is satisfied; that is, the light is totally reflected back into the glass with no light escaping from the glass surface. (This is an idealized situation. In practice, there is always some tunneling of optical energy through the interface. This can be explained in terms of the electromagnetic wave theory of light, which is presented in Sec. 2.4.)

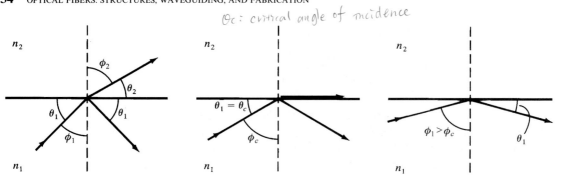

θ_c: critical angle of incidence

FIGURE 2-7
Representation of the critical angle and total internal reflection at a glass–air interface.

As an example, consider the glass–air interface shown in Fig. 2-7. When the light ray in air is parallel to the glass surface, then $\phi_2 = 90°$ so that $\sin \phi_2 = 1$. The critical angle in the glass is thus

$$\sin \phi_c = \frac{n_2}{n_1} \tag{2-18}$$

Example 2-1. Using $n_1 = 1.50$ for glass and $n_2 = 1.00$ for air, ϕ_c is about 52°. Any light in the glass incident on the interface at an angle ϕ_1 greater than 52° is totally reflected back into the glass.

In addition, when light is totally internally reflected, a phase change δ occurs in the reflected wave. This phase change depends on the angle $\theta_1 < \pi/2 - \phi_c$ according to the relationships[1]

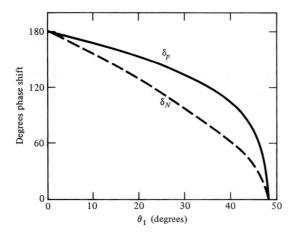

FIGURE 2-8
Phase shifts occurring from the reflection of wave components normal (δ_N) and parallel (δ_p) to the plane of incidence.

$$\tan \frac{\delta_N}{2} = \frac{\sqrt{n^2 \cos^2 \theta_1 - 1}}{n \sin \theta_1} \qquad (2\text{-}19a)$$

$$\tan \frac{\delta_p}{2} = \frac{n\sqrt{n^2 \cos^2 \theta_1 - 1}}{\sin \theta_1} \qquad (2\text{-}19b)$$

Here, δ_N and δ_p are the phase shifts of the electric-field wave components normal and parallel to the plane of incidence, respectively, and $n = n_1/n_2$. These phase shifts are shown in Fig. 2-8 for a glass–air interface ($n = 1.5$ and $\phi_c = 52°$). The values range from zero immediately at the critical angle to $\pi/2 - \phi_c$ when $\phi_c = 90°$.

These basic optical principles will now be used to illustrate how optical power is transmitted along a fiber.

2.3 OPTICAL FIBER MODES AND CONFIGURATIONS

Before going into details on optical fiber characteristics in Sec. 2.3.3, we first present a brief overview of the underlying concepts of optical fiber modes and optical fiber configurations.

2.3.1 Fiber Types

An optical fiber is a dielectric waveguide that operates at optical frequencies. This fiber waveguide is normally cylindrical in form. It confines electromagnetic energy in the form of light to within its surfaces and guides the light in a direction parallel to its axis. The transmission properties of an optical waveguide are dictated by its structural characteristics, which have a major effect in determining how an optical signal is affected as it propagates along the fiber. The structure basically establishes the information-carrying capacity of the fiber and also influences the response of the waveguide to environmental perturbations.

The propagation of light along a waveguide can be described in terms of a set of guided electromagnetic waves called the *modes* of the waveguide. These guided modes are referred to as the *bound* or *trapped* modes of the waveguide. Each guided mode is a pattern of electric and magnetic field distributions that is repeated along the fiber at equal intervals. Only a certain discrete number of modes are capable of propagating along the guide. As will be seen in Sec. 2.4, these modes are those electromagnetic waves that satisfy the homogeneous wave equation in the fiber and the boundary condition at the waveguide surfaces.

Although many different configurations of the optical waveguide have been discussed in the literature,[3] the most widely accepted structure is the single solid dielectric cylinder of radius a and index of refraction n_1 shown in Fig. 2-9. This cylinder is known as the *core* of the fiber. The core is surrounded by a solid dielectric *cladding* which has a refractive index n_2 that is less than n_1. Although, in principle, a cladding is not necessary for light to propagate along the core of the

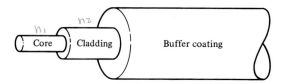

FIGURE 2-9
Schematic of a single-fiber structure. A circular solid core of refractive index n_1 is surrounded by a cladding having a refractive index $n_2 < n_1$. An elastic plastic buffer encapsulates the fiber.

fiber, it serves several purposes. The cladding reduces scattering loss that results from dielectric discontinuities at the core surface, it adds mechanical strength to the fiber, and it protects the core from absorbing surface contaminants with which it could come in contact.

In low- and medium-loss fibers the core material is generally glass and is surrounded by either a glass or a plastic cladding. Higher-loss plastic-core fibers with plastic claddings are also widely in use. In addition, most fibers are encapsulated in an elastic, abrasion-resistant plastic material. This material adds further strength to the fiber and mechanically isolates or buffers the fibers from small geometrical irregularities, distortions, or roughnesses of adjacent surfaces. These perturbations could otherwise cause scattering losses indicated by random microscopic bends that can arise when the fibers are incorporated into cables or supported by other structures.

Variations in the material composition of the core give rise to the two commonly used fiber types shown in Fig. 2-10. In the first case, the refractive index of the core is uniform throughout and undergoes an abrupt change (or step) at the cladding boundary. This is called a *step-index fiber*. In the second case, the core refractive index is made to vary as a function of the radial distance from the center of the fiber. This type is a *graded-index fiber*.

Both the step- and the graded-index fibers can be further divided into single-mode and multimode classes. As the name implies, a single-mode fiber sustains only one mode of propagation, whereas multimode fibers contain many hundreds of modes. A few typical sizes of single- and multimode fibers are given in Fig. 2-10 to provide an idea of the dimensional scale. Multimode fibers offer several advantages compared with single-mode fibers. As we shall see in Chap. 5, the larger core radii of multimode fibers make it easier to launch optical power into the fiber and facilitate the connecting together of similar fibers. Another advantage is that light can be launched into a multimode fiber using a light-emitting-diode (LED) source, whereas single-mode fibers must generally be excited with laser diodes. Although LEDs have less optical output power than laser diodes (as we shall discuss in Chap. 4), they are easier to make, are less expensive, require less complex circuitry, and have longer lifetimes than laser diodes, thus making them more desirable in certain applications.

A disadvantage of multimode fibers is that they suffer from intermodal dispersion. We shall describe this effect in detail in Chap. 3. Briefly, intermodal dispersion can be described as follows. When an optical pulse is launched into a fiber, the optical power in the pulse is distributed over all (or most) of the modes

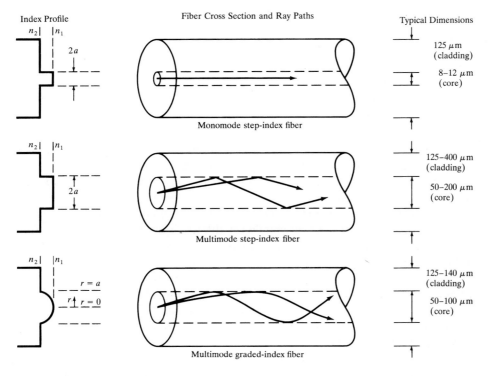

Index Profile

Fiber Cross Section and Ray Paths

Typical Dimensions

n_2| |n_1

2a

Monomode step-index fiber

125 μm
(cladding)

8–12 μm
(core)

n_2| |n_1

2a

Multimode step-index fiber

125–400 μm
(cladding)

50–200 μm
(core)

n_2| |n_1

$r = a$

r↑ $r = 0$

Multimode graded-index fiber

125–140 μm
(cladding)

50–100 μm
(core)

FIGURE 2-10
Comparison of single-mode and multimode step-index and graded-index optical fibers.

of the fiber. Each of the modes that can propagate in a multimode fiber travels at a slightly different velocity. This means that the modes in a given optical pulse arrive at the fiber end at slightly different times, thus causing the pulse to spread out in time as it travels along the fiber. This effect, which is known as *intermodal dispersion* or *intermodal distortion*, can be reduced by using a graded-index profile in a fiber core. This allows graded-index fibers to have much larger bandwidths (data rate transmission capabilities) then step-index fibers. Even higher bandwidths are possible in single-mode fibers, where intermodal dispersion effects are not present.

2.3.2 Rays and Modes

The electromagnetic light field that is guided along an optical fiber can be represented by a superposition of bound or trapped modes. Each of these guided modes consists of a set of simple electromagnetic field configurations. For mono- chromatic light fields of radian frequency ω, a mode traveling in the positive z direction (i.e., along the fiber axis) has a time and z dependence given by

$$e^{j(\omega t - \beta z)}$$

The factor β is the z component of the wave propagation constant $k = 2\pi/\lambda$ and is the main parameter of interest in describing fiber modes. For guided modes, β can assume only certain discrete values, which are determined from the requirement that the mode field must satisfy Maxwell's equations and the electric and magnetic field boundary conditions at the core–cladding interface. This is described in detail in Sec. 2.4.

Another method for theoretically studying the propagation characteristics of light in an optical fiber is the geometrical optics or ray-tracing approach. This method provides a good approximation to the light acceptance and guiding properties of optical fibers when the ratio of the fiber radius to the wavelength is large. This is known as the *small-wavelength limit*. Although the ray approach is strictly valid only in the zero-wavelength limit, it is still relatively accurate and extremely valuable for nonzero wavelengths when the number of guided modes is large; that is, for multimode fibers. The advantage of the ray approach is that, compared with the exact electromagnetic wave (modal) analysis, it gives a more direct physical interpretation of the light propagation characteristics in an optical fiber.

Since the concept of a light ray is very different from that of a mode, let us see qualitatively what the relationship is between them. (The mathematical details of this relationship are beyond the scope of this book but can be found in the literature.[4-6]) A guided mode traveling in the z direction (along the fiber axis) can be decomposed into a family of superimposed plane waves that collectively form a standing-wave pattern in the direction transverse to the fiber axis. That is, the phases of the plane waves are such that the envelope of the collective set of waves remains stationary.[2b] Since with any plane wave we can associate a light ray that is perpendicular to the phase front of the wave, the family of plane waves corresponding to a particular mode forms a set of rays called a *ray congruence*. Each ray of this particular set travels in the fiber at the same angle relative to the fiber axis. We note here that, since only a certain number M of discrete guided modes exist in a fiber, the possible angles of the ray congruences corresponding to these modes are also limited to the same number M. Although a simple ray picture appears to allow rays at any angle greater than the critical angle to propagate in a fiber, the allowable quantized propagation angles result when the phase condition for standing waves is introduced into the ray picture. This is discussed further in Sec. 2.3.5.

Despite the usefulness of the approximate geometrical optics method, a number of limitations and discrepancies exist between it and the exact modal analysis. An important case is the analysis of single-mode or few-mode fibers, which must be dealt with by using electromagnetic theory. Problems involving coherence or interference phenomena must also be solved with an electromagnetic approach. In addition, a modal analysis is necessary when a knowledge of the field distribution of individual modes is required. This arises, for example, when analyzing the excitation of an individual mode or when analyzing the coupling of power between modes at waveguide imperfections (which we shall discuss in Chap. 3).

Another discrepancy between the ray optics approach and the modal analysis occurs when an optical fiber is uniformly bent with a constant radius of curvature. As we shall show in Chap. 3, wave optics correctly predicts that every mode of the curved fiber experiences some radiation loss. Ray optics, on the other hand, erroneously predicts that some ray congruences can undergo total internal reflection at the curve and, consequently, can remain guided without loss.

2.3.3 Step-Index Fiber Structure

We begin our discussion of light propagation in an optical waveguide by considering the step-index fiber. In practical step-index fibers the core of radius a has a refractive index n_1 which is typically equal to 1.48. This is surrounded by a cladding of lightly lower index n_2, where

$$n_2 = n_1(1 - \Delta) \tag{2-20}$$

The parameter Δ is called the *core-cladding index difference* or simply the *index difference*. Values of n_2 are chosen such that Δ is nominally 0.01. Typical values range from 1 to 3 percent for multimode fibers and from 0.2 to 1.0 percent for single-mode fibers. Since the core refractive index is larger than the cladding index, electromagnetic energy at optical frequencies is made to propagate along the fiber waveguide through internal reflection at the core–cladding interface.

2.3.4 Ray Optics Representation

Since the core size of multimode fibers is much larger than the wavelength of the light we are interested in (which is approximately $1\,\mu$m), an intuitive picture of the propagation mechanism in an ideal multimode step-index optical waveguide is most easily seen by a simple ray (geometrical) optics representation.[6-11] For simplicity, in this analysis we shall consider only a particular ray belonging to a ray congruence which represents a fiber mode. The two types of rays that can propagate in a fiber are meridional rays and skew rays. *Meridional rays* are confined to the meridian planes of the fiber, which are the planes that contain the axis of symmetry of the fiber (the core axis). Since a given meridional ray lies in a single plane, its path is easy to track as it travels along the fiber. Meridional rays can be divided into two general classes: bound rays that are trapped in the core and propagate along the fiber axis according to the laws of geometrical optics, and unbound rays that are refracted out of the fiber core.

Skew rays are not confined to a single plane, but instead tend to follow a helical-type path along the fiber as shown in Fig. 2-11. These rays are more difficult to track as they travel along the fiber, since they do not lie in a single plane. Although skew rays constitute a major portion of the total number of guided rays, their analysis is not necessary to obtain a general picture of rays propagating in a fiber. The examination of meridional rays will suffice for this purpose. However, a detailed inclusion of skew rays will change such expressions

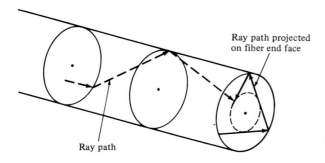

FIGURE 2-11
Ray optics representation of skew rays traveling in a step-index optical fiber core.

as the light-acceptance ability of the fiber and power losses of light traveling along a waveguide.[6,10]

A greater power loss arises when skew rays are included in the analyses, since many of the skew rays that geometric optics predicts to be trapped in the fiber are actually leaky rays.[5,12,13] These leaky rays are only partially confined to the core of the circular optical fiber and attenuate as the light travels along the optical waveguide. This partial reflection of leaky rays cannot be described by pure ray theory alone. Instead, the analysis of radiation loss arising from these types of rays must be described by mode theory. This is explained further in Sec. 2.4.

The meridional ray is shown in Fig. 2-12 for a step-index fiber. The light ray enters the fiber core from a medium of refractive index n at an angle θ_0 with respect to the fiber axis and strikes the core–cladding interface at a normal angle ϕ. If it strikes this interface at such an angle that it is totally internally reflected, then the meridional ray allows a zigzag path along the fiber core, passing through the axis of the guide after each reflection.

From Snell's law, the minimum angle ϕ_{min} that supports total internal reflection for the meridional ray is given by

$$\sin \phi_{min} = \frac{n_2}{n_1} \tag{2-21}$$

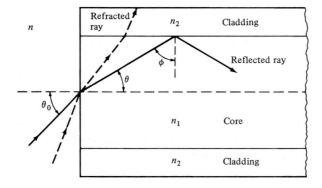

FIGURE 2-12
Meridional ray optics representation of the propagation mechanism in an ideal step-index optical waveguide.

Rays striking the core–cladding interface at angles less than ϕ_{min} will refract out of the core and be lost in the cladding. By applying Snell's law to the air–fiber face boundary, the condition of Eq. (2-21) can be related to the maximum entrance angle $\theta_{0,max}$ through the relationship

$$n \sin \theta_{0,max} = n_1 \sin \theta_c = (n_1^2 - n_2^2)^{1/2} \tag{2-22}$$

where $\theta_c = \pi/2 - \phi_c$. Thus, those rays having entrance angles θ_0 less than $\theta_{0,max}$ will be totally internally reflected at the core–cladding interface.

Equation (2-22) also defines the *numerical aperture* (NA) of a step-index fiber for meridional rays:

$$NA = n \sin \theta_{0,max} = (n_1^2 - n_2^2)^{1/2} \approx n_1 \sqrt{2\Delta} \tag{2-23}$$

The approximation on the right-hand side is valid for the typical case where Δ, as defined by Eq. (2-20), is much less than 1. Since the numerical aperture is related to the maximum acceptance angle, it is commonly used to describe the light acceptance or gathering capability of a fiber and to calculate source-to-fiber optical power coupling efficiencies. This is detailed in Chap. 5. The numerical aperture is a dimensionless quantity which is less than unity, with values normally ranging from 0.14 to 0.50.

> **Example 2.2.** Preferred sizes of multimode glass optical fibers and their corresponding numerical apertures are as follows:

Core diameter (μm)	Clad diameter (μm)	Numerical aperture
50	125	0.19–0.25
62.5	125	0.27–0.31
85	125	0.25–0.30
100	140	0.25–0.30

2.3.5 Wave Representation in a Dielectric Slab Waveguide

Referring to Fig. 2-12, the ray theory appears to allow rays at any angle ϕ greater than the critical angle ϕ_c to propagate along the fiber. However, when the interference effect due to the phase of the plane wave associated with the ray is taken into account, it is seen that only waves at certain discrete angles greater than or equal to ϕ_c are capable of propagating along the fiber.

To see this, let us consider wave propagation in an infinite dielectric slab waveguide of thickness d. Its refractive index n_1 is greater than the index n_2 of the material above and below the slab. A wave will thus propagate in this guide through multiple reflections, provided that the angle of incidence with respect to the upper and lower surfaces satisfies the condition given in Eq. (2-22).

Figure 2-13 show the geometry of the waves reflecting at the material interfaces. Here, we consider two rays, designated ray 1 and ray 2, associated with the same wave. The rays are incident on the material interface at an angle $\theta < \theta_c = \pi/2 - \phi_c$. The ray paths in Fig. 2-13 are denoted by solid lines and their associated constant-phase fronts by dashed lines.

The condition required for wave propagation in the dielectric slab is that all points on the same phase front of a plane wave must be in phase. This means that the phase change occurring in ray 1 when traveling from point A to point B minus the phase change in ray 2 between points C and D must differ by an integer multiple of 2π. As the wave travels through the material, it undergoes a phase shift Δ given by

$$\Delta = k_1 s = n_1 k s = n_1 2\pi s/\lambda$$

where $k_1 =$ the propagation constant in the medium of refractive index n_1
$k = k_1/n_1$ is the free-space propagation constant
$s =$ the distance the wave has traveled in the material

The phase of the wave changes not only as the wave travels but also upon reflection from a dielectric interface, as described in Sec. 2.2.

In going from point A to point B, ray 1 travels a distance $s_1 = d/\sin\theta$ in the material, and undergoes two phase changes δ at the reflection points. Ray 2 does not incur any reflections in going from point C to point D. To determine its phase change, first note that the distance from point A to point D is $\overline{AD} = (d/\tan\theta) - d\tan\theta$. Thus, the distance between points C and D is

$$s_2 = \overline{AD}\cos\theta = (\cos^2\theta - \sin^2\theta)d/\sin\theta$$

The requirement for wave propagation can then be written as

$$\frac{2\pi n_1}{\lambda}(s_1 - s_2) + 2\delta = 2\pi m \qquad (2\text{-}24a)$$

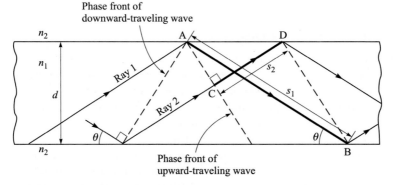

FIGURE 2-13
Light wave propagating along a fiber waveguide. Phase changes occur both as the wave travels through the fiber medium and at the reflection points.

where $m = 0, 1, 2, 3, \ldots$. Substituting the expressions for s_1 and s_2 into Eq. (2-24a) then yields

$$\frac{2\pi n_1}{\lambda}\left\{\frac{d}{\sin\theta} - \left[\frac{(\cos^2\theta - \sin^2\theta)d}{\sin\theta}\right]\right\} + 2\delta = 2\pi m \qquad (2\text{-}24b)$$

which can be reduced to

$$\frac{2\pi n_1 d \sin\theta}{\lambda} + \delta = \pi m \qquad (2\text{-}24c)$$

Considering only electric waves with components normal to the plane of incidence, we have from Eq. (2-19a) that the phase shift upon reflection is

$$\delta = -2\arctan\left[\frac{\sqrt{\cos^2\theta - (n_2^2/n_1^2)}}{\sin\theta}\right] \qquad (2\text{-}25)$$

The negative sign is needed here since the wave in the medium must be a decaying and not a growing wave. Substituting this expression into Eq. (2-24c) yields

$$\frac{2\pi n_1 d \sin\theta}{\lambda} - \pi m = 2\arctan\left[\frac{\sqrt{\cos^2\theta - (n_2^2/n_1^2)}}{\sin\theta}\right] \qquad (2\text{-}26a)$$

or

$$\tan\left(\frac{\pi n_1 d \sin\theta}{\lambda} - \frac{\pi m}{2}\right) = \left[\frac{\sqrt{n_1^2 \cos^2\theta - n_2^2}}{n_1 \sin\theta}\right] \qquad (2\text{-}26b)$$

Thus, only waves that have those angles θ which satisfy the condition in Eq. (2-26) will propagate in the dielectric slab waveguide (see Prob. 2-12).

2.4 MODE THEORY FOR CIRCULAR WAVEGUIDES

To attain a more detailed understanding of the optical power propagation mechanism in a fiber, it is necessary to solve Maxwell's equations subject to the cylindrical boundary conditions at the interface between the core and the cladding of the fiber. This has been done in extensive detail in a number of works.[7-10,14-18] Since a complete treatment is beyond the scope of this book, only a general outline of a simplified (but still complex) analysis will be given here.

Before going into the details of mode theory in circular optical fibers, in Sec. 2.4.1 we first give a qualitative overview of the concepts of modes in a waveguide. Next, Sec. 2.4.2 presents a brief summary of the fundamental results obtained from the detailed analyses in Secs. 2.4.3 through 2.4.9, so that those who are not familiar with Maxwell's equations can skip over those sections designated by a star (★) without loss of continuity.

When solving Maxwell's equations for hollow metallic waveguides, only transverse electric (TE) modes and transverse magnetic (TM) modes are found. However, in optical fibers the core–cladding boundary conditions lead to a coupling between the electric and magnetic field components. This gives rise to hybrid modes, which makes optical waveguide analysis more complex than metallic waveguide analysis. The hybrid modes are designated as HE or EH modes, depending on whether the transverse electric field (the E field) or the transverse magnetic field (the H field) is larger for that mode. The two lowest-order modes are designated by HE_{11} and TE_{01}, where the subscripts refer to possible modes of propagation of the optical field.

Although the theory of light propagation in optical fibers is well understood, a complete description of the guided and radiation modes is rather complex since it involves six-component hybrid electromagnetic fields that have very involved mathematical expressions. A simplification[19-23] of these expressions can be carried out, in practice, since fibers usually are constructed so that the difference in the core and cladding indices of refraction is very small; that is, $n_1 - n_2 \ll 1$. With this assumption, only four field components need to be considered and their expressions become significantly simpler. The field components are called *linearly polarized* (LP) modes and are labeled LP_{jm} where j and m are integers designating mode solutions. In this scheme for the lowest-order modes, each LP_{0m} mode is derived from an HE_{1m} mode and each LP_{1m} mode comes from TE_{0m}, TM_{0m}, and HE_{0m} modes. Thus, the fundamental LP_{01} mode corresponds to an HE_{11} mode.

Although the analysis required for even these simplifications is still fairly involved, this material is key to understanding the principles of optical fiber operation. In Secs. 2.4.3 through 2.4.9 we first will solve Maxwell's equations for a circular step-index waveguide and then will describe the resulting solutions for some of the lower-order modes.

2.4.1 Overview of Modes

Before we progress with a discussion of mode theory in circular optical fibers, let us qualitatively examine the appearance of modal fields in the planar dielectric slab waveguide shown in Fig. 2-14. This waveguide is composed of a dielectric slab of refractive index $n_1 < n_2$, which we shall call the cladding. This represents the simplest form of an optical waveguide and can serve as a model to gain an understanding of wave propagation in optical fibers. In fact, a cross-sectional view of the slab waveguide looks the same as the cross-sectional view of an optical fiber cut along its axis. Figure 2-14 shows the field patterns of several of the lower-order transverse electric (TE) modes (which are solutions of Maxwell's equations for the slab waveguide[7-10]). The *order* of a mode is equal to the number of field zeros across the guide. The order of the mode is also related to the angle that the ray congruence corresponding to this mode makes with the plane of the waveguide (or the axis of a fiber); that is, the steeper the angle, the higher the order of the mode. The plots show that the electric fields of the guided modes are not

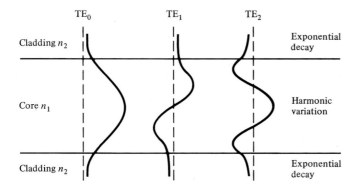

FIGURE 2-14
Electric field distributions for several of the lower-order guided modes in a symmetrical-slab waveguide.

completely confined to the central dielectric slab (i.e., they do not go to zero at the guide–cladding interface), but, instead, they extend partially into the cladding. The fields vary harmonically in the guiding region of refractive index n_1 and decay exponentially outside of this region. For low-order modes the fields are tightly concentrated near the center of the slab (or the axis of an optical fiber), with little penetration into the cladding region. On the other hand, for higher-order modes the fields are distributed more toward the edges of the guide and penetrate further into the cladding region.

Solving Maxwell's equations shows that, in addition to supporting a finite number of guided modes, the optical fiber waveguide has an infinite continuum of *radiation modes* that are not trapped in the core and guided by the fiber but are still solutions of the same boundary-value problem. The radiation field basically results from the optical power that is outside the fiber acceptance angle being refracted out of the core. Because of the finite radius of the cladding, some of this radiation gets trapped in the cladding, thereby causing cladding modes to appear. As the core and cladding modes propagate along the fiber, mode coupling occurs between the cladding modes and the higher-order core modes. This coupling occurs because the electric fields of the guided core modes are not completely confined to the core but extend partially into the cladding (see Fig. 2-14) and likewise for the cladding modes. A diffusion of power back and forth between the core and cladding modes thus occurs; this generally results in a loss of power from the core modes. In practice, the cladding modes will be suppressed by a lossy coating which covers the fiber or they will scatter out of the fiber after traveling a certain distance because of roughness on the cladding surface.

In addition to bound and refracted modes, a third category of modes called *leaky modes*[5,6,10,12,13] is present in optical fibers. These leaky modes are only partially confined to the core region, and attenuate by continuously radiating their power out of the core as they propagate along the fiber. This power radiation out of the waveguide results from a quantum mechanical phenomenon known as the *tunnel effect*. Its analysis is fairly lengthy and beyond the scope of this book. However, it is essentially based on the upper and lower bounds that the boundary conditions for the solutions of Maxwell's equations impose on the propagation factor β. A mode remains guided as long as β satisfies the condition

$$n_2 k < \beta < n_1 k$$

where n_1 and n_2 are the refractive indices of the core and cladding, respectively, and $k = 2\pi/\lambda$. The boundary between truly guided modes and leaky modes is defined by the *cutoff condition* $\beta = n_2 k$. As soon as β becomes smaller than $n_2 k$, power leaks out of the core into the cladding region. Leaky modes can carry significant amounts of optical power in short fibers. Most of these modes disappear after a few centimeters, but a few have sufficiently low losses to persist in fiber lengths of a kilometer.

2.4.2 Summary of Key Modal Concepts

An important parameter connected with the cutoff condition is the *V number* defined by

$$V = \frac{2\pi a}{\lambda}(n_1^2 - n_2^2)^{1/2} = \frac{2\pi a}{\lambda}\text{NA} \tag{2-27}$$

This is a dimensionless number that determines how many modes a fiber can support. Except for the lowest-order HE_{11} mode, each mode can exist only for values of V that exceed a certain limiting value (with each mode having a different V limit). The modes are cut off when $\beta = n_2 k$. This occurs when $V \leq 2.405$. The HE_{11} mode has not cutoff and ceases to exist only when the core diameter is zero. This is the principle on which single-mode fibers are based. The details for these and other modes are shown later in Fig. 2-18.

The V number can also be used to express the number of modes M in a multimode fiber when V is large. For this case, an estimate of the total number of modes supported in a fiber is

$$M \approx \frac{1}{2}\left(\frac{2\pi a}{\lambda}\right)^2 (n_1^2 - n_2^2) = \frac{V^2}{2} \tag{2-28}$$

Since the field of a guided mode extends partly into the cladding, as shown in Fig. 2-14, a final quantity of interest for a step-index fiber is the fractional power flow in the core and cladding for a given mode. As the V number approaches cutoff for any particular mode, more of the power of that mode is in the cladding. At the cutoff point, the mode becomes radiative with all the optical power of the mode residing in the cladding. Far from cutoff—that is, for large values of V—the fraction of the average optical power residing in the cladding can be estimated by

$$\frac{P_{\text{clad}}}{P} \approx \frac{4}{3\sqrt{M}} \tag{2-29}$$

where P is the total optical power in the fiber. The details for the power distribution between the core and the cladding of various LP_{jm} modes are shown later in Fig. 2-22. Note that since M is proportional to V^2, the power flow in the cladding decreases as V increases. However, this increases the number of modes in the fiber, which is not desirable for a high-bandwidth capability.

2.4.3 Maxwell's Equations★

To analyze the optical waveguide we need to consider Maxwell's equations that give the relationships between the electric and magnetic fields. Assuming a linear, isotropic dielectric material having no currents and free charges, these equations take the form[2]

$$\nabla \times \mathbf{E} = -\frac{\partial \mathbf{B}}{\partial t} \tag{2-30a}$$

$$\nabla \times \mathbf{H} = \frac{\partial \mathbf{D}}{\partial t} \tag{2-30b}$$

$$\nabla \cdot \mathbf{D} = 0 \tag{2-30c}$$

$$\nabla \cdot \mathbf{B} = 0 \tag{2-30d}$$

where $\mathbf{D} = \epsilon \mathbf{E}$ and $\mathbf{B} = \mu \mathbf{H}$. The parameter ϵ is the permittivity (or dielectric constant) and μ is permeability of the medium.

A relationship defining the wave phenomena of the electromagnetic fields can be derived from Maxwell's equations. Taking the curl of Eq. (2.30a) and making use of Eq. (2.30b) yields

$$\nabla \times (\nabla \times \mathbf{E}) = -\mu \frac{\partial}{\partial t}(\nabla \times \mathbf{H}) = -\epsilon \mu \frac{\partial^2 \mathbf{E}}{\partial t^2} \tag{2-31a}$$

Using the vector identity (see App. B),

$$\nabla \times (\nabla \times \mathbf{E}) = \nabla(\nabla \cdot \mathbf{E}) - \nabla^2 \mathbf{E}$$

and using Eq. (2-30c) (i.e., $\nabla \cdot \mathbf{E} = 0$), Eq. (2-31a) becomes

$$\nabla^2 \mathbf{E} = \epsilon \mu \frac{\partial^2 \mathbf{E}}{\partial t^2} \tag{2-31b}$$

Similarly, by taking the curl of Eq. (2-30b), it can be shown that

$$\nabla^2 \mathbf{H} = \epsilon \mu \frac{\partial^2 \mathbf{H}}{\partial t^2} \tag{2-31c}$$

Equations (2-31b) and (2-31c) are the standard *wave equations*.

2.4.4 Waveguide Equations★

Consider electromagnetic waves propagating along the cylindrical fiber shown in Fig. 2-15. For this fiber, a cylindrical coordinate system $\{r, \phi, z\}$ is defined with the z axis lying along the axis of the waveguide. If the electromagnetic waves are to propagate along the z axis, they will have a functional dependence of the form

$$\mathbf{E} = \mathbf{E}_0(r, \phi)e^{j(\omega t - \beta z)} \tag{2-32a}$$

$$\mathbf{H} = \mathbf{H}_0(r, \phi)e^{j(\omega t - \beta z)} \tag{2-32b}$$

which are harmonic in time t and coordinate z. The parameter β is the z component of the propagation vector and will be determined by the boundary conditions on the electromagnetic fields at the core–cladding interface described in Sec. 2.4.6.

When Eqs. (2-32a) and (2-32b) are substituted into Maxwell's curl equations, we have, from Eq. (2.30a)

$$\frac{1}{r}\left(\frac{\partial E_z}{\partial \phi} + jr\beta E_\phi\right) = -j\omega\mu H_r \tag{2-33a}$$

$$j\beta E_r + \frac{\partial E_z}{\partial r} = j\omega\mu H_\phi \tag{2-33b}$$

$$\frac{1}{r}\left[\frac{\partial}{\partial r}(rE_\phi) - \frac{\partial E_r}{\partial \phi}\right] = -j\mu\omega H_z \tag{2-33c}$$

and, from Eq. (2-30b),

$$\frac{1}{r}\left(\frac{\partial H_z}{\partial \phi} + jr\beta H_\phi\right) = j\epsilon\omega E_r \tag{2-34a}$$

$$j\beta H_r + \frac{\partial H_z}{\partial r} = -j\epsilon\omega E_\phi \tag{2-34b}$$

$$\frac{1}{r}\left[\frac{\partial}{\partial r}(rH_\phi) - \frac{\partial H_r}{\partial \phi}\right] = j\epsilon\omega E_z \tag{2-34c}$$

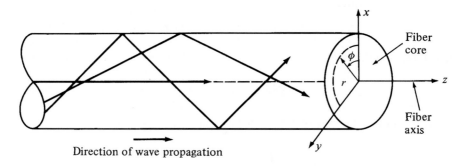

Direction of wave propagation

FIGURE 2-15
Cylindrical coordinate system used for analyzing electromagnetic wave propagation in an optical fiber.

By eliminating variables these equations can be rewritten such that, when E_z and H_z are known, the remaining transverse components E_r, E_ϕ, H_r, and H_ϕ can be determined. For example, E_ϕ or H_r can be eliminated from Eqs. (2-33a) and (2-34b) so that the component H_ϕ or E_r, respectively, can be found in terms of E_z or H_z. Doing so yields

$$E_r = -\frac{j}{q^2}\left(\beta\frac{\partial E_z}{\partial r} + \frac{\mu\omega}{r}\frac{\partial H_z}{\partial \phi}\right) \tag{2-35a}$$

$$E_\phi = -\frac{j}{q^2}\left(\frac{\beta}{r}\frac{\partial E_z}{\partial \phi} - \mu\omega\frac{\partial H_z}{\partial r}\right) \tag{2-35b}$$

$$H_r = \frac{-j}{q^2}\left(\beta\frac{\partial H_z}{\partial r} - \frac{\omega\epsilon}{r}\frac{\partial E_z}{\partial \phi}\right) \tag{2-35c}$$

$$H_\phi = \frac{-j}{q^2}\left(\frac{\beta}{r}\frac{\partial H_z}{\partial \phi} + \omega\epsilon\frac{\partial E_z}{\partial r}\right) \tag{2-35d}$$

where $q^2 = \omega^2\epsilon\mu - \beta^2 = k^2 - \beta^2$.

Substitution of Eqs. (2-35c) and (2-35d) into Eq. (2-34c) results in the wave equation in cylindrical coordinates,

$$\frac{\partial^2 E_z}{\partial r^2} + \frac{1}{r}\frac{\partial E_z}{\partial r} + \frac{1}{r^2}\frac{\partial^2 E_z}{\partial \phi^2} + q^2 E_z = 0 \tag{2-36}$$

and substitution of Eqs. (2-35a) and (2-35b) into Eq. (2-33c) leads to

$$\frac{\partial^2 H_z}{\partial r^2} + \frac{1}{r}\frac{\partial H_z}{\partial r} + \frac{1}{r^2}\frac{\partial^2 H_z}{\partial \phi^2} + q^2 H_z = 0 \tag{2-37}$$

It is interesting to note that Eqs. (2-36) and (2-37) each contain either only E_z or only H_z. This appears to imply that the longitudinal components of **E** and **H** are uncoupled and can be chosen arbitrarily provided that they satisfy Eqs. (2-36) and (2-37). However, in general, coupling of E_z and H_z is required by the boundary conditions of the electromagnetic field components described in Sec. 2.4.6. If the boundary conditions do not lead to coupling between the field components, mode solutions can be obtained in which either $E_z = 0$ or $H_z = 0$. When $E_z = 0$ the modes are called *transverse electric* or TE modes, and when $H_z = 0$ they are called *transverse magnetic* or TM modes. *Hybrid* mode exist if both E_z and H_z are nonzero. These are designated as HE or EH modes, depending on whether H_z or E_z, respectively, makes a larger contribution to the transverse field. The fact that the hybrid modes are present in optical waveguides makes their analysis more complex than in the simpler case of hollow metallic waveguides where only TE and TM modes are found.

2.4.5 Wave Equations for Step-Index Fibers★

We now use the above results to find the guided modes in a step-index fiber. A standard mathematical procedure for solving equations such as Eq. (2-36) is to use the separation-of-variables method, which assumes a solution of the form

$$E_z = AF_1(r)F_2(\phi)F_3(z)F_4(t) \tag{2-38}$$

As was already assumed, the time- and z-dependent factors are given by

$$F_3(z)F_4(t) = e^{j(\omega t - \beta z)} \tag{2-39}$$

since the wave is sinusoidal in time and propagates in the z direction. In addition, because of the circular symmetry of the waveguide, each field component must not change when the coordinate ϕ is increased by 2π. We thus assume a periodic function of the form

$$F_2(\phi) = e^{jv\phi} \tag{2-40}$$

The constant v can be positive or negative, but it must be an integer since the fields must be periodic in ϕ with a period of 2π.

Substituting Eq. (2-40) into Eq. (2-38), the wave equation for E_z [Eq. 2-36] becomes

$$\frac{\partial^2 F_1}{\partial r^2} + \frac{1}{r}\frac{\partial F_1}{\partial r} + \left(q^2 - \frac{v^2}{r^2}\right)F_1 = 0 \tag{2-41}$$

which is a well-known differential equation for Bessel functions.[24-26] An exactly identical equation can be derived for H_z.

For the configuration of the step-index fiber we consider a homogeneous core of refractive index n_1 and radius a, which is surrounded by an infinite cladding of index n_2. The reason for assuming an infinitely thick cladding is that the guided modes in the core have exponentially decaying fields outside the core and these must have insignificant values at the outer boundary of the cladding. In practice, optical fibers are designed with claddings that are sufficiently thick so that the guided-mode field does not reach the outer boundary of the cladding. To get an idea of the field patterns, the electric field distributions for several of the lower-order guided modes in a symmetrical-slab waveguide were shown in Fig. 2-14. The fields vary harmonically in the guiding region of refractive index n_1 and decay exponentially outside of this region.

Equation (2-41) must now be solved for the regions inside and outside the core. For the inside region the solutions for the guided modes must remain finite as $r \to 0$, whereas on the outside the solutions must decay to zero as $r \to \infty$. Thus, for $r < a$ the solutions are Bessel functions of the first kind of order v. For these functions we use the common designation $J_v(ur)$. Here, $u^2 = k_1^2 - \beta^2$ with $k_1 = 2\pi n_1/\lambda$. The expressions for E_z and H_z inside the core are thus

$$E_z(r < a) = AJ_v(ur)e^{jv\phi}e^{j(\omega t - \beta z)} \tag{2-42}$$

$$H_z(r < a) = BJ_v(ur)e^{jv\phi}e^{j(\omega t - \beta z)} \tag{2-43}$$

where A and B are arbitrary constants.

Outside of the core the solutions to Eq. (2-41) are given by modified Bessel functions of the second kind, $K_\nu(wr)$, where $w^2 = \beta^2 - k_2^2$ with $k_2 = 2\pi n_2/\lambda$. The expressions for E_z and H_z outside the core are therefore

$$E_z(r > a) = CK_\nu(wr)e^{j\nu\phi}e^{j(\omega t - \beta z)} \tag{2-44}$$

$$H_z(r > a) = DK_\nu(wr)e^{j\nu\phi}e^{j(\omega t - \beta z)} \tag{2-45}$$

with C and D being arbitrary constants.

The definitions of $J_\nu(ur)$ and $K_\nu(wr)$ and various recursion relations are given in App. C. From the definition of the modified Bessel function, it is seen that $K_\nu(wr) \to e^{-wr}$ as $wr \to \infty$. Since $K_\nu(wr)$ must go to zero as $r \to \infty$, it follows that $w > 0$. This, in turn, implies that $\beta \geq k_2$, which represents a cutoff condition. The *cutoff condition* is the point at which a mode is no longer bound to the core region. A second condition on β can be deduced from the behavior of $J_\nu(ur)$. Inside the core the parameter u must be real for F_1 to be real, from which it follows that $k_1 \geq \beta$. The permissible range of β for bound solutions is therefore

$$n_2 k = k_2 \leq \beta \leq k_1 = n_1 k \tag{2-46}$$

where $k = 2\pi/\lambda$ is the free-space propagation constant.

2.4.6 Modal Equation★

The solutions for β must be determined from the boundary conditions. The boundary conditions require that the tangential components E_ϕ and E_z of **E** inside and outside of the dielectric interface at $r = a$ must be the same, and similarly for the tangential components H_ϕ and H_z. Consider first the tangential components of **E**. For the z component we have, from Eq. (2-42) at the inner core–cladding boundary ($E_z = E_{z1}$) and from Eq. (2–44) at the outside of the boundary ($E_z = E_{z2}$), that

$$E_{z1} - E_{z2} = AJ_\nu(ua) - CK_\nu(wa) = 0 \tag{2-47}$$

The ϕ component is found from Eq. (2-35b). Inside the core the factor q^2 is given by

$$q^2 = u^2 = k_1^2 - \beta^2 \tag{2-48}$$

where $k_1 = 2\pi n_1/\lambda = \omega\sqrt{\epsilon_1\mu}$, while outside the core

$$w^2 = \beta^2 - k_2^2 \tag{2-49}$$

with $k_2 = 2\pi n_2/\lambda = \omega\sqrt{\epsilon_2\mu}$. Substituting Eqs. (2-42) and (2-43) into Eq. (2-35b) to find $E_{\phi 1}$, and, similarly, using Eqs. (2-44) and (2-45) to determine $E_{\phi 2}$, yields, at $r = a$,

$$E_{\phi 1} - E_{\phi 2} = -\frac{j}{u^2}\left[A\frac{jv\beta}{a}J_v(ua) - B\omega\mu uJ_v'(ua)\right]$$

$$-\frac{j}{w^2}\left[C\frac{jv\beta}{a}K_v(wa) - D\omega\mu wK_v'(wa)\right] = 0 \tag{2-50}$$

where the prime indicates differentiation with respect to the argument.

Similarly, for the tangential components of **H** it is readily shown that, at $r = a$,

$$H_{z1} - H_{z2} = BJ_v(ua) - DK_v(wa) = 0 \tag{2-51}$$

and

$$H_{\phi 1} - H_{\phi 2} = -\frac{j}{u^2}\left[B\frac{jv\beta}{a}J_v(ua) + A\omega\epsilon_1 uJ_v'(ua)\right]$$

$$-\frac{j}{w^2}\left[D\frac{jv\beta}{a}K_v(wa) + C\omega\epsilon_2 wK_v'(wa)\right] = 0 \tag{2-52}$$

Equations (2-47), (2-50), (2-51), and (2-52) are a set of four equations with four unknown coefficients, A, B, C, and D. A solution to these equations exists only if the determinant of these coefficients is zero:

$$\begin{vmatrix} J_v(ua) & 0 & -K_v(wa) & 0 \\ \frac{\beta v}{au^2}J_v(ua) & \frac{j\omega\mu}{u}J_v'(ua) & \frac{\beta v}{aw^2}K_v(wa) & \frac{j\omega\mu}{w}K_v'(wa) \\ 0 & J_v(ua) & 0 & -K_v(wa) \\ -\frac{j\omega\epsilon_1}{u}J_v'(ua) & \frac{\beta v}{au^2}J_v(ua) & -\frac{j\omega\epsilon_2}{w}K_v'(wa) & \frac{\beta v}{aw^2}K_v(wa) \end{vmatrix} = 0 \tag{2-53}$$

Evaluation of this determinant yields the following eigenvalue equation for β:

$$(\mathcal{J}_v + \mathcal{K}_v)(k_1^2\mathcal{J}_v + k_2^2\mathcal{K}_v) = \left(\frac{\beta v}{a}\right)^2\left(\frac{1}{u^2} + \frac{1}{w^2}\right)^2 \tag{2-54}$$

where

$$\mathcal{J}_v = \frac{J_v'(ua)}{uJ_v(ua)} \qquad \text{and} \qquad \mathcal{K}_v = \frac{K_v'(wa)}{wK_v(wa)}$$

Upon solving Eq. (2-54) for β, it will be found that only discrete values restricted to the range given by Eq. (2-46) will be allowed. Although Eq. (2-54) is a complicated transcendental equation which is generally solved by numerical techniques, its solution for any particular mode will provide all the characteristics of that mode. We shall now consider this equation for some of the lowest-order modes of a step-index waveguide.

2.4.7 Modes in Step-Index Fibers★

To help describe the modes, we shall first examine the behavior of the J-type Bessel functions. These are plotted in Fig. 2-16 for the first three orders. The J-type Bessel functions are similar to harmonic functions since they exhibit oscillatory behavior for real k, as is the case for sinusoidal functions. Because of the oscillatory behavior of J_ν, there will be m roots of Eq. (2-54) for a given ν value. These roots will be designated by $\beta_{\nu m}$, and the corresponding modes are either $\text{TE}_{\nu m}$, $\text{TM}_{\nu m}$, $\text{EH}_{\nu m}$, or $\text{HE}_{\nu m}$. Schematics of the transverse electric field patterns for the four lowest-order modes over the cross section of a step-index fiber are shown in Fig. 2-17.

For the dielectric fiber waveguide, all modes are hybrid modes except those for which $\nu = 0$. When $\nu = 0$ the right-hand side of Eq. (2-54) vanishes and two different eigenvalue equations result. These are

$$\mathcal{J}_0 + \mathcal{K}_0 = 0 \qquad (2\text{-}55a)$$

or, using the relations for J_ν' and K_ν' in App. C,

$$\frac{J_1(ua)}{uJ_0(ua)} + \frac{K_1(wa)}{wK_0(wa)} = 0 \qquad (2\text{-}55b)$$

which corresponds to TE_{0m} modes ($E_z = 0$), and

$$k_1^2 \mathcal{J}_0 + k_2^2 \mathcal{K}_0 = 0 \qquad (2\text{-}56a)$$

or

$$\frac{k_1^2 J_1(ua)}{uJ_0(ua)} + \frac{k_2^2 K_1(wa)}{wK_0(wa)} = 0 \qquad (2\text{-}56b)$$

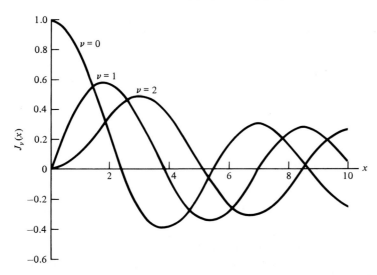

FIGURE 2-16
Variation of the Bessel function $J_\nu(x)$ for the first three orders ($\nu = 0, 1, 2$) plotted as a function of x.

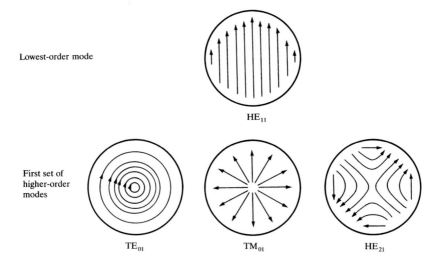

Lowest-order mode

HE_{11}

First set of higher-order modes

TE_{01} TM_{01} HE_{21}

FIGURE 2-17
Cross-sectional views of the transverse electric field vectors for the four lowest-order modes in a step-index fiber.

which corresponds to TM_{0m} modes ($H_z = 0$). The proof of this is left as an exercise (see Prob. 2-15).

When $\nu \neq 0$ the situation is more complex and numerical methods are needed to solve Eq. (2-54) exactly. However, simplified and highly accurate approximations based on the principle that the core and cladding refractive indices are nearly the same have been derived by Snyder[19] and Gloge.[20,27] The condition that $n_1 - n_2 \ll 1$ was referred to by Gloge as giving rise to *weakly guided* modes. A treatment of these derivations is given in Sec. 2.4.8.

Let us examine the cutoff conditions for fiber modes. As was mentioned in relation to Eq. (2-46), a mode is referred to as being cut off when it is not longer bound to the core of the fiber, so that its field no longer decays on the outside of the core. The cutoffs for the various modes are found by solving Eq. (2-54) in the limit $w^2 \rightarrow 0$. This is, in general, fairly complex, so that only the results,[14,16] which are listed in Table 2-1, will be given here.

An important parameter connected with the cutoff condition is the *normalized frequency V* (also called the *V number* or *V parameter*) defined by

$$V^2 = (u^2 + w^2)a^2 = \left(\frac{2\pi a}{\lambda}\right)^2 (n_1^2 - n_2^2) = \left(\frac{2\pi a}{\lambda}\right)^2 NA^2 \qquad (2\text{-}57)$$

which is a dimensionless number that determines how many modes a fiber can support. The number of modes that can exist in a waveguide as a function of V may be conveniently represented in terms of a *normalized propagation constant b* defined by [20]

TABLE 2-1
Cutoff conditions for some lower-order modes

v	Mode	Cutoff condition
0	TE_{0m}, TM_{0m}	$J_0(ua) = 0$
1	HE_{1m}, EH_{1m}	$J_1(ua) = 0$
≥ 2	EH_{vm}	$J_v(ua) = 0$
	HE_{vm}	$\left(\dfrac{n_1^2}{n_2^2} + 1\right)J_{v-1}(ua) = \dfrac{ua}{v-1}J_v(ua)$

$$b = \frac{a^2 w^2}{V^2} = \frac{(\beta/k)^2 - n_2^2}{n_1^2 - n_2^2}$$

A plot of b (in terms of β/k) as a function of V is shown in Fig. 2-18 for a few of the low-order modes. This figure shows that each mode can exist only for values of V that exceed a certain limiting value. The modes are cut off when $\beta/k = n_2$. The HE_{11} mode has no cutoff and ceases to exist only when the core diameter is zero. This is the principle on which the single-mode fiber is based. By appropriately choosing a, n_1, and n_2 so that

$$V = \frac{2\pi a}{\lambda}(n_1^2 - n_2^2)^{1/2} \leq 2.405 \tag{2-58}$$

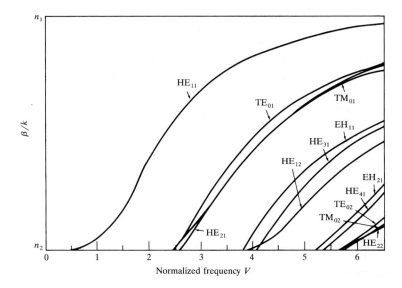

FIGURE 2-18
Plots of the propagation constant (in terms of β/k) as a function of V for a few of the lowest-order modes.

which is the value at which the lowest-order Bessel function $J_0 = 0$ (see Fig. 2-16), all modes except the HE_{11} mode are cut off.

Example 2-3. A step-index fiber has a normalized frequency $V = 26.6$ at a 1300-nm wavelength. If the core radius is 25 μm, let us find the numerical aperture. From Eqs. (2-23) and (2-50) we have

$$V = \frac{2\pi a}{\lambda} NA$$

or

$$NA = V \frac{\lambda}{2\pi a} = 26.6 \frac{1.3}{2\pi(25)} = 0.22$$

The parameter V can also be related to the number of modes M in a multimode fiber when M is large. An approximate relationship for step-index fibers can be derived from ray theory. A ray congruence incident on the end of a fiber will be accepted by the fiber if it lies within an angle θ defined by the numerical aperture as given in Eq. (2-23):

$$NA = \sin\theta = (n_1^2 - n_2^2)^{1/2} \tag{2-59}$$

For practical numerical apertures, $\sin\theta$ is small so that $\sin\theta \simeq \theta$. The solid acceptance angle for the fiber is therefore

$$\Omega = \pi\theta^2 = \pi(n_1^2 - n_2^2) \tag{2-60}$$

For electromagnetic radiation of wavelength λ emanating from a laser or a waveguide, the number of modes per unit solid angle is given by $2A/\lambda^2$, where A is the area the mode is leaving or entering.[28] The area A in this case is the core cross section πa^2. The factor 2 comes from the fact that the plane wave can have two polarization orientations. The total number of modes M entering the fiber is thus given by

$$M \simeq \frac{2A}{\lambda^2}\Omega = \frac{2\pi^2 a^2}{\lambda^2}(n_1^2 - n_2^2) = \frac{V^2}{2} \tag{2-61}$$

2.4.8   Linearly Polarized Modes★

As may be apparent by now, the exact analysis for the modes of a fiber is mathematically very complex. However, a simpler but highly accurate approximation can be used, based on the principle that in a typical step-index fiber the difference between the indices of refraction of the core and cladding is very small; that is, $\Delta \ll 1$. This is the basis of the *weakly guiding fiber approximation*.[7,19,20,27] In this approximation the electromagnetic field patterns and the propagation constants of the mode pairs $HE_{v+1,m}$ and $EH_{v-1,m}$ are very similar. This holds likewise for the three modes TE_{0m}, TM_{0m}, and HE_{2m}. This can be seen from Fig. 2-18 with $(v, m) = (0, 1)$ and $(2, 1)$ for the mode groupings

$\{HE_{11}\}$, $\{TE_{01}, TM_{01}, HE_{21}\}$, $\{HE_{31}, EH_{11}\}$, $\{HE_{12}\}$, $\{HE_{41}, EH_{21}\}$, and $\{TE_{02}, TM_{02}, HE_{22}\}$. The result is that only four field components need to be considered instead of six, and the field description is further simplified by the use of caretesian instead of cylindrical coordinates.

When $\Delta \ll 1$ we have that $k_1^2 \simeq k_2^2 \simeq \beta^2$. Using these approximations, Eq. (2-54) becomes

$$\mathcal{J}_\nu + \mathcal{K}_\nu = \pm \frac{\nu}{a}\left(\frac{1}{u^2} + \frac{1}{w^2}\right) \tag{2-62}$$

Thus, Eq. (2-55b) for TE_{0m} modes is the same as Eq. (2-56b) for TM_{0m} modes. Using the recurrence relations for J_ν' and K_ν' given in App. C, we get two sets of equations for Eq. (2-62) for the positive and negative signs. The positive sign yields

$$\frac{J_{\nu+1}(ua)}{uJ_\nu(ua)} + \frac{K_{\nu+1}(wa)}{wK_\nu(wa)} = 0 \tag{2-63}$$

The solution of this equation gives a set of modes called the EH modes. For the negative sign in Eq. (2-62) we get

$$\frac{J_{\nu-1}(ua)}{uJ_\nu(ua)} - \frac{K_{\nu-1}(wa)}{wK_\nu(wa)} = 0 \tag{2-64a}$$

or, alternatively, taking the inverse of Eq. (2-64a) and using the first expressions for $J_\nu(ua)$ and $K_\nu(wa)$ from Sec. C.1.2 and Sec. C.2.2,

$$-\frac{uJ_{\nu-2}(ua)}{J_{\nu-1}(ua)} = \frac{wK_{\nu-2}(wa)}{K_{\nu-1}(wa)} \tag{2-64b}$$

This results in a set of modes called the HE modes.

If we define a new parameter

$$j = \begin{cases} 1 & \text{for TE and TM modes} \\ \nu + 1 & \text{for EH modes} \\ \nu - 1 & \text{for HE modes} \end{cases} \tag{2-65}$$

then Eqs. (2-55b), (2-63), and (2-64b) can be written in the unified form

$$\frac{uJ_{j-1}(ua)}{J_j(ua)} = -\frac{wK_{j-1}(wa)}{K_j(wa)} \tag{2-66}$$

Equations (2-65) and (2-66) show that within the weakly guiding approximation all modes characterized by a common set of j and m satisfy the same characteristic equation. This means that these modes are degenerate. Thus, if an $HE_{\nu+1,m}$ mode is degenerate with an $EH_{\nu-1,m}$ mode (i.e., if HE and EH modes of corresponding radial order m and equal circumferential order ν form degenerate pairs), then any combination of an $HE_{\nu+1,m}$ mode with an $EH_{\nu-1,m}$ mode will likewise constitute a guided mode of the fiber.

Gloge[20] proposed that such degenerate modes be called *linearly polarized* (LP) modes, and be designated LP$_{jm}$ modes regardless of their TM, TE, EH, or

HE field configuration. The normalized propagation constant b as a function of V is given for various LP_{jm} modes in Fig. 2-19. In general, we have the following:

1. Each LP_{0m} mode is derived from an HE_{1m} mode.
2. Each LP_{1m} mode comes from TE_{0m}, TM_{0m}, and HE_{2m} modes.
3. Each LP_{vm} mode ($v \geq 2$) is from an $HE_{v+1,m}$ and an $EH_{v-1,m}$ mode.

The correspondence between the ten lowest LP modes (i.e., those having the lowest cutoff frequencies) and the traditional TM, TE, EH, and HE modes is given in Table 2-2. This table also shows the number of degenerate modes.

A very useful feature of the LP-mode designation is the ability to readily visualize a mode. In a complete set of modes only one electric and one magnetic field component are significant. The electric field vector **E** can be chosen to lie along an arbitrary axis, with the magnetic field vector **H** being perpendicular to it. In addition, there are equivalent solutions with the field polarities reversed. Since each of the two possible polarization directions can be coupled with either a $\cos j\phi$ or a $\sin j\phi$ azimuthal dependence, four discrete mode patterns can be obtained from a single LP_{jm} label. As an example, the four possible electric and magnetic field directions and the corresponding intensity distributions for the LP_{11} mode are shown in Fig. 2-20. Figures 2-21a and 2-21b illustrate how two LP_{11} modes are composed from the exact HE_{21} plus TE_{01} and the exact HE_{21} plus TM_{01} modes, respectively.

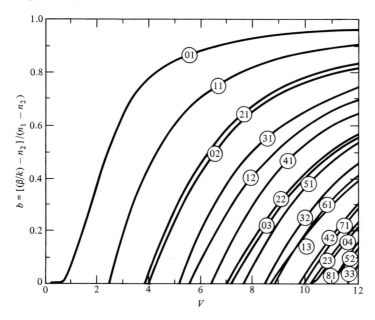

FIGURE 2-19
Plots of the propagation constant b as a function of V for various LP_{jm} modes. (Reproduced with permission from Gloge.[20])

TABLE 2-2
Composition of the lower-order linearly polarized modes

LP-mode designation	Traditional-mode designation and number of modes	Number of degenerate modes
LP_{01}	$HE_{11} \times 2$	2
LP_{11}	TE_{01}, TM_{01}, $HE_{21} \times 2$	4
LP_{21}	$EH_{11} \times 2$, $HE_{31} \times 2$	4
LP_{02}	$HE_{12} \times 2$	2
LP_{31}	$EH_{21} \times 2$, $HE_{41} \times 2$	4
LP_{12}	TE_{02}, TM_{02}, $HE_{22} \times 2$	4
LP_{41}	$EH_{31} \times 2$, $HE_{51} \times 2$	4
LP_{22}	$EH_{12} \times 2$, $HE_{32} \times 2$	4
LP_{03}	$HE_{13} \times 2$	2
LP_{51}	$EH_{41} \times 2$, $HE_{61} \times 2$	4

2.4.9 Power Flow in Step-Index Fibers★

A final quantity of interest for step-index fibers is the fractional power flow in the core and cladding for a given mode. As illustrated in Fig. 2-14, the electromagnetic field for a given mode does not go to zero at the core–cladding interface, but changes from an oscillating form in the core to an exponential decay in the cladding. Thus, the electromagnetic energy of a guided mode is carried partly in the core and partly in the cladding. The further away a mode is from its cutoff frequency, the more concentrated its energy is in the core. As cutoff is

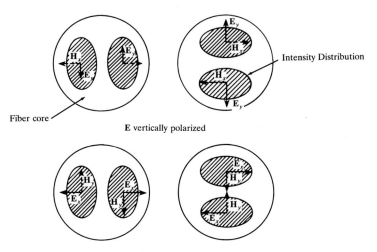

FIGURE 2-20
The four possible transverse electric field and magnetic field directions and the corresponding intensity distributions for the LP_{11} mode.

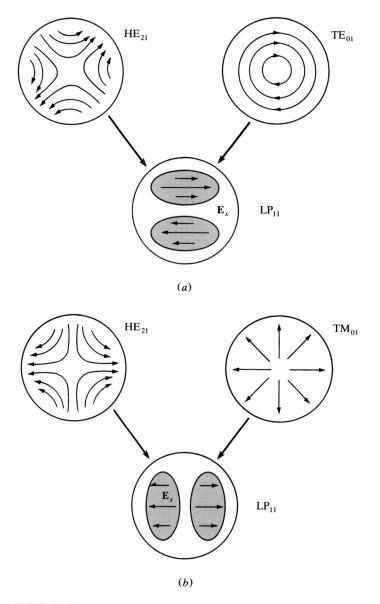

FIGURE 2-21
Composition of two LP_{11} modes from exact modes and their transverse electric field and intensity distributions.

approached, the field penetrates further into the cladding region and a greater percentage of the energy travels in the cladding. At cutoff the field no longer decays outside the core and the mode now becomes a fully radiating mode.

The relative amounts of power flowing in the core and the cladding can be obtained by integrating the Poynting vector in the axial direction,

$$S_z = \tfrac{1}{2} \operatorname{Re}(\mathbf{E} \times \mathbf{H}^*) \cdot \mathbf{e}_z \tag{2-67}$$

over the fiber cross section. Thus, the powers in the core and cladding, respectively, are given by

$$P_{\text{core}} = \tfrac{1}{2} \int_0^a \int_0^{2\pi} r(E_x H_y^* - E_y H_x^*) d\phi \, dr \tag{2-68}$$

$$P_{\text{clad}} = \tfrac{1}{2} \int_a^\infty \int_0^{2\pi} r(E_x H_y^* - E_y H_x^*) d\phi \, dr \tag{2-69}$$

where the asterisk denotes the complex conjugate. Gloge[20,29] has shown that, based on the weakly guided mode approximation, which has an accuracy on the order of the index difference Δ between the core and cladding, the relative core and cladding powers for a particular mode ν are given by

$$\frac{P_{\text{core}}}{P} = \left(1 - \frac{u^2}{V^2}\right)\left[1 - \frac{J_\nu^2(ua)}{J_{\nu+1}(ua)J_{\nu-1}(ua)}\right] \tag{2-70}$$

and

$$\frac{P_{\text{clad}}}{P} = 1 - \frac{P_{\text{core}}}{P} \tag{2-71}$$

where P is the total power in the mode ν. The relationships between P_{core} and P_{clad} are plotted in Fig. 2-22 in terms of the fractional powers P_{core}/P and P_{clad}/P for various LP_{jm} modes. In addition, far from cutoff the average total power in the cladding has been derived for fibers in which many modes can propagate. Because of this large number of modes, those few modes that are appreciably close to cutoff can be ignored to a reasonable approximation. The derivation assumes an incoherent source, such as a tungsten filament lamp or a light-emitting diode, which, in general, excites every fiber mode with the same amount of power. The total average cladding power is thus approximated by[20]

$$\left(\frac{P_{\text{clad}}}{P}\right)_{\text{total}} = \tfrac{4}{3} M^{-1/2} \tag{2-72}$$

where, from Eq. (2-61), M is the total number of modes entering the fiber. From Fig. 2-22 and Eq. (2-72) it can be seen that, since M is proportional to V^2, the power flow in the cladding decreases as V increases.

Example 2-4. As an example, consider a fiber having a core radius of 25 μm, a core index of 1.48, and $\Delta = 0.01$. At an operating wavelength of 0.84 μm the value of V is 39 and there are 760 modes in the fiber. From Eq. (2-72), approximately 5 percent of the power propagates in the cladding. If Δ is decreased to, say, 0.003 in order to

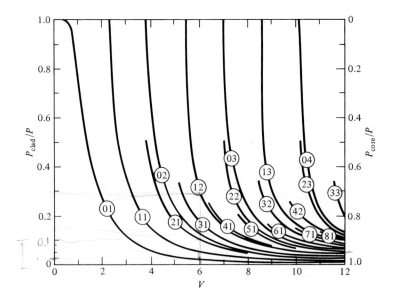

FIGURE 2-22
Fractional power flow in the cladding of a step-index optical fiber as a function of V. When $\nu \neq 1$, the curve numbers νm designate the $HE_{\nu+1,m}$ and $EH_{\nu-1,m}$ modes. For $\nu = 1$, the curve numbers νm give the HE_{2m}, TM_{0m}, and TM_{0m} modes. (Reproduced with permission from Gloge.[20])

decrease signal dispersion (see Chap. 3), then 242 modes propagate in the fiber and about 9 percent of the power resides in the cladding. For the case of the single-mode fiber, considering the LP_{01} mode (the HE_{11} mode) in Fig. 2-22, it is seen that for $V = 1$ about 70 percent of the power propagates in the cladding, whereas for $V = 2.405$, which is where the LP_{11} mode (the TE_{01} mode) begins, approximately 84 percent of the power is now within the core.

2.5   SINGLE-MODE FIBERS

Single-mode fibers are constructed by letting the dimensions of the core diameter be a few wavelengths (usually 8–12) and by having small index differences between the core and the cladding. From Eq. (2-27) or (2-58) with $V = 2.4$, it can be seen that single-mode propagation is possible for fairly large variations in values of the physical core size a and the core–cladding index differences Δ. However, in practical designs of single-mode fibers,[27] the core–cladding index difference varies between 0.2 and 1.0 percent, and the core diameter should be chosen to be just below the cutoff of the first higher-order mode; that is, for V slightly less than 2.4. For example, a typical single-mode fiber may have a core radius of 3 μm and a numerical aperture of 0.1 at a wavelength of 0.8 μm. From Eqs. (2-23) and (2-57) [or Eq. (2-27)], this yields $V = 2.356$.

2.5.1 Mode-Field Diameter

For single-mode fibers the geometric distribution of light in the propagating mode (rather than the core diameter and the numerical aperture) is what is important when predicting the performance characteristics of these fibers. Thus, a fundamental parameter of a single-mode fiber is the *mode-field diameter* (MFD).[30-35] This parameter can be determined from the mode-field distribution of the fundamental LP_{01} mode. The mode-field diameter is analogous to the core diameter in multimode fibers, except that in single-mode fibers not all the light that propagates through the fiber is carried in the core. This is illustrated in Fig. 2-23.

A variety of models for characterizing and measuring the MFD have been proposed.[30-33,36-40]. The main consideration in all these methods is how to approximate the electric field distribution. First, let us assume the distribution to be gaussian:[21]

$$E(r) = E_0 \exp(-r^2/W_0^2) \tag{2-73}$$

where r is the radius, E_0 is the field at zero radius, and W_0 is the width of the electric field distribution. Then, one method is take the width $2W_0$ of the MFD to be twice the e^{-1} radius of the optical electric field (which is equivalent to the e^{-2}

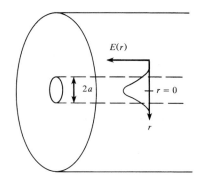

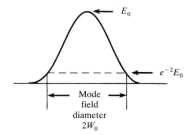

FIGURE 2-23
Distribution of light in a single-mode fiber above its cutoff wavelength. For a gaussian distribution, the MFD is given by the $1/e^2$ width of the optical power.

radius of the optical power) given in Eq. (2-73). The MFD width $2W_0$ of the LP_{01} mode can then be defined as

$$2W_0 = 2\left[\frac{2\int_0^\infty r^3 E^2(r)dr}{\int_0^\infty r E^2(r)dr}\right]^{1/2} \tag{2-74}$$

where $E(r)$ denotes the field distribution of the LP_{01} mode. This definition is not unique, and several others have been proposed.[31] Also note that, in general, the mode field varies with the refractive index profile and thus deviates from a gaussian distribution.

2.5.2   Propagation Modes in Single-Mode Fibers

As described in Sec. 2.4.8, in any ordinary single-mode fiber there are actually two independent, degenerate propagation modes.[34,41-43]. These modes are very similar but their polarization planes are orthogonal. These may be chosen arbitrarily as the horizontal (H) and the vertical (V) polarizations as shown in Fig. 2-24. Either one of these two polarization modes constitutes the fundamental HE_{11} mode. In general, the electric field of the light propagating along the fiber is a linear super-position of these two polarization modes and depends on the polarization of the light at the launching point into the fiber.

  Suppose we arbitrarily choose one of the modes to have its transverse electric field polarized along the x direction and the other independent orthogonal mode to be polarized in the y direction as shown in Fig. 2-24. In ideal fibers with perfect rotational symmetry, the two modes are degenerate with equal propagation constants ($k_x = k_y$), and any polarization state injected into the fiber will propagate unchanged. In actual fibers there are imperfections, such as asymmetrical lateral stresses, noncircular cores, and variations in refractive-index profiles. These imperfections break the circular symmetry of the ideal fiber and lift the degeneracy of the two modes. The modes propagate with different phase velocities, and the difference between their effective refractive indices is called the fiber *birefringence*,

$$B_f = n_y - n_x \tag{2-75}$$

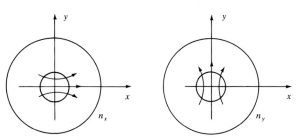

FIGURE 2-24
Two polarizations of the fundamental HE_{11} mode in a single-mode fiber.

Horizontal mode   Vertical mode

Equivalently, we may define the birefringence as

$$\beta = k_0(n_y - n_x) \tag{2-76}$$

where $k_0 = 2\pi/\lambda$ is the free-space propagation constant.

If light is injected into the fiber so that both modes are excited, then one will be delayed in phase relative to the other as they propagate. When this phase difference is an integral multiple of 2π, the two modes will beat at this point and the input polarization state will be reproduced. The length over which this beating occurs is the *fiber beat length*,

$$L_p = 2\pi/\beta \tag{2-77}$$

Example 2-5. A single mode optical fiber has a beat length of 8 cm at 1300 nm. From Eqs. (2-75) to (2-77) we have that the modal birefringence is

$$B_f = n_y - n_x = \frac{\lambda}{L_p} = \frac{1.3 \times 10^{-6}\,\text{m}}{8 \times 10^{-2}\,\text{m}} = 1.63 \times 10^{-5}$$

or, alternatively,

$$\beta = \frac{2\pi}{L_p} = \frac{2\pi}{0.08\,\text{m}} = 78.5\,\text{m}^{-1}$$

This indicates an intermediate-type fiber, since birefringences can vary from $B_f = 1 \times 10^{-3}$ (a typical high-birefringence fiber) to $B_f = 1 \times 10^{-8}$ (a typical low-birefringence fiber).

2.6 GRADED-INDEX FIBER STRUCTURE

In the graded-index fiber design the core refractive index decreases continuously with increasing radial distance r from the center of the fiber, but is generally constant in the cladding. The most commonly used construction for the refractive-index variation in the core is the power law relationship

$$n(r) = \begin{cases} n_1\left[1 - 2\Delta\left(\dfrac{r}{a}\right)^\alpha\right]^{1/2} & \text{for} \quad 0 \le r \le a \\[2mm] n_1(1 - 2\Delta)^{1/2} \simeq n_1(1 - \Delta) = n_2 & \text{for} \quad r \ge a \end{cases} \tag{2-78}$$

Here, r is the radial distance from the fiber axis, a is the core radius, n_1 is the refractive index at the core axis, n_2 is the refractive index of the cladding, and the dimensionless parameter α defines the shape of the index profile. The index difference Δ for the graded-index fiber is given by

$$\Delta = \frac{n_1^2 - n_2^2}{2n_1^2} \simeq \frac{n_1 - n_2}{n_1} \tag{2-79}$$

The approximation on the right-hand side of this equation reduces the expression for Δ to that of the step-index fiber given by Eq. (2-20). Thus, the same symbol is used in both cases. For $\alpha = \infty$, Eq. (2-78) reduces to the step-index profile $n(r) = n_1$.

Determining the NA for graded-index fibers is more complex than for step-index fibers, since it is a function of position across the core end face. This is in contrast to the step-index fiber, where the NA is constant across the core. Geometrical optics considerations show that light incident on the fiber core at position r will propagate as a guided mode only if it is within the local numerical aperture NA(r) at that point. The local numerical aperture is defined as[44]

$$\text{NA}(r) = \begin{cases} [n^2(r) - n_2^2]^{1/2} \simeq \text{NA}(0)\sqrt{1 - (r/a)^\alpha} & \text{for} \quad r \leq a \\ 0 & \text{for} \quad r > a \end{cases} \quad (2\text{-}80a)$$

where the axial numerical aperture is defined as

$$\text{NA}(0) = [n^2(0) - n_2^2]^{1/2} = (n_1^2 - n_2^2)^{1/2} \simeq n_1\sqrt{2\Delta} \quad (2\text{-}80b)$$

Thus, the NA of a graded-index fiber decreases from NA(0) to zero as r moves from the fiber axis to the core–cladding boundary. A comparison of the numerical apertures for fibers having various α profiles is shown in Fig. 2-25. The number of bound modes in a graded-index fiber is[45-48]

$$M = \frac{\alpha}{\alpha + 2} a^2 k^2 n_1^2 \Delta \quad (2\text{-}81)$$

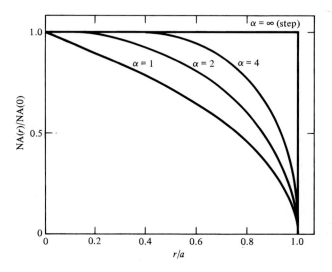

FIGURE 2-25
A comparison of the numerical apertures for fibers having various α profiles.

2.7 FIBER MATERIALS

In selecting materials for optical fibers, a number of requirements must be satisfied. For example:

1. It must be possible to make long, thin, flexible fibers from the material.
2. The material must be transparent at a particular optical wavelength in order for the fiber to guide light efficiently.
3. Physically compatible materials that have slightly different refractive indices for the core and cladding must be available.

Materials that satisfy these requirements are glasses and plastics.

The majority of fibers are made of glass consisting of either silica (SiO_2) or a silicate. The variety of available glass fibers ranges from high-loss glass fibers with large cores used for short-transmission distances to very transparent (low-loss) fibers employed in long-haul applications. Plastic fibers are less widely used because of their substantially higher attenuation than glass fibers. The main use of plastic fibers is in short-distance applications (several hundred meters) and in abusive environments, where the greater mechanical strength of plastic fibers offers an advantage over the use of glass fibers.

2.7.1 Glass Fibers

Glass is made by fusing mixtures of metal oxides, sulfides, or selenides.[49-54] The resulting material is a randomly connected molecular network rather than a well-defined ordered structure as found in crystalline materials. A consequence of this random order is that glasses do not have well-defined melting points. When glass is heated up from room temperature, it remains a hard solid up to several hundred degrees centigrade. As the temperature increases further, the glass gradually begins to soften until at very high temperatures it becomes a viscous liquid. The expression "melting temperature" is commonly used in glass manufacture. This term refers only to an extended temperature range in which the glass becomes fluid enough to free itself fairly quickly of gas bubbles.

The largest category of optically transparent glasses from which optical fibers are made consists of the oxide glasses. Of these, the most common is silica (SiO_2), which has a refractive index of 1.458 at 850 nm. To produce two similar materials that have slightly different indices of refraction for the core and cladding, either fluorine or various oxides (referred to as *dopants*), such as B_2O_3, GeO_2, or P_2O_5, are added to the silica. As shown in Fig. 2-26 the addition of GeO_2 or P_2O_5 increases the refractive index, whereas doping the silica with fluorine or B_2O_3 decreases it. Since the cladding must have a lower index than the core, examples of fiber compositions are

1. GeO_2–SiO_2 core; SiO_2 cladding
2. P_2O_5–SiO_2 core; SiO_2 cladding

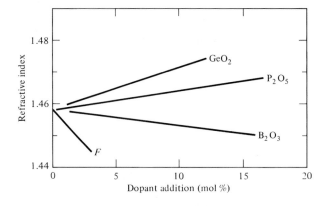

FIGURE 2-26
Variation in refractive index as a function of doping concentration in silica glass.

3. SiO_2 core; B_2O_3–SiO_2 cladding
4. GeO_2–B_2O_3–SiO_2 core; B_2O_3–SiO_2 cladding

Here, the notation GeO_2–SiO_2, for example, denotes a GeO_2-doped silica glass.

The principal raw material for silica is sand. Glass composed of pure silica is referred to as either *silica glass, fused silica,* or *vitreous silica.* Some of its desirable properties are a resistance to deformation at temperatures as high as 1000°C, a high resistance to breakage from thermal shock because of its low thermal expansion, good chemical durability, and high transparency in both the visible and infrared regions of interest to fiber optic communication systems. Its high melting temperature is a disadvantage if the glass is prepared from a molten state. However, this problem is partially avoided when using vapor deposition techniques.

2.7.2 Halide Glass Fibers

In 1975 researchers at the Université de Rennes[55] discovered fluoride glasses that have extremely low transmission losses at mid-infrared wavelengths (0.2–8 μm, with the lowest loss being around 2.55 μm). Fluoride glasses belong to a general family of halide glasses in which the anions are from elements in group VII of the periodic table, namely fluorine, chlorine, bromine, and iodine.

The material that researchers have concentrated on is a *heavy metal fluoride glass*, which uses ZrF_4 as the major component and glass network former. Several other constituents need to be added to make a glass that has moderate resistance to crystallization.[56,57] Table 2-3 lists the constituents and their molecular percentages of a particular fluoride glass referred to as ZBLAN (after its elements ZrF_4, BaF_2, LaF_3, AlF_3 and NaF). This material forms the core of a glass fiber. To make a lower-refractive-index glass, one partially replaces ZrF_4 by HaF_4 to get a ZHBLAN cladding.

Although these glasses potentially offer intrinsic minimum losses of 0.01–0.001 dB/km, fabricating long lengths of these fibers is difficult. First, ultrapure

TABLE 2-3
Molecular composition of a ZBLAN fluoride glass

Material	Molecular percentage
ZrF_4	54
BaF_2	20
LaF_3	4.5
AlF_3	3.5
NaF	18

materials must be used to reach this low loss level. Second, fluoride glass is prone to devitrification. Fiber-making techniques have to take this into account to avoid the formation of microcrystallites, which have a drastic effect on scattering losses.

2.7.3 Active Glass Fibers

Incorporating rare-earth elements (atomic numbers 57–71) into a normally passive glass gives the resulting material new optical and magnetic properties. These new properties allow the material to perform amplification, attenuation, and phase retardation on the light passing through it.[58-60] Doping can be carried out for both silica and halide glasses.

Two commonly used materials for fiber lasers are erbium and neodymium. The ionic concentrations of the rare-earth elements are low (on the order of 0.005–0.05 mole percent) to avoid clustering effects. By examining the absorption and fluorescence spectra of these materials, one can use an optical source which emits at an absorption wavelength to excite electrons to higher energy levels in the rare-earth dopants. When these excited electrons drop to lower energy levels, they emit light in a narrow optical spectrum at the fluorescence wavelength. Chapter 11 discusses the applications of erbium-doped fibers to optical amplifiers.

2.7.4 Chalgenide Glass Fibers

In addition to allowing the creation of optical amplifiers, the nonlinear properties of glass fibers can be exploited for other applications, such as all-optical switches and fiber lasers. Chalgenide glass is one candidate for these uses because of its high optical nonlinearity and its long interaction length.[61,62] These glasses contain at least one chalcogen element (S, Se, or Te) and typically one other element such as P, I, Cl, Br, Cd, Ba, Si, or Tl for tailoring the thermal, mechanical, and optical properties of the glass. Among the various chalgenide glasses, As_2S_3 is one of the most well-known materials. Single-mode fibers have been made using $As_{40}S_{58}Se_2$ and As_2S_3 for the core and cladding materials, respectively. Losses in these glasses typically range around 1 dB/m.

TABLE 2-4
Sample characteristics of PMMA and PFP polymer optical fibers

Characteristic	PMMA POF	PFP POF
Core diameter	0.4 mm	0.125–0.30 mm
Cladding diameter	1.0 mm	0.25–0.60 mm
Numerical aperture	0.25	0.20
Attenuation	150 dB/km at 650 nm	60–80 dB/km at 650–1300 nm
Bandwidth	2.5 Gb/s over 100 m	2.5 Gb/s over 300 m

2.7.5   Plastic Optical Fibers

The growing demand for delivering high-speed services directly to the workstation has led fiber developers to create high-bandwidth graded-index polymer (plastic) optical fibers (POF) for use in a customer premises.[63-65] The core of these fibers is either polymethylmethacrylate or a perfluorinated polymer. These fibers are hence referred to as PMMA POF and PFP POF, respectively. Although they exhibit considerably greater optical signal attenuations than glass fibers, they are tough and durable. For example, since the modulus of these polymers is nearly two orders of magnitude lower than that of silica, even a 1-mm-diameter graded-index POF is sufficiently flexible to be installed in conventional fiber cable routes. Compared with silica fibers, the core diameters of plastic fibers are 10–20 times larger, which allows a relaxation of connector tolerances without sacrificing optical coupling efficiencies. Thus, inexpensive plastic injection-molding technologies can be used to fabricate connectors, splices, and transceivers.

Table 2-4 gives sample characteristics of PMMA and PFP polymer optical fibers.

2.8   FIBER FABRICATION

Two basic techniques[66-68] are used in the fabrication of all-glass optical waveguides. These are the vapor-phase oxidation process and the direct-melt methods. The direct-melt method follows traditional glass-making procedures in that optical fibers are made directly from the molten state of purified components of silicate glasses. In the vapor-phase oxidation process, highly pure vapors of metal halides (e.g., $SiCl_4$ and $GeCl_4$) react with oxygen to form a white powder of SiO_2 particles. The particles are then collected on the surface of a bulk glass by one of four different commonly used processes and are sintered (transformed to a homogeneous glass mass by heating without melting) by one of a variety of techniques to form a clear glass rod or tube (depending on the process). This rod or tube is called a *preform*. It is typically around 10–25 mm in diameter and 60–120 cm long. Fibers are made from the preform[69-72] by using the equipment shown in Fig. 2-27. The preform is precision-fed into a circular heater called the *drawing furnace*. Here, the preform end is softened to the point where it can be

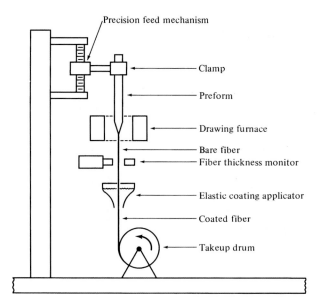

FIGURE 2-27
Schematic of a fiber-drawing apparatus.

drawn into a very thin filament, which becomes the optical fiber. The turning speed of the takeup drum at the bottom of the draw tower determines how fast the fiber is drawn. This, in turn, will determine the thickness of the fiber, so that a precise rotation rate must be maintained. An optical fiber thickness monitor is used in a feedback loop for this speed regulation. To protect the bare glass fiber from external contaminants, such as dust and water vapor, an elastic coating is applied to the fiber immediately after it is drawn.

2.8.1   Outside Vapor-Phase Oxidation

The first fiber to have a loss of less than 20 dB/km was made at the Corning Glass Works[73-75] by the *outside vapor-phase oxidation* (OVPO) process. This method is illustrated in Fig. 2-28. First, a layer of SiO_2 particles called a *soot* is deposited from a burner onto a rotating graphite or ceramic mandrel. The glass soot adheres to this bait rod and, layer by layer, a cylindrical, porous glass preform is built up. By properly controlling the constituents of the metal halide vapor stream during the deposition process, the glass compositions and dimensions desired for the core and cladding can be incorporated into the preform. Either step- or graded-index preforms can thus be made.

When the deposition process is completed, the mandrel is removed and the porous tube is then vitrified in a dry atmosphere at a high temperature (above 1400°) to a clear glass preform. This clear preform is subsequently mounted in a fiber-drawing tower and made into a fiber, as shown in Fig. 2-27. The central hole in the tube preform collapses during this drawing process.

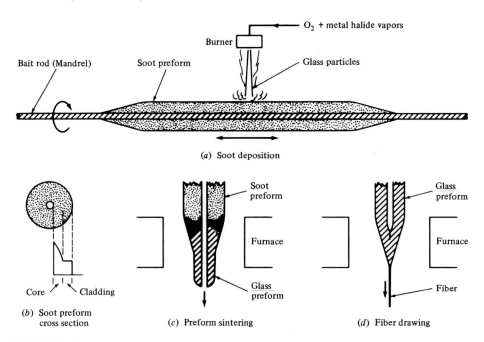

FIGURE 2-28
Basic steps in preparing a preform by the OVPO process. (*a*) Bait rod rotates and moves back and forth under the burner to produce a uniform deposition of glass soot particles along the rod; (*b*) profiles can be step or graded index; (*c*) following deposition, the soot preform is sintered into a clear glass preform; (*d*) fiber is drawn from the glass preform. (Reproduced with permission from Schultz.[66])

2.8.2 Vapor-Phase Axial Deposition

The OVPO process described in Sec. 2.8.1 is a lateral deposition method. Another OVPO-type process is the *vapor-phase axial deposition* (VAD) method,[76,77] illustrated in Fig. 2-29. In this method, the SiO_2 particles are formed in the same way as described in the OVPO process. As these particles emerge from the torches, they are deposited onto the end surface of a silica glass rod which acts as a seed. A porous preform is grown in the axial direction by moving the rod upward. The rod is also continuously rotated to maintain cylindrical symmetry of the particle deposition. As the porous preform moves upward, it is transformed into a solid, transparent rod preform by zone melting (heating in a narrow localized zone) with the carbon ring heater shown in Fig. 2-29. The resultant preform can then be drawn into a fiber by heating it in another furnace, as shown in Fig. 2-27.

Both step- and graded-index fibers in either multimode or single-mode varieties can be made by the VAD method. The advantages of the VAD method are (1) the preform has no central hole as occurs with the OVPO process, (2) the preform can be fabricated in continuous lengths which can affect process costs and product yields, and (3) the fact that the deposition chamber and the zone-melting ring heater are tightly connected to each other in the same enclosure allows the achievement of a clean environment.

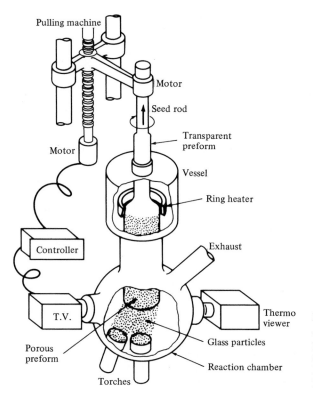

FIGURE 2-29
Apparatus for the VAD (vapor-phase axial deposition) process. (Reproduced with permission from Izawa and Inagaki,[76] © 1980, IEEE.)

2.8.3 Modified Chemical Vapor Deposition

The *modified chemical vapor deposition* (MCVD) process shown in Fig. 2-30 was pioneered at Bell Laboratories,[53,78] and widely adopted elsewhere to produce very low-loss graded-index fibers. The glass vapor particles, arising from the reaction of the constituent metal halide gases and oxygen, flow through the inside of a revolving silica tube. As the SiO_2 particles are deposited, they are sintered to a clear glass layer by an oxyhydrogen torch which travels back and forth along the tube. When the desired thickness of glass has been deposited, the vapor flow is shut off and the tube is heated strongly to cause it to collapse into a solid rod preform. The fiber that is subsequently drawn from this preform rod will have a core that consists of the vapor-deposited material and a cladding that consists of the original silica tube.

2.8.4 Plasma-Activated Chemical Vapor Deposition

Scientists at Philips Research invented the *plasma-activated chemical vapor deposition* (PCVD) process.[79-81] As shown in Fig. 2-31, the PCVD method is similar to the MCVD process in that deposition occurs within a silica tube. However, a

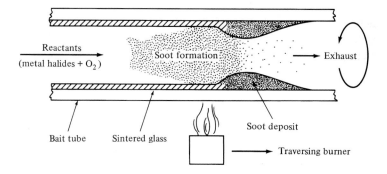

FIGURE 2-30
Schematic of MCVD (modified chemical vapor deposition) process. (Reproduced with permission from Schultz.[66])

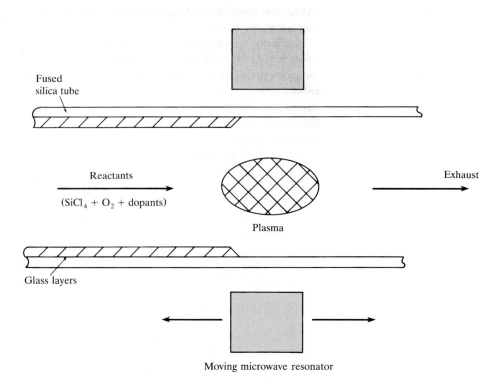

FIGURE 2-31
Schematic of PCVD (plasma-activated chemical vapor deposition) process.

nonisothermal microwave plasma operating at low pressure initiates the chemical reaction. With the silica tube held at temperatures in the range of 1000–1200°C to reduce mechanical stresses in the growing glass films, a moving microwave resonator operating at 2.45 GHz generates a plasma inside the tube to activate the chemical reaction. This process deposits clear glass material directly on the tube wall; there is no soot formation. Thus, no sintering is required. When one has deposited the desired glass thickness, the tube is collapsed into a preform just as in the MCVD case.

2.8.5 Double-Crucible Method

Silica, chalgenide, and halide glass fibers can all be made using a direct-melt *double-crucible* technique.[8,57] In this method, glass rods for the core and cladding materials are first made separately by melting mixtures of purified powders to make the appropriate glass composition. These rods are then used as feedstock for each of two concentric crucibles, as shown in Fig. 2-32. The inner crucible contains molten core glass and the outer one contains the cladding glass. The fibers are drawn from the molten state through orifices in the bottom of the two concentric crucibles in a continuous production process.

Although this method has the advantage of being a continuous process, careful attention must be paid to avoid contaminants during the melting. The main sources of contamination arise from the furnace environment and from the crucible. Silica crucibles are normally used in preparing the glass feed rods, whereas the double concentric crucibles used in the drawing furnace are made from platinum. A detailed description of the crucible design and an analysis of the fiber-drawing process is given by Midwinter.[8]

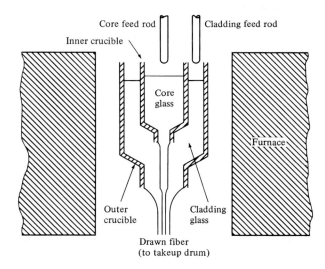

FIGURE 2-32
Double-crucible arrangement for drawing fibers from molten glass.

2.9 MECHANICAL PROPERTIES OF FIBERS

In addition to the transmission properties of optical waveguides, their mechanical characteristics play a very important role when they are used as the transmission medium in optical communication systems.[82-85] Fibers must be able to withstand the stresses and strains that occur during the cabling process and the loads induced during the installation and service of the cable. During cable manufacture and installation, the loads applied to the fiber can be either impulsive or gradually varying. Once the cable is in place, the service loads are usually slowly varying loads, which can arise from temperature variations or a general settling of the cable following installation.

Strength and *static fatigue* are the two basic mechanical characteristics of glass optical fibers. Since the sight and sound of shattering glass are quite familiar, one intuitively suspects that glass is not a very strong material. However, the longitudinal breaking stress of pristine glass fibers is comparable to that of metal wires. The cohesive bond strength of the constituent atoms of a glass fiber governs its theoretical intrinsic strength. Maximum tensile strengths of 14 GPa (2×10^6 lb/in.^{2}) have been observed in short-gauge-length glass fibers. This is close to the 20-GPa tensile strength of steel wire. The difference between glass and metal is that, under an applied stress, glass will extend elastically up to its breaking strength, whereas metals can be stretched plastically well beyond their true elastic range. Copper wires, for example, can be elongated plastically by more than 20 percent before they fracture. For glass fibers, elongations of only about 1 percent are possible before fracture occurs.

In practice, the existence of stress concentrations at surface flaws or microcracks limits the median strength of long glass fibers to the 700-to-3500-MPa ($1-5 \times 10^5$ lb/in.^{2}) range. The fracture strength of a given length of glass fiber is determined by the size and geometry of the severest flaw (the one that produces the largest stress concentration) in the fiber. A hypothetical, physical flaw model is shown in Fig. 2-33. This elliptically shaped crack is generally referred to as a *Griffith microcrack*.[86] It has a width w, a depth χ, and a tip radius ρ. The strength of the crack for silica fibers follows the relation

$$K = Y\chi^{1/2}\sigma \qquad (2-82)$$

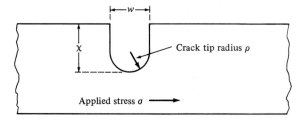

FIGURE 2-33
A hypothetical model of a microcrack in an optical fiber.

where the stress intensity factor K is given in terms of the stress σ in megapascals applied to the fiber, the crack depth χ in millimeters, and a dimensionless constant Y that depends on flaw geometry. For surface flaws, which are the most critical in glass fibers, $Y = \sqrt{\pi}$. From this equation the maximum crack size allowable for a given applied stress level can be calculated. The maximum values of K depend upon the glass composition but tend to be in the range 0.6–$0.9 \, \text{MN/m}^{3/2}$.

Since an optical fiber generally contains many flaws that have a random distribution of size, the fracture strength of a fiber must be viewed statistically. If $F(\sigma, L)$ is defined as the cumulative probability that a fiber of length L will fail below a stress level σ, then, under the assumption that the flaws are independent and randomly distributed in the fiber and that the fracture will occur at the most severe flaw, we have

$$F(\sigma, L) = 1 - e^{-LN(\sigma)} \qquad (2\text{-}83)$$

where $N(\sigma)$ is the cumulative number of flaws per unit length with a strength less than σ. A widely used form for $N(\sigma)$ is the empirical expression proposed by Weibull,[87]

$$N(\sigma) = \frac{1}{L_0} \left(\frac{\sigma}{\sigma_0} \right)^m \qquad (2\text{-}84)$$

where m, σ_0, and L_0 are constants related to the initial inert strength distribution. This leads to the so-called *Weibull expression*

$$F(\sigma, L) = 1 - \exp\left[-\left(\frac{\sigma}{\sigma_0} \right)^m \frac{L}{L_0} \right] \qquad (2\text{-}85)$$

A plot of Weibull expression is shown in Fig. 2-34 for measurements performed on long-fiber samples.[85,88] These data were obtained by testing to destruction a large number of fiber samples. The fact that a single curve can be drawn through the data indicates that the failures arise from a single type of flaw. Earlier works[89] showed a double-curve Weibull distribution with different slopes for short and long fibers. This is indicative of flaws arising from two sources: one from the fiber manufacturing process and the other from fundamental flaws occurring in the glass preform and the fiber. By careful environmental control of the fiber-drawing furnace, numerous 1-km lengths of silica fiber with a single failure distribution and a maximum strength of 3500 MPa have been fabricated.

In contrast to strength, which relates to instantaneous failure under an applied load, *static fatigue* relates to the slow growth of pre-existing flaws in the glass fiber under humid conditions and tensile stress.[82,85] This gradual flaw growth causes the fiber to fail at a lower stress level than that which could be reached under a strength test. A flaw such as the one shown in Fig. 2-33 propagates through the fiber because of chemical erosion of the fiber material at the flaw tip. The primary cause of this erosion is the presence of water in the environment, which reduces the strength of the SiO_2 bonds in the glass. The speed of the growth reaction is increased when the fiber is put under stress. Certain fiber

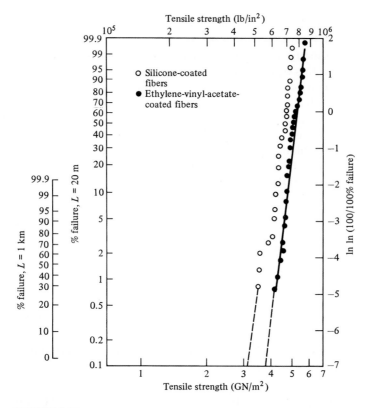

FIGURE 2-34

A Weibull-type plot showing the cumulative probability that fibers of 20-m and 1-km lengths will fracture at the indicated applied stress. (Reproduced with permission from Miller, Hart, Vroom, and Bowden.[88])

materials are more resistant to static fatigue than others, with fused silica being the most resistant of the glasses in water. In general, coatings that are applied to the fiber immediately during the manufacturing process afford a good degree of protection against environmental corrosion.[90]

Another important factor to consider is *dynamic fatigue*. When an optical cable is being installed in a duct, it experiences repeated stress owing to surging effects. The surging is caused by varying degrees of friction between the optical cable and the duct or guiding tool in a manhole on a curved route. Varying stresses also arise in aerial cables that are set into transverse vibration by the wind. Theoretical and experimental investigations[91,92] have shown that the time to failure under these conditions is related to the maximum allowable stress by the same lifetime parameters that are found from the cases of static stress and stress that increases at a constant rate.

A high assurance of fiber reliability can be provided by proof testing.[35,93,94] In this method, an optical fiber is subjected to a tensile load greater than that expected at any time during the cable manufacturing, installation, and service. Any fibers which do not pass the proof test are rejected. Empirical studies of slow crack growth show that the growth rate $d\chi/dt$ is approximately proportional to a power of the stress intensity factor; that is,

$$\frac{d\chi}{dt} = AK^b \tag{2-86}$$

Here, A and b are material constants and the stress intensity factor is given by Eq. (2-82). For most glasses, b ranges between 15 and 50.

If a proof test stress σ_p is applied for a time t_p, then from Eq. (2-86) we have

$$B(\sigma_i^{b-2} - \sigma_p^{b-2}) = \sigma_p^b t_p \tag{2-87}$$

where σ_i is the initial inert strength and

$$B = \frac{2}{b-2}\left(\frac{K}{Y}\right)^{2-b}\frac{1}{AY^b} \tag{2-88}$$

When this fiber is subjected to a static stress σ_s after proof testing, the time to failure t_s is found from Eq. (2-86) to be

$$B(\sigma_p^{b-2} - \sigma_s^{b-2}) = \sigma_s^b t_s \tag{2-89}$$

Combining Eqs. (2-87) and (2-89) yields

$$B(\sigma_i^{b-2} - \sigma_s^{b-2}) = \sigma_p^b t_p + \sigma_s^b t_s \tag{2-90}$$

To find the failure probability F_s of a fiber after a time t_s after proof testing, we first define $N(t, \sigma)$ to be the number of flaws per unit length which when fail in a time t under an applied stress σ. Assuming that $N(\sigma_i) \gg N(\sigma_s)$, then

$$N(t_s, \sigma_s) \simeq N(\sigma_i) \tag{2-91}$$

Solving Eq. (2-90) for σ_i and substituting into Eq. (2-84), we have, from Eq. (2-91),

$$N(t_s, \sigma_s) = \frac{1}{L_0}\left\{\frac{\left[(\sigma_p^b t_p + \sigma_s^b t_s)/B + \sigma_s^{b-2}\right]^{1/(b-2)}}{\sigma_0}\right\}^m \tag{2-92}$$

The failure number $N(t_p, \sigma_p)$ per unit length during proof testing is found from Eq. (2-92) by setting $\sigma_s = \sigma_p$ and letting $t_s = 0$, so that

$$N(t_p, \sigma_p) = \frac{1}{L_0}\left[\frac{\left(\sigma_p^b t_p/B + \sigma_p^{b-2}\right)^{1/(b-2)}}{\sigma_0}\right]^m \tag{2-93}$$

Letting $N(t_x, \sigma_x) = N_x$, the failure probability F_s for a fiber after it has been proof tested is given by

$$F_s = 1 - e^{-L(N_s - N_p)} \tag{2-94}$$

Substituting Eqs. (2-92) and (2-93) into Eq. (2-94), we have

$$F_s = 1 - \exp\left(-N_p L\left\{\left[\left(1 + \frac{\sigma_s^b t_s}{\sigma_p^b t_p}\right)\frac{1}{1 + C}\right]^{m/(b-2)} - 1\right\}\right) \tag{2-95}$$

where $C = B/(\sigma_p^2 t_p)$, and where we have ignored the term

$$\left(\frac{\sigma_s}{\sigma_p}\right)^b \frac{B}{\sigma_s^2 t_p} \ll 1 \tag{2-96}$$

This holds because typical values of the parameters in this term are $\sigma_s/\sigma_p \simeq 0.3$–$0.4$, $t_p \simeq 10$ s, $b > 15$, $\sigma_p = 350\,\text{MN/m}^2$, and $B \simeq 0.05$–$0.5\,(\text{MN/m}^2)^2 \cdot \text{s}$.

The expression for F_s given by Eq. (2-95) is valid only when the proof stress is unloaded immediately, which is not the case in actual proof testing of optical fibers. When the proof stress is released within a finite duration, the C value should be rewritten as

$$C = \gamma \frac{B}{\sigma_p^2 t_p} \tag{2-97}$$

where γ is a coefficient of slow-crack-growth effect arising during the unloading period.

2.10   FIBER OPTIC CABLES

In any practical application of optical waveguide technology, the fibers need to be incorporated in some type of cable structure.[95-98] The cable structure will vary greatly, depending on whether the cable is to be pulled into underground or intrabuilding ducts, buried directly in the ground, installed on outdoor poles, or submerged under water. Different cable designs are required for each type of application, but certain fundamental cable design principles will apply in every case. The objectives of cable manufacturers have been that the optical fiber cables should be installable with the same equipment, installation techniques, and precautions as those used for conventional wire cables. This requires special cable designs because of the mechanical properties of glass fibers.

One important mechanical property is the maximum allowable axial load on the cable, since this factor determines the length of cable that can be reliably installed. In copper cables the wires themselves are generally the principal load-bearing members of the cable, and elongations of more than 20 percent are possible without fracture. On the other hand, extremely strong optical fibers tend to break at 4-percent elongation, whereas typical good-quality fibers exhibit long-length breaking elongations of about 0.5–1.0 percent. Since static fatigue occurs very quickly at stress levels above 40 percent of the permissible elongation and very slowly below 20 percent, fiber elongations during cable manufacture and installation should be limited to 0.1–0.2 percent.

Steel wire which has a Young's modulus of 2×10^4 MPa has been extensively used for reinforcing conventional electric cables and can also be employed for optical fiber cables. For some applications it is desirable to use nonmetallic constructions, either to avoid the effects of electromagnetic induction or to reduce cable weight. In this case, plastic strength members and high-tensile-strength organic yarns such as Kevlar® (a product of the DuPont Chemical Corporation) are used. With good fabrication practices, the optical fibers are isolated from other cable components, they are kept close to the neutral axis of the cable, and room is provided for the fibers to move when the cable is flexed or stretched.

Another factor to consider is fiber brittleness. Since glass fibers do not deform plastically, they have a low tolerance for absorbing energy from impact loads. Hence, the outer sheath of an optical cable must be designed to protect the glass fibers inside from impact forces. In addition, the outer sheath should not crush when subjected to side forces, and it should provide protection from corrosive environmental elements. In underground installations, a heavy-gauge-metal outer sleeve may also be required to protect against potential damage from burrowing rodents, such as gophers.

In designing optical fiber cables, several types of fiber arrangements are possible and a large variety of components could be included in the construction. The simplest designs are one- or two-fiber cables intended for indoor use. In the hypothetical two-fiber design shown in Fig. 2-35, a fiber is first coated with a buffer material and placed loosely in a tough, oriented polymer tube, such as polyethylene. For strength purposes this tube is surrounded by strands of aramid yarn and, in turn, is encapsulated in a polyurethane jacket. A final outer jacket of polyurethane, polyethylene, or nylon binds the two encapsulated fiber units together.

Larger cables can be created by stranding several basic fiber building blocks (as shown in Fig. 2-35) around a central strength member. This is illustrated in Fig. 2-36 for a six-fiber cable. The fiber units are bound onto the strength member with paper or plastic binding tape, and then surrounded by an outer jacket. If

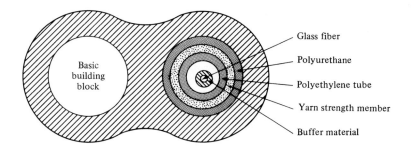

FIGURE 2-35
A hypothetical two-fiber cable design. The basic building block on the left is identical to that shown for the right-hand fiber.

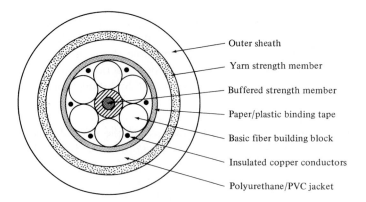

Outer sheath

Yarn strength member

Buffered strength member

Paper/plastic binding tape

Basic fiber building block

Insulated copper conductors

Polyurethane/PVC jacket

FIGURE 2-36
A typical six-fiber cable created by stranding six basic fiber-building blocks around a central strength member.

repeaters are required along the route where the cable is to be installed, it may be advantageous to include wires within the cable structure for powering these repeaters. The wires can also be used for fault isolation or as an engineering order wire for voice communications during cable installations.

PROBLEMS

2-1. Consider an electric field by the expression

$$\mathbf{E} = \left[100e^{j30°}\mathbf{e}_x + 20e^{-j50°}\mathbf{e}_y + 40e^{j210°}\mathbf{e}_z\right]e^{j\omega t}$$

Express this as a measurable electric field as described by Eq. (2-2) at a frequency of 100 MHz.

2-2. A wave is specified by $y = 8\cos 2\pi(2t - 0.8z)$, where y is expressed in micrometers and the propagation constant is given in μm^{-1}. Find (a) the amplitude, (b) the wavelength, (c) the angular frequency, and (d) the displacement at time $t = 0$ and $z = 4\,\mu m$.

2-3. What are the energies in electron volts (eV) of light at wavelengths of 820 nm, 1320 nm, and 1550 nm? What are the values of the propagation constant k at these wavelengths?

2-4. Consider two plane waves X_1 and X_2 traveling in the same direction. If they have the same frequency ω but different amplitudes a_i and phases δ_i, then we can represent them by

$$X_1 = a_1\cos(\omega t - \delta_1)$$
$$X_2 = a_2\cos(\omega t - \delta_2)$$

According to the principle of superposition, the resultant wave X is simply the sum of X_1 and X_2. Show that X can be written in the form

$$X = A\cos(\omega t - \phi)$$

where

$$A^2 = a_1^2 + a_2^2 + 2a_1 a_2 \cos(\delta_1 - \delta_2)$$

and

$$\tan\phi = \frac{a_1 \sin\delta_1 + a_2 \sin\delta_2}{a_1 \cos\delta_1 + a_2 \cos\delta_2}$$

2-5. Elliptically polarized light can be represented by the two orthogonal waves given in Eqs. (2-2) and (2-3). Show that elimination of the $(\omega t - kz)$ dependence between them yields

$$\left(\frac{E_x}{E_{0x}}\right)^2 + \left(\frac{E_y}{E_{0y}}\right)^2 - 2\frac{E_x}{E_{0x}}\frac{E_y}{E_{0y}}\cos\delta = \sin^2\delta$$

which is the equation of an ellipse making an angle α with the x axis, where α is given by Eq. (2-8).

2-6. Let $E_{0x} = E_{0y} = 1$ in Eq. (2-7). Using a computer or a graphical calculator, write a program to plot this equation for values of $\delta = (n\pi)/8$, where $n = 0, 1, 2, \ldots, 16$. What does this show about the state of polarization as the angle δ changes?

2-7. Show that any linearly polarized wave may be considered as the superposition of left and right circularly polarized waves which are in phase and have equal amplitudes and frequencies.

2-8. Light traveling in air strikes a glass plate at an angle $\theta_1 = 33°$, where θ_1 is measured between the incoming ray and the glass surface. Upon striking the glass, part of the beam is reflected and part is refracted. If the refracted and reflected beams make an angle of $90°$ with each other, what is the refractive index of the glass? What is the critical angle for this glass?

2-9. A point source of light is 12 cm below the surface of a large body of water ($n = 1.33$ for water). What is the radius of the largest circle on the water surface through which the light can emerge?

2-10. A $45°$–$45°$–$90°$ prism is immersed in alcohol ($n = 1.45$). What is the minimum refractive index the prism must have if a ray incident normally on one of the short faces is to be totally reflected at the long face of the prism?

2-11. Calculate the numerical aperture of a step-index fiber having $n_1 = 1.48$ and $n_2 = 1.46$. What is the maximum entrance angle $\theta_{0,\max}$ for this fiber if the outer medium is air with $n = 1.00$?

2-12. Consider a dielectric slab having a thickness $d = 10$ mm and index of refraction $n_1 = 1.50$. Let the medium above and below the slab be air, in which $n_2 = 1$. Let the wavelength be $\lambda = 10$ mm (equal to the thickness of the waveguide).

(a) What is the critical angle for the slab waveguide?

(b) Solve Eq. (2-26b) graphically to show that there are three angles of incidence which satisfy this equation.

(c) What happens to the number of angles as the wavelength is decreased?

2-13. Derive the approximation of the right-hand side of Eq. (2-23) for $\Delta \ll 1$. What is the difference in the approximate and exact expressions for the value of NA if $n_1 = 1.49$ and $n_2 = 1.48$?

2-14. Using the expressions in Eqs. (2-33) and (2-34) derived from Maxwell's curl equations, derive the radial and transverse electric and magnetic field components given in Eqs. (2-35a) to (2-35d). Show that these expressions lead to Eqs. (2-36) and (2-37).

2-15. Show that for $v = 0$, Eq. (2-55b) corresponds to TE_{0m} modes ($E_z = 0$) and that Eq. (2-56b) corresponds to TM_{0m} modes ($H_z = 0$).

2-16. Verify that $k_1^2 \approx k_2^2 \approx \beta^2$ when $\Delta \ll 1$, where k_1 and k_2 are the core and cladding propagation constants, respectively, as defined in Eq. (2-46).

2-17. Replot Fig. 2-16 for $J_0(x)$ and $J_1(x)$. On the resulting graph, indicate the range of values of x for the lower-order LP_{1m} modes and the exact lower-order HE_{1m}, TE_{0m}, and TM_{0m} modes.

2-18. A step-index multimode fiber with a numerical aperture of 0.20 supports approximately 1000 modes at an 850-nm wavelength.
(a) What is the diameter of its core?
(b) How many modes does the fiber support at 1320 nm?
(c) How any modes does the fiber support at 1550 nm?

2-19. (a) Determine the normalized frequency at 820 nm for a step-index fiber having a 25-μm core radius, $n_1 = 1.48$, and $n_2 = 1.46$.
(b) How many modes propagate in this fiber at 820 nm?
(c) How many modes propagate in this fiber at 1320 nm?
(d) How many modes propagate in this fiber at 1550 nm?
(e) What percent of the optical power flows in the cladding in each case?

2-20. Consider a fiber with a 25-μm core radius, a core index $n_1 = 1.48$, and $\Delta = 0.01$.
(a) If $\lambda = 1320$ nm, what is the value of V and how many modes propagate in the fiber?
(b) What percent of the optical power flows in the cladding?
(c) If the core–cladding difference is reduced to $\Delta = 0.003$, how many modes does the fiber support and what fraction of the optical power flows in the cladding?

2-21. Find the core radius necessary for single-mode operation at 1320 nm of a step-index fiber with $n_1 = 1.480$ and $n_2 = 1.478$. What are the numerical aperture and maximum acceptance angle of this fiber?

2-22. A manufacturer wishes to make a silica-core, step-index fiber with $V = 75$ and a numerical aperture NA $= 0.30$ to be used at 820 nm. If $n_1 = 1.458$, what should the core size and cladding index be?

2-23. Draw a design curve of the fractional refractive-index difference Δ versus the core radius a for a silica-core ($n_1 = 1.458$), single-mode fiber to operate at 1300 nm. Suppose the fiber we select from this curve has a 5-μm core radius. Is this fiber still single-mode at 820 nm? Which modes exist in the fiber at 820 nm?

2-24. Using the following approximation for W_0 given by Marcuse,[99]

$$W_0 = a(0.65 + 1.619V^{-3/2} + 2.879V^{-6})$$

evaluate and plot $E(r)/E_0$ with r ranging from 0 to 3 for values of $V = 1.0, 1.4, 1.8, 2.2, 2.6$, and 3.0.

2-25. Commonly available single-mode fibers have beat lengths in the range 10 cm $< L_p$ < 2 m. What range of refractive-index differences does this correspond to for $\lambda = 1300$ nm?

2-26. Plot the refractive-index profiles from n_1 to n_2 as a function of radial distance $r \leq a$ for graded-index fibers that have α values of 1, 2, 4, 8, and ∞ (step index). Assume the fibers have a 25-μm core radius, $n_1 = 1.48$, and $\Delta = 0.01$.

2-27. Calculate the number of modes at 820 nm and 1.3 μm in a graded-index fiber having a parabolic-index profile ($\alpha = 2$), a 25-μm core radius, $n_1 = 1.48$, and $n_2 = 1.46$. How does this compare to a step-index fiber?

2-28. Calculate the numerical apertures of (*a*) a plastic step-index fiber having a core refractive index of $n_1 = 1.60$ and a cladding index of $n_2 = 1.49$, (*b*) a step-index fiber having a silica core ($n_1 = 1.458$) and a silicone resin cladding ($n_2 = 1.405$).

2-29. When a preform is drawn into a fiber, the principle of conservation of mass must be satisfied under steady-state drawing conditions. Show that for a solid rod preform this is represented by the expression

$$s = S\left(\frac{D}{d}\right)^2$$

where D and d are the preform and fiber diameters, and S and s are the preform feed and fiber-draw speeds, respectively. A typical drawing speed is 1.2 m/s for a 125-μm outer-diameter fiber. What is the preform feed rate in cm/min for a 9-mm-diameter preform?

2-30. A silica tube with inside and outside radii of 3 and 4 mm, respectively, is to have a certain thickness of glass deposited on the inner surface. What should the thickness of this glass deposition be if a fiber having a core diameter of 50 μm and an outer cladding diameter of 125 μm is to be drawn from this preform?

2-31. (*a*) The density of fused silica is 2.6 g/cm^3. How many grams are needed for a 1-km-long 50-μm-diameter fiber core?

(*b*) If the core material is to be deposited inside of a glass tube at a 0.5-g/min deposition rate, how long does it take to make the preform for this fiber?

2-32. During fabrication of optical fibers, dust particles incorporated into the fiber surface are prime examples of surface flaws that can lead to reduced fiber strength. What size dust particles are tolerable if a glass fiber having a 20-N/mm$^{3/2}$ stress intensity factor is to withstand a 700-MN/m^2 stress?

2-33. Static fatigue in a glass fiber refers to the condition where a fiber is stressed to a level σ_a, which is much less than the fracture stress associated with the weakest flaw. Initially, the fiber will not fail but, with time, cracks in the fiber will grow as a result of chemical erosion at the crack tip. One model for the growth rate of a crack of depth χ assumes a relation of the form given in Eq. (2-86).

(*a*) Using this equation, show that the time required for a crack of initial depth χ_i to grow to its failure size χ_f is given by

$$t = \frac{2}{(b-2)A(Y\sigma)^b}(\chi_i^{(2-b)/2} - \chi_f^{(2-b)/2})$$

(*b*) For long, static fatigue times (on the order of 20 years), $K_i^{2-b} \ll K_f^{2-b}$ for large values of b. Show that under this condition the failure time is

$$t = \frac{2K_i^{2-b}}{(b-2)A\sigma^2 Y^2}$$

2-34. Derive Eq. (2-90) by starting with Eq. (2-86).

2-35. Derive Eq. (2-95) by using the expressions given in Eqs. (2-92) and (2-93) for the number of flaws per unit length failing in a time t. Verify the relationship given in Eq. (2-96).

2-36. Consider two similar fiber samples of lengths L_1 and L_2 subjected to stress levels of σ_1 and σ_2, respectively. If σ_{1c} and σ_{2c} are the corresponding fast-fracture stress levels for equal failure probability, show that

$$\frac{\sigma_{1c}}{\sigma_{2c}} = \left(\frac{L_2}{L_1}\right)^{1/m}$$

From Fig. 2-34 estimate the value of m for a 10-percent failure probability of these particular ethylene-vinyl-acetate-coated fibers.

REFERENCES

1. See any general physics book or introductory optics book; for example:
   (a) M. V. Klein and T. E. Furtak, *Optics*, Wiley, New York, 2nd ed., 1986.
   (b) E. Hecht and A. Zajac, *Optics*, Addison-Wesley, Reading, MA, 2nd ed., 1987.
   (c) F. A. Jenkins and H. E. White, *Fundamentals of Optics*, McGraw-Hill, New York, 4th ed., 1976.
2. See any introductory electromagnetics book; for example:
   (a) W. H. Hayt, Jr., *Engineering Electromagnetics*, McGraw-Hill, New York, 5th ed., 1989.
   (b) J. D. Kraus, *Electromagnetics*, McGraw-Hill, New York, 4th ed., 1992.
   (c) C. R. Paul and S. R. Nasar, *Introduction to Electromagnetic Fields*, McGraw-Hill, New York, 2nd ed., 1987.
3. E. A. J. Marcatili, "Objectives of early fibers: Evolution of fiber types," in S.E. Miller and A. G. Chynoweth, eds., *Optical Fiber Telecommunications*, Academic, New York, 1979.
4. (a) S. J. Maurer and L. B. Felsen, "Ray-optical techniques for guided waves," *Proc. IEEE*, vol. 55, pp. 1718–1729, Oct. 1967.
   (b) L. B. Felsen, "Rays and modes in optical fibers," *Electron. Lett.*, vol. 10, pp. 95–96, Apr. 1974.
5. A. W. Snyder and D. J. Mitchell, "Leaky rays on circular optical fibers," *J. Opt. Soc. Amer.*, vol. 64, pp. 599–607, May 1974.
6. A. W. Snyder and J. D. Love, *Optical Waveguide Theory*, Chapman & Hall, New York, 1983.
7. D. Marcuse, *Theory of Dielectric Optical Waveguides*, Academic, New York, 2nd ed., 1991.
8. J. Midwinter, *Optical Fibers for Transmission*, Wiley, New York, 1979.
9. R. Syms and J. Cozens, *Optical Guided Waves and Devices*, McGraw-Hill, New York, 1992.
10. H. G. Unger, *Planar Optical Waveguides and Fibers*, Clarendon, Oxford, 1977.
11. J. A. Buck, *Fundamentals of Optical Fibers*, Wiley, New York, 1995.
12. R. Olshansky, "Leaky modes in graded index optical fibers," *Appl. Opt.*, vol. 15, pp. 2773–2777, Nov. 1976.
13. A Tomita and L. G. Cohen, "Leaky-mode loss of the second propagation mode in single-mode fibers with index well profiles," *Appl. Opt.*, vol. 24, pp. 1704–1707, 1985.
14. E. Snitzer, "Cylindrical dielectric waveguide modes," *J. Opt. Soc. Amer.*, vol. 51, pp. 491–498, May 1961.
15. M. Koshiba, *Optical Waveguide Analysis*, McGraw-Hill, New York, 1992.
16. D. Marcuse, *Light Transmission Optics*, Van Nostrand Reinhold, New York, 2nd ed., 1982.
17. R. Olshansky, "Propagation in glass optical waveguides," *Rev. Mod. Phys.*, vol. 51, pp. 341–367, Apr. 1979.
18. D. Gloge, "The optical fiber as a transmission medium," *Rep. Prog. Phys.*, vol. 42, pp. 1777–1824, Nov. 1979.
19. A. W. Snyder, "Asymptotic expressions for eigenfunctions and eigenvalues of a dielectric or optical waveguide," *IEEE Trans. Microwave Theory Tech.*, vol. MTT-17, pp. 1130–1138, Dec. 1969.
20. D. Gloge, "Weakly guiding fibers," *Appl. Opt.*, vol. 10, pp. 2252–2258, Oct. 1971.

21. D. Marcuse, "Gaussian approximation of the fundamental modes of graded index fibers," *J. Opt. Soc. Amer.*, vol. 68, pp. 103–109, Jan. 1978.

22. H. M. DeRuiter, "Integral equation approach to the computation of modes in an optical waveguide," *J. Opt. Soc. Amer.*, vol. 70, pp. 1519–1524, Dec. 1980.

23. A. W. Snyder, "Understanding monomode optical fibers," *Proc. IEEE.*, vol. 69, pp. 6–13, Jan. 1981.

24. C. E. Pearson, *Handbook of Applied Mathematics*, Chapman & Hall, New York, 1990.

25. M. Kurtz, *Handbook of Applied Mathematics for Engineers and Scientists*, McGraw-Hill, New York, 1991.

26. D. Zwillinger, ed., *Standard Mathematical Tables and Formulae*, CRC Press, Boca Raton, FL, 30th ed., 1995.

27. D. Marcuse, D. Gloge, and E. A. J. Marcatili, "Guiding properties of fibers," in S. E. Miller and A. G. Chynoweth, eds., *Optical Fiber Telecommunications*, Academic, New York, 1979.

28. R. M. Gagliardi and S. Karp, *Optical Communications*, Wiley, New York, 2nd ed., 1995.

29. D. Gloge, "Propagation effects in optical fibers," *IEEE Trans. Microwave Theory Tech.*, vol. MTT-23, pp. 106–120, Jan. 1975.

30. M. Artiglia, G. Coppa, P. DiVita, M. Potenza, and A. Sharma, "Mode field diameter measurements in single-mode optical fibers," *J. Lightwave Tech.*, vol. 7, pp. 1139–1152, Aug. 1989.

31. T. J. Drapela, D. L. Franzen, A. H. Cherin, and R. J. Smith, "A comparison of far-field methods for determining mode field diameter of single-mode fibers using both gaussian and Petermann definitions," *J. Lightwave Tech.*, vol. 7, pp. 1153–1157, Aug. 1989.

32. K. Petermann, "Constraints for fundamental mode spot size for broadband dispersion-compensated single-mode fibers," *Electron. Lett.*, vol. 19, pp. 712–714, Sept. 1983.

33. M. Ohashi, K.-I. Kitayama, and S. Seikai, "Mode field diameter measurement conditions for fibers by transmitted field pattern methods," *J. Lightwave Tech.*, vol. 4, pp. 109–115, Feb. 1986.

34. L. B. Jeunhomme, *Single-Mode Fiber Optics*, Dekker, New York, 2nd ed., 1989.

35. ITU-T Recommendation G.650, *Definition and Test Methods for the Relevant Parameters of Single-Mode Fibers*, Mar. 1993.

36. TIA/EIA FOTP-164A, *Single Mode Fiber, Measurement of Mode Field Diameter by Far-Field Scanning*, May 1991.

37. F. Kapron, "Fiber-optic test methods," in F. Allard, ed., *Fiber Optics Handbook for Engineers and Scientists*, McGraw-Hill, Chicago, 1990.

38. D. L. Franzen and R. Srivastava, "Determining the mode-field diameter of single-mode optical fiber: An interlaboratory comparison," *J. Lightwave Tech.*, vol. LT-3, pp. 1073–1077, Oct. 1985.

39. W. T. Anderson, V. Shah, L. Curtis, A. J. Johnson, and J. P. Kilmer, "Mode-field diameter measurements for single-mode fibers with non-gaussian field profiles," *J. Lightwave Tech.*, vol. LT-5, pp. 211–217, Feb. 1987.

40. W. T. Anderson and D. L. Philen, "Spot-size measurements for single-mode fibers—A comparison of four techniques," *J. Lightwave Tech.*, vol. LT-1, pp. 20–26, Mar. 1983.

41. (a) I. P. Kaminow, "Polarization in optical fibers," *IEEE J. Quantum Electron.*, vol. QE-17, pp. 15–22, Jan. 1981.

    (b) I. P. Kaminow, "Polarization maintaining fibers," *Appl. Scientific Research*, vol. 41, pp. 257–270, 1984.

42. S. C. Rashleigh, "Origins and control of polarization effects in single-mode fibers," *J. Lightwave Tech.*, vol. LT-1, pp. 312–331, June 1983.

43. (a) X.-H. Zheng, W. M. Henry, and A. W. Snyder, "Polarization characteristics of the fundamental mode of optical fibers," *J. Lightwave Tech.*, vol. LT-6, pp. 1300–1305, Aug. 1988.

    (b) S.-Y. Huang, J. N. Blake, and B. Y. Kim, "Perturbation effects on mode propagation in highly elliptical core two-mode fibers," *J. Lightwave Tech.*, vol. 8, pp. 23–33, Jan. 1990.

    (c) S. J. Garth and C. Pask, "Polarization rotation in nonlinear bimodal optical fibers," *J. Lightwave Tech.*, vol. 8, pp. 129–137, Feb. 1990.

44. D. Gloge and E. Marcatili, "Multimode theory of graded core fibers," *Bell Sys. Tech. J.*, vol. 52, pp. 1563–1578, Nov. 1973.

45. E. Merzbacher, *Quantum Mechanics*, Wiley, New York, 3rd ed., 1988.

46. R. Srivastava, C. K. Kao, and R. V. Ramaswamy, "WKB analysis of planar surface waveguides with truncated index profiles." *J. Lightwave Tech.*, vol. LT-5, pp. 1605–1609, Nov. 1987.

47. B. E. A. Saleh and M. Teich, *Fundamentals of Photonics*, Wiley, New York, 1991, Chap. 8.

48. M. Born and E. Wolf, *Principles of Optics*, Pergamon, Oxford, 6th ed., 1980.

49. N. P. Bansal and R. H. Doremus, *Handbook of Glass Properties*, Academic, New York, 1986.

50. I. Fanderlilk, *Optical Properties of Glass*, Elsevier, New York, 1983.

51. J. C. Phillips, "The physics of glass," *Phys. Today*, vol. 35, pp. 27–33, Feb. 1982.

52. B. C. Bagley, C. R. Kurkjian, J. W. Mitchell, G. E. Peterson, and A. R. Tynes, "Materials, properties, and choices," in S. E. Miller and A. G. Chynoweth, eds., *Optical Fiber Telecommunications*, Academic Press, New York, 1979.

53. S. R. Nagel, "Fiber materials and fabrication methods," in S. E. Miller and I. P. Kaminow, eds., *Optical Fiber Telecommunications—II*, Academic, New York, 1988.

54. H. Rawson, *Properties and Applications of Glass*, Elsevier, New York, 1980.

55. M. Poulain, U. Poulain, J. Lucas, and P. Bran, "Verres fluores au tetrafluorure de zirconium: Propriétés optiques d'un verre dopé au Nd^{3+}." *Mater. Res. Bull.*, vol. 10, no. 4, pp. 243–246, 1975.

56. (a) D. C. Tran, G. H. Siegel, Jr., and B. Bendow, "Heavy metal fluoride glasses and fibers: A review," *J. Lightwave Tech.*, vol. 2, pp. 566–586, Oct. 1984.

    (b) J. Lucas, "Review: Fluoride Glasses," *J. Mater. Sci.*, vol. 24, pp. 1–13, Jan. 1989.

57. P. W. France, *Fluoride Glass Optical Fibers*, CRC Press, Boca Raton, FL, 1990.

58. B. J. Ainslie, "A review of the fabrication and properties of erbium-doped fibers for optical amplifiers," *J. Lightwave Tech.*, vol. 9, pp. 220–227, Feb. 1991.

59. W. Miniscalco, "Erbium-doped glasses for fiber amplifiers at 1500 nm," *J. Lightwave Tech.*, vol. 9, pp. 234–250, Feb. 1991.

60. J. R. Simpson, "Rare earth doped fiber fabrication: Techniques and physical properties," in M. J. F. Digonnet, ed., *Rare Earth Doped Fiber Lasers and Amplifiers*, Dekker, New York, 1993.

61. M. Asobe, "Nonlinear optical properties of chalcogenide glass fibers and their application to all-optical switching," *Opt. Fiber Technol.*, vol. 3, pp. 142–148, Apr. 1997.

62. R. Mossadegh, J. S. Sanghera, D. Schaafsma, B. J. Cole, V. Q. Nguyen, R. E. Miklos, and I. Aggarwal, "Fabrication of single-mode chalcogenide optical fiber," *J. Lightwave Tech.*, vol. 16, pp. 214–217, Feb. 1998.

63. A. N. Sinha, M. Groten, and G. D. Khoe, "A flexible polymer fiber infrastructure for an evolutionary customer premises network," *Intl. J. Commun. Systems*, vol. 10, pp. 23–30, Jan./Feb. 1997.

64. T. Ishigure, M. Satoh, O. Takanashi, E. Nihei, T. Nyu, S. Yamazaki, and Y. Koike, "Formation of the refractive index profile in graded-index polymer optical fiber for gigabit data transmission," *J. Lightwave Tech.*, vol. 15, pp. 2095–2100, Nov. 1997.

65. N. Yoshihara, "Low-loss, high-bandwidth fluorinated POF for visible to 1.3-μm wavelengths," *OSA/IEEE Proc. OFC-98*, Paper ThM4, p. 308, Feb. 1998.

66. P. C. Schultz, "Progress in optical waveguide process and materials," *Appl. Opt.*, vol. 18, pp. 3684–3693, Nov. 1979.

67. W. G. French, R. E. Jaeger, J. B. MacChesney, S. R. Nagel, K. Nassau, and A. D. Pearson, "Fiber preform preparation," in S. E. Miller and A. G. Chynoweth, eds., *Optical Fiber Telecommunications*, Academic, New York, 1979.

68. R. Dorn, A. Baumgärtner, A. Gutu-Nelle, J. Koppenborg, W. Rehm, R. Schneider, and S. Schneider, "Mechanical shaping of preforms for low loss at low cost," *J. Opt. Commun.*, vol. 10, pp. 2–5, Mar. 1989.

69. R. E. Jaeger, A. D. Pearson, J. C. Williams, and H. M. Presby, "Fiber drawing and control," in S. E. Miller and A. G. Chynoweth, eds., *Optical Fiber Telecommunications*, Academic, New York, 1979.

70. U. C. Paek, "High-speed high-strength fiber drawing," *J. Lightwave Tech.*, vol. LT-4, pp. 1048–1060, Aug. 1986.

71. C. Brehm, P. Dupont, G. Lavanant, P. Ledoux, C. LeSergent, C. Reinaudo, J. M., Saugrain, M. Carratt, and R. Jocteur, "Improved drawing conditions for very low loss 1.55-μm dispersion-shifted fiber," *Fiber and Integrated Opt.*, vol. 7, no. 4, pp. 333–341, 1988.

72. P. L. Chu, T. Whitbread, and P. M. Allen, "An on-line fiber drawing tension and diameter measurement device," *J. Lightwave Tech.*, vol. 7, pp. 255–261, Feb. 1989.

73. F. P. Kapron, D. B. Keck, and R. D. Maurer, "Radiation losses in glass optical waveguides," *Appl. Phys. Lett.*, vol. 17, pp. 423–425, Nov. 1970.

74. P. C. Schultz, "Fabrication of optical waveguides by the outside vapor deposition process," *Proc. IEEE*, vol. 68, pp. 1187–1190, Oct. 1980.

75. R. V. VanDewoestine and A. J. Morrow, "Developments in optical waveguide fabrication by the outside vapor deposition process," *J. Lightwave Tech.*, vol. LT-4, pp. 1020–1025, Aug. 1986.

76. T. Izawa and N. Inagaki, "Materials and processes for fiber preform fabrication: Vapor-phase axial deposition," *Proc. IEEE*, vol. 68, pp. 1184–1187, Oct. 1980.

77. H. Murata, "Recent developments in vapor phase axial deposition," *J. Lightwave Tech.*, vol. LT-4, pp. 1026–1033, Aug. 1986.

78. S. R. Nagel, J. B. MacChesney, and K. L. Walker, "Modified chemical vapor deposition," in T. Li, ed., *Optical Fiber Communications, Vol. 1, Fiber Fabrication*, Academic, New York, 1985.

79. (a) P. Geittner, D. Küppers, and H. Lydtin, "Low loss optical fibers prepared by plasma-activated chemical vapor deposition (PCVD)," *Appl. Phys. Lett.*, vol 28, pp. 645–646, June 1976.
    (b) P. Geittner and H. Lydtin, "Manufacturing optical fibers by the PCVD process," *Philips Tech. Rev. (Netherlands)*, vol. 44, pp. 241–249, May 1989.

80. T. Hünlich, H. Bauch, R. T. Kersten, V. Paquet, and G. F. Weidmann, "Fiber-preform fabrication using plasma technology: A review," *J. Opt. Commun.*, vol. 8, pp. 122–129, Dec. 1987.

81. H. Lydtin, "PCVD: A technique suitable for large-scale fabrication of optical fibers," *J. Lightwave Tech.*, vol. LT-4, pp. 1034–1038, Aug. 1986.

82. R. D. Maurer, "Behavior of flaws in fused silcia fibers," in C. R. Kurkjian, ed., *Strength of Organic Glasses*, Plenum, New York, 1986.

83. D. Kalish, D. L. Key, C. R. Kurkjian, B. K. Tariyal, and T. T. Wang, "Fiber characterization— mechanical," in S. E. Miller and A. G. Chynoweth, eds., *Optical Fiber Telecommunications*, Academic, New York, 1979.

84. M. J. Matthewson, C. R. Kurkjian, and J. R. Hamblin, "Acid stripping of fused silica optical fibers without strength degradation," *J. Lightwave Tech.*, vol. 15, pp. 490–497, Mar. 1997.

85. C. R. Kurkjian, J. T. Krause, and M. J. Matthewson, "Strength and fatigue of silica optical fibers," *J. Lightwave Tech.*, vol. 7, pp. 1360–1370, Sept. 1989.

86. A. A. Griffith, "The phenomena of rupture and flow in solids," *Philos. Trans. Roy. Soc. (London)*, vol. 221A, pp. 163–198, Oct. 1920.

87. W. Weibull, "A statistical theory of the strength of materials," *Ing. Vetenskaps. Akad. Handl. (Proc. Roy. Swed. Inst. Eng. Res.)*, no. 151, 1939; "The phenomenon of rupture in solids," *ibid.*, no. 153, 1939.

88. T. J. Miller, A. C. Hart, W. I. Vroom, and M. J. Bowden, "Silicone and ethylene-vinyl-acetate-coated laser-drawn silica fibers with tensile strengts > 3.5 GN/m^2 (500 kpsi) in > 3 km lengths," *Electron. Lett.*, vol. 14, pp. 603–605, Aug. 1978.

89. H. Schonhorn, C. R. Kurkjian, R. E. Jaeger, H. N. Vazirani, R. V. Albarino, and F. V. DiMarcello, "Epoxy-acrylate coated fused silcia fibers with tensile strengths greater than 500 kpsi (3.5 GN/m^2) in 1-km gauge lengths," *Appl. Phys. Lett.*, vol. 29, pp. 712–714, Dec. 1976.

90. (a) K. E. Lu, G. S. Glaesemann, R. V. VanDewoestine, and G. Kar, "Recent developments in hermetically coated optical fiber," *J. Lightwave Tech.*, vol. 6, pp. 240–244, Feb. 1988.
    (b) K. E. Lu, "Hermetic coatings," *Technical Digest: IEEE/OSA Optical Fiber Commun. Conf.*, p. 174, San Francisco, CA. Jan. 1990.

91. Y. Katsuyama, Y. Mitsunaga, H. Kobayashi, and Y. Ishida, "Dynamic fatigue of optical fiber under repeated stress," *J. Appl. Phys.*, vol. 53, pp. 318–321, Jan. 1982.

92. V. Annovazzi-Lodi, S. Donati, S. Merlo, and G. Zapelloni, "Statistical analysis of fiber failures under bending-stress fatigue," *J. Lightwave Tech.*, vol. 15, pp. 288–293, Feb. 1997.

93. TIA/EIA FOTP-31C, *Proof Testing Optical Fibers by Tension*, Sept. 1994.

94. TIA/EIA FOTP-76, *Method for Measuring Dynamic Fatigue of Optical Fibers by Tension*, June 1993.

95. B. Wiltshire and M. H. Reeve, "A review of the environmental factors affecting optical cable design," *J. Lightwave Tech.*, vol. 6, pp. 179–185, Feb. 1988.

96. C. H. Gartside, III, P. D. Patel, and M. R. Santana, "Optical fiber cables," in S. E. Miller and I. P. Kaminow, eds., *Optical Fiber Telecommunications—II*, Academic, New York, 1988.

97. K. Hogari, S. Furukawa, Y. Nakatsuji, S. Koshio, and K. Nishizawa, "Optical fiber cables for residential and business premises," *J. Lightwave Tech.*, vol. 16, pp. 207–213, Feb. 1998.

98. G. Mahlke and P. Goessing, *Fiber Optic Cables: Fundamentals, Cable Engineering, System Planning*, Wiley, New York, 1997.

99. D. Marcuse, "Loss analysis of single-mode fiber splices," *Bell Sys. Tech. J.*, vol. 56, pp. 703–718, May/June 1977.

CHAPTER
3

SIGNAL DEGRADATION IN OPTICAL FIBERS

In Chap. 2 we showed the structure of optical fibers and examined the concepts of how light propagates along a cylindrical dielectric optical waveguide. Here, we shall continue the discussion of optical fibers by answering two very important questions:

1. What are the loss or signal attenuation mechanisms in a fiber?
2. Why and to what degree do optical signals get distorted as they propagate along a fiber?

Signal attenuation (also known as *fiber loss* or *signal loss*) is one of the most important properties of an optical fiber, because it largely determines the maximum unamplified or repeaterless separation between a transmitter and a receiver. Since amplifiers and repeaters are expensive to fabricate, install, and maintain, the degree of attenuation in a fiber has large influence on system cost. Of equal importance is signal distortion. The distortion mechanisms in a fiber cause optical signal pulses to broaden as they travel along a fiber. If these pulses travel sufficiently far, they will eventually overlap with neighboring pulses, thereby creating errors in the receiver output. The signal distortion mechanisms thus limit the information-carrying capacity of a fiber.

Section 3.3 on pulse broadening in graded-index fibers and Sec. 3.4 on mode coupling encompass advanced material that can be skipped without loss of continuity; these sections are designated by a star (⋆). In relation to signal-degradation effects, App. F shows the origin of several factors contributing to fiber dispersion.

3.1 ATTENUATION

Attenuation of a light signal as it propagates along a fiber is an important consideration in the design of an optical communication system, since it plays a major role in determining the maximum transmission distance between a transmitter and a receiver or an in-line amplifier. The basic attenuation mechanisms in a fiber are absorption, scattering, and radiative losses of the optical energy.[1-5] Absorption is related to the fiber material, whereas scattering is associated both with the fiber material and with structural imperfections in the optical waveguide. Attenuation owing to radiative effects originates from perturbations (both microscopic and macroscopic) of the fiber geometry.

In this section we shall first discuss the units in which fiber losses are measured and then present the physical phenomena giving rise to attenuation.

3.1.1 Attenuation Units

As light travels along a fiber, its power decreases exponentially with distance. If $P(0)$ is the optical power in a fiber at the origin (at $z = 0$), then the power $P(z)$ at a distance z further down the fiber is

$$P(z) = P(0)e^{-\alpha_p z} \tag{3-1a}$$

where

$$\alpha_p = \frac{1}{z} \ln\left[\frac{P(0)}{P(z)}\right] \tag{3-1b}$$

is the fiber *attenuation coefficient* given in units of, for example, km^{-1}. Note that the units for $2z\alpha_p$ can also be designated by *nepers* (see App. D).

For simplicity in calculating optical signal attenuation in a fiber, the common procedure is to express the attenuation coefficient in units of *decibels per kilometer*, denoted by dB/km. Designating this parameter by α, we have

$$\alpha \, (dB/km) = \frac{10}{z} \log\left[\frac{P(0)}{P(z)}\right] = 4.343\alpha_p \, (km^{-1}) \tag{3-1c}$$

This parameter is generally referred to as the *fiber loss* or the *fiber attenuation*. It depends on several variables, as is shown in the following sections, and it is a function of the wavelength, as is illustrated by the general attenuation curve in Fig. 3-1.

Example 3-1. An ideal fiber would have no loss so that $P_{out} = P_{in}$. This corresponds to a 0-dB/km attenuation, which, in practice, is impossible. An actual low-loss fiber may have a 3-dB/km average loss at 900 nm, for example. This means that the optical signal power would decrease by 50 percent over a 1-km length and would decrease by 75 percent (a 6-dB loss) over a 2-km length, since loss contributions expressed in decibels are additive.

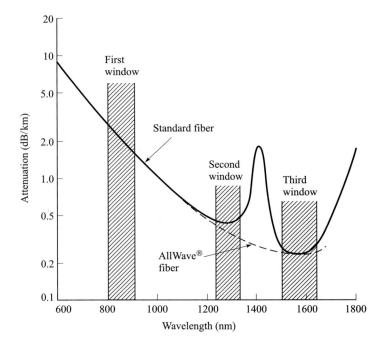

FIGURE 3-1
Optical fiber attenuation as a function of wavelength yields nominal values of 0.5 dB/km at 1300 nm and 0.3 dB/km at 1550 nm for standard single-mode fiber (solid curve). This fiber shows an attenuation peak around 1400 nm resulting from absorption by water molecules. The dashed curve is for a water-free AllWave® fiber (data courtesy of Lucent Technologies).

Note that App. D contains a review of decibels, which are used to facilitate calculations of power budgets in a light-wave link. As described therein, optical powers are commonly expressed in units of *dBm*, which is the decibel power level referred to 1 mW.

Example 3-2. Consider a 30-km long optical fiber that has an attenuation of 0.8 dB/km at 1300 nm. Suppose we want to find the optical output power P_{out} if 200 μW of optical power is launched into the fiber. We first use Eq. (D-2) to express the input power in dBm units:

$$P_{in}(dBm) = 10 \log\left[\frac{P_{in}(W)}{1\,mW}\right] = 10 \log\left[\frac{200 \times 10^{-6}\,W}{1 \times 10^{-3}\,W}\right] = -7.0\,dBm$$

From Eq. (3-1c) we then have that the output power level (in dBm) at $z = 30$ km is

$$P_{out}(dBm) = 10 \log\left[\frac{P_{out}(W)}{1\,mW}\right] = 10 \log\left[\frac{P_{in}(W)}{1\,mW}\right] - \alpha z$$

$$= -7.0\,dBm - (0.8\,dB/km)(30\,km) = -31.0\,dBm$$

In units of watts, from Eq. (D-2) the output power is

$$P(30\,\text{km}) = 10^{-31.0/10}(1\,\text{mW}) = 0.79 \times 10^{-3}\,\text{mW} = 0.79\,\mu\text{W}$$

3.1.2 Absorption

Absorption is caused by three different mechanisms:

1. Absorption by atomic defects in the glass composition.
2. Extrinsic absorption by impurity atoms in the glass material.
3. Intrinsic absorption by the basic constituent atoms of the fiber material.

Atomic defects are imperfections in the atomic structure of the fiber material. Examples are missing molecules, high-density clusters of atom groups, or oxygen defects in the glass structure. Usually, absorption losses arising from these defects are negligible compared with intrinsic and impurity absorption effects. However, they can be significant if the fiber is exposed to ionizing radiation, as might occur in a nuclear reactor environment, in medical radiation therapies, in space missions that pass through the earth's Van Allen belts, or in accelerator instrumentation.[6-9] In such applications, high radiation doses may be accumulated over several years.

Radiation damages a material by changing its internal structure. The damage effects depend on the energy of the ionizing particles or rays (e.g., electrons, neutrons, or gamma rays), the radiation flux (dose rate), and the fluence (particles per square centimeter). The total dose a material receives is expressed in units of rad(Si), which is a measure of radiation absorbed in bulk silicon. This unit is defined as

$$1\,\text{rad(Si)} = 100\,\text{erg/g} = 0.01\,\text{J/kg}$$

The basic response of a fiber to ionizing radiation is an increase in attenuation owing to the creation of atomic defects, or attenuation centers, that absorb optical energy. The higher the radiation level, the larger the attenuation, as Fig. 3-2a illustrates. However, the attenuation centers will relax or anneal out with time, as shown in Fig. 3-2b. Thus, the specific radiation-induced loss in fibers is complex, and some guidelines for application have been recommended.[7]

The dominant absorption factor in fibers prepared by the direct-melt method is the presence of impurities in the fiber material. Impurity absorption results predominantly from transition metal ions, such as iron, chromium, cobalt, and copper, and from OH (water) ions. The transition metal impurities which are present in the starting materials used for direct-melt fibers range between 1 and 10 parts per billion (ppb), causing losses from 1 to 10 dB/km. The impurity levels in vapor-phase deposition processes are usually one to two orders of magnitude lower. Impurity absorption losses occur either because of electronic transitions between the energy levels associated with the incompletely filled inner subshell of these ions or because of charge transitions from one ion to another. The absorp-

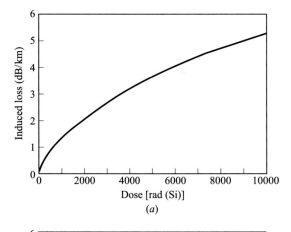

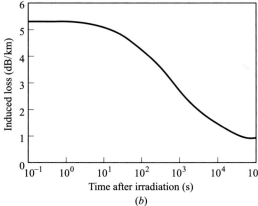

FIGURE 3-2

Effects of ionizing radiation on optical fiber attenuation. (*a*) Loss increase during steady irradiation to a total dose of 10^4 rad(Si). (*b*) Subsequent recovery as a function of time after radiation has stopped. (Modified with permission from West et al.,[7c] © 1994, IEEE.)

tion peaks of the various transition metal impurities tend to be broad, and several peaks may overlap, which further broadens the absorption region.

The presence of OH (water) ion impurities in fiber preforms results mainly from the oxyhydrogen flame used for the hydrolysis reaction of the $SiCl_4$, $GeCl_4$, and $POCl_3$ starting materials. Water impurity concentrations of less than a few parts per billion are required if the attenuation is to be less than 20 dB/km. Early optical fibers had high levels of OH ions which resulted in large absorption peaks occurring at 1400, 950, and 725 nm. These are the first, second, and third overtones, respectively, of the fundamental absorption peak of water near 2.7 μm, as shown in Fig. 1-7. Between these absorption peaks there are regions of low attenuation.

The peaks and valleys in the attenuation curve resulted in the designation of various "transmission windows" to optical fibers. By reducing the residual OH content of fibers to around 1 ppb, standard commercially available single-mode fibers have nominal attenuations of 0.5 dB/km in the 1300-nm window and 0.3 dB/km in the 1550-nm window, as shown by the solid curve in Fig. 3-1. An effectively complete elimination of water molecules from the fiber results in the dashed curve shown in Fig. 3-1. This is for an AllWave® fiber made by Lucent Technologies.

Intrinsic absorption is associated with the basic fiber material (e.g., pure SiO_2) and is the principal physical factor that defines the transparency window of a material over a specified spectral region. It occurs when the material is in a perfect state with no density variations, impurities, material inhomogeneities, and so on. Intrinsic absorption thus sets the fundamental lower limit on absorption for any particular material.

Intrinsic absorption results from electronic absorption bands in the ultraviolet region and from atomic vibration bands in the near-infrared region. The electronic absorption bands are associated with the band gaps of the amorphous glass materials. Absorption occurs when a photon interacts with an electron in the valence band and excites it to a higher energy level, as is described in Sec. 2.1. The ultraviolet edge of the electron absorption bands of both amorphous and crystalline materials follow the empirical relationship[1,3]

$$\alpha_{uv} = Ce^{E/E_0} \tag{3-2a}$$

which is known as Urbach's rule. Here, C and E_0 are empirical constants and E is the photon energy. The magnitude and characteristic exponential decay of the ultraviolet absorption are shown in Fig. 3-3. Since E is inversely proportional to the wavelength λ, ultraviolet absorption decays exponentially with increasing wavelength. In particular, the ultraviolet loss contribution in dB/km at any wavelength can be expressed empirically as a function of the mole fraction x of GeO_2 as [10,11]

$$\alpha_{uv} = \frac{154.2x}{46.6x + 60} \times 10^{-2} \exp\left(\frac{4.63}{\lambda}\right) \tag{3-2b}$$

As shown in Fig. 3-3, the ultraviolet loss is small compared with scattering loss in the near-infrared region.

In the near-infrared region above $1.2\,\mu m$, the optical waveguide loss is predominantly determined by the presence of OH ions and the inherent infrared absorption of the constituent material. The inherent infrared absorption is associated with the characteristic vibration frequency of the particular chemical bond between the atoms of which the fiber is composed. An interaction between the vibrating bond and the electromagnetic field of the optical signal results in a transfer of energy from the field to the bond, thereby giving rise to absorption. This absorption is quite strong because of the many bonds present in the fiber. An empirical expression for the infrared absorption in dB/km for GeO_2–SiO_2 glass is [10,11]

$$\alpha_{IR} = 7.81 \times 10^{11} \times \exp\left(\frac{-48.48}{\lambda}\right) \tag{3-3}$$

These mechanisms result in a wedge-shaped spectral-loss characteristic. Within this wedge, losses as low as 0.154 dB/km at $1.55\,\mu m$ in a single-mode

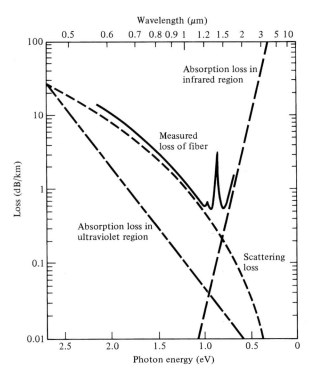

FIGURE 3-3
Optical fiber attenuation characteristics and their limiting mechanisms for a GeO_2-doped low-loss low-OH-content silica fiber. (Reproduced with permission from Osanai et al.[13])

fiber have been measured.[12] A comparison[13] of the infrared absorption induced by various doping materials in low-water content fibers is shown in Fig. 3-4. This indicates that for operation at longer wavelengths GeO_2-doped fiber material is the most desirable. Note that the absorption curve shown in Fig. 3-3 is for a GeO_2-doped fiber.

3.1.3 Scattering Losses

Scattering losses in glass arise from microscopic variations in the material density, from compositional fluctuations, and from structural inhomogeneities or defects occurring during fiber manufacture. As we saw in Sec. 2.7, glass is composed of a randomly connected network of molecules. Such a structure naturally contains regions in which the molecular density is either higher or lower than the average density in the glass. In addition, since glass is made up of several oxides, such as SiO_2, GeO_2, and P_2O_5 compositional fluctuations can occur. These two effects give rise to refractive-index variations which occur within the glass over distances that are small compared with the wavelength. These index variations cause a Rayleigh-type scattering of the light. Rayleigh scattering in glass is the same phenomenon that scatters light from the sun in the atmosphere, thereby giving rise to a blue sky.

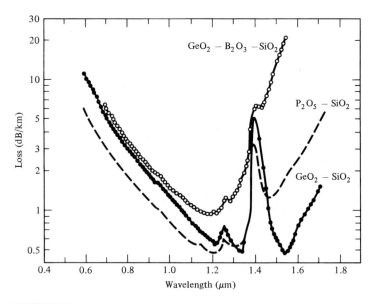

FIGURE 3-4
A comparison of the infrared absorption induced by various doping materials in low-loss silica fibers. (Reproduced with permission from Osanai et al.[13])

The expressions for scattering-induced attenuation are fairly complex owing to the random molecular nature and the various oxide constituents of glass. For single-component glass the scattering loss at a wavelength λ resulting from density fluctuations can be approximated by[3,14] (in base e units)

$$\alpha_{\text{scat}} = \frac{8\pi^3}{3\lambda^4}(n^2 - 1)^2 k_B T_f \beta_T \qquad (3\text{-}4a)$$

Here, n is the refractive index, k_B is Boltzmann's constant, β_T is the isothermal compressibility of the material, and the fictive temperature T_f is the temperature at which the density fluctuations are frozen into the glass as it solidifies (after having been drawn into a fiber). Alternatively, the relation[3,15] (in base e units)

$$\alpha_{\text{scat}} = \frac{8\pi^3}{3\lambda^4}n^8 p^2 k_B T_f \beta_T \qquad (3\text{-}4b)$$

has been derived, where p is the photoelastic coefficient. A comparison of Eqs. (3-4a) and (3-4b) is given in Prob. 3-6. Note that Eqs. (3-4a) and (3-4b) are given in units of *nepers* (that is, base e units). As shown in Eq. (3-1), to change this to decibels for optical power attenuation calculations, multiply these equations by $10 \log e = 4.343$.

For multicomponent glasses the scattering is given by[3]

$$\alpha = \frac{8\pi^3}{3\lambda^4}(\delta n^2)^2 \, \delta V \tag{3-5}$$

where the square of the mean-square refractive-index fluctuation $(\delta n^2)^2$ over a volume of δV is

$$(\delta n^2)^2 = \left(\frac{\partial n}{\partial \rho}\right)^2 (\delta \rho)^2 + \sum_{i=1}^{m} \left(\frac{\partial n^2}{\partial C_i}\right)^2 (\delta C_i)^2 \tag{3-6}$$

Here, $\delta\rho$ is the density fluctuation and δC_i is the concentration fluctuation of the ith glass component. The magnitudes of the composition and density fluctuations are generally not known and must be determined from experimental scattering data. Once they are known the scattering loss can be calculated.

Structural inhomogeneities and defects created during fiber fabrication can also cause scattering of light out of the fiber. These defects may be in the form of trapped gas bubbles, unreacted starting materials, and crystallized regions in the glass. In general, the preform manufacturing methods that have evolved have minimized these extrinsic effects to the point where scattering that results from them is negligible compared with the intrinsic Rayleigh scattering.

Since Rayleigh scattering follows a characteristic λ^{-4} dependence, it decreases dramatically with increasing wavelength, as is shown in Fig. 3-3. For wavelengths below about 1 μm it is the dominant loss mechanisms in a fiber and gives the attenuation-versus-wavelength plots their characteristic downward trend with increasing wavelength. At wavelengths longer than 1 μm, infrared absorption effects tend to dominate optical signal attenuation.

Combining the infrared, ultraviolet, and scattering losses, we get the results shown in Fig. 3-5 for multimode fibers and Fig. 3-6 for single-mode fibers.[16] Both of these figures are for typical commercial-grade silica fibers. The losses of multimode fibers are generally higher than those of single-mode fibers. This is a result of higher dopant concentrations and the accompanying larger scattering loss due

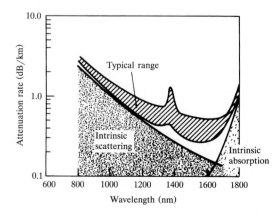

FIGURE 3-5
Typical spectral attenuation range for production-run graded-index multimode fibers. (Reproduced with permission from Keck,[16] © 1985, IEEE.)

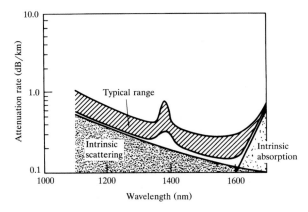

FIGURE 3-6

Typical spectral attenuation range for production-run single-mode fibers. (Reproduced with permission from Keck,[16] © 1985, IEEE.)

to greater compositional fluctuation in multimode fibers. In addition, multimode fibers are subject to higher-order-mode losses owing to perturbations at the core–cladding interface.

3.1.4 Bending Losses

Radiative losses occur whenever an optical fiber undergoes a bend of finite radius of curvature.[17-25] Fibers can be subject to two types of bends: (*a*) macroscopic bends having radii that are large compared with the fiber diameter, for example, such as those that occur when a fiber cable turns a corner, and (*b*) random microscopic bends of the fiber axis that can arise when the fibers are incorporated into cables.

Let us first examine large-curvature radiation losses, which are known as *macrobending losses* or simply *bending losses*. For slight bends the excess loss is extremely small and is essentially unobservable. As the radius of curvature decreases, the loss increases exponentially until at a certain critical radius the curvature loss becomes observable. If the bend radius is made a bit smaller once this threshold point has been reached, the losses suddenly become extremely large.

Qualitatively, these curvature loss effects can be explained by examining the modal electric field distributions shown in Fig. 2-14. Recall that this figure shows that any bound core mode has an evanescent field tail in the cladding which decays exponentially as a function of distance from the core. Since this field tail moves along with the field in the core, part of the energy of a propagating mode travels in the fiber cladding. When a fiber is bent, the field tail on the far side of the center of curvature must move faster to keep up with the field in the core, as is shown in Fig. 3-7 for the lowest-order fiber mode. At a certain critical distance x_c from the center of the fiber, the field tail would have to move faster than the speed of light to keep up with the core field. Since this is not possible the optical energy in the field tail beyond x_c radiates away.

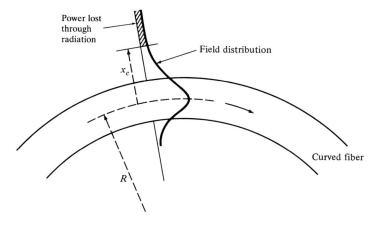

FIGURE 3-7
Sketch of the fundamental mode field in a curved optical waveguide. (Reproduced with permission from E. A. J. Marcatili and S. E. Miller, *Bell Sys. Tech. J.,* vol. 48, p. 2161, Sept. 1969, © 1969, AT&T.)

The amount of optical radiation from a bent fiber depends on the field strength at x_c and on the radius of curvature R. Since higher-order modes are bound less tightly to the fiber core than lower-order modes, the higher-order modes will radiate out of the fiber first. Thus, the total number of modes that can be supported by a curved fiber is less than in a straight fiber. Gloge[18] has derived the following expression for the effective number of modes N_{eff} that are guided by a curved multimode fiber of radius a:

$$N_{\text{eff}} = N_\infty \left\{ 1 - \frac{\alpha+2}{2\alpha\Delta} \left[\frac{2a}{R} + \left(\frac{3}{2n_2 kR} \right)^{2/3} \right] \right\} \qquad (3\text{-}7)$$

where α defines the graded-index profile, Δ is the core–cladding index difference, n_2 is the cladding refractive index, $k = 2\pi/\lambda$ is the wave propagation constant, and

$$N_\infty = \frac{\alpha}{\alpha+2} (n_1 ka)^2 \Delta \qquad (3\text{-}8)$$

is the total number of modes in a straight fiber [see Eq. (2-81)].

> **Example 3-3.** As an example, let us find the radius of curvature R at which the number of modes decreases by 50 percent in a graded-index fiber. For this fiber, let $\alpha = 2$, $n_2 = 1.5$, $\Delta = 0.01$, $a = 25\,\mu\text{m}$, and let the wavelength of the guided light be $1.3\,\mu\text{m}$. Solving Eq. (3-7) yields $R = 1.0$ cm.

Another form of radiation loss in optical waveguides results from mode coupling caused by random microbends of the optical fiber.[25-30] *Microbends* are repetitive small-scale fluctuations in the radius of curvature of the fiber axis, as is illustrated in Fig. 3-8. They are caused either by nonuniformities in the manufac-

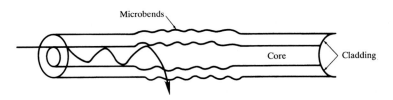

Microbends

Core   Cladding

Power loss from higher-order modes

Power coupling to higher-order modes

FIGURE 3-8
Small-scale fluctuations in the radius of curvature of the fiber axis leads to microbending losses. Microbends can shed higher-order modes and can cause power from low-order modes to couple to higher-order modes.

turing of the fiber or by nonuniform lateral pressures created during the cabling of the fiber. The latter effect is often referred to as *cabling* or *packaging losses*. An increase in attenuation results from microbending because the fiber curvature causes repetitive coupling of energy between the guided modes and the leaky or nonguided modes in the fiber.

One method of minimizing microbending losses is by extruding a compressible jacket over the fiber. When external forces are applied to this configuration, the jacket will be deformed but the fiber will tend to stay relatively straight, as shown in Fig. 3-9. For a multimode graded-index fiber having a core radius a, outer radius b (excluding the jacket), and index difference Δ, the microbending loss α_M of a jacketed fiber is reduced from that of an unjacketed fiber by a factor[31]

$$F(\alpha_M) = \left[1 + \pi \Delta^2 \left(\frac{b}{a} \right)^4 \frac{E_f}{E_j} \right]^{-2}$$ (3-9)

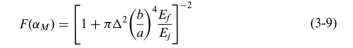

External force

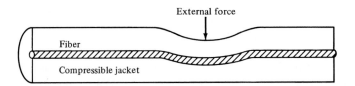

Fiber

Compressible jacket

FIGURE 3-9
A compressible jacket extruded over a fiber reduces microbending resulting from external forces.

Here, E_j and E_f are the Young's moduli of the jacket and fiber, respectively. The Young's modulus of common jacket materials ranges from 20 to 500 MPa. The Young's modulus of fused silica glass is about 65 GPa.

3.1.5 Core and Cladding Losses

Upon measuring the propagation losses in an actual fiber, all the dissipative and scattering losses will be manifested simultaneously. Since the core and cladding have different indices of refraction and therefore differ in composition, the core and cladding generally have different attenuation coefficients, denoted α_1 and α_2, respectively. If the influence of modal coupling is ignored,[32] the loss for a mode of order (v, m) for a step-index waveguide is

$$\alpha_{vm} = \alpha_1 \frac{P_{core}}{P} + \alpha_2 \frac{P_{clad}}{P} \qquad (3\text{-}10a)$$

where the fractional powers P_{core}/P and P_{clad}/P are shown in Fig. 2-22 for several low-order modes. Using Eq. (2-71), this can be written as

$$\alpha_{vm} = \alpha_1 + (\alpha_2 - \alpha_1) \frac{P_{clad}}{P} \qquad (3\text{-}10b)$$

The total loss of the waveguide can be found by summing over all modes weighted by the fractional power in that mode.

For the case of a graded-index fiber the situation is much more complicated. In this case, both the attenuation coefficients and the modal power tend to be functions of the radial coordinate. At a distance r from the core axis the loss is [32]

$$\alpha(r) = \alpha_1 + (\alpha_2 - \alpha_1) \frac{n^2(0) - n^2(r)}{n^2(0) - n_2^2} \qquad (3\text{-}11)$$

where α_1 and α_2 are the axial and cladding attenuation coefficients, respectively, and the n terms are defined by Eq. (2-78). The loss encountered by a given mode is then

$$\alpha_{gi} = \frac{\int_0^\infty \alpha(r) p(r) r \, dr}{\int_0^\infty p(r) r \, dr} \qquad (3\text{-}12)$$

where $p(r)$ is the power density of that mode at r. The complexity of the multimode waveguide has prevented an experimental correlation with a model. However, it has generally been observed that the loss increases with increasing mode number.[25,33]

3.2 SIGNAL DISTORTION IN OPTICAL WAVEGUIDES

An optical signal becomes increasingly distorted as it travels along a fiber. This distortion is a consequence of intramodal dispersion and intermodal delay effects. These distortion effects can be explained by examining the behavior of the group velocities of the guided modes, where the *group velocity* is the speed at which energy in a particular mode travels along the fiber.

Intramodal dispersion or *chromatic dispersion* is pulse spreading that occurs within a single mode. The spreading arises from the finite spectral emission width of an optical source. This phenomenon is also known as *group velocity dispersion* (GVD), since the dispersion is a result of the group velocity being a function of the wavelength. Because intramodal dispersion depends on the wavelength, its effect on signal distortion increases with the spectral width of the optical source. This spectral width is the band of wavelengths over which the source emits light. It is normally characterized by the root-mean-square (rms) spectral width σ_λ (see Fig. 4-12). For light-emitting diodes (LEDs) the rms spectral width is approximately 5 percent of a central wavelength. For example, if the peak emission wavelength of an LED source is 850 nm, a typical source spectral width would be 40 nm; that is, the source emits most of its optical power in the 830-to-870-nm wavelength band. Laser diode optical sources have much narrower spectral widths, with typical values being 1–2 nm for multimode lasers and 10^{-4} nm for single-mode lasers.

The two main causes of intramodal dispersion are as follows:

1. *Material dispersion*, which arises from the variation of the refractive index of the core material as a function of wavelength. (Material dispersion is sometimes referred to as *chromatic dispersion*, since this is the same effect by which a prism spreads out a spectrum.) This causes a wavelength dependence of the group velocity of any given mode; that is, pulse spreading occurs even when different wavelengths follow the same path.

2. *Waveguide dispersion*, which occurs because a single-mode fiber confines only about 80 percent of the optical power to the core. Dispersion thus arises, since the 20 percent of the light propagating in the cladding travels faster than the light confined to the core. The amount of waveguide dispersion depends on the fiber design, since the modal propagation constant β is a function of a/λ (the optical fiber dimension relative to the wavelength λ; here, a is the core radius).

The other factor giving rise to pulse spreading is *intermodal delay*, which is a result of each mode having a different value of the group velocity at a single frequency.

Of these three, waveguide dispersion usually can be ignored in multimode fibers. However, this effect is significant in single-mode fibers. The full effects of these three distortion mechanisms are seldom observed in practice, since they tend to be mitigated by other factors, such as nonideal index profiles, optical power-launching conditions (different amounts of optical power launched into the various modes), nonuniform mode attenuation, and mode mixing in the fiber and in splices; and by statistical variations in these effects along the fiber. In this section we shall first discuss the general effects of signal distortion and then examine the various dispersion mechanisms. By using a Taylor series expansion of the propagation constant β, App. F shows the origin of several factors contributing to dispersion.

3.2.1 Information Capacity Determination

A result of the dispersion-induced signal distortion is that a light pulse will broaden as it travels along the fiber. As shown in Fig. 3-10, this pulse broadening will eventually cause a pulse to overlap with neighboring pulses. After a certain amount of overlap has occurred, adjacent pulses can no longer be individually distinguished at the receiver and errors will occur. Thus, the dispersive properties determine the limit of the information capacity of the fiber.

A measure of the information capacity of an optical waveguide is usually specified by the *bandwidth–distance product* in MHz·km. For a step-index fiber the various distortion effects tend to limit the bandwidth-distance product to about 20 MHz·km. In graded-index fibers the radial refractive-index profile can be carefully selected so that pulse broadening is minimized at a specific operating wavelength. This had led to bandwidth–distance products as high as 2.5 GHz·km. Single-mode fibers can have capacities well in excess of this. A comparison of the information capacities of various optical fibers with the capacities of typical coaxial cables used for UHF and VHF transmission is shown in

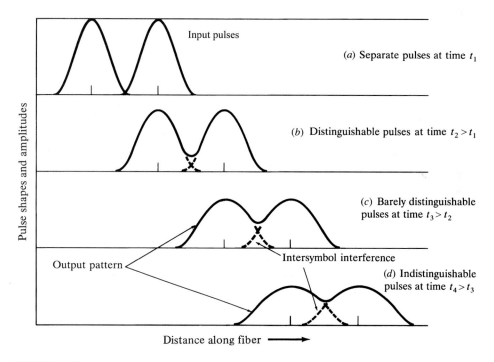

FIGURE 3-10
Broadening and attenuation of two adjacent pulses as they travel along a fiber. (*a*) Originally, the pulses are separate; (*b*) the pulses overlap slightly and are clearly distinguishable; (*c*) the pulses overlap significantly and are barely distinguishable; (*d*) eventually, the pulses strongly overlap and are indistinguishable.

Fig. 3-11. The curves are shown in terms of signal attenuation versus data rate. The flatness of the attenuation curves for the fibers extend up to the microwave spectrum.

The information-carrying capacity can be determined by examining the deformation of short light pulses propagating along the fiber. The following discussion on signal distortion is thus carried out primarily from the standpoint of pulse broadening, which is representative of digital transmission.

3.2.2 Group Delay

Let us examine a signal that modulates an optical source. We shall assume that the modulated optical signal excites all modes equally at the input end of the fiber. Each mode thus carries an equal amount of energy through the fiber. Furthermore, each mode contains all the spectral components in the wavelength band over which the source emits. The signal may be considered as modulating each of these spectral components in the same way. As the signal propagates along the fiber, each spectral component can be assumed to travel independently, and to undergo a time delay or *group delay* per unit length in the direction of propagation given by[34]

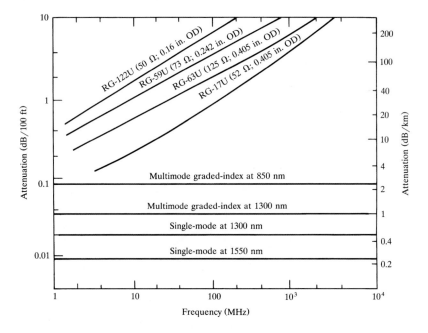

FIGURE 3-11
A comparison of the attenuation as a function of frequency or data rate of various coaxial cables and several types of high-bandwidth optical fibers.

$$\frac{\tau_g}{L} = \frac{1}{V_g} = \frac{1}{c}\frac{d\beta}{dk} = -\frac{\lambda^2}{2\pi c}\frac{d\beta}{d\lambda} \qquad (3\text{-}13)$$

Here, L is the distance traveled by the pulse, β is the propagation constant along the fiber axis, $k = 2\pi/\lambda$, and the *group velocity*

$$V_g = c\left(\frac{d\beta}{dk}\right)^{-1} = \left(\frac{\partial\beta}{\partial\omega}\right)^{-1} \qquad (3\text{-}14)$$

is the velocity at which the energy in a pulse travels along a fiber.

Since the group delay depends on the wavelength, each spectral component of any particular mode takes a different amount of time to travel a certain distance. As a result of this difference in time delays, the optical signal pulse spreads out with time as it is transmitted over the fiber. Thus, the quantity we are interested in is the amount of pulse spreading that arises from the group delay variation.

If the spectral width of the optical source is not too wide, the delay difference per unit wavelength along the propagation path is approximately $d\tau_g/d\lambda$. For spectral components which are $\delta\lambda$ apart and which lie $\delta\lambda/2$ above and below a central wavelength λ_0, the total delay difference $\delta\tau$ over a distance L is

$$\delta\tau = \frac{d\tau_g}{d\lambda}\delta\lambda = -\frac{L}{2\pi c}\left(2\lambda\frac{d\beta}{d\lambda} + \lambda^2\frac{d^2\beta}{d\lambda^2}\right)\delta\lambda \qquad (3\text{-}15a)$$

In terms of the angular frequency ω, this is written as

$$\delta\tau = \frac{d\tau_g}{d\omega}\delta\omega = \frac{d}{d\omega}\left(\frac{L}{V_g}\right)\delta\omega = L\left(\frac{d^2\beta}{d\omega^2}\right)\delta\omega \qquad (3\text{-}15b)$$

The factor $\beta_2 \equiv d^2\beta/d\omega^2$ is the GVD parameter, which determines how much a light pulse broadens as it travels along an optical fiber (see App. F).

If the spectral width $\delta\lambda$ of an optical source is characterized by its rms value (see Fig. 4-12), then the pulse spreading can be approximated by the rms pulse width,

$$\sigma_g \approx \left|\frac{d\tau_g}{d\lambda}\right|\sigma_\lambda = \frac{L\sigma_\lambda}{2\pi c}\left|2\lambda\frac{d\beta}{d\lambda} + \lambda^2\frac{d^2\beta}{d\lambda^2}\right| \qquad (3\text{-}16)$$

The factor

$$D = \frac{1}{L}\frac{d\tau_g}{d\lambda} = \frac{d}{d\lambda}\left(\frac{1}{V_g}\right) = -\frac{2\pi c}{\lambda^2}\beta_2 \qquad (3\text{-}17)$$

is designated as the *dispersion*. It defines the pulse spread as a function of wavelength and is measured in picoseconds per kilometer per nanometer [ps/ (nm · km)]. It is a result of material and waveguide dispersion. In many theoretical treatments of intramodal dispersion it is assumed, for simplicity, that material dispersion and waveguide dispersion can be calculated separately and then added to give the total dispersion of the mode. In reality, these two mechanisms are intricately related, since the dispersive properties of the refractive index (which gives rise to material dispersion) also effects the waveguide dispersion. However, an examination[35] of the interdependence of material and waveguide dispersion has shown that, unless a very precise value is desired, a good estimate of the total intramodal dispersion can be obtained by calculating the effect of signal distortion arising from one type of dispersion in the absence of the other. Thus, to a very good approximation, D can be written as the sum of the material dispersion D_{mat} and the waveguide dispersion D_{wg}. Material dispersion and waveguide dispersion are therefore considered separately in the next two sections.

3.2.3   Material Dispersion

Material dispersion occurs because the index of refraction varies as a function of the optical wavelength. This is exemplified in Fig. 3-12 for silica.[36] As a consequence, since the group velocity V_g of a mode is a function of the index of refraction, the various spectral components of a given mode will travel at different speeds, depending on the wavelength.[37] Material dispersion is, therefore, an intramodal dispersion effect, and is of particular importance for single-mode waveguides and for LED system (since an LED has a broader output spectrum than a laser diode).

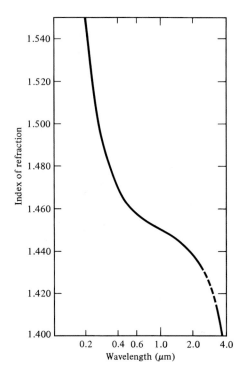

FIGURE 3-12

Variations in the index of refraction as a function of the optical wavelength for silica. (Reproduced with permission from I. H. Malitson, *J. Opt. Soc. Amer.*, vol. 55, pp. 1205–1209, Oct. 1965.)

To calculate material-induced dispersion, we consider a plane wave propagating in an infinitely extended dielectric medium that has a refractive index $n(\lambda)$ equal to that of the fiber core. The propagation constant β is thus given as

$$\beta = \frac{2\pi n(\lambda)}{\lambda} \tag{3-18}$$

Substituting this expression for β into Eq. (3-13) with $k = 2\pi/\lambda$ yields the group delay τ_{mat} resulting from material dispersion:

$$\tau_{mat} = \frac{L}{c}\left(n - \lambda \frac{dn}{d\lambda}\right) \tag{3-19}$$

Using Eq. (3-16), the pulse spread σ_{mat} for a source of spectral width σ_λ is found by differentiating this group delay with respect to wavelength and multiplying by σ_λ to yield

$$\sigma_{mat} \approx \left|\frac{d\tau_{mat}}{d\lambda}\right|\sigma_\lambda = \frac{\sigma_\lambda L}{c}\left|\lambda \frac{d^2 n}{d\lambda^2}\right| = \sigma_\lambda L |D_{mat}(\lambda)| \tag{3-20}$$

where $D_{mat}(\lambda)$ is the *material dispersion*.

A plot of the material dispersion for unit length L and unit optical source spectral width σ_λ is given in Fig. 3-13 for the silica material shown in Fig. 3-12. From Eq. (3-20) and Fig. 3-13 it can be seen that material dispersion can be reduced either by choosing sources with narrower spectral output widths (reducing σ_λ) or by operating at longer wavelengths.[38]

Example 3-4. As an example, consider a typical GaAlAs LED having a spectral width of 40 nm at an 800-nm peak output so that $\sigma_\lambda/\lambda = 5$ percent. As can be seen from Fig. 3-13 and Eq. (3-20), this produces a pulse spread of 4.4 ns/km. Note that material dispersion goes to zero at 1.27 μm for pure silica.

3.2.4 Waveguide Dispersion

The effect of waveguide dispersion on pulse spreading can be approximated by assuming that the refractive index of the material is independent of wavelength. Let us first consider the group delay—that is, the time required for a mode to travel along a fiber of length L. To make the results independent of fiber configuration,[37] we shall express the group delay in terms of the normalized propagation constant b defined as

$$b = 1 - \left(\frac{ua}{V}\right)^2 = \frac{\beta^2/k^2 - n_2^2}{n_1^2 - n_2^2} \tag{3-21}$$

For small values of the index difference $\Delta = (n_1 - n_2)/n_1$, Eq. (3-21) can be approximated by

$$b \approx \frac{\beta/k - n_2}{n_1 - n_2} \tag{3-22}$$

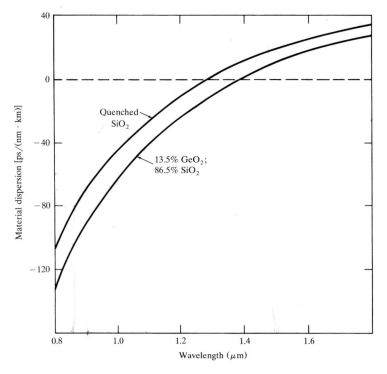

FIGURE 3-13
Material dispersion as a function of optical wavelength for pure silica and 13.5 percent GeO$_2$/86.5 percent SiO$_2$. (Reproduced with permission from J. W. Fleming, *Electron. Lett.*, vol. 14, pp. 326–328, May 1978.)

Solving Eq. (3-22) for β, we have

$$\beta \approx n_2 k(b\Delta + 1) \tag{3-23}$$

With this expression for β and using the assumption that n_2 is not a function of wavelength, we find that the group delay τ_{wg} arising from waveguide dispersion is

$$\tau_{wg} = \frac{L}{c}\frac{d\beta}{dk} = \frac{L}{c}\left[n_2 + n_2\Delta\frac{d(kb)}{dk}\right] \tag{3-24}$$

The modal propagation constant β is obtained from the eigenvalue equation expressed by Eq. (2-54), and is generally given in terms of the normalized frequency V defined by Eq. (2-57). We shall therefore use the approximation

$$V = ka(n_1^2 - n_2^2)^{1/2} \simeq kan_2\sqrt{2\Delta}$$

which is valid for small values of Δ, to write the group delay in Eq. (3-24) in terms of V instead of k, yielding

$$\tau_{wg} = \frac{L}{c}\left[n_2 + n_2\Delta\frac{d(Vb)}{dV}\right] \qquad (3\text{-}25)$$

The first term in Eq. (3-25) is a constant and the second term represents the group delay arising from waveguide dispersion. The factor $d(Vb)/dV$ can be expressed as[37]

$$\frac{d(Vb)}{dV} = b\left[1 - \frac{2J_\nu^2(ua)}{J_{\nu+1}(ua)J_{\nu-1}(ua)}\right]$$

where u is defined by Eq. (2-48) and a is the fiber radius. This factor is plotted in Fig. 3-14 as a function of V for various LP modes. The plots show that, for a fixed value of V, the group delay is different for every guided mode. When a light pulse is launched into a fiber, it is distributed among many guided modes. These various modes arrive at the fiber end at different times depending on their group delay, so that a pulse spreading results. For multimode fibers the waveguide dispersion is generally very small compared with material dispersion and can therefore be neglected.

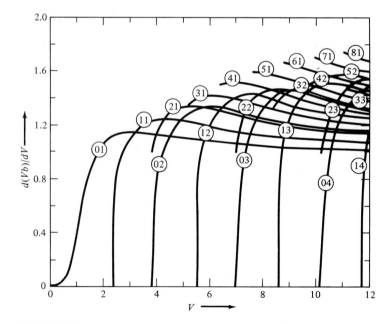

FIGURE 3-14
The group delay arising from waveguide dispersion as a function of the V number for a step-index optical fiber. The curve numbers jm designate the LP_{jm} modes. (Reproduced with permission from Gloge.[37])

3.2.5 Signal Distortion in Single-Mode Fibers

For single-mode fibers, waveguide dispersion is of importance and can be of the same order of magnitude as material dispersion. To see this, let us compare the two dispersion factors. The pulse spread σ_{wg} occurring over a distribution of wavelengths σ_λ is obtained from the derivative of the group delay with respect to wavelength:[37]

$$\sigma_{wg} \approx \left|\frac{d\tau_{wg}}{d\lambda}\right|\sigma_\lambda = L|D_{wg}(\lambda)|\sigma_\lambda$$

$$= \frac{V}{\lambda}\left|\frac{d\tau_{wg}}{d\lambda}\right|\sigma_\lambda = \frac{n_2 L \Delta \sigma_\lambda}{c\lambda} V \frac{d^2(Vb)}{dV^2} \tag{3-26}$$

where $D_{wg}(\lambda)$ is the *waveguide dispersion*.

To see the behavior of the waveguide dispersion, consider the expression of the factor ua for the lowest-order mode (i.e., the HE_{11} mode or, equivalently, the LP_{01} mode) in the normalized propagation constant. This can be approximated by[37]

$$ua = \frac{(1 + \sqrt{2})V}{1 + (4 + V^4)^{1/4}} \tag{3-27a}$$

Substituting this into Eq. (3-21) yields, for the HE_{11} mode,

$$b(V) = 1 - \frac{(1 + \sqrt{2})^2}{[1 + (4 + V^4)^{1/4}]^2} \tag{3-27b}$$

Figure 3-15 shows plots of this expression for b and its derivatives $d(Vb)/dV$ and $V\, d^2(Vb)/dV^2$ as functions of V.

Example 3-5. From Eq. (3-26) we have that the waveguide dispersion is

$$D_{wg}(\lambda) = -\frac{n_2\Delta}{c}\frac{1}{\lambda}\left[V\frac{d^2(Vb)}{dV^2}\right]$$

Let $n_2 = 1.48$ and $\Delta = 0.2$ percent. At $V = 2.4$, from Fig. 3-15 the expression in square brackets is 0.26. Choosing $\lambda = 1320$ nm, we then have $D_{wg}(\lambda) = -1.9$ ps/ (nm·km).

Figure 3-16 gives examples of the magnitudes of material and waveguide dispersion for a fused-silica-core single-mode fiber having $V = 2.4$. Comparing the waveguide dispersion with the material dispersion, we see that for a standard non-dispersion-shifted fiber, waveguide dispersion is important around 1320 nm. At this point, the two dispersion factors cancel to give a zero total dispersion. However, material dispersion dominates waveguide dispersion at shorter and longer wavelengths; for example, at 900 nm and 1550 nm. This figure used the approximation that material and waveguide dispersions are additive.

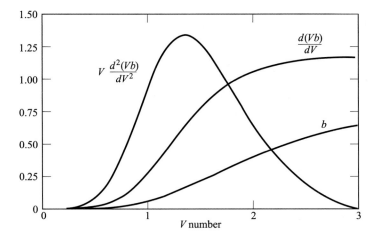

FIGURE 3-15
The waveguide parameter b and its derivatives $d(Vb)/dV$ and $V\ d^2(Vb)/dV^2$ plotted as a function of the V number for the HE_{11} mode.

3.2.6 Polarization-Mode Dispersion

The effects of fiber birefringence on the polarization states of an optical signal are another source of pulse broadening. This is particularly critical for high-rate long-haul transmission links (e.g., 10 Gb/s over tens of kilometers) that are designed to operate near the zero-dispersion wavelength of the fiber. Birefringence can result from intrinsic factors such as geometric irregularities of the fiber core or internal stresses on it. Deviations of less than 1 percent in the circularity of the core can already have a noticeable effect in a high-speed lightwave system. In addition, external factors, such as bending, twisting, or pinching of the fiber, can also lead to birefringence. Since all these mechanisms exist to some extent in any field-installed fiber, there will be a varying birefringence along its length.

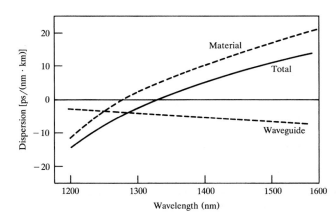

FIGURE 3-16
Examples of the magnitudes of material and waveguide dispersion as a function of optical wavelength for a single-mode fused-silica-core fiber. (Reproduced with permission from Keck,[16] © 1985, IEEE.)

A fundamental property of an optical signal is its polarization state. *Polarization* refers to the electric-field orientation of a light signal, which can vary significantly along the length of a fiber. As shown in Fig. 3-17, signal energy at a given wavelength occupies two orthogonal polarization modes. A varying birefringence along its length will cause each polarization mode to travel at a slightly different velocity and the polarization orientation will rotate with distance. The resulting difference in propagation times $\Delta\tau$ between the two orthogonal polarization modes will result in pulse spreading. This is the *polarization-mode dispersion* (PMD).[38,39] If the group velocities of the two orthogonal polarization modes are v_{gx} and v_{gy}, then the differential time delay $\Delta\tau_{pol}$ between the two polarization components during propagation of the pulse over a distance L is

$$\Delta\tau_{pol} = \left| \frac{L}{v_{gx}} - \frac{L}{v_{gy}} \right| \tag{3-28}$$

An important point to note is that, in contrast to chromatic dispersion, which is a relatively stable phenomenon along a fiber, PMD varies randomly along a fiber. A principal reason for this is that the perturbations causing the birefringence effects vary with temperature. In practice, this shows up as a random, time-varying fluctuation in the value of the PMD at the fiber output. Thus, $\Delta\tau_{pol}$ given in Eq. (3-28) cannot be used directly to estimate PMD. Instead, statistical predictions are needed to account for its effects.

A useful means of characterizing PMD for long fiber lengths is in terms of the mean value of the differential group delay (see Chap. 13 for PMD measurement techniques). This can be calculated according to the relationship

$$\langle \Delta\tau_{pol} \rangle \approx D_{PMD}\sqrt{L} \tag{3-29}$$

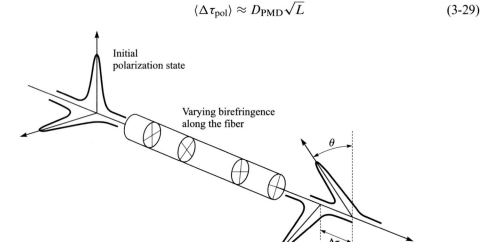

FIGURE 3-17
Variation in the polarization states of an optical pulse as it passes through a fiber with varying birefringence along its length.

where D_{PMD}, which is measured in $ps/\sqrt{km}$, is the average PDM parameter. Typical values of D_{PMD} range from 0.1 to 1.0 $ps/\sqrt{km}$. As an example, one experiment measured values of PMD for three different types of cable installations that were subjected to different environments.[40] The setups were a 36-km spooled fiber in a temperature-controlled chamber, a 48.8-km buried cable, and a 48-km aerial cable. Over a 12- to 15-h period, the average PMD parameters were measured as 0.028, 0.29, and 1.28 $ps/\sqrt{km}$, respectively. The larger value of PMD for the aerial cable is caused by sudden changes in temperature or because of movement of the fiber due to wind. In contrast to the instantaneous value $\Delta\tau_{pol}$, which varies over time and type of source, the mean value does not change from day to day or from source to source.

3.2.7 Intermodal Distortion

The final factor giving rise to signal degradation is intermodal distortion, which is a result of different values of the group delay for each individual mode at a single frequency. To see this pictorially, consider the meridional ray picture given for the step-index fiber in Fig. 2-12. The steeper the angle of propagation of the ray congruence, the higher is the mode number and, consequently, the slower the axial group velocity. This variation in the group velocities of the different modes results in a group delay spread of intermodal distortion. This distortion mechanism is eliminated by single-mode operation, but is important in multimode fibers. The maximum pulse broadening arising from intermodal distortion is the difference between the travel time T_{max} of the longest ray congruence paths (the highest-order mode) and the travel time T_{min} of the shortest ray congruence paths (the fundamental mode). This is simply obtained from ray tracing and is given by

$$\delta T_{mod} = T_{max} - T_{min} = \frac{n_1 \Delta L}{c} \tag{3-30}$$

Note that this simple derivation considers only pulse broadening owing to meridional rays and does not take into account skew rays.

3.3 PULSE BROADENING IN GRADED-INDEX WAVEGUIDES★

The analysis of pulse broadening in graded-index waveguides is more involved owing to the radial variation in core refractive index. The feature of this grading of the refractive-index profile is that it offers multimode propagation in a relatively large core together with the possibility of very low intermodal delay distortion. This combination allows the transmission of high data rates over long distances while still maintaining a reasonable degree of light launching and coupling ease. The reason for this low intermodal distortion can be seen by examining the light ray congruence propagation paths shown in Fig. 2-10. Since the index of refraction is lower at the outer edges of the core, light rays will travel faster in this region than in the center of the core where the refractive index is higher. This can be seen from the fundamental relationship $v = c/n$, where v is the speed of light in a medium of refractive index n. Thus, the ray congruence characterizing the higher-order mode will tend to travel further than the fundamental ray congruence, but at a faster rate. The higher-order mode will thereby tend to keep up with the lower-order mode, which, in turn, reduces the spread in the modal delay.

The root-mean-square (rms) pulse broadening σ in a graded-index fiber can be obtained from the sum[41]

$$\sigma = (\sigma_{intermodal}^2 + \sigma_{intramodal}^2)^{1/2} \tag{3-31}$$

where $\sigma_{intermodal}$ is the rms pulse width resulting from intermodal delay distortion and $\sigma_{intramodal}$ is the rms pulse width resulting from pulse broadening within each mode. To find the intermodal delay distortion, we use the relationship connecting intermodal delay to pulse broadening derived by Personick,[41]

$$\sigma_{intermodal} = \left(\langle \tau_g^2 \rangle - \langle \tau_g \rangle^2\right)^{1/2} \tag{3-32}$$

where the group delay τ_g of a mode as given by Eq. (3-13) in general depends on the order (v, m) of the mode, as described in Sec. 2.4.5, since the parameter β_{vm} denotes the roots of the modal equation given by Eq. (2-54). That is, in general,

$$\frac{\tau_g(v, m)}{L} = \frac{1}{c} \frac{\partial \beta_{vm}}{\partial k}$$

The quantities $\langle \tau_g^2 \rangle$ and $\langle \tau_g \rangle$ are then defined as the averages of τ_g^2 and τ_g, respectively, over the mode distribution; that is

$$\langle \tau_g^2 \rangle = \sum_{v,m} \frac{P_{vm} \tau_g^2(v, m)}{M} \tag{3-33a}$$

and

$$\langle \tau_g \rangle = \sum_{v, m} \frac{P_{vm} \tau_g(v, m)}{M} \tag{3-33b}$$

where P_{vm} is the optical power contained in the mode of order (v, m) and M is the number of fiber modes. For simplicity of notation, we will omit the subscripts v and m, since we will later assume that all modes are excited equally.

The group delay

$$\tau_g = \frac{L}{c} \frac{\partial \beta}{\partial k} \tag{3-34a}$$

is the time it takes energy in a mode having a propagation constant β to travel a distance L. To evaluate τ_g we use the following expression[42] for β:

$$\beta = \left[k^2 n_1^2 - 2\left(\frac{\alpha + 2}{\alpha} \frac{m}{a^2}\right)^{\alpha/(\alpha+2)} (n_1^2 k^2 \Delta)^{2/(\alpha+2)} \right]^{1/2} \tag{3-34b}$$

or, equivalently,

$$\beta = kn_1 \left[1 - 2\Delta \left(\frac{m}{M} \right)^{\alpha/(\alpha+2)} \right]^{1/2}$$

where m is the number of guided modes having propagation constants between $n_1 k$ and β, and M is the total number of possible guided modes given by Eq. (2-81). Substituting Eq. (3-34b) into Eq. (3-34a), keeping in mind that n_1 and Δ also depend on k, we obtain

$$\begin{aligned}
\tau_g &= \frac{L}{c} \frac{kn_1}{\beta} \left[N_1 - \frac{4\Delta}{\alpha+2} \left(\frac{m}{M} \right)^{\alpha/(\alpha+2)} \left(N_1 + \frac{n_1 k}{2\Delta} \frac{\partial\Delta}{\partial k} \right) \right] \\
&= \frac{LN_1}{c} \frac{kn_1}{\beta} \left[1 - \frac{\Delta}{\alpha+2} \left(\frac{m}{M} \right)^{\alpha/(\alpha+2)} (4+\epsilon) \right]
\end{aligned}$$

(3-35)

where we have used Eq. (2-81) for M and have defined the quantities

$$N_1 = n_1 + k \frac{\partial n_1}{\partial k}$$

(3-36a)

$$\epsilon = \frac{2n_1 k}{N_1 \Delta} \frac{\partial\Delta}{\partial k}$$

(3-36b)

As we noted in Eq. (2-46), guided modes only exist for values of β lying between kn_2 and kn_1. Since n_1 differs very little from n_2, that is,

$$n_2 = n_1(1 - \Delta)$$

where $\Delta \ll 1$ is the core-cladding index difference, it follows that $\beta \simeq n_1 k$. Thus, we can use the relationship

$$y = \Delta \left(\frac{m}{M} \right)^{\alpha/(\alpha+2)} \ll 1$$

(3-37)

in order to expand Eq. (3-35) in a power series in y. Using the approximation

$$\frac{kn_1}{\beta} = (1 - 2y)^{-1/2} \simeq 1 + y + \frac{3y^2}{2}$$

(3-38)

then Eq. (3-35) becomes

$$\begin{aligned}
\tau_g = \frac{N_1 L}{c} &\left[1 + \frac{\alpha - 2 - \epsilon}{\alpha + 2} \Delta \left(\frac{m}{M} \right)^{\alpha/(\alpha+2)} \right. \\
&\left. + \frac{3\alpha - 2 - 2\epsilon}{2(\alpha + 2)} \Delta^2 \left(\frac{m}{M} \right)^{2\alpha/(\alpha+2)} + O(\Delta^3) \right]
\end{aligned}$$

(3-39)

Equation (3-39) shows that to first order in Δ, the group delay difference between the modes is zero if

$$\alpha = 2 + \epsilon$$

(3-40)

Since ϵ is generally small, this indicates that minimum intermodal distortion will result from core refractive-index profiles which are nearly parabolic: that is, $\alpha \simeq 2$.

If we assume that all modes are equally excited (i.e., $P_{vm} = P$ for all modes), and if the number of fiber modes is assumed to be large, then the summation in Eqs. (3-33) can be replaced by an integral. Using these assumptions, Eq. (3-39) can be substituted into Eq. (3-32) to yield[42]

$$\sigma_{\text{intermodal}} = \frac{LN_1\Delta}{2c} \frac{\alpha}{\alpha+1} \left(\frac{\alpha+2}{3\alpha+2}\right)^{1/2}$$

$$\times \left[c_1^2 + \frac{4c_1c_2(\alpha+1)\Delta}{2\alpha+1} + \frac{16\Delta^2c_2^2(\alpha+1)^2}{(5\alpha+2)(3\alpha+2)}\right]^{1/2} \tag{3-41}$$

where we have used the abbreviations

$$c_1 = \frac{\alpha-2-\epsilon}{\alpha+2}$$

$$c_2 = \frac{3\alpha-2-2\epsilon}{2(\alpha+2)} \tag{3-42}$$

To find the intramodal pulse broadening, we use the definition[42]

$$\sigma_{\text{intramodal}}^2 = \left(\frac{\sigma_\lambda}{\lambda}\right)^2 \left\langle \left(\lambda \frac{d\tau_g}{d\lambda}\right)^2 \right\rangle \tag{3-43}$$

where σ_λ is the rms spectral width of the optical source. Equation (3-39) can be used to evaluate $\lambda\, d\tau_g/d\lambda$. If we neglect all terms of second and higher order in Δ, we obtain

$$\lambda\frac{d\tau_g}{d\lambda} = -\frac{L}{c}\lambda^2\frac{d^2n_1}{d\lambda^2} + \frac{N_1L\Delta}{c}\frac{\alpha-2-\epsilon}{\alpha+2}\frac{2\alpha}{\alpha+2}\left(\frac{m}{M}\right)^{\alpha/(\alpha+2)} \tag{3-44}$$

Here, we have kept only the largest terms; that is, terms involving factors such as $d\Delta/d\lambda$ and $\Delta dn_1/d\lambda$ are negligibly small. Both terms in Eq. (3-44) contribute to $\lambda d\tau_g/d\lambda$ for large values of α, since $\lambda^2 d^2n_1/d\lambda^2$ and Δ are the same order of magnitude. However, the second term in Eq. (3-44) is small compared with the first term when α is close to 2.

To evaluate $\sigma_{\text{intramodal}}$ we again assume that all modes are equally excited and that the summation in Eqs. (3-33) can be replaced by an integral. Thus, substituting Eq. (3-44) into Eq. (3-43) we have[42]

$$\sigma_{\text{intramodal}} = \frac{L}{c}\frac{\sigma_\lambda}{\lambda}\left[\left(-\lambda^2\frac{d^2n_1}{d\lambda^2}\right)^2\right.$$

$$\left.-N_1c_1\Delta\left(2\lambda^2\frac{d^2n_1}{d\lambda^2}\frac{\alpha}{\alpha+1} - N_1c_1\Delta\frac{4\alpha^2}{(\alpha+2)(3\alpha+2)}\right)\right]^{1/2} \tag{3-45}$$

Olshansky and Keck[42a] have evaluated σ as a function of α at $\lambda = 900$ nm for a titania-doped silica fiber having a numerical aperture of 0.16. This is shown in Fig. 3-18. Here the uncorrected curve assumes $\epsilon = 0$ and includes only inter-modal dispersion (no material dispersion). The inclusion of the effect of ϵ shifts the curve to higher values of α. The effect of the spectral width of the optical source on the rms pulse width is clearly demonstrated in Fig. 3-18. The light sources shown are an LED, an injection laser diode, and a distributed-feedback laser having rms spectral widths of 15, 1, and 0.2 nm, respectively. The data transmission capacities of these sources are approximately 0.13, 2, and $10 \,(\text{Gb} \cdot \text{km})/\text{s}$, respectively.

The value of α which minimizes pulse distortion depends strongly on wavelength. To see this, let us examine the structure of a graded-index fiber. A simple model of this structure is to consider the core to be composed of concentric cylindrical layers of glass, each of which has a different material composition. For each layer, the refractive index has a different variation with wavelength λ, since the glass composition is different in each layer. Consequently, a fiber with a given index profile α will exhibit different pulse spreading according to the source wavelength used. This is generally called *profile dispersion*. An example of this is given in Fig. 3-19 for a GeO_2–SiO_2 fiber.[43] This shows that the optimum value of α decreases with increasing wavelength. Suppose one wishes to transmit at 900 nm. A fiber having an optimum profile α_{opt} at 900 nm should exhibit a sharp

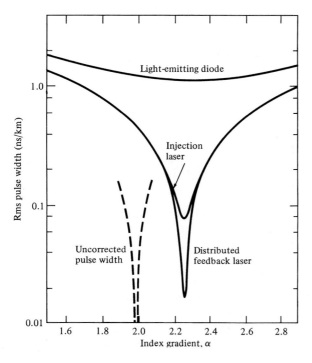

FIGURE 3-18
Calculated rms pulse spreading in a graded-index fiber versus the index parameter α at 900 nm. The uncorrected pulse curve is for $\epsilon = 0$ and assumes mode dispersion only. The other curves include material dispersion for an LED, an injection laser diode, and a distributed-feedback laser having spectral widths of 15, 1, and 0.2 nm, respectively. (Reproduced with permission from Olshansky and Keck.[42a])

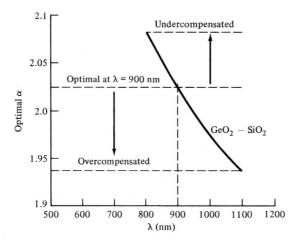

FIGURE 3-19

Profile dispersion effect on the optimum value of α as a function of wavelength for a GeO_2–SiO_2 graded index fiber. (Reproduced with permission from Cohen, Kaminow, Astle, and Stulz,[43] © 1978, IEEE.)

bandwidth peak at that wavelength. Fibers with undercompensated profiles, characterized by $\alpha > \alpha_{opt}$ (900 nm), tend to have a peak bandwidth at a shorter wavelength. On the other hand, overcompensated fibers which have an index profile $\alpha < \alpha_{opt}$ (900 nm) become optimal at a longer wavelength.

If the effect of material dispersion is ignored (i.e., $dn_1/d\lambda = 0$), an expression for the optimum index profile can be found from the minimum of Eq. (3-41) as a function of α. This occurs at[42]

$$\alpha_{opt} = 2 + \epsilon - \Delta \frac{(4 + \epsilon)(3 + \epsilon)}{5 + 2\epsilon} \tag{3-46}$$

If we take $\epsilon = 0$ and $dn_1/d\lambda = 0$, then Eq. (3-41) reduces to

$$\sigma_{opt} = \frac{n_1 \Delta^2 L}{20\sqrt{3}c} \tag{3-47}$$

This can be compared with the dispersion in a step-index fiber by setting $\alpha = \infty$ and $\epsilon = 0$ in Eq. (3-41), yielding

$$\sigma_{step} = \frac{n_1 \Delta L}{c} \frac{1}{2\sqrt{3}} \left(1 + 3\Delta + \frac{12\Delta^2}{5}\right)^{1/2} \simeq \frac{n_1 \Delta L}{2\sqrt{3}c} \tag{3-48}$$

Thus, under the assumptions made in Eqs. (3-47) and (3-48),

$$\frac{\sigma_{step}}{\sigma_{opt}} = \frac{10}{\Delta} \tag{3-49}$$

Hence, since typical values of Δ are 0.01, Eq. (3-49) indicates that the capacity of a graded-index fiber is about three orders of magnitude larger than that of a step-index fiber. For $\Delta = 1$ percent, the rms pulse spreading in a step-index fiber is about 14 ns/km, whereas that for a graded-index fiber is calculated to be 0.014 ns/km. In practice, these values are greater because of manufacturing difficulties. For

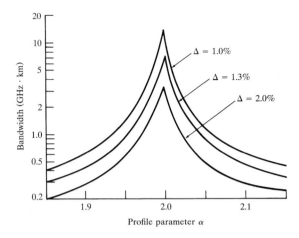

FIGURE 3-20
Variations in bandwidth resulting from slight deviations in the refractive-index profile for a graded-index fiber with $\Delta = 1$, 1.3, and 2 percent. (Reproduced with permission from Marcuse and Presby.[44])

example, although theory predicts a bandwidth of about $8\,\text{GHz} \cdot \text{km}$, it has been shown that, in practice, very slight deviations of the refractive-index profile from its optimum shape, owing to unavoidable manufacturing tolerances, can decrease the fiber bandwidth dramatically. This is illustrated in Fig. 3-20 for fibers with $\Delta = 1$, 1.3, and 2 percent. A change in α of a few percent can decrease the bandwidth by an order of magnitude.

3.4 MODE COUPLING★

In real systems, pulse distortion will increase less rapidly after a certain initial length of fiber because of mode coupling and differential mode loss[45,46] In this initial length of fiber, coupling of energy from one mode to another arises because of structural imperfections, fiber diameter and refractive-index variations, and cabling-induced microbends. The mode coupling tends to average out the propagation delays associated with the modes, thereby reducing intermodal dispersion. Associated with this coupling is an additional loss, which is designated by h and which has units of dB/km. The result of this phenomenon is that, after a certain coupling length L_c, the pulse distortion will change from an L dependence to a $(L_c L)^{1/2}$ dependence.

The improvement in pulse spreading caused by mode coupling over the distance $Z < L_c$ is related to the excess loss hZ incurred over this distance by the equation

$$hZ\left(\frac{\sigma_c}{\sigma_0}\right)^2 = C \tag{3-50}$$

Here, C is a constant, σ_0 is the pulse width increase in the absence of mode coupling, σ_c is the pulse broadening in the presence of strong mode coupling, and hZ is the excess attenuation resulting from mode coupling. The constant C in Eq. (3-50) is independent of all dimensional quantities and refractive indices. It

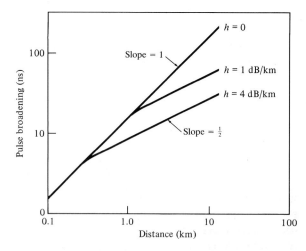

FIGURE 3-21
Mode-coupling effects on pulse distortion in long fibers for various coupling losses.

depends only on the fiber profile shape, the mode-coupling strength, and the modal attenuations.

The effect of mode coupling on pulse distortion can be significant for long fibers, as is shown in Fig. 3-21 for various coupling losses in a graded-index fiber. The parameters of this fiber are $\Delta = 1$ percent, $\alpha = 4$, and $C = 1.1$. The coupling loss h must be determined experimentally, since a calculation would require a detailed knowledge of the mode coupling introduced by the various waveguide perturbations. Measurements of bandwidth as a function of distance have produced values of L_c ranging from about 100 to 550 m.

An important point to note is that extensive mode coupling and power distribution can occur at connectors, splices, and other passive components in an optical link; this can have a significant effect on the overall system bandwidth.[47-49]

3.5 DESIGN OPTIMIZATION OF SINGLE-MODE FIBERS

Since telecommunication companies use single-mode fibers as the principal optical transmission medium in their networks, and because of the importance of single-mode fibers in microwave-speed localized applications,[50] this section addresses their basic design and operational properties. Some of the attributes of single-mode fibers include a long expected installation lifetime, very low attenuation, high-quality signal transfer because of the absence of modal noise, and the largest available bandwidth–distance product. Here, we shall examine design-optimization characteristics, cutoff wavelength, dispersion, mode-field diameter, and bending loss.

3.5.1 Refractive-Index Profiles

In the design of single-mode fibers, dispersion behavior is a major distinguishing feature, since this is what limits long-distance and very high-speed transmission. Comparing Figs. 3-3 and 3-16, we see that whereas the dispersion of a single-mode silica fiber is lowest at 1300 nm, its attenuation is a minimum at 1550 nm, where the dispersion is higher. Ideally, for achieving a maximum transmission distance of a high-capacity link, the dispersion null should be at the wavelength of minimum attenuation. To achieve this, one can adjust the basic fiber parameters to shift the zero-dispersion minimum to longer wavelengths.

The basic material dispersion is hard to alter significantly, but it is possible to modify the waveguide dispersion by changing from a simple step-index core profile design to more complicated index profiles.[13,51-57] Researchers have thus examined a variety of core and cladding refractive-index configurations for altering the behavior of single-mode fibers. Figure 3-22 shows representative refractive-index profiles of the four main categories: 1300-nm-optimized fibers, dispersion-shifted fibers, dispersion-flattened fibers, and large-effective-core-area fibers. To get a better feeling of their geometry, Fig. 3-23 shows the three-dimensional index profiles for several different types of single-mode fibers.

The most popular single-mode fibers used in telecommunication networks are near-step-index fibers, which are dispersion-optimized for operation at 1300 nm. These *1300-nm-optimized single-mode fibers* are of either the *matched-cladding*[13,51,52] or the *depressed-cladding*[53,54] design, as shown in Figs. 3-22a, 3-23a and 3-23b. Matched-cladding fibers have a uniform refractive index throughout the cladding. Typical mode-field diameters are 9.5 μm and core-to-cladding index differences are around 0.37 percent. In depressed-cladding fibers the cladding portion next to the core has a lower index than the outer cladding region. Mode-field diameters are around 9 μm, and typical positive and negative index differences are 0.25 and 0.12 percent, respectively.

As we saw from Eqs. (3-20) and (3-26), whereas material dispersion depends only on the composition of the material, waveguide dispersion is a function of the core radius, the refractive-index difference, and the shape of the refractive-index profile. Thus, the waveguide dispersion can vary dramatically with the fiber design parameters. By creating a fiber with a larger negative waveguide dispersion and assuming the same values for material dispersion as in a standard single-mode fiber, the addition of waveguide and material dispersion can then shift the zero-dispersion point to longer wavelengths. The resulting optical fibers are known as *dispersion-shifted fibers*.[52,55-58] Examples of refractive-index profiles for dispersion-shifted fibers are shown in Figs. 3-22b and 3-23c. A typical waveguide dispersion curve for this type of fiber is depicted in Fig. 3-24a. The resultant total dispersion curve is shown in Fig. 3-24b for fibers with a zero-dispersion wavelength at 1550 nm.

An alternative is to reduce fiber dispersion by spreading the dispersion minimum out over a wider range. This approach is known as *dispersion flattening*.[59,60] Dispersion-flattened fibers are more complex to design than dis-

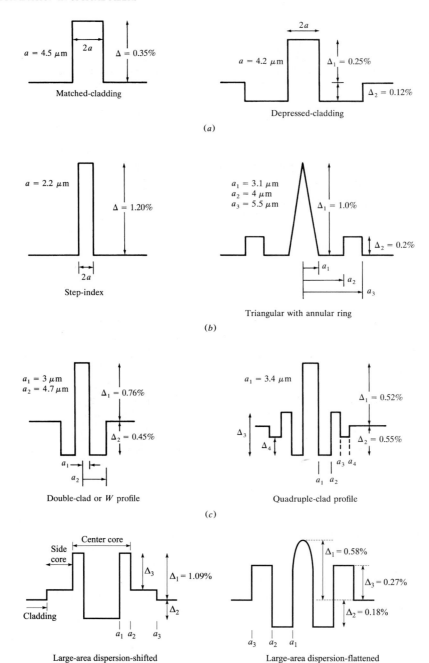

FIGURE 3-22
Representative cross sections of index profiles for (*a*) 1300-nm-optimized, (*b*) dispersion-shifted, (*c*) dispersion-flattened, and (*d*) large-effective-core-area fibers.

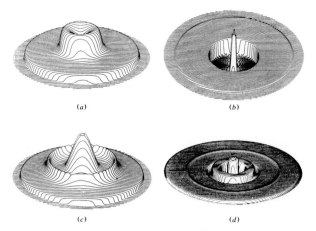

FIGURE 3-23
Three-dimensional refractive index profiles for (a) matched-cladding 1300-nm-optimized, (b) depressed-cladding 1300-nm-optimized, (c) triangular dispersion-shifted, and (d) quadruple-clad dispersion-flattened single-mode fibers. [(a) and (c) courtesy of Corning, Inc.; (b) courtesy of York Technology; (d) reproduced with permission from H. Lydtin, *J. Lightwave Tech.*, vol. LT-4, pp. 1034–1038, Aug. 1986, © 1986, IEEE.]

persion-shifted fibers, because dispersion must be considered over a much broader range of wavelengths. However, they offer desirable characteristics over a wide span of wavelengths. Figures 3-22c and 3-23d show typical cross-sectional and three-dimensional refractive-index profiles, respectively. A typical waveguide dispersion curve for this type of fiber is depicted in Fig. 3-24a. Figure 3-24b gives the resultant total dispersion-flattened characteristic.

The advent of optical fiber amplifiers for operation in the 1550-nm region (see Chap. 11) and the accompanying demand for long-distance high-capacity links led to the development of a single-mode optical fiber with a larger effective core area.[61-64] The impetus for larger core areas is the need to reduce the effects of fiber nonlinearities, which limit system capacities, as is detailed in Chap. 12. Figure 3-22d gives two examples of the index profile for these large-effective-area (LEA) fibers. Whereas standard single-mode fibers have effective core areas of about $55 \, \mu m^2$, these profiles yield values greater than $100 \, \mu m^2$.

3.5.2 Cutoff Wavelength

The cutoff wavelength of the first higher-order mode (LP_{11}) is an important transmission parameter for single-mode fibers, since it separates the single-mode from the multimode regions.[65-67] As we saw from Eq. (2-58), single-mode operation occurs above the theoretical cutoff wavelength given by

$$\lambda_{c;th} = \frac{2\pi a}{V}(n_1^2 - n_2^2)^{1/2} \tag{3-51}$$

with $V = 2.405$ for step-index fibers. At this wavelength, only the LP_{01} mode (i.e., the HE_{11} mode) should propagate in the fiber.

Since in the cutoff region the field of the LP_{11} mode is widely spread across the fiber cross section (i.e., it is not tightly bound to the core), its attenuation is strongly affected by fiber bends, length, and cabling. Recommendation G.650 of

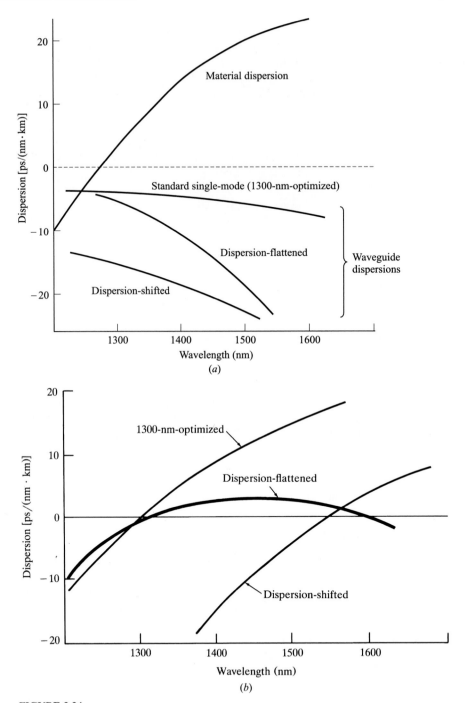

FIGURE 3-24
(*a*) Typical waveguide dispersions and the common material dispersion of three different single-mode fiber designs; (*b*) resultant total dispersions.

the ITU-T[66] and the EIA-455-80A Standard[67] specify methods for determining an effective cutoff wavelength λ_c. The setup consists of a 2-m length of fiber that contains a single 14-cm-radius loop or several 14-cm-radius curvatures that add up to one complete loop. Using a tunable light source that has a full-width half-maximum linewidth not exceeding 10 nm, light is launched into the fiber so that both the LP_{01} and the LP_{11} modes are uniformly excited.

First, the output power $P_1(\lambda)$ is measured as a function of wavelength in a sufficiently wide range around the expected cutoff wavelength. Next, the output power $P_2(\lambda)$ is measured over the same wavelength range when a loop of sufficiently small radius is included in the test fiber to filter the LP_{11} mode. A typical radius for this loop is 30 mm. With this method, the logarithmic ratio $R(\lambda)$ between the two transmitted powers $P_1(\lambda)$ and $P_2(\lambda)$ is calculated as

$$R(\lambda) = 10 \log \left[\frac{P_1(\lambda)}{P_2(\lambda)} \right] \tag{3-52}$$

Figure 3-25 gives a typical curve of the result. The effective cutoff wavelength λ_c is defined as the largest wavelength at which the higher-order LP_{11} mode power relative to the fundamental LP_{01} mode power is reduced to 0.1 dB; that is, when $R(\lambda) = 0.1$ dB, as is shown in Fig. 3-25. Recommended values of λ_c range from 1100 to 1280 nm, to avoid modal noise and dispersion problems.

3.5.3 Dispersion Calculations

As noted in Sec. 3.5.1, the total dispersion in single-mode fibers consists mainly of material and waveguide dispersions. The resultant intramodal or chromatic dispersion is represented by[58,66,68-70]

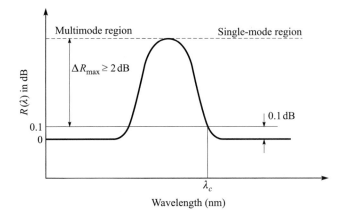

FIGURE 3-25
Typical attenuation-ratio versus wavelength plot for determining cutoff wavelength using the bend-reference (or single-mode-reference) transmission method. The peak ratio should be at least 2 dB above the cutoff level.

$$D(\lambda) = \frac{d\tau}{d\lambda} \tag{3-53}$$

where τ is the measured group delay per unit length of fiber (thus the factor of L difference between this equation and the expressions given in Sec. 3.2). The dispersion is commonly expressed in ps/(nm·km). The broadening σ of an optical pulse over a fiber of length L is given by

$$\sigma = D(\lambda)L\sigma_\lambda \tag{3-54}$$

where σ_λ is the half-power spectral width of the optical source. To measure the dispersion, one examines the pulse delay over a desired wavelength range.

As illustrated in Fig. 3-24, the dispersion behavior varies with wavelength and also with fiber type. Thus, the EIA and the ITU-T have recommended different formulas to calculate the chromatic dispersion for specific fiber types operating in a given wavelength region. To calculate the dispersion for a non-dispersion-shifted fiber (called a Class IVa fiber by the EIA) in the 1270-to-1340-nm region, the standards recommend fitting the measured group delay per unit length to a three-term Sellmeier equation of the form[66]

$$\tau = A + B\lambda^2 + C\lambda^{-2} \tag{3-55}$$

to the measured pulse data. Here, A, B, and C are the curve-fitting parameters. An equivalent expression is

$$\tau = \tau_0 + \frac{S_0}{8}\left(\lambda - \frac{\lambda_0^2}{\lambda}\right)^2 \tag{3-56}$$

where τ_0 is the relative delay minimum at the zero-dispersion wavelength λ_0, and S_0 is the value of the *dispersion slope* $S(\lambda) = dD/d\lambda$ at λ_0, which is given in ps/(nm^2 · km). Using Eq. (3-53), the dispersion for a non-dispersion-shifted fiber is

$$D(\lambda) = \frac{\lambda S_0}{4}\left[1 - \left(\frac{\lambda_0}{\lambda}\right)^4\right] \tag{3-57}$$

To calculate the dispersion for a dispersion-shifted fiber (called a Class IVb fiber by the EIA) in the 1500-to-1600-nm region, the standards recommend using the quadratic expression.[66]

$$\tau = \tau_0 + \frac{S_0}{2}(\lambda - \lambda_0)^2 \tag{3-58}$$

which results in the dispersion expression

$$D(\lambda) = (\lambda - \lambda_0)S_0 \tag{3-59}$$

Note from App. F that the *third-order dispersion* β_3 can be given as

$$\beta_3 = \frac{\lambda^2}{(2\pi c)^2}[\lambda^2 S_0 + 2\lambda D] \tag{3-60}$$

When measuring a set of fibers, one will get values of λ_0 ranging from $\lambda_{0,min}$ to $\lambda_{0,max}$. Figure 3-26 shows the range of expected dispersion values for a set of

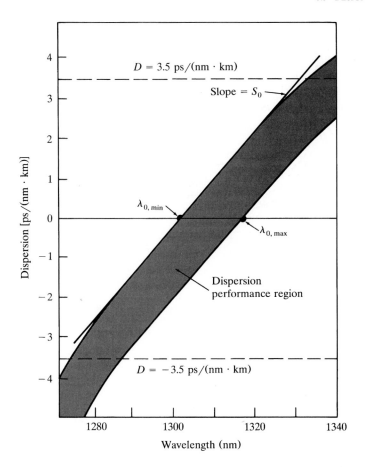

FIGURE 3-26

Example of a dispersion performance curve for a set of single-mode fibers. The two slightly curved lines are found by solving Eq. (3-57). S_0 is the slope of $D(\lambda)$ at the zero-dispersion wavelength λ_0.

non-dispersion-shifted fibers in the 1270-to-1340-nm region. Typical values of S_0 are 0.092 ps/(nm² · km) for standard non-dispersion-shifted fibers, and are between 0.06 and 0.08 ps/(nm² · km) for dispersion-shifted fibers. Alternatively, the ITU-T Rec. G.652 has specified this as a maximum dispersion of 3.5 ps/ (nm · km) in the 1285-to-1330-nm region, as denoted by the dashed lines in Fig. 3-26.

Figure 3-27 illustrates the importance of controlling dispersion in single-mode fibers. As optical pulses travel down a fiber, temporal broadening occurs because material and waveguide dispersion cause different wavelengths in the optical pulse to propagate with different velocities. Thus, as Eq. (3-54) implies, the broader the spectral width σ_λ of the source, the greater the pulse dispersion will be. This effect is clearly seen in Fig. 3-27.

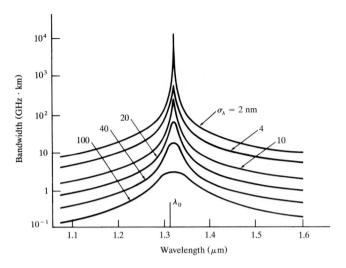

FIGURE 3-27
Examples of bandwidth versus wavelength for different source spectral widths σ_λ in a single-mode fiber having a dispersion minimum at 1300 nm. (Reproduced with permission from Reed, Cohen, and Shang,[56] © 1987, AT&T.)

3.5.4 Mode-Field Diameter

Section 2.5.1 gives the definition of the mode-field diameter in single-mode fibers. One uses the mode-field diameter in describing the functional properties of a single-mode fiber, since it takes into account the wavelength-dependent field penetration into the cladding. This is shown in Fig. 3-28 for 1300-nm-optimized, dispersion-shifted, and dispersion-flattened single-mode fibers.

3.5.5 Bending Loss

Macrobending and microbending losses are important in the design of single-mode fibers.[19-25] These losses are principally evident in the 1550-nm region, and show up as a rapid increase in attenuation when the fiber is bent smaller than a certain bend radius. The lower the cutoff wavelength relative to the oper-ating wavelength, the more susceptible single-mode fibers are to bending. For example, in a fiber which is optimized for operation at 1300 nm, both the micro-bending and macrobending losses are greater at 1550 nm that at 1300 nm by a factor of 3 to 5, as Fig. 3-29 illustrates. A fiber thus might be transmitting at 1300 nm but have a significant loss at 1550 nm.

The bending losses are primarily a function of the mode-field diameter. Generally, the smaller the mode-field diameter (i.e., the tighter the confinement of the mode to the core), the smaller the bending loss. This is true for both matched-clad and depressed-clad fibers, as Fig. 3-30 shows.

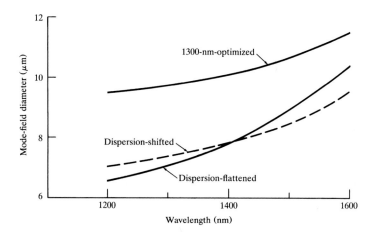

FIGURE 3-28
Typical mode-field diameter variations with wavelength for (*a*) 1300-nm-optimized, (*b*) dispersion-shifted, and (*c*) dispersion-flattened single-mode fibers.

In examining the bending loss, early theories assume a simple model of a fiber with an infinitely extending cladding. This results in the prediction of a smooth exponential increase of bending loss with increasing wavelength or radius of curvature. In an actual fiber, oscillations in the bend loss versus both the wavelength and bending radius are observed. These oscillations can be attributed to coherent coupling between the field propagating in the core and the fraction of the radiated field that is reflected at the boundary between the cladding and the

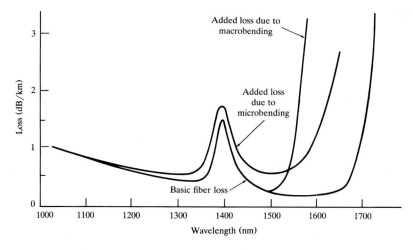

FIGURE 3-29
Representative increases in single-mode fiber attenuation owing to microbending and macrobending effects. (Reproduced with permission from Kalish and Cohen,[57] © 1987, AT&T.)

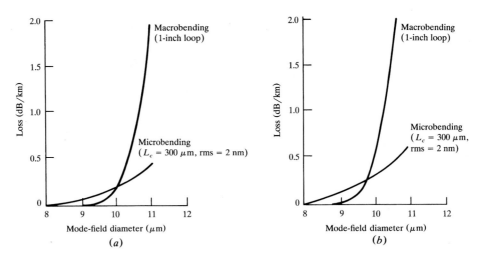

FIGURE 3-30
Calculated increase in attenuation at 1310 nm from microbending and macrobending effects as a function of mode-field diameter for (*a*) depressed-cladding single-mode fiber ($V = 2.514$) and (*b*) matched-cladding single-mode fiber ($V = 2.373$). The microbending calculations assume a correlation length L_c (microbending repetition rate) of 300 nm and a 2-nm deformation amplitude. (Reproduced with permission from Kalish and Cohen,[57] © 1987, AT&T.)

fiber-coating material. Figure 3-31 gives an example of calculated bend loss as a function of bend radius at a 1300-nm wavelength. The fiber parameters were core radius $a = 3.6\,\mu$m, cladding radius $b = 60\,\mu$m, $(n_1 - n_2)/n_2 = 3.56 \times 10^{-3}$ and $(n_3 - n_2)/n_2 = 0.07$, where n_1, n_2, and n_3 are the core, cladding, and coating indices of refraction, respectively.

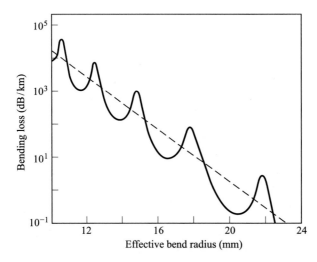

FIGURE 3-31
Calculated bend loss as a function of bend radius at 1300 nm. The dashed line represents the infinite-cladding case; that is, $n_2 = n_3$. (Modified with permission from Renner,[22] © 1992, IEEE.)

By specifying bend-radius limitations, one can largely avoid high macro-bending losses. Manufacturers usually recommend a minimum fiber or cable bend diameter of 40–50 mm (1.6–2.0 in.). This is consistent with typical bend diameters of 50–75 mm found in fiber-splice enclosures, in equipment bays, or on optoelectronic packages. Since single-mode fibers are designed to have little or no additional attenuation at 1550 nm from bend diameters greater than 50 mm, bending loss should not be a limiting performance factor in correctly installed cables.

PROBLEMS

3-1. Verify the expression given in Eq. (3-1c) that relates α, which is in units of dB/km, to α_p, which is in units of km^{-1}.

3-2. A certain optical fiber has an attenuation of 0.6 dB/km at 1300 nm and 0.3 dB/km at 1550 nm. Suppose the following two optical signals are launched simultaneously into the fiber: an optical power of 150 μW at 1300 nm and an optical power of 100 μW at 1550 nm. What are the power levels in μW of these two signals at (a) 8 km and (b) 20 km?

3-3. An optical signal at a specific wavelength has lost 55 percent of its power after traversing 3.5 km of fiber. What is the attenuation in dB/km of this fiber?

3-4. A continuous 12-km-long optical fiber link has a loss of 1.5 dB/km.
   (a) What is the minimum optical power level that must be launched into the fiber to maintain an optical power level of 0.3 μW at the receiving end?
   (b) What is the required input power if the fiber has a loss of 2.5 dB/km?

3-5. Consider a step-index fiber with a SiO_2–GeO_2 core having a mole fraction 0.08 of GeO_2. Plot Eqs. (3-2b) and (3-3) from 500 nm to 5 μm, and compare the results with the curves in Fig. 3-5.

3-6. The optical power loss resulting from Rayleigh scattering in a fiber can be calculated from either Eq. (3-4a) or Eq. (3-4b). Compare these two equations for silica ($n = 1.460$ at 630 nm), given that the fictive temperature T_f is 1400 K, the isothermal compressibility β_T is 6.8×10^{-12} cm^2/dyn, and the photoelastic coefficient is 0.286. How does this agree with measured values ranging from 3.9 to 4.8 dB/km at 633 nm?

3-7. Using a computer, solve Eq. (3-7) to make plots of the radius of bend curvature versus the fiber core radius a for values of $N_{\text{eff}}/N_\infty = 10$, 50, and 75 percent at wavelengths of 1300 nm and 1550 nm. Let a be in the range 5 μm $\leq a \leq 30$ μm.

3-8. Consider graded-index fibers having index profiles $\alpha = 2.0$, cladding refractive indices $n_2 = 1.50$, and index differences $\Delta = 0.01$. Using Eq. (3-7), plot the ratio N_{eff}/N_∞ for bend radii less than 10 cm at $\lambda = 1$ μm for fibers having core radii of 4, 25, and 100 μm.

3-9. Three common fiber jacket materials are Elvax® 265 ($E_j = 21$ MPa) and Hytrel® 4056 ($E_j = 58$ MPa), both made by DuPont, and Versalon® 1164 ($E_j = 104$ MPa) made by General Mills. If the Young's modulus of a glass fiber is 64 GPa, plot the reduction in microbending loss as a function of the index difference Δ when fibers are coated with these materials. Make these plots for Δ values ranging from 0.1 to 1.0 percent and for a fiber cladding-to-core ratio of $b/a = 2$.

3-10. Assume that a step-index fiber has a V number of 6.0.

(a) Using Fig. 2-22, estimate the fractional power P_{clad}/P traveling in the cladding for the six lowest-order LP modes.

(b) If the fiber in (a) is a glass-core glass-clad fiber having core and cladding attenuations of 3.0 and 4.0 dB/km, respectively, find the attenuations for each of the six lowest-order modes.

(c) Suppose the fiber in (a) is a glass-core polymer-clad fiber having core and cladding attenuations of 5 and 1000 dB/km, respectively. Find the attenuations for each of the six lowest-order modes.

3-11. Assume a given mode in a graded-index fiber has a power density $p(r) = P_0 \exp(-Kr^2)$, where the factor K depends on the modal power distribution.

(a) Letting $n(r)$ in Eq. (3-11) be given by Eq. (2-78) with $\alpha = 2$, show that the loss in this mode is

$$\alpha_{gi} = \alpha_1 + \frac{\alpha_2 - \alpha_1}{Ka^2}$$

Since $p(r)$ is a rapidly decaying function of r and since $\Delta \ll 1$, for ease of calculation assume that the top relation in Eq. (2-78) holds for all values of r.

(b) Choose K such that $p(a) = 0.1P_0$; that is, 10 percent of the power flows in the cladding. Find α_{gi} in terms of α_1 and α_2.

3-12. For wavelengths less than $1.0 \, \mu\text{m}$ the refractive index n satisfies a Sellmeier relation of the form[71]

$$n^2 = 1 + \frac{E_0 E_d}{E_0^2 - E^2}$$

where $E = hc/\lambda$ is the photon energy and E_0 and E_d are, respectively, material oscillator energy and dispersion energy parameters. In SiO_2 glass, $E_0 = 13.4$ eV and $E_d = 14.7$ eV. Show that, for wavelengths between 0.20 and $1.0 \, \mu\text{m}$, the values of n found from the Sellmeier relation are in good agreement with those shown in Fig. 3-12.

3-13. (a) An LED operating at 850 nm has a spectral width of 45 nm. What is the pulse spreading in ns/km due to material dispersion? What is the pulse spreading when a laser diode having a 2-nm spectral width is used?

(b) Find the material-dispersion-induced pulse spreading at 1550 nm for an LED with a 75-nm spectral width.

3-14. (a) Using Eqs. (2-48), (2-49), and (2-57) show that the normalized propagation constant defined by Eq. (3-21) can be written in the form

$$b = \frac{\beta^2/k^2 - n_2^2}{n_1^2 - n_2^2}$$

(b) For small core–cladding refractive-index differences show that the expression for b derived in (a) reduces to

$$b = \frac{\beta/k - n_2}{n_1 - n_2}$$

from which it follows that

$$\beta = n_2 k(b\Delta + 1)$$

3-15. Using a computer (preferably), verify the plots for b, $d(Vb)/dV$, and $V\,d^2(Vb)/dV^2$ shown in Fig. 3-15. Use the expression for b given by Eq. (3-27b).

3-16. Derive Eq. (3-30) by using a ray-tracing method.

3-17. Consider a step-index fiber with core and cladding diameters of 62.5 and 125 μm, respectively. Let the core index $n_1 = 1.48$ and let the index difference $\Delta = 1.5$ percent. Compare the modal dispersion in units of ns/km of this fiber as given by Eq. (3-30) with the more exact expression

$$\frac{\sigma_{\text{mod}}}{L} = \frac{n_1 - n_2}{c}\left(1 - \frac{\pi}{V}\right)$$

where L is the length of the fiber and n_2 is the cladding index.

3-18. Verify that Eq. (3-41) reduces to Eq. (3-48) for the step-index case when $\alpha = \infty$ and $\epsilon = 0$.

3-19. Show that, when the effect of material dispersion is ignored and for $\epsilon = 0$, Eq. (3-41) reduces to Eq. (3-47).

3-20. Make a plot on log–log paper of the rms pulse broadening in a parabolic graded-index fiber ($\alpha = 2$) as a function of the optical source spectral width σ_λ in the range 0.10–100 nm for peak operating wavelengths of 850 nm and 1300 nm. Let $\Delta = 0.01$, $N_1 = 1.46$, and $\epsilon = 0$ at both wavelengths. Assume the factor $\lambda^2 d^2 n/d\lambda^2$ is 0.025 at 850 nm and 0.004 at 1300 nm.

3-21. Repeat Prob. 3-20 for a graded-index single-mode fiber with $\Delta = 0.001$.

3-22. Derive Eq. (3-35) by substituting Eq. (3-34b) into Eq. (3-34a).

3-23. Derive Eq. (3-44) from Eq. (3-39).

3-24. Verify the expression given in Eq. (3-45).

3-25. Consider a standard non-dispersion-shifted single-mode optical fiber that has a zero-dispersion wavelength at 1310 nm with a dispersion slope of $S_0 = 0.090\,\text{ps}/(\text{nm}^2 \cdot \text{km})$. Plot the dispersion in the wavelength range 1270 nm $\leq \lambda \leq$ 1340 nm.

3-26. A typical dispersion-shifted single-mode optical fiber has a zero-dispersion wavelength at 1550 nm with a dispersion slope of $S_0 = 0.070\,\text{ps}/(\text{nm}^2 \cdot \text{km})$.

    (a) Plot the dispersion in the wavelength range 1500 nm $\leq \lambda \leq$ 1600 nm.

    (b) Compare the dispersion at 1500 nm with the dispersion value for the non-dispersion-shifted fiber described in Prob. 3-25.

3-27. Compare the rms pulse broadening per kilometer for the following three fibers:

    (a) A multimode step-index fiber with core index $n_1 = 1.49$ and relative index difference $\Delta = 1$ percent.

    (b) A graded-index fiber having an optimum parabolic index profile and the same core index ($n_1 = 1.49$) and relative index difference ($\Delta = 1$ percent) as the step-index fiber in (a).

    (c) The same type of graded-index fiber as in (b) but with $\Delta = 0.5$ percent.

3-28. Consider an optical link consisting of a 5-km-long step-index fiber with core index $n_1 = 1.49$ and relative index difference $\Delta = 1$ percent.

    (a) Find the delay difference at the fiber end between the slowest and fastest modes.

    (b) Find the rms pulse broadening caused by intermodal dispersion.

    (c) Calculate the maximum bit rate B_T that can be transmitted over the fiber without significant errors, which is given by $B_T = 0.2/\sigma_{\text{step}}$ (see Chap. 8).

(*d*) Assuming the maximum bit rate equals the bandwidth, what is the bandwidth–distance product of this fiber?

3-29. The net delay σ_{inter} of a graded-index fiber between the lowest-order and highest-order modes can be written as[34]

$$
\sigma_{inter} =
\begin{cases}
\dfrac{n_1 \Delta L}{c}\left(\dfrac{\alpha - \alpha_{opt}}{\alpha + 2}\right) & \text{for } \alpha \neq \alpha_{opt} \\[2ex]
\dfrac{n_1 \Delta^2 L}{c} & \text{for } \alpha = \alpha_{opt}
\end{cases}
$$

If $\alpha_{opt} = 2.0$, use a computer to plot the ratio of σ_{inter} (at $\alpha = \alpha_{opt}$) to σ_{inter} (at $\alpha \neq \alpha_{opt}$) for index differences of $\Delta = 0.5$, 1.0, and 2.0 percent over the index-profile range $0.90 \leq \alpha \leq 1.10$; that is, for values of α ranging from 90 to 110 percent of α_{opt}.

3-30. Calculate the waveguide dispersion at 1320 nm in units of [ps/(nm · km)] for a single-mode fiber with core and cladding diameters of $9\,\mu$m and $125\,\mu$m, respectively. Let the core index $n_1 = 1.48$ and let the index difference $\Delta = 0.22$ percent.

3-31. Starting with Eq. (3-55), derive the dispersion expression given in Eq. (3-57).

3-32. Renner[22] derived a simplified approximation to describe the bend losses of single-mode optical fibers. This expression for the bending loss is

$$
\alpha_{simp} = \alpha_{conv}\frac{2(Z_3 Z_2)^{1/2}}{(Z_3 + Z_2) - (Z_3 - Z_2)\cos(2\Theta)}
$$

where the conventional bending loss is

$$
\alpha_{conv} = \frac{1}{2}\left(\frac{\pi}{\gamma^3 R}\right)^{1/2}\frac{\kappa^2}{V^2 K_1^2(\gamma a)}\exp\left(-\frac{2\gamma^3 R}{3\beta_0^3}\right)
$$

where V is given by Eq. (2-57), β_0 is the propagation constant in a straight fiber with an infinite cladding given by Eq. (2-46), K_1 is the modified Bessel function (see App. C), and

$$
Z_q \approx k^2 n_q^2(1 + 2b/R) - \beta_0^2 \approx k^2 n_q^2(1 + 2b/R) - k^2 n_2^2 \qquad \text{for } q = 2, 3
$$

$$
\Theta = \frac{\gamma^3 R}{3k^2 n_2^2}\left(\frac{R_c}{R} - 1\right)^{3/2}
$$

$$
\gamma = \left(\beta_0^2 - k^2 n_2^2\right)^{1/2} \approx k\left(n_1^2 - n_2^2\right)^{1/2}
$$

$$
\kappa^2 = k^2 n_1^2 - \beta_0^2 \approx k^2\left(n_1^2 - n_2^2\right)
$$

$$
R_c = 2k^2 n_2^2 b/\gamma^2 = \text{the critical bend radius}
$$

Using a computer, (*a*) verify the plot given in Fig. 3-31 at 1300 nm, and (*b*) calculate and plot the bend loss as a function of wavelength for $800\,\text{nm} \leq \lambda \leq 1600\,\text{nm}$ at several different bend radii (e.g., 15 and 20 mm). Let $n_1 = 1.480$, $n_2 = 1.475$, $n_3 = 1.07 n_2 = 1.578$, and $b = 60\,\mu$m.

3-33. Faustini and Martini[24] developed a more detailed formula for describing the oscillatory behavior of bend loss as a function of the bend radius and wavelength. Using a computer, use their formulation to reproduce the three-dimensional plots of bend loss versus radius of curvature and wavelength given in Fig. 5 of their paper (*J. Lightwave Tech.*, vol. 15, pp. 671–679, Apr. 1997).

REFERENCES

1. B. C. Bagley, C. R. Kurkjian, J. W. Mitchell, G. E. Peterson, and A. R. Tynes, "Materials, properties, and choices," in S. E. Miller and A. G. Chynoweth, eds., *Optical Fiber Telecommunications*, Academic, New York, 1979.

2. P. Kaiser and D. B. Keck, "Fiber types and their status," in S. E. Miller and I. P. Kaminow, eds., *Optical Fiber Telecommunications—II*, Academic, New York, 1988.

3. R. Olshansky, "Propagation in glass optical waveguides," *Rev. Mod. Phys.*, vol. 51, pp. 341–367, Apr. 1979.

4. D. Gloge, "The optical fibre as a transmission medium," *Rep. Prog. Phys.*, vol. 42, pp. 1777–1824, Nov. 1979.

5. A. W. Snyder and J. D. Love, *Optical Waveguide Theory*, Chapman & Hall, New York, 1983.

6. A. Iino and J. Tamura, "Radiation resistivity in silica optical fibers," *J. Lightwave Tech.*, vol. 6, pp. 145–149, Feb. 1988.

7. (a) E. W. Taylor, E. J. Friebele, H. Henschel, R. H. West, J. A. Krinsky, and C. E. Barnes, "Interlaboratory comparison of radiation-induced attenuation in optical fibers. Part II: Steady state," *J. Lightwave Tech.*, vol. 8, pp. 967–976, June 1990.

   (b) E. J. Friebele et al., "Interlaboratory comparison of radiation-induced attenuation in optical fibers. Part III: Transient exposures," *J. Lightwave Tech.*, vol. 8, pp. 977–989, June 1990.

   (c) R. H. West, H. Buker, E. J. Friebele, H. Henschel, and P. B. Lyons, "The use of optical time-domain reflectometers to measure radiation-induced losses in optical fibers," *J. Lightwave Tech.*, vol. 12, pp. 614–620, Apr. 1994.

8. H. Henschel and E. Baumann, "Effect of natural radioactivity on optical fibers of undersea cables," *J. Lightwave Tech.*, vol. 14, pp. 724–731, May 1996.

9. J. Söderqvist et al., "Radiation hardness evaluation of an analog optical link for operation at cryogenic temperatures," *IEEE Trans. Nucl. Sci.*, vol. 44, pp. 861–865, June 1997.

10. V. Miya, Y. Terunuma, T. Hosaka, and T. Miyashita, "Ultra low loss single-mode fibers at $1.55\,\mu m$," *Electron. Lett.*, vol. 15, pp. 106–108, 1979.

11. (a) S. R. Nagel, J. B. MacChesney, and K. L. Walker, "An overview of the MCVD process and performance," *IEEE J. Quantum Electron.*, vol. QE-18, pp. 459–476, Apr. 1982.

    (b) S. R. Nagel, "Fiber materials and fabrication methods," in S. E. Miller and I. P. Kaminow, eds., *Optical Fiber Telecommunications—II*, Academic, New York, 1988.

12. M. Ohashi, K. Shiraki, and K. Tajima, "Optical loss property of silica-based single-mode fibers," *J. Lightwave Tech.*, vol. 10, pp. 539–543, May 1992.

13. H. Osanai, T. Shioda, T. Moriyama, S. Araki, M. Horiguchi, T. Izawa, and H. Takata, "Effects of dopants on transmission loss of low OH content optical fibers," *Electron. Lett.*, vol. 12, pp. 549–550, Oct. 1976.

14. R. Maurer, "Glass fibers for optical communications," *Proc. IEEE*, vol. 61, pp. 452–462, Apr. 1973.

15. D. A. Pinnow, T. C. Rich, F. W. Ostermeyer, and M. DiDomenico, Jr., "Fundamental optical attenuation limits in the liquid and gassy state with application to fiber optical waveguide material," *Appl. Phys. Lett.*, vol. 22, pp. 527–529, May 1973.

16. D. B. Keck, "Fundamentals of optical waveguide fibers," *IEEE Commun. Mag.*, vol. 23, pp. 17–22, May 1985.

17. D. Marcuse, "Curvature loss formula for optical fibers," *J. Opt. Soc. Amer.*, vol. 66, pp. 216–220, Mar. 1976.

18. D. Gloge, "Bending loss in multimode fibers with graded and ungraded core index," *Appl. Opt.*, vol. 11, pp. 2506–2512, Nov. 1972.

19. A. J. Harris and P. F. Castle, "Bend loss measurements on high numerical aperture single-mode fibers as a function of wavelength and bend radius," *J. Lightwave Tech.*, vol. LT-4, pp. 34–40, Jan. 1986.

20. G. L. Tangonan, H. P. Hsu, V. Jones, and J. Pikulski, "Bend loss measurements of small mode field diameter fibers," *Electron. Lett.*, vol. 25, pp. 142–143, Jan. 19, 1989.

21. N. Kamikawa and C.-T. Chang, "Losses in small-radius bends in single-mode fibers," *Electron. Lett.*, vol. 25, pp. 947–949, July 20, 1989.

22. H. Renner, "Bending loses of coated single-mode fibers: A simple approach," *J. Lightwave Tech.*, vol. 10, pp. 544–551, May 1992.

23. F. Wilczewski, "Determination of the 'Field radius' from bending loss measurements of optical fibers with arbitrary index profile," *IEEE Photonics Tech. Lett.*, vol. 8, pp. 90–91, Jan. 1996.

24. L. Faustini and G. Martini, "Bend loss in single-mode fibers," *J. Lightwave Tech.*, vol. 15, pp. 671–679, Apr. 1997.

25. J. D. Love, "Application of low-loss criterion to optical waveguides and devices," *IEE Proc.*, vol. 136, pt. J, pp. 225–228, Aug. 1989.

26. W. B. Gardner, "Microbending loss in optical fibers," *Bell Sys. Tech. J.*, vol. 54, pp. 457–465, Feb. 1975.

27. J. Sakai and T. Kimura, "Practical microbending loss formula for single mode optical fibers," *IEEE J. Quantum Electron.*, vol. QE-15, pp. 497–500, June 1979.

28. S.-T. Shiue and Y.-K. Tu, "Design of single-coated optical fibers to minimize thermally and mechanically induced microbending losses," *J. Opt. Commun.*, vol. 15, pp. 16–19, Jan. 1994.

29. C. Unger and W. Stöcklein, "Investigation of the microbending sensitivity of fibers," *J. Lightwave Tech.*, vol. 12, pp. 591–596, Apr. 1994.

30. V. Arya, K. A. Murphy, A. Wang, and R. O. Claus, "Microbend losses in single-mode optical fibers: Theoretical and experimental investigation," *J. Lightwave Tech.*, vol. 13, pp. 1998–2002, Oct. 1995.

31. D. Gloge, "Optical fiber packaging and its influence on fiber straightness and loss," *Bell Sys. Tech. J.*, vol. 54, pp. 245–262, Feb. 1975.

32. D. Gloge, "Propagation effects in optical fibers," *IEEE Trans. Microwave Theory Tech.*, vol. MTT-23, pp. 106–120, Jan. 1975.

33. D. Marcuse, *Theory of Dielectric Optical Waveguides*, Academic, New York, 2nd ed., 1991.

34. D. Gloge, E. A. J. Marcatili, D. Marcuse, and S. D. Personick, "Dispersion properties of fibers," in S. E. Miller and A. G. Chynoweth, eds., *Optical Fiber Telecommunications*, Academic, New York, 1979.

35. D. Marcuse, "Interdependence of waveguide and material dispersion," *Appl. Opt.*, vol. 18, pp. 2930–2932, Sept. 1979.

36. F. P. Kapron and D. B. Keck, "Pulse transmission through a dielectric optical waveguide," *Appl. Opt.*, vol. 10, pp. 1519–1523, July 1971.

37. D. Gloge, "Weakly guiding fibers," *Appl. Opt.*, vol. 10, pp. 2252–2258, Oct. 1971; "Dispersion in weakly guiding fibers," *Appl. Opt.*, vol. 10, pp. 2442–2445, Nov. 1971.

38. C. D. Poole and J. Nagel, "Polarization effects in lightwave systems," in I. P. Kaminow and T. L. Koch, eds., *Optical Fiber Telecommunications—III*, vol. A, Academic, New York, 1997, chap. 6, pp. 114–161.

39. C. De Angelis, A. Galtarossa, G. Gianello, F. Matera, and M. Schiano, "Time evolution of polarization mode dispersion in long terrestrial links," *J. Lightwave Tech.*, vol. 10, pp. 552–555, May 1992.

40. J. Cameron, L. Chen, X. Bao, and J. Stears, "Time evolution of polarization mode dispersion in optical fibers," *IEEE Photonics Tech. Lett.*, vol. 10, pp. 1265–1267, Sept. 1998.

41. S. D. Personick, "Receiver design for digital fiber optic communication systems," *Bell Sys. Tech. J.*, vol. 52, pp. 843–874, July/Aug. 1973.

42. (a) R. Olshansky and D. Keck, "Pulse broadening in graded index optical fibers," *Appl. Opt.*, vol. 15, pp. 483–491, Feb. 1976.
   (b) G. Einarsson, "Pulse broadening in graded index optical fibers: Correction," *Appl. Opt.*, vol. 25, p. 1030, Apr. 1986.

43. L. Cohen, I. Kaminow, H. Astle, and L. Stulz, "Profile dispersion effects on transmission bandwidths in graded index optical fibers," *IEEE J. Quantum Electron.*, vol. QE-14, pp. 37–41, Jan. 1978.

44. D. Marcuse and H. M. Presby, "Effects of profile deformation on fiber bandwidth," *Appl. Opt.*, vol. 18, pp. 3758–3763, Nov. 1979; *Appl. Opt.*, vol. 19, p. 188, Jan. 1980.

45. (*a*) R. Olshansky, "Mode coupling effects in graded index optical fibers," *Appl. Opt.*, vol. 14, pp. 935–945, Apr. 1975.

   (*b*) S. Geckeler, "Pulse broadening in optical fibers with mode mixing," *Appl. Opt.*, vol. 18, pp. 2192–2198, July 1979.

   (*c*) M. J. Hackert, "Evolution of power distributions in fiber optic systems: Development of a measurement strategy," *Fiber & Integrated Optics*, vol. 8, pp. 163–167, 1989.

46. D. Marcuse, *Principles of Optical Fiber Measurements*, Academic, New York, 1981.

47. Q. Yu, P. H. Zongo, and P. Facq, "Refractive-index profile influences on mode coupling effects at optical fiber splices and connectors," *J. Lightwave Tech.*, vol. 11, pp. 1270–1273, Aug. 1993.

48. D. Rice and G. Keiser, "Short-haul fiber-optic link connector loss," *33rd International Wire & Cable Symp.*, Reno, NV, Nov. 1984, pp. 190–192.

49. (*a*) A. R. Michelson, M. Ericsrud, S. Aamlid, and N. Ryen, "Role of the fusion splice in the concatenation problem," *J. Lightwave Tech.*, vol. LT-2, pp. 126–138, Apr. 1984.

   (*b*) P. J. W. Severin and W. H. Bardoel, "Differential mode loss and mode conversion in passive fiber optic components," *J. Lightwave Tech.*, vol. LT-4, pp. 1640–1646, Nov. 1986.

50. (*a*) Special Issue on "Broadband Lightwave Video Transmission," *J. Lightwave Tech.*, vol. 11, Jan 1993.

   (*b*) Special Issues on "Microwave and Millimeter-wave Photonics," *IEEE Trans. Microwave Theory Tech.*, vol. 43, Sept. 1995; vol. 45, pt. II, Aug. 1997.

51. J. C. Lapp, V. A. Bhagavatula, and A. J. Morrow, "Segmented-core single-mode fiber optimized for bending performance," *J. Lightwave Tech.*, vol. 6, pp. 1462–1465, Oct. 1988.

52. B. J. Ainsle and C. R. Day, "A review of single-mode fibers with modified dispersion characteristics," *J. Lightwave Tech.*, vol. LT-4, pp. 967–979, Aug. 1986.

53. D. P. Jablonowski, U. C. Paek, and L. S. Watkins, "Optical fiber manufacturing techniques," *AT&T Tech. J.*, vol. 66, pp. 33–44, Jan./Feb. 1987.

54. H. J. Hagemann, H. Lade, J. Warnier, and D. H. Wiechert, "The performance of depressed-cladding single-mode fibers with different *b/a* ratios," *J. Lightwave Tech.*, vol. 9, pp. 689–694, June 1991.

55. Y. W. Li, C. D. Hussey, and T. A. Birks, "Triple-clad single-mode fibers for dispersion-shifting," *J. Lightwave Tech.*, vol. 11, pp. 1812–1819, Nov. 1993.

56. W. A. Reed, L. G. Cohen, and H. T. Shang, "Tailoring optical characteristics of dispersion-shifted lightguides for applications near $1.55\,\mu m$," *AT&T Tech. J.*, vol. 65, pp. 105–122, Sept./Oct. 1986.

57. D. Kalish and L. G. Cohen, "Single-mode fiber: From research and development to manufacturing," *AT&T Tech. J.*, vol. 66, pp. 19–32, Jan./Feb. 1987.

58. ITU-T Recommendation G.653, *Characteristics of a Dispersion-Shifted Single-Mode Optical Fiber Cable*, Mar. 1993. (Note: This recommendation is for fibers having a zero-dispersion wavelength around 1550 nm).

59. P. K. Bachmann, D. Leers, H. Wehr, D. U. Wiechert, J. A. Van Steenwijk, D. L. A. Tjaden, and E. R. Wehrhahn, "Dispersion-flattened single-mode fibers prepared with PCVD: Performance, limitations, design optimization," *J. Lightwave Tech.*, vol. LT-4, pp. 858–863, July 1986.

60. V. A. Bhagavatula, M. S. Spotz, W. F. Love, and D. B. Keck, "Segmented-core single-mode fibers with low loss and low dispersions," *Electron. Lett.*, vol. 19, pp. 317–318, Apr. 28, 1983.

61. S. F. Mahmoud and A. M. Kharbat, "Transmission characteristics of a coaxial optical fiber line," *J. Lightwave Tech.*, vol. 11, pp. 1717–1720, Nov. 1993.

62. C. Weistein, "Fiber design improves long-haul performance," *Laser Focus World*, vol. 33, pp. 215–220, May 1997.

63. M. Kato, K. Kurokawa, and Y. Miyajima, "A new design for dispersion-shifted fiber with an effective core area larger than $100\,\mu m^2$ and good bending characteristics," *1998 OSA Tech. Digest—Opt. Fiber Comm. Conf. (OFC 98)*, pp. 301–302, Feb. 1998.

64. H. Hatayama, T. Kato, M. Onishi, E. Sasaoka, and M. Nishimura, "Dispersion-flattened fiber with large-effective-core area more than $50\,\mu m^2$," *1998 OSA Tech. Digest—Opt. Fiber Comm. Conf.* (OFC 98), pp. 304–305, Feb. 1998.

65. D. L. Franzen, "Determining the effective cutoff wavelength of single-mode fibers: An interlaboratory comparison," *J. Lightwave Tech.*, vol. 3, pp. 128–134, Feb. 1985.

66. ITU-T Recommendation G.650, *Definition and Test Methods for the Relevant Parameters of Single-Mode Fibers*, Mar. 1993.

67. TIA/EIA-455-80A, *Measuring Cutoff Wavelength of Uncabled Single-Mode Fiber by Transmitted Power*, Feb. 1996.

68. A. J. Barlow, R. S. Jones, and K. W. Forsyth, "Technique for direct measurement of single-mode fiber chromatic dispersion," *J. Lightwave Tech.*, vol. 5, pp. 1207–1217, Sept. 1987.

69. TIA/EIA-455-168A, *Chromatic Dispersion Measurement of Multimode Graded-Index and Single-Mode Optical Fibers by Spectral Group Delay Measurement in the Time Domain*, Mar. 1992.

70. ITU-T Recommendation G.652, *Characteristics of a Single-Mode Optical Fiber Cable*, Mar. 1993. (Note: This recommendation is for fibers having a zero-dispersion wavelength around 1310 nm).

71. M. DiDomenico, Jr., "Material dispersion in optical fiber waveguides," *Appl. Opt.*, vol. 11, pp. 652–654, Mar. 1972.

CHAPTER
4

OPTICAL
SOURCES

The principal light sources used for fiber optic communications applications are heterojunction-structured semiconductor *laser diodes* (also referred to as *injection laser diodes* or ILDs) and *light-emitting diodes* (LEDs). A *heterojunction* consists of two adjoining semiconductor materials with different band-gap energies. These devices are suitable for fiber transmission systems because they have adequate output power for a wide range of applications, their optical power output can be directly modulated by varying the input current to the device, they have a high efficiency, and their dimensional characteristics are compatible with those of the optical fiber. Comprehensive treatments of the major aspects of LEDs and laser diodes are presented in various books.[1-6] Review articles and book chapters covering the operating principles of these devices are also available,[7-12] and the reader is referred to these for details.

The intent of this chapter is to give an overview of the pertinent character-istic of fiber-compatible luminescent sources. The first section discusses semicon-ductor material fundamentals that are relevant to light source operation. The next two sections present the output and operating characteristics of LEDs and laser diodes, respectively. These are followed by sections discussing the temperature responses of optical sources, their linearity characteristics, and their reliability under various operating conditions.

We shall see in this chapter that the light-emitting region of both LEDs and laser diodes consists of a *pn* junction constructed of direct-band-gap III–V semi-conductor materials. When this junction is forward biased, electrons and holes are injected into the *p* and *n* regions, respectively. These injected minority carriers can recombine either radiatively, in which case a photon of energy $h\nu$ is emitted, or

141

nonradiatively, whereupon the recombination energy is dissipated in the form of heat. This *pn* junction is thus known as the *active* or *recombination region*.

A major difference between LEDs and laser diodes is that the optical output from an LED is incoherent, whereas that from a laser diode is coherent. In a coherent source, the optical energy is produced in an optical resonant cavity. The optical energy released from this cavity has spatial and temporal coherence, which means it is highly monochromatic and the output beam is very directional. In an incoherent LED source, no optical cavity exists for wavelength selectivity. The output radiation has a broad spectral width, since the emitted photon energies range over the energy distribution of the recombining electrons and holes, which usually lie between 1 and $2k_BT$ (k_B is Boltzmann's constant and T is the absolute temperature at the *pn* junction). In addition, the incoherent optical energy is emitted into a hemisphere according to a cosine power distribution and thus has a large beam divergence.

In choosing an optical source compatible with the optical waveguide, various characteristics of the fiber, such as its geometry, its attenuation as a function of wavelength, its group delay distortion (bandwidth), and its modal characteristics, must be taken into account. The interplay of these factors with the optical source power, spectral width, radiation pattern, and modulation capability needs to be considered. The spatially directed coherent optical output from a laser diode can be coupled into either single-mode or multimode fibers. In general, LEDs are used with multimode fibers, since normally it is only into a multimode fiber that the incoherent optical power from an LED can be coupled in sufficient quantities to be useful. However, LEDs have been employed in high-speed local-area applications in which one wants to transmit several wavelengths on the same fiber. Here, a technique called *spectral slicing* is used.[13-15] This entails using a passive device such as a waveguide grating array (see Chap. 10) to split the broad spectral emission of the LED into narrow spectral slices. Since these slices are each centered at a different wavelength, they can be individually modulated externally with independent data streams and simultaneously sent on the same fiber.

4.1   TOPICS FROM SEMICONDUCTOR PHYSICS

Since the material in this chapter assumes a rudimentary knowledge of semiconductor physics, various relevant definitions are given here for semiconductor material properties, including the concepts of energy bands, intrinsic and extrinsic materials, *pn* junctions, and direct and indirect band gaps. Further details can be found in Refs. 16–18.

4.1.1   Energy Bands

Semiconductor materials have conduction properties that lie somewhere between those of metals and insulators. As an example material, we consider silicon (Si), which is located in the fourth column (group IV) of the periodic table of elements.

A Si atom has four electrons in its outer shell, by which it makes covalent bonds with its neighboring atoms in a crystal.

The conduction properties can be interpreted with the aid of the *energy-band diagrams* shown in Fig. 4-1a. In a pure crystal at low temperatures, the *conduction band* is completely empty of electrons and the *valence band* is completely full. These two bands are separated by an *energy gap*, or *band gap*, in which no energy levels exist. As the temperature is raised, some electrons are thermally excited across the band gap. For Si this excitation energy must be greater than 1.1 eV, which is the band-gap energy. This gives rise to a concentration n of free electrons in the conduction band, which leaves behind an equal concentration p of vacancies, or *holes*, in the valence band, as is shown schematically in Fig. 4-1b. Both the free electrons and the holes are mobile within the material, so that both can contribute to electrical conductivity; that is, an electron in the valance band can move into a vacant hole. This action makes the hole move in the opposite direction to the electron flow, as is shown in Fig. 4-1a.

The concentration of electrons and holes is known as the *intrinsic carrier concentration n_i*, and for a perfect material with no imperfections or impurities it is given by

$$n = p = n_i = K \exp\left(-\frac{E_g}{2k_BT}\right) \qquad (4\text{-}1)$$

where

$$K = 2(2\pi k_BT/h^2)^{3/2}(m_e m_h)^{3/4}$$

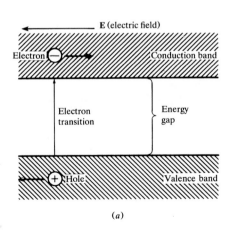

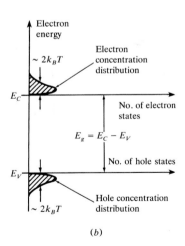

(a) (b)

FIGURE 4-1
(a) Energy-level diagrams showing the excitation of an electron from the valence band (of energy E_V) to the conduction band (of energy E_C). The resultant free electron and free hole move under the influence of an external electric field E; (b) equal electron and hole concentrations in an intrinsic semiconductor created by the thermal excitation of electrons across the band gap.

is a constant that is characteristic of the material. Here, T is the temperature in degrees Kelvin, k_B is Boltzmann's constant, h is Planck's constant, and m_e and m_h are the effective masses of the electrons and holes, respectively, which can be smaller by a factor of 10 or more than the free-space electron rest mass of 9.11×10^{-31} kg.

Example 4-1. Given the following parameter values for GaAs at 300 K:

$$\text{Electron rest mass } m = 9.11 \times 10^{-31} \text{ kg}$$

$$\text{Effective electron mass } m_e = 0.068m = 6.19 \times 10^{-32} \text{ kg}$$

$$\text{Effective hole mass } m_h = 0.56m = 5.10 \times 10^{-31} \text{ kg}$$

$$\text{Band-gap energy } E_g = 1.42 \text{ eV}$$

then from Eq. (4-1) we find that the intrinsic carrier concentration is

$$n_i = 2.62 \times 10^{12} \text{m}^{-3} = 2.62 \times 10^6 \text{ cm}^{-3}$$

The conduction can be greatly increased by adding traces of impurities from the group V elements (e.g., P, As, Sb). This process is called *doping* and the doped semiconductor is called an *extrinsic material*. These elements have five electrons in the outer shell. When they replace a Si atom, four electrons are used for covalent bonding, and the fifth, loosely bound electron is available for conduction. As shown in Fig. 4-2a, this gives rise to an occupied level, just below the conduction band, called the *donor level*. The impurities are called *donors* because they can give up an electron to the conduction band. This is reflected by the increase in the free-

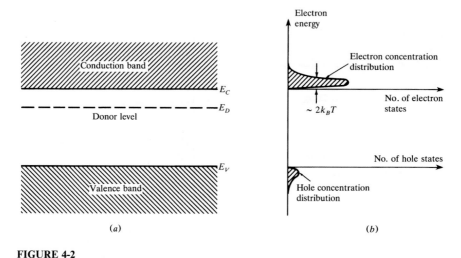

FIGURE 4-2

(a) Donor level in an *n*-type material; (b) the ionization of donor impurities creates an increased electron concentration distribution.

electron concentration in the conduction band, as shown in Fig. 4-2b. Since in this type of material the current is carried by (negative) electrons (because the electron concentration is much higher than that of holes), it is called *n-type* material.

The conduction can also be increased by adding group III elements, which have three electrons in the outer shell. In this case, three electrons make covalent bonds, and a hole with properties identical to that of the donor electron is created. As shown in Fig. 4-3a, this gives rise to an unoccupied level just above the valence band. Conduction occurs when electrons are excited from the valence band to this *acceptor level* (so called because the impurity atoms have accepted electrons from the valence band). Correspondingly, the free-hole concentration increases in the valence band, as shown in Fig. 4-3b. This is called *p*-type material because the conduction is a result of (positive) hole flow.

4.1.2 Intrinsic and Extrinsic Material

A perfect material containing no impurities is called an *intrinsic material*. Because of thermal vibrations of the crystal atoms, some electrons in the valence band gain enough energy to be excited to the conduction band. This *thermal generation process* produces free electron–hole pairs, since every electron that moves to the conduction band leaves behind a hole. Thus, for an intrinsic material the number of electrons and holes are both equal to the intrinsic carrier density, as denoted by Eq. (4-1). In the opposite *recombination process*, a free electron releases its energy and drops into a free hole in the valence band. For an extrinsic semiconductor, the increase of one type of carrier reduces the number of the other type. In this case, the product of the two types of carriers remains constant at a given temperature. This gives rise to the *mass-action law*

$$pn = n_i^2 \qquad\qquad (4\text{-}2)$$

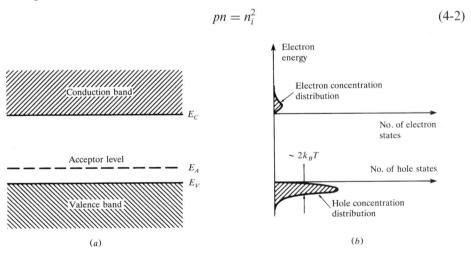

(a) (b)

FIGURE 4-3
(a) Acceptor level in p-type material; (b) the ionization of acceptor impurities creates an increased hole concentration distribution.

which is valid for both intrinsic and extrinsic materials under thermal equilibrium.

Since the electrical conductivity is proportional to the carrier concentration, two types of charge carriers are defined for this material:

1. *Majority carriers* refer either to electrons in *n*-type material or to holes in *p*-type material.
2. *Minority carriers* refer either to holes in *n*-type material or to electrons in *p*-type material.

The operation of semiconductor devices is essentially based on the *injection* and *extraction* of minority carriers.

Example 4-2. Consider an *n*-type semiconductor which has been doped with a net concentration of N_D donor impurities. Let n_N and p_N be the electron and hole concentrations, respectively, where the subscript N is used to denote *n*-type semiconductor characteristics. In this case, holes are created exclusively by thermal ionization of intrinsic atoms. This process generates equal concentrations of electrons and holes, so that the hole concentration in an *n*-type semiconductor is

$$p_N = p_i = n_i$$

Since conduction electrons are generated by both impurity and intrinsic atoms, the total conduction-electron concentration n_N is

$$n_N = N_D + n_i = N_D + p_N$$

Substituting Eq. (4-2) for p_N (which states that, in equilibrium, the product of the electron and hole concentrations equals the square of the intrinsic carrier density, so that $p_N = n_i^2/n_N$), we have

$$n_N = \frac{N_D}{2}\left(\sqrt{1 + \frac{4n_i^2}{N_D^2}} + 1\right)$$

If $n_i \ll N_D$, which is generally the case, then to a good approximation

$$n_N = N_D \qquad \text{and} \qquad p_N = n_i^2/N_D$$

4.1.3 The *pn* Junctions

Doped *n*- or *p*-type semiconductor material by itself serves only as a conductor. To make devices out of these semiconductors, it is necessary to use both types of materials (in a single, continuous crystal structure). The junction between the two material regions, which is known as the *pn junction*, is responsible for the useful electrical characteristics of a semiconductor device.

When a *pn* junction is created, the majority carriers diffuse across it. This causes electrons to fill holes in the *p* side of the junction and causes holes to appear on the *n* side. As a result, an electric field (or *barrier potential*) appears across the junction, as is shown in Fig. 4-4. This field prevents further net movements of charges once equilibrium has been established. The junction area now

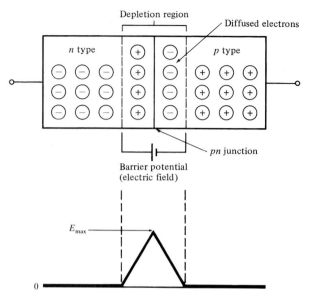

FIGURE 4-4

Electron diffusion across a *pn* junction creates a barrier potential (electric field) in the depletion region.

has no mobile carriers, since its electrons and holes are locked into a covalent bond structure. This region is called either the *depletion region* or the *space charge region*.

When an external battery is connected to the *pn* junction with its positive terminal to the *n*-type material and its negative terminal to the *p*-type material, the junction is said to be *reverse-biased*. This is shown in Fig. 4-5. As a result of the reverse bias, the width of the depletion region will increase on both the *n* side and the *p* side. This effectively increases the barrier potential and prevents any majority carriers from flowing across the junction. However, minority carriers can move with the field across the junction. The minority carrier flow is small at normal temperatures and operating voltages, but it can be significant when excess carriers are created as, for example, in an illuminated photodiode.

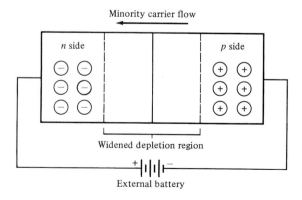

FIGURE 4-5

A reverse bias widens the depletion region, but allows minority carriers to move freely with the applied field.

When the *pn* junction is *forward-biased*, as shown in Fig. 4-6, the magnitude of the barrier potential is reduced. Conduction-band electrons on the *n* side and valence-band holes on the *p* side are, thereby, allowed to diffuse across the junction. Once across, they significantly increase the minority carrier concentrations, and the excess carriers then recombine with the oppositely charged majority carriers. The recombination of excess minority carriers is the mechanisms by which optical radiation is generated.

4.1.4 Direct and Indirect Band Gaps

In order for electron transitions to take place to or from the conduction band with the absorption or emission of a photon, respectively, both energy and momentum must be conserved. Although a photon can have considerable energy, its momentum $h\nu/c$ is very small.

Semiconductors are classified as either *direct-band-gap* or *indirect-band-gap* materials depending on the shape of the band gap as a function of the momentum k, as shown in Fig. 4-7. Let us consider recombination of an electron and a hole, accompanied by the emission of a photon. The simplest and most probable recombination process will be that where the electron and hole have the same momentum value (see Fig. 4-7*a*). This is a direct-band-gap material.

For indirect-band-gap materials, the conduction-band minimum and the valence-band maximum energy levels occur at different values of momentum, as shown in Fig. 4-7*b*. Here, band-to-band recombination must involve a third particle to conserve momentum, since the photon momentum is very small. *Phonons* (i.e., crystal lattice vibrations) serve this purpose.

4.1.5 Semiconductor Device Fabrication

In fabricating semiconductor devices, the crystal structure of the various material regions must be carefully taken into account. In any crystal structure, single atoms (e.g., Si or Ge) or groups of atoms (e.g., NaCl or GaAs) are arranged in a repeated pattern in space. This periodic arrangement defines a *lattice*, and the

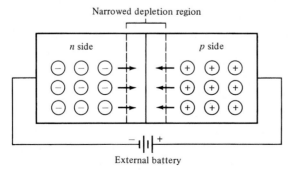

FIGURE 4-6
Lowering the barrier potential with a forward bias allows majority carriers to diffuse across the junction.

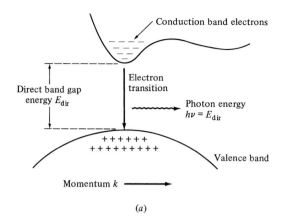

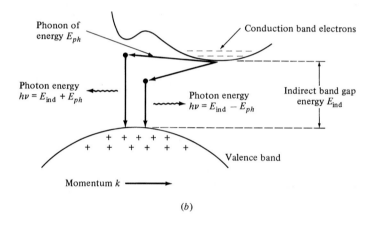

FIGURE 4-7
(*a*) Electron recombination and the associated photon emission for a direct-band-gap material; (*b*) electron recombination for indirect-band-gap materials requires a phonon of energy E_{ph} and momentum k_{ph}.

spacing between the atoms or groups of atoms is called the *lattice spacing* or the *lattice constant*. Typical lattice spacings are a few angstroms.

Semiconductor devices are generally fabricated by starting with a crystalline substrate which provides mechanical strength for mounting the device and for making electric contacts. A technique of crystal growth by chemical reaction is then used to grow thin layers of semiconductor materials on the substrate. These materials must have lattice structures that are identical to those of the substrate crystal. In particular, the lattice spacings of adjacent materials should be closely matched to avoid temperature-induced stresses and strains at the material interfaces. This type of growth is called *epitaxial*, which is derived from the Greek words *epi* meaning "on" and *taxis* meaning "arrangement"; that is, it is an arrangement of atoms from one material on another material. An important characteristic of epitaxial growth is that it is relatively simple to change the impurity concentration of successive material layers, so that a layered semiconductor device can be fabricated in a continuous process. Epitaxial layers can be formed by growth techniques of either vapor phase, liquid phase, or molecular beam.[16,19,20]

4.2 LIGHT-EMITTING DIODES (LEDs)

For optical communication systems requiring bit rates less than approximately 100–200 Mb/s together with multimode fiber-coupled optical power in the tens of microwatts, semiconductor light-emitting diodes (LEDs) are usually the best light source choice. These LEDs require less complex drive circuitry than laser diodes since no thermal or optical stabilization circuits are needed (see Sec. 4.3.6), and they can be fabricated less expensively with higher yields.

4.2.1 LED Structures

To be useful in fiber transmission applications an LED must have a high radiance output, a fast emission response time, and a high quantum efficiency. Its *radiance* (or *brightness*) is a measure, in watts, of the optical power radiated into a unit solid angle per unit area of the emitting surface. High radiances are necessary to couple sufficiently high optical power levels into a fiber, as shown in detail in Chap. 5. The emission response time is the time delay between the application of a current pulse and the onset of optical emission. As we discuss in Secs. 4.2.4 and 4.3.7, this time delay is the factor limiting the bandwidth with which the source can be modulated directly by varying the injected current. The quantum efficiency is related to the fraction of injected electron–hole pairs that recombine radiatively. This is defined and described in detail in Sec. 4.2.3.

To achieve a high radiance and a high quantum efficiency, the LED structure must provide a means of confining the charge carriers and the stimulated optical emission to the active region of the *pn* junction where radiative recombination takes place. Carrier confinement is used to achieve a high level of radiative recombination in the active region of the device, which yields a high quantum efficiency. Optical confinement is of importance for preventing absorption of the emitted radiation by the material surrounding the *pn* junction.

To achieve carrier and optical confinement, LED configurations such as homojunctions and single and double heterojunctions have been widely investigated. The most effective of these structures is the configuration shown in Fig. 4-8. This is referred to as a *double-heterostructure* (or *heterojunction*) device because of the two different alloy layers on each side of the active region. This configuration evolved from studies on laser diodes. By means of this sandwich structure of differently composed alloy layers, both the carriers and the optical field are confined in the central active layer. The band-gap differences of adjacent layers confine the charge carriers (Fig. 4-8*b*), while the differences in the indices of

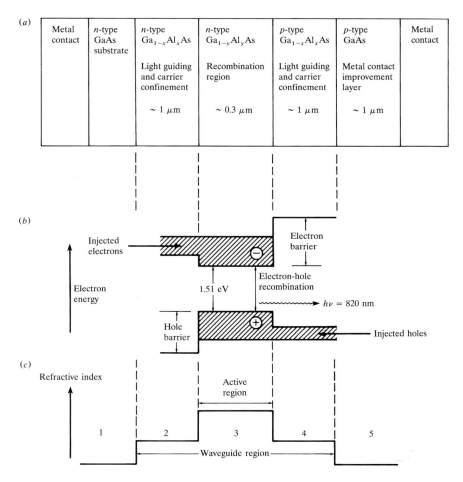

FIGURE 4-8

(*a*) Cross-section drawing (not to scale) of a typical GaAlAs double-heterostructure light emitter. In this structure, $x > y$ to provide for both carrier confinement and optical guiding. (*b*) Energy-band diagram showing the active region, and the electron and hole barriers which confine the charge carriers to the active layer. (*c*) Variations in the refractive index; the lower index of refraction of the material in regions 1 and 5 creates an optical barrier around the waveguide because of the higher band-gap energy of this material.

refraction of adjoining layers confine the optical field to the central active layer (Fig. 4-8*c*). This dual confinement leads to both high efficiency and high radiance. Other parameters influencing the device performance include optical absorption in the active region (self-absorption), carrier recombination at the heterostructure interfaces, doping concentration of the active layer, injection carrier density, and active-layer thickness. We shall see the effects of these parameters in the following sections.

The two basic LED configurations being used for fiber optics are *surface emitters* (also called *Burrus* or *front emitters*) and *edge emitters*.[21] In the surface emitter, the plane of the active light-emitting region is oriented perpendicularly to the axis of the fiber, as shown in Fig. 4-9. In this configuration, a well is etched through the substrate of the device, into which a fiber is then cemented in order to accept the emitted light. The circular active area in practical surface emitters is nominally 50 μm in diameter and up to 2.5 μm thick. The emission pattern is essentially isotropic with a 120° half-power beam width.

This isotropic pattern from a surface emitter is called a *lambertian pattern*. In this pattern, the source is equally bright when viewed from any direction, but the power diminishes as $\cos\theta$, where θ is the angle between the viewing direction and the normal to the surface (this is because the projected area one sees decreases as $\cos\theta$). Thus, the power is down to 50 percent of its peak when $\theta = 60°$, so that the total half-power beam width is 120°.

The edge emitter depicted in Fig. 4-10 consists of an active junction region, which is the source of the incoherent light, and two guiding layers. The guiding layers both have a refractive index which is lower than that of the active region but higher than the index of the surrounding material. This structure forms a waveguide channel that directs the optical radiation toward the fiber core. To match the typical fiber-core diameters (50–100 μm), the contact stripes for the edge emitter are 50–70 μm wide. Lengths of the active regions usually range from 100 to 150 μm. The emission pattern of the edge emitter is more directional than that of the surface emitter, as is illustrated in Fig. 4-10. In the plane parallel to the junction, where there is no waveguide effect, the emitted beam is lambertian (varying as $\cos\theta$) with a half-power width of $\theta_\parallel = 120°$. In the plane perpendicular to the junction, the half-power beam width $\theta_\perp$ has been made as small as 25–35° by a proper choice of the waveguide thickness.[21]

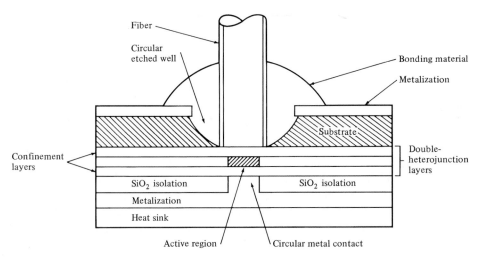

FIGURE 4-9
Schematic (not to scale) of a high-radiance surface-emitting LED. The active region is limited to a circular section that has an area compatible with the fiber-core end face.

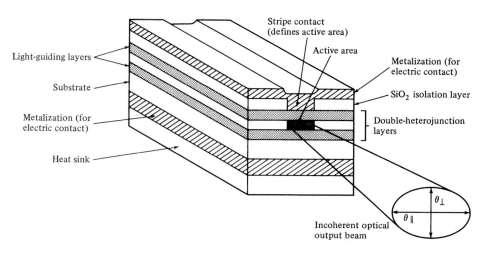

FIGURE 4-10
Schematic (not to scale) of an edge-emitting double-heterojunction LED. The output beam is lamber-tian in the plane of the *pn* junction ($\theta_\parallel = 120°$) and highly directional perpendicular to the *pn* junction ($\theta_\perp \approx 30°$).

4.2.2 Light Source Materials

The semiconductor material that is used for the active layer of an optical source must have a direct band gap. In a direct-band-gap semiconductor, electrons and holes can recombine directly across the band gap without needing a third particle to conserve momentum. Only in direct-band-gap material is the radiative recombination sufficiently high to produce an adequate level of optical emission. Although none of the normal single-element semiconductors are direct-gap materials, many binary compounds are. The most important of these are the so-called III–V materials. These are made from compounds of a group III element (e.g., Al, Ga, or In) and a group V element (e.g., P, As, or Sb). Various ternary and quaternary combinations of binary compounds of these elements are also direct-gap materials and are suitable candidates for optical sources.

For operation in the 800-to-900-nm spectrum, the principal material used is the ternary alloy $Ga_{1-x}Al_xAs$. The ratio x of aluminum arsenide to gallium arsenide determines the band gap of the alloy and, correspondingly, the wavelength of the peak emitted radiation. This is illustrated in Fig. 4-11. The value of x for the active-area material is usually chosen to give an emission wavelength of 800–850 nm. An example of the emission spectrum of a $Ga_{1-x}Al_xAs$ LED with $x = 0.08$ is shown in Fig. 4-12. The peak output power occurs at 810 nm. The width of the spectral pattern at its half-power point is known as the *full-width half-maximum* (FWHM) spectral width. As shown in Fig. 4-12, this FWHM spectral width σ_λ is 36 nm.

At longer wavelengths the quaternary alloy $In_{1-x}Ga_xAs_yP_{1-y}$ is one of the primary material candidates. By varying the mole fractions x and y in the active

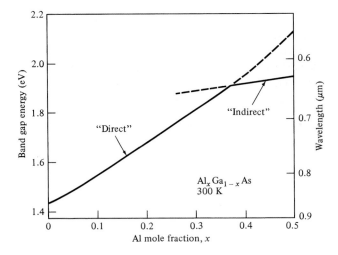

FIGURE 4-11
Band-gap energy and output wavelength as a function of aluminum mole fraction x for $Al_xGa_{1-x}As$ at room temperature. (Reproduced with permission from Miller, Marcatili, and Lee, *Proc. IEEE*, vol. 61, pp. 1703–1751, Dec. 1973, © 1973, IEEE.)

area, LEDs with peak output powers at any wavelength between 1.0 and 1.7 μm can be constructed. For simplicity, the notations GaAlAs and InGaAsP are generally used unless there is an explicit need to know the values of x and y. Other notations such as AlGaAs, (Al,Ga)As, (GaAl)As, GaInPAs, and $In_xGa_{1-x}As_yP_{1-y}$ are also found in the literature. From the last notation, it is obvious that, depending on the preference of the particular author, the values of x and $1 - x$ for the same material could be interchanged in different articles in the literature.

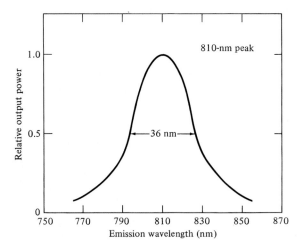

FIGURE 4-12
Spectral emission pattern of a representative $Ga_{1-x}Al_xAs$ LED with $x = 0.08$. The width of the spectral pattern at its half-power point is 36 nm.

The alloys GaAlAs and InGaAsP are chosen to make semiconductor light sources because it is possible to match the lattice parameters of the heterostructure interfaces by using a proper combination of binary, ternary, and quaternary materials. A very close match between the crystal lattice parameters of the two adjoining heterojunctions is required to reduce interfacial defects and to minimize strains in the device as the temperature varies. These factors directly affect the radiative efficiency and lifetime of a light source. Using the fundamental quantum-mechanical relationship between energy E and frequency ν,

$$E = h\nu = \frac{hc}{\lambda}$$

the peak emission wavelength λ in micrometers can be expressed as a function of the band-gap energy E_g in electron volts by the equation

$$\lambda(\mu\text{m}) = \frac{1.240}{E_g \text{ (eV)}} \tag{4-3}$$

The relationships between the band-gap energy E_g and the crystal lattice spacing (or lattice constant) a_0 for various III–V compounds are plotted in Fig. 4-13.

A heterojunction with matching lattice parameters is created by choosing two material compositions that have the same lattice constant but different band-gap energies (the band-gap differences are used to confine the charge carriers). In the ternary alloy GaAlAs the band-gap energy E_g and the crystal lattice spacing a_0 are determined by the dashed line in Fig. 4-13 that connects the materials GaAs ($E_g = 1.43$ eV and $a_0 = 5.64$ Å) and AlAs ($E_g = 2.16$ eV and $a_0 = 5.66$ Å). The

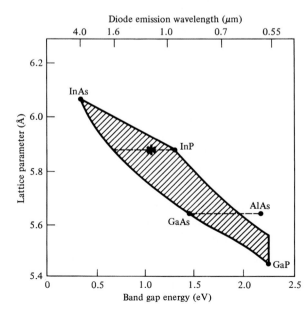

FIGURE 4-13
Relationships between the crystal lattice spacing, energy gap, and diode emission wavelength at room temperature. The shaded area is for the quaternary alloy InGaAsP. The asterisk (*) is for $In_{0.8}Ga_{0.2}As_{0.35}P_{0.65}$ ($E_g \approx 1.1$ eV) lattice-matched to InP. (Used with permission from *Optical Fibre Communications* by Tech. Staff of CSELT, © 1980, McGraw-Hill Book Company.)

energy gap in electron volts for values of x between zero and 0.37 (the direct-band-gap region) can be found from the empirical equation[1]

$$E_g = 1.424 + 1.266x + 0.266x^2 \tag{4-4}$$

Given the value of E_g in electron volts, the peak emission wavelength in micrometers is found from Eq. (4-3).

■■■ **Example 4-3.** Consider a $Ga_{1-x}Al_xAs$ laser with $x = 0.07$. From Eq. (4-4), we have $E_g = 1.51$ eV, so that Eq. (4-3) yields $\lambda = 0.82\,\mu m$.

The band-gap energy and lattice-constant range for the quaternary alloy InGaAsP are much larger, as shown by the shaded area in Fig. 4-13. These materials are generally grown on an InP substrate, so that lattice-matched configurations are obtained by selecting a compositional point along the top dashed line in Fig. 4-13, which passes through the InP point. Along this line, the compositional parameters x and y follow the relationship $y \simeq 2.20x$ with $0 \le x \le 0.47$. For $In_{1-x}Ga_xAs_yP_{1-y}$ compositions that are lattice-matched to InP, the band gap in eV varies as

$$E_g = 1.35 - 0.72y + 0.12y^2 \tag{4-5}$$

Band-gap wavelengths from 0.92 to $1.65\,\mu m$ are covered by this material system.

■■■ **Example 4-4.** Consider the alloy $In_{0.74}Ga_{0.26}As_{0.57}P_{0.43}$ (i.e., $x = 0.26$ and $y = 0.57$). Then, from Eq. (4-5), we have $E_g = 0.97$ eV, so that Eq. (4-3) yields $\lambda = 1.27\,\mu m$.

Whereas the FWHM power spectral widths of LEDs in the 800-nm region are around 35 nm, this increases in longer-wavelength materials. For devices operating in the 1300-to-1600-nm region, the spectral widths vary from around 70 to 180 nm. Figure 4-14 shows an example for devices emitting at 1300 nm. In addition, as Fig. 4-14 shows, the output spectral widths of surface-emitting LEDs tend to be broader than those of edge-emitting LEDs because of different internal-absorption effects of the emitted light in the two device structures.

4.2.3   Quantum Efficiency and LED Power

An excess of electrons and holes in p- and n-type material, respectively (referred to as *minority carriers*) is created in a semiconductor light source by carrier injection at the device contacts. The excess densities of electrons n and holes p are equal, since the injected carriers are formed and recombine in pairs in accordance with the requirement for charge neutrality in the crystal. When carrier injection stops, the carrier density returns to the equilibrium value. In general, the excess carrier density decays exponentially with time according to the relation

$$n = n_0 e^{-t/\tau} \tag{4-6}$$

where n_0 is the initial injected excess electron density and the time constant τ is the carrier lifetime. This lifetime is one of the most important operating parameters of

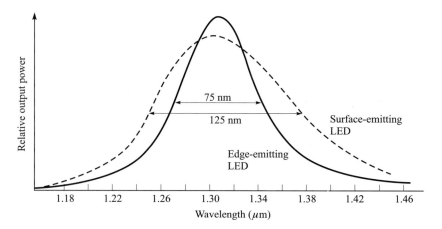

FIGURE 4-14
Typical spectral patterns for edge-emitting and surface-emitting LEDs at 1310 nm. The patterns broaden with increasing wavelength and are wider for surface emitters.

an electro-optic device. Its value can range from milliseconds to fractions of a nanosecond depending on material composition and device defects.

The excess carriers can recombine either radiatively or nonradiatively. In radiative recombination a photon of energy $h\nu$, which is approximately equal to the band-gap energy, is emitted. Nonradiative recombination effects include optical absorption in the active region (self-absorption), carrier recombination at the heterostructure interfaces, and the Auger process in which the energy released during an electron–hole recombination is transferred to another carrier in the form of kinetic energy.

When there is a constant current flow into an LED, an equilibrium condition is established. That is, the excess density of electrons n and holes p is equal since the injected carriers are created and recombined in pairs such that charge neutrality is maintained within the device. The total rate at which carriers are generated is the sum of the externally supplied and the thermally generated rates. The externally supplied rate is given by J/qd, where J is the current density in A/cm^2, q is the electron charge, and d is the thickness of the recombination region. The thermal generation rate is given by n/τ. Hence, the rate equation for carrier recombination in an LED can be written as

$$\frac{dn}{dt} = \frac{J}{qd} - \frac{n}{\tau} \qquad (4\text{-}7)$$

The equilibrium condition is found by setting Eq. (4-7) equal to zero, yielding

$$n = \frac{J\tau}{qd} \qquad (4\text{-}8)$$

This relationship gives the steady-state electron density in the active region when a constant current is flowing through it.

The *internal quantum efficiency* in the active region is the fraction of the electron–hole pairs that recombine radiatively. If the radiative recombination rate is R_r and the nonradiative recombination rate is R_{nr}, then the internal quantum efficiency η_{int} is the ratio of the radiative recombination rate to the total recombination rate:

$$\eta_{int} = \frac{R_r}{R_r + R_{nr}} \tag{4-9}$$

For exponential decay of excess carriers, the radiative recombination lifetime is $\tau_r = n/R_r$ and the nonradiative recombination lifetime is $\tau_{nr} = n/R_{nr}$. Thus, the internal quantum efficiency can be expressed as

$$\eta_{int} = \frac{1}{1 + \tau_r/\tau_{nr}} = \frac{\tau}{\tau_r} \tag{4-10}$$

where the *bulk recombination lifetime* τ is

$$\frac{1}{\tau} = \frac{1}{\tau_r} + \frac{1}{\tau_{nr}} \tag{4-11}$$

In general, τ_r and τ_{nr} are comparable for direct-band-gap semiconductors, such as GaAlAs and InGaAsP. This also means that R_r and R_{nr} are similar in magnitude, so that the internal quantum efficiency is about 50 percent for simple homojunction LEDs. However, LEDs having double-heterojunction structures can have quantum efficiencies of 60–80 percent. This high efficiency is achieved because the thin active regions of these devices mitigate the self-absorption effects, which reduces the nonradiative recombination rate.

If the current injected into the LED is I, then the total number of recombinations per second is

$$R_r + R_{nr} = I/q \tag{4-12}$$

Substituting Eq. (4-12) into Eq. (4-9) then yields $R_r = \eta_{int}I/q$. Noting that R_r is the total number of photons generated per second and that each photon has an energy $h\nu$, then the optical power generated internally to the LED is

$$P_{int} = \eta_{int} \frac{I}{q} h\nu = \eta_{int} \frac{hcI}{q\lambda} \tag{4-13}$$

Example 4-5. A double-heterojunction InGaAsP LED emitting at a peak wavelength of 1310 nm has radiative and nonradiative recombination times of 30 and 100 ns, respectively. The drive current is 40 mA. From Eq. (4-11), the bulk recombination lifetime is

$$\tau = \frac{\tau_r \tau_{nr}}{\tau_r + \tau_{nr}} = \frac{30 \times 100}{30 + 100} \text{ ns} = 23.1 \text{ ns}$$

Using Eq. (4-10), the internal quantum efficiency is

$$\eta_{\text{int}} = \frac{\tau}{\tau_r} = \frac{23.1}{30} = 0.77$$

Substituting this into Eq. (4-13) yields an internal power level of

$$P_{\text{int}} = \eta_{\text{int}} \frac{hcI}{q\lambda} = 0.77 \frac{(6.6256 \times 10^{-34}\,\text{J} \cdot \text{s})(3 \times 10^8\,\text{m/s})(0.040\,\text{A})}{(1.602 \times 10^{-19}\,\text{C})(1.31 \times 10^{-6}\,\text{m})} = 29.2\,\text{mW}$$

Not all internally generated photons will exit the device. To find the emitted power, one needs to consider the *external quantum efficiency* η_{ext}. This is defined as the ratio of the photons emitted from the LED to the number of internally generated photons. To find the external quantum efficiency, we need to take into account reflection effects at the surface of the LED. As shown in Fig. 4-15 and described in Sec. 2.2, at the interface of a material boundary only that fraction of light falling within a cone defined by the critical angle $\phi_c = \pi/2 - \theta_c$ will cross the interface. From Eq. (2-18), we have that $\phi_c = \sin^{-1}(n_2/n_1)$. Here, n_1 is the refractive index of the semiconductor material and n_2 is the refractive index of the outside material, which nominally is air with $n_2 = 1.0$. The external quantum efficiency can then be calculated from the expression

$$\eta_{\text{ext}} = \frac{1}{4\pi} \int_0^{\phi_c} T(\phi)(2\pi \sin \phi)d\phi \tag{4-14}$$

where $T(\phi)$ is the *Fresnel transmission coefficient* or *Fresnel transmissivity*. This factor depends on the incidence angle ϕ, but, for simplicity, we can use the expression for normal incidence, which is[22]

$$T(0) = \frac{4n_1 n_2}{(n_1 + n_2)^2} \tag{4-15}$$

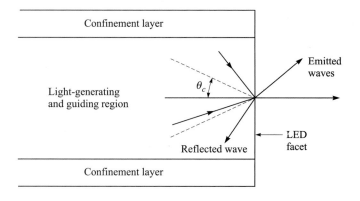

FIGURE 4-15
Only light falling within a cone defined by the critical angle will be emitted from an optical source.

Assuming the outside medium is air and letting $n_1 = n$, we have $T(0) = 4n/(n+1)^2$. The external quantum efficiency is then approximately given by

$$\eta_{\text{ext}} \approx \frac{1}{n(n+1)^2} \tag{4-16}$$

From this, it follows that the optical power emitted from the LED is

$$P = \eta_{\text{ext}} P_{\text{int}} = \frac{P_{\text{int}}}{n(n+1)^2} \tag{4-17}$$

Example 4-6. Assuming a typical value of $n = 3.5$ for the refractive index of an LED material, then from Eq. (4-16) we obtain $\eta_{\text{ext}} = 1.41$ percent. This shows that only a small fraction of the internally generated optical power is emitted from the device.[23]

4.2.4 Modulation of an LED

The frequency response of an LED is largely determined by the following three factors: the doping level in the active region, the injected carrier lifetime τ_i in the recombination region, and the parasitic capacitance of the LED. If the drive current is modulated at a frequency ω, the optical output power of the device will vary as[24,25]

$$P(\omega) = P_0 \left[1 + (\omega\tau_i)^2\right]^{-1/2} \tag{4-18}$$

where P_0 is the power emitted at zero modulation frequency. The parasitic capacitance can cause a delay of the carrier injection into the active junction, and, consequently, could delay the optical output.[26,27] This delay is negligible if a small, constant forward bias is applied to the diode. Under this condition, Eq. (4-18) is valid and the modulation response is limited only by the carrier recombination time.

The modulation bandwidth of an LED can be defined in either electrical or optical terms. Normally, electrical terms are used since the bandwidth is actually determined via the associated electrical circuitry. Thus, the modulation bandwidth is defined as the point where the electrical signal power, designated by $p(\omega)$, has dropped to half its constant value resulting from the modulated portion of the optical signal. This is the electrical 3-dB point; that is, the frequency at which the output electrical power is reduced by 3 dB with respect to the input electrical power, as is illustrated in Fig. 4-16.

Since an optical source exhibits a linear relationship between light power and current, currents rather than voltages (which are used in electrical systems) are compared in optical systems. Thus, since $p(\omega) = I^2(\omega)/R$, the ratio of the output electrical power at the frequency ω to the power at zero modulation is

$$\text{Ratio}_{\text{elec}} = 10 \log \left[\frac{p(\omega)}{p(0)}\right] = 10 \log \left[\frac{I^2(\omega)}{I^2(0)}\right] \tag{4-19}$$

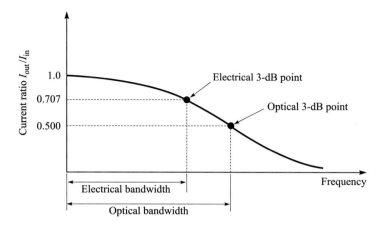

FIGURE 4-16
Frequency response of an optical source showing the electrical and optical 3-dB-bandwidth points.

where $I(\omega)$ is the electrical current in the detection circuitry. The electrical 3-dB point occurs at that frequency point where the detected electrical power $p(\omega) = p(0)/2$. This happens when

$$\frac{I^2(\omega)}{I^2(0)} = \frac{1}{2} \tag{4-20}$$

or $I(\omega)/I(0) = 1/\sqrt{2} = 0.707$.

Sometimes, the modulation bandwidth of an LED is given in terms of the 3-dB bandwidth of the modulated optical power $P(\omega)$; that is, it is specified at the frequency where $P(\omega) = P_0/2$. In this case, the 3-dB bandwidth is determined from the ratio of the optical power at frequency ω to the unmodulated value of the optical power. Since the detected current is directly proportional to the optical power, this ratio is

$$\text{Ratio}_{\text{optical}} = 10 \log\left[\frac{P(\omega)}{P(0)}\right] = 10 \log\left[\frac{I(\omega)}{I(0)}\right] \tag{4-21}$$

The optical 3-dB point occurs at that frequency where the ratio of the currents is equal to 1/2. As shown in Fig. 4-16, this gives an inflated value of the modulation bandwidth, which corresponds to an electrical power attenuation of 6 dB.

4.3   LASER DIODES

Lasers come in many forms with dimensions ranging from the size of a grain of salt to one that will occupy an entire room. The lasing medium can be a gas, a liquid, an insulating crystal (solid state), or a semiconductor. For optical fiber systems the laser sources used almost exclusively are semiconductor laser diodes. They are similar to other lasers, such as the conventional solid-state and gas lasers, in that the emitted radiation has spatial and temporal coherence; that is, the output radiation is highly monochromatic and the light beam is very directional.

Despite their differences, the basic principle of operation is the same for each type of laser. Laser action is the result of three key processes: photon absorption, spontaneous emission, and stimulated emission. These three processes are represented by the simple two-energy-level diagrams in Fig. 4-17, where E_1 is the ground-state energy and E_2 is the excited-state energy. According to Planck's law, a transition between these two states involves the absorption or emission of a photon of energy $h\nu_{12} = E_2 - E_1$. Normally, the system is in the group state. When a photon of energy $h\nu_{12}$ impinges on the system, an electron in state E_1 can absorb the photon energy and be excited to state E_2, as shown in Fig. 4-17a. Since this is an unstable state, the electron will shortly return to the ground state, thereby emitting a photon of energy $h\nu_{12}$. This occurs without any external stimulation and is called *spontaneous emission*. These emissions are isotropic and of random phase, and thus appear as a narrowband gaussian output.

The electron can also be induced to make a downward transition from the excited level to the ground-state level by an external stimulation. As shown in Fig. 4-17c, if a photon of energy $h\nu_{12}$ impinges on the system while the electron is still in its excited state, the electron is immediately stimulated to drop to the ground state and give off a photon of energy $h\nu_{12}$. This emitted photon is in phase with the incident photon, and the resultant emission is known as *stimulated emission*.

In thermal equilibrium the density of excited electrons is very small. Most photons incident on the system will therefore be absorbed, so that stimulated emission is essentially negligible. Stimulated emission will exceed absorption only if the population of the excited states is greater than that of the ground state. This condition is known as *population inversion*. Since this is not an equilibrium condition, population inversion is achieved by various "pumping" techniques. In a semiconductor laser, population inversion is accomplished by injecting electrons into the material at the device contacts to fill the lower energy states of the conduction band.

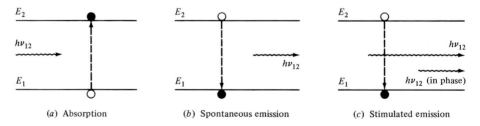

(a) Absorption   (b) Spontaneous emission   (c) Stimulated emission

FIGURE 4-17
The three key transition processes involved in laser action. The open circle represents the initial state of the electron and the filled circle represents the final state. Incident photons are shown on the left of each diagram and emitted photons are shown on the right.

4.3.1 Laser Diode Modes and Threshold Conditions

For optical fiber communication systems requiring bandwidths greater than approximately 200 MHz, the semiconductor injection laser diode is preferred over the LED. Laser diodes typically have response times less than 1 ns, have optical bandwidths of 2 nm or less, and, in general, are capable of coupling several tens of milliwatts of useful luminescent power into optical fibers with small cores and small mode-field diameters. Virtually all laser diodes in use are multilayered heterojunction devices. As mentioned in Sec. 4.2, the double-heterojunction LED configuration evolved from the successful demonstration of both carrier and optical confinement in heterojunction injection laser diodes. The more rapid evolvement and utilization of LEDs as compared with laser diodes lies in the inherently simpler construction, the smaller temperature dependence of the emitted optical power, and the absence of catastrophic degradation in LEDs (see Sec. 4.5). The construction of laser diodes is more complicated, mainly because of the additional requirement of current confinement in a small lasing cavity.

Stimulated emission in semiconductor lasers arises from optical transitions between distributions of energy states in the valence and conduction bands. This differs from gas and solid-state lasers, in which radiative transitions occur between discrete isolated atomic or molecular levels. The radiation in the laser diode is generated within a Fabry-Perot resonator cavity,[1-3] shown in Fig. 4-18, as

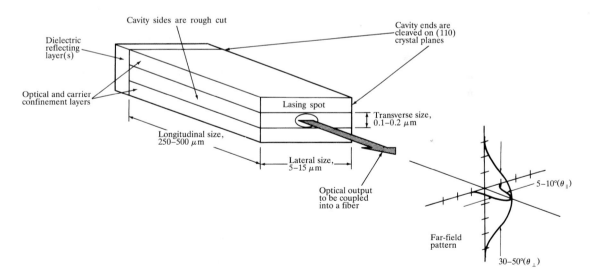

FIGURE 4-18
Fabry-Perot resonator cavity for a laser diode. The cleaved crystal ends function as partially reflecting mirrors. The unused end (the rear facet) can be coated with a dielectric reflector to reduce optical loss in the cavity. Note that the light beam emerging from the laser forms a vertical ellipse, even though the lasing spot at the active-area facet is a horizontal ellipse.

in most other types of lasers. However, this cavity is much smaller, being approximately 250–500 μm long, 5–15 μm wide, and 0.1–0.2 μm thick. These dimensions are commonly referred to as the *longitudinal, lateral,* and *transverse dimensions* of the cavity, respectively.

In the laser diode Fabry-Perot resonator, a pair of flat, partially reflecting mirrors are directed toward each other to enclose the cavity. The mirror facets are constructed by making two parallel clefts along natural cleavage planes of the semiconductor crystal. The purpose of these mirrors is to provide strong *optical feedback* in the longitudinal direction, thereby converting the device into an oscillator with a gain mechanism that compensates for optical losses in the cavity. The laser cavity can have many resonant frequencies. The device will oscillate (thereby emitting light) at those resonant frequencies for which the gain is sufficient to overcome the losses. The sides of the cavity are simply formed by roughening the edges of the device to reduce unwanted emissions in these directions.

In another laser diode type, commonly referred to as the *distributed-feedback* (DFB) *laser,*[1,8,28-31] the cleaved facets are not required for optical feedback. A typical DFB laser configuration is shown in Fig. 4-19. The fabrication of this device is similar to the Fabry-Perot types, except that the lasing action is obtained from Bragg reflectors (gratings) or periodic variations of the refractive index (called *distributed-feedback corrugations*), which are incorporated into the multilayer structure along the length of the diode. This is discussed in more detail in Sec. 4.3.6.

In general, the full optical output is needed only from the front facet of the laser—that is, the one to be aligned with an optical fiber. In this case, a dielectric reflector can be deposited on the rear laser facet to reduce the optical loss in the cavity, to reduce the threshold current density (the point at which lasing starts), and to increase the external quantum efficiency. Reflectivities greater than 98 percent have been achieved with a six-layer reflector.[32]

The optical radiation within the resonance cavity of a laser diode sets up a pattern of electric and magnetic field lines called the *modes of the cavity* (see Secs.

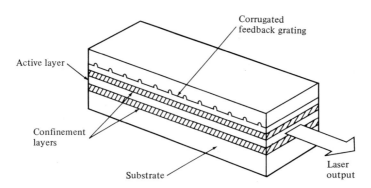

FIGURE 4-19
Structure of a distributed-feedback (DFB) laser diode.

2.3 and 2.4 for details on modes). These can conveniently be separated into two independent sets of transverse electric (TE) and transverse magnetic (TM) modes. Each set of modes can be described in terms of the longitudinal, lateral, and transverse half-sinusoidal variations of the electromagnetic fields along the major axes of the cavity. The *longitudinal modes* are related to the length L of the cavity and determine the principal structure of the frequency spectrum of the emitted optical radiation. Since L is much larger than the lasing wavelength of approximately 1 μm, many longitudinal modes can exist. *Lateral modes* lie in the plane of the *pn* junction. These modes depend on the side wall preparation and the width of the cavity, and determine the shape of the lateral profile of the laser beam. *Transverse modes* are associated with the electromagnetic field and beam profile in the direction perpendicular to the plane of the *pn* junction. These modes are of great importance, since they largely determine such laser characteristics as the radiation pattern (the angular distribution of the optical output power) and the threshold current density.

To determine the lasing conditions and the resonant frequencies, we express the electromagnetic wave propagating in the longitudinal direction (along the axis normal to the mirrors) in terms of the electric field phasor

$$E(z, t) = I(z)\, e^{j(\omega t - \beta z)} \tag{4-22}$$

where $I(z)$ is the optical field intensity, ω is the optical radian frequency, and β is the propagation constant (see Sec. 2.3.2).

Lasing is the condition at which light amplification becomes possible in the laser diode. The requirement for lasing is that a population inversion be achieved. This condition can be understood by considering the fundamental relationship between the optical field intensity I, the absorption coefficient α_λ, and the gain coefficient g in the Fabry-Perot cavity. The stimulated emission rate into a given mode is proportional to the intensity of the radiation in that mode. The radiation intensity at a photon energy $h\nu$ varies exponentially with the distance z that it traverses along the lasing cavity according to the relationship

$$I(z) = I(0)\exp\big\{[\Gamma g(h\nu) - \bar{\alpha}(h\nu)]z\big\} \tag{4-23}$$

where $\bar{\alpha}$ is the effective absorption coefficient of the material in the optical path and Γ is the *optical-field confinement factor*—that is, the fraction of optical power in the active layer (see Prob. 4-11 concerning details of transverse and lateral optical-field confinement factors).

Optical amplification of selected modes is provided by the feedback mechanism of the optical cavity. In the repeated passes between the two partially reflecting parallel mirrors, a portion of the radiation associated with those modes that have the highest optical gain coefficient is retained and further amplified during each trip through the cavity.

Lasing occurs when the gain of one or several guided modes is sufficient to exceed the optical loss during one roundtrip through the cavity; that is, $z = 2L$. During this roundtrip, only the fractions R_1 and R_2 of the optical radiation are reflected from the two laser ends 1 and 2, respectively, where R_1 and R_2 are the mirror reflectivities or Fresnel reflection coefficients, which are given by

$$R = \left(\frac{n_1 - n_2}{n_1 + n_2}\right)^2 \tag{4-24}$$

for the reflection of light at an interface between two materials having refractive indices n_1 and n_2. From this lasing condition, Eq. (4-23) becomes

$$I(2L) = I(0)R_1 R_2 \exp\{2L[\Gamma g(h\nu) - \bar{\alpha}(h\nu)]\} \tag{4-25}$$

At the lasing threshold, a steady-state oscillation takes place, and the magnitude and phase of the returned wave must be equal to those of the original wave. This gives the conditions

$$I(2L) = I(0) \tag{4-26}$$

for the amplitude and

$$e^{-j2\beta L} = 1 \tag{4-27}$$

for the phase. Equation (4-27) gives information concerning the resonant frequencies of the Fabry-Perot cavity. This is discussed further in Sec. 4.3.2. From Eq. (4-26), we can find which modes have sufficient gain for sustained oscillation and we can find the amplitudes of these modes. The condition to just reach the lasing threshold is the point at which the optical gain is equal to the total loss α_t in the cavity. From Eq. (4-26), this condition is

$$\Gamma g_{th} = \alpha_t = \bar{\alpha} + \frac{1}{2L}\ln\left(\frac{1}{R_1 R_2}\right) = \bar{\alpha} + \alpha_{end} \tag{4-28}$$

where α_{end} is the mirror loss in the lasing cavity. Thus, for lasing to occur, we must have the gain $g \geq g_{th}$. This means that the pumping source that maintains the population inversion must be sufficiently strong to support or exceed all the energy-consuming mechanisms within the lasing cavity.

Example 4-7. For GaAs, $R_1 = R_2 = 0.32$ for uncoated facets (i.e., 32 percent of the radiation is reflected at a facet) and $\bar{\alpha} \simeq 10\,\text{cm}^{-1}$. This yields $\Gamma g_{th} = 33\,\text{cm}^{-1}$ for a laser diode of length $L = 500\,\mu\text{m}$.

The mode that satisfies Eq. (4-28) reaches threshold first. Theoretically, at the onset of this condition, all additional energy introduced into the laser should augment the growth of this particular mode. In practice, various phenomena lead to the excitation of more than one mode.[1] Studies on the conditions of longitudinal single-mode operation have shown that important factors are thin active regions and a high degree of temperature stability.

The relationship between optical output power and diode drive current is presented in Fig. 4-20. At low diode currents, only spontaneous radiation is emitted. Both the spectral range and the lateral beam width of this emission are broad like that of an LED. A dramatic and sharply defined increase in the power

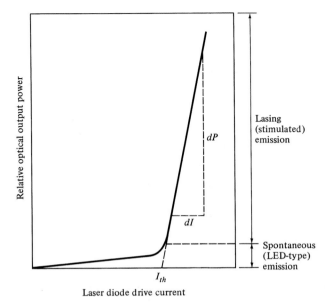

Relative optical output power

dP

Lasing
(stimulated)
emission

dI

I_{th}

Laser diode drive current

Spontaneous
(LED-type)
emission

FIGURE 4-20
Relationship between optical output power and laser diode drive current. Below the lasing threshold, the optical output is a spontaneous LED-type emission.

output occurs at the lasing threshold. As this transition point is approached, the spectral range and the beam width both narrow with increasing drive current. The final spectral width of approximately 1 nm and the fully narrowed lateral beam width of nominally 5–10° are reached just past the threshold point. The *threshold current* I_{th} is conventionally defined by extrapolation of the lasing region of the power-versus-current curve, as shown in Fig. 4-20. At high power outputs, the slope of the curve decreases because of junction heating.

For laser structures that have strong carrier confinement, the *threshold current density* for stimulated emission J_{th} can to a good approximation be related to the lasing-threshold optical gain by

$$g_{th} = \beta J_{th} \tag{4-29}$$

where β is a constant that depends on the specific device construction (see Prob. 4-15).

4.3.2 Laser Diode Rate Equations

The relationship between optical output power and the diode drive current can be determined by examining the rate equations that govern the interaction of photons and electrons in the active region. As noted earlier, the total carrier population is determined by carrier injection, spontaneous recombination, and stimulated emission. For a *pn* junction with a carrier-confinement region of depth *d*, the *rate equations* are given by

$$\frac{d\Phi}{dt} = Cn\Phi + R_{sp} - \frac{\Phi}{\tau_{ph}}$$

$$\text{(4-30)}$$

$$= \text{stimulated emission} + \text{spontaneous emission} + \text{photon loss}$$

which governs the number of photons Φ, and

$$\frac{dn}{dt} = \frac{J}{qd} - \frac{n}{\tau_{sp}} - Cn\Phi$$

$$\text{(4-31)}$$

$$= \text{injection} + \text{spontaneous recombination} + \text{stimulated emission}$$

which governs the number of electrons n. Here, C is a coefficient describing the strength of the optical absorption and emission interactions, R_{sp} is the rate of spontaneous emission into the lasing mode (which is much smaller than the total spontaneous-emission rate), τ_{ph} is the photon lifetime, τ_s is the spontaneous-recombination lifetime, and J is the injection-current density.

Equations (4-30) and (4-31) may be balanced by considering all the factors that affect the number of carriers in the laser cavity. The first term in Eq. (4-30) is a source of photons resulting from stimulated emission. The second term, describing the number of photons produced by spontaneous emission, is relatively small compared with the first term. The third term in Eq. (4-30) indicates the decay in the number of photons caused by loss mechanisms in the lasing cavity. In Eq. (4-31), the first term represents the increase in the electron concentration in the conduction band as current flows into the device. The second and third terms give the number of electrons lost from the conduction band owing to spontaneous and stimulated transitions, respectively.

Solving these two equations for a steady-state condition will yield an expression for the output power. The steady state is characterized by the left-hand sides of Eqs. (4-30) and (4-31) being equal to zero. First, from Eq. (4-30), assuming R_{sp} is negligible and noting that $d\Phi/dt$ must be positive when Φ is small, we have

$$Cn - \frac{1}{\tau_{ph}} \geq 0$$

$$\text{(4-32)}$$

This shows that n must exceed a threshold value n_{th} in order for Φ to increase. Using Eq. (4-31), this threshold value can be expressed in terms of the threshold current J_{th} needed to maintain an inversion level $n = n_{th}$ in the steady state when the number of photons $\Phi = 0$:

$$\frac{n_{th}}{\tau_{sp}} = \frac{J_{th}}{qd}$$

$$\text{(4-33)}$$

This expression defines the current required to sustain an excess electron density in the laser when spontaneous emission is the only decay mechanism.

Next, consider the photon and electron rate equations in the steady-state condition at the lasing threshold. Respectively, Eqs. (4-30) and (4-31) become

$$0 = Cn_{\text{th}}\Phi_s + R_{\text{sp}} - \frac{\Phi_s}{\tau_{\text{ph}}} \tag{4-34}$$

and

$$0 = \frac{J}{qd} - \frac{n_{\text{th}}}{\tau_{\text{sp}}} - Cn_{\text{th}}\Phi_s \tag{4-35}$$

where Φ_s is the steady-state photon density. Adding Eqs. (4-34) and (4-35), using Eq. (4-33) for the term $n_{\text{th}}/\tau_{\text{sp}}$, and solving for Φ_s yields the number of photons per unit volume:

$$\Phi_s = \frac{\tau_{\text{ph}}}{qd}(J - J_{\text{th}}) + \tau_{\text{ph}}R_{\text{sp}} \tag{4-36}$$

The first term in Eq. (4-36) is the number of photons resulting from stimulated emission. The power from these photons is generally concentrated in one or a few modes. The second term gives the spontaneously generated photons. The power resulting from these photons is not mode-selective, but is spread over all the possible modes of the volume, which are on the order of 10^8 modes.

4.3.3 External Quantum Efficiency

The *external differential quantum efficiency* η_{ext} is defined as the number of photons emitted per radiative electron–hole pair recombination above threshold. Under the assumption that above threshold the gain coefficient remains fixed at g_{th}, η_{ext} is given by[1]

$$\eta_{\text{ext}} = \frac{\eta_i(g_{\text{th}} - \bar{\alpha})}{g_{\text{th}}} \tag{4-37}$$

Here, η_i is the internal quantum efficiency. This is not a well-defined quantity in laser diodes, but most measurements show that $\eta_i \simeq 0.6$–0.7 at room temperature. Experimentally, η_{ext} is calculated from the straight-line portion of the curve for the emitted optical power P versus drive current I, which gives

$$\eta_{\text{ext}} = \frac{q}{E_g}\frac{dP}{dI} = 0.8065\lambda \ (\mu\text{m})\frac{dP \ (\text{mW})}{dI \ (\text{mA})} \tag{4-38}$$

where E_g is the band-gap energy in electron volts, dP is the incremental change in the emitted optical power in milliwatts for an incremental change dI in the drive current (in milliamperes), and λ is the emission wavelength in micrometers. For standard semiconductor lasers, external differential quantum efficiencies of 15–20 percent per facet are typical. High-quality devices have differential quantum efficiencies of 30–40 percent.

4.3.4 Resonant Frequencies

Now let us return to Eq. (4-27) to examine the resonant frequencies of the laser. The condition in Eq. (4-27) holds when

$$2\beta L = 2\pi m \tag{4-39}$$

where m is an integer. Using $\beta = 2\pi n/\lambda$ for the propagation constant from Eq. (2-46), we have

$$m = \frac{L}{\lambda/2n} = \frac{2Ln}{c} \nu \tag{4-40}$$

where $c = \nu\lambda$. This states that the cavity resonates (i.e., a standing-wave pattern exists within it) when an integer number m of half-wavelengths spans the region between the mirrors.

Since in all lasers the gain is a function of frequency (or wavelength, since $c = \nu\lambda$), there will be a range of frequencies (or wavelengths) for which Eq. (4-40) holds. Each of these frequencies corresponds to a mode of oscillation of the laser. Depending on the laser structure, any number of frequencies can satisfy Eqs. (4-26) and (4-27). Thus, some lasers are single-mode and some are multimode. The relationship between gain and frequency can be assumed to have the gaussian form

$$g(\lambda) = g(0)\exp\left[-\frac{(\lambda - \lambda_0)^2}{2\sigma^2}\right] \tag{4-41}$$

where λ_0 is the wavelength at the center of the spectrum, σ is the spectral width of the gain, and the maximum gain $g(0)$ is proportional to the population inversion.

Let us now look at the frequency, or wavelength, spacing between the modes of a multimode laser. Here, we consider only the longitudinal modes. Note, however, that for each longitudinal mode there may be several transverse modes that arise from one or more reflections of the propagating wave at the sides of the resonator cavity.[1,3] To find the frequency spacing, consider two successive modes of frequencies ν_{m-1} and ν_m represented by the integers $m-1$ and m. From Eq. (4-40), we have

$$m - 1 = \frac{2Ln}{c}\nu_{m-1} \tag{4-42}$$

and

$$m = \frac{2Ln}{c}\nu_m \tag{4-43}$$

Subtracting these two equations yields

$$1 = \frac{2Ln}{c}(\nu_m - \nu_{m-1}) = \frac{2Ln}{c}\Delta\nu \tag{4-44}$$

from which we have the frequency spacing

$$\Delta\nu = \frac{c}{2Ln} \tag{4-45}$$

This can be related to the wavelength spacing $\Delta\lambda$ through the relationship $\Delta\nu/\nu = \Delta\lambda/\lambda$, yielding

$$\Delta\lambda = \frac{\lambda^2}{2Ln} \qquad (4\text{-}46)$$

Thus, given Eqs. (4-41) and (4-46), the output spectrum of a multimode laser follows the typical gain-versus-frequency plot given in Fig. 4-21, where the exact number of modes, their heights, and their spacings depend on the laser construction.

Example 4-8. A GaAs laser operating at 850 nm has a 500-μm length and a refractive index $n = 3.7$. What are the frequency and wavelength spacings. If, at the half-power point, $\lambda - \lambda_0 = 2$ nm, what is the spectral width σ of the gain?

From Eq. (4-45) we have $\Delta\nu = 81$ GHz, and from Eq. (4-46) we find that $\Delta\lambda = 0.2$ nm. Using Eq. (4-41) with $g(\lambda) = 0.5g(0)$ yields $\sigma = 1.70$ nm.

4.3.5 Laser Diode Structures and Radiation Patterns

A basic requirement for efficient operation of laser diodes is that, in addition to transverse optical and carrier confinement between heterojunction layers, the current flow must be restricted laterally to a narrow stripe along the length of the laser. Numerous novel methods of achieving this, with varying degrees of success, have been proposed, but all strive for the same goals of limiting the

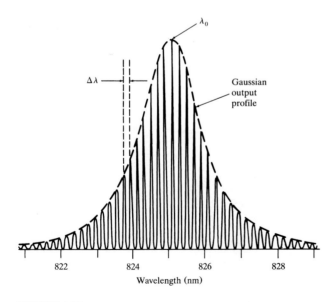

FIGURE 4-21
Typical spectrum from a gain-guided GaAlAs/GaAs laser diode. (Reproduced with permission from K. Peterman and G. Arnold, *IEEE J. Quantum Electron.*, vol. 18, pp. 543–555, Apr. 1982, © 1982, IEEE.)

number of lateral modes so that lasing is confined to a single filament, stabilizing the lateral gain, and ensuring a relatively low threshold current.

Figure 4-22 shows the three basic *optical-confinement methods* used for bounding laser light in the lateral direction.[8] In the first structure, a narrow electrode stripe (less than $8\,\mu m$ wide) runs along the length of the diode. The injection of electrons and holes into the device alters the refractive index of the active layer directly below the stripe. The profile of these injected carriers creates a weak, complex waveguide that confines the light laterally. This type of device is commonly referred to as a *gain-guided laser*. Although these lasers can emit optical powers exceeding 100 mW, they have strong instabilities and can have highly astigmatic, two-peaked beams as shown in Fig. 4-22*a*.

More stable structures use the configurations shown in Fig. 4-22*b* and *c*. Here, dielectric waveguide structures are fabricated in the lateral direction. The variations in the real refractive index of the various materials in these structures control the lateral modes in the laser. Thus, these devices are called *index-guided lasers*. If a particular index-guided laser supports only the fundamental transverse mode and the fundamental longitudinal mode, it is known as a *single-mode laser*. Such a device emits a single, well-collimated beam of light that has an intensity profile which is a bell-shaped gaussian curve.

Index-guided lasers can have either positive-index or negative-index wave-confining structures. In a *positive-index waveguide*, the central region has a higher refractive index than the outer regions. Thus, all of the guided light is reflected at

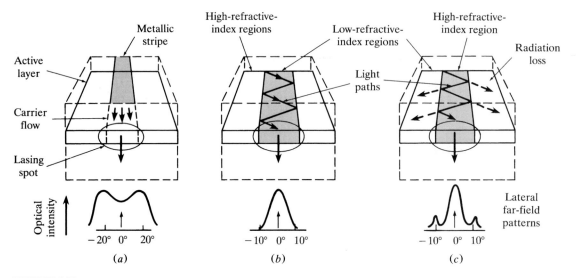

FIGURE 4-22

Three fundamental structures for confining optical waves in the lateral direction: (*a*) in the gain-induced guide, electrons injected via a metallic stripe contact alter the index of refraction of the active layer; (*b*) the positive-index waveguide has a higher refractive index in the central portion of the active region; (*c*) the negative-index waveguide has a lower refractive index in the central portion of the active region. (Reproduced with permission from Botez,[8a] © 1985, IEEE.)

the dielectric boundary, just as it is at the core–cladding interface in an optical fiber. By proper choice of the change in refractive index and the width of the higher-index region, one can make a device that supports only the fundamental lateral mode.

In a *negative-index waveguide*, the central region of the active layer has a lower refractive index than the outer regions. At the dielectric boundaries, part of the light is reflected and the rest is refracted into the surrounding material and is thus lost. This radiation loss appears in the far-field radiation pattern as narrow side lobes to the main beam, as shown in Fig. 4-22c. Since the fundamental mode in this device has less radiation loss than any other mode, it is the first to lase. The positive-index laser is the more popular of these two structures.

Index-guided lasers can be made using any one of four fundamental structures. These are the buried heterostructure, a selectively diffused construction, a varying-thickness structure, and a bent-layer configuration. To make the *buried heterostructure* (BH) laser shown in Fig. 4-23, one etches a narrow mesa stripe (1–2 μm wide) in double-heterostructure material. The mesa is then embedded in high-resistivity lattice-matched n-type material with an appropriate band gap and low refractive index. This material is GaAlAs in 800-to-900-nm lasers with a GaAs active layer, and is InP for 1300-to-1600-nm lasers with an InGaAsP active layer. This configuration thus strongly traps generated light in a lateral waveguide. A number of variations of this fundamental structure have been used to fabricate high-performing laser diodes.[6,7]

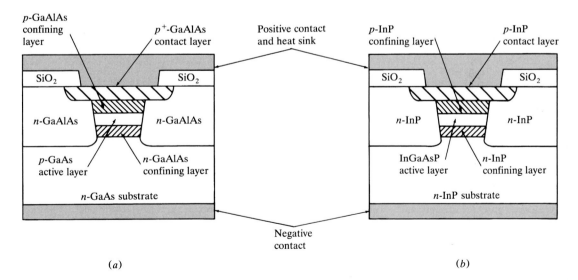

FIGURE 4-23
(*a*) Short-wavelength (800–900 nm) GaAlAs and (*b*) long-wavelengths (1300–1600 nm) InGaAsP buried-heterostructure laser diodes.

The *selectively diffused construction* is shown in Fig. 4-24*a*. Here, a chemical dopant, such as zinc for GaAlAs lasers and cadmium for InGaAsP lasers, is diffused into the active layer immediately below the metallic contact stripe. The dopant changes the refractive index of the active layer to form a lateral waveguide channel. In the *varying-thickness structure* shown in Fig. 4-24*b*, a channel (or other topological configuration, such as a mesa or terrace) is etched into the substrate. Layers of crystal are then regrown into the channel using liquid-phase epitaxy. This process fills in the depressions and partially dissolves the protrusions, thereby creating variations in the thicknesses of the active and confining layers. When an optical wave encounters a local increase in the thickness, as shown in Fig. 4-24*b*, the thicker area acts as a positive-index waveguide of higher-index material. In the *bent-layer structure*, a mesa is etched into the substrate as shown in Fig. 4-24*c*. Semiconductor material layers are grown onto this structure using vapor-phase epitaxy to exactly replicate the mesa configuration. The active layer has a constant thickness with lateral bends. As an optical wave travels along the flat top of the mesa in the active area, the lower-index material outside of the bends confines the light along this lateral channel.

In addition to confining the optical wave to a narrow lateral stripe to achieve continuous high optical output power, one also needs to restrict the drive current tightly to the active layer so more than 60 percent of the current contributes to lasing. Figure 4-25 shows the four basic *current-confinement methods*. In each method, the device architecture blocks current on both sides of the lasing region. This is achieved either by high-resistivity regions or by reverse-biased *pn* junctions, which prevent the current from flowing while the device is forward-biased under normal conditions. For structures with a continuous active layer, the current can be confined either above or below the lasing region. The diodes are forward-biased so that current travels from the *p*-type to the *n*-type regions. In the *preferential-dopant diffusion* method, partially diffusing a *p*-type

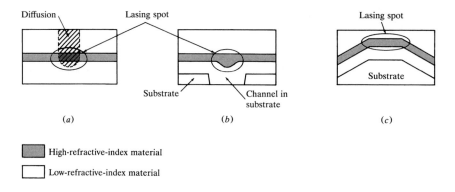

(a)   *(b)*   *(c)*

High-refractive-index material

Low-refractive-index material

FIGURE 4-24
Positive-index optical-wave-confining structure of the (*a*) selectively diffused, (*b*) varying-thickness, and (*c*) bent-layer types. (Adapted with permission from Botez,[8a] © 1985, IEEE.)

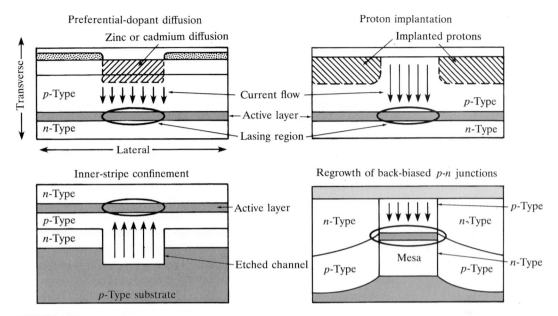

FIGURE 4-25

Four basic methods for achieving current confinement in laser diodes: (*a*) preferential-dopant diffusion, (*b*) proton implantation, (*c*) inner-stripe confinement, and (*d*) regrowth of back-biased *pn* junctions. (Adapted with permission from Botez,[8a] © 1985, IEEE.)

dopant (Zn or Cd) through an *n*-type capping layer establishes a narrow path for the current, since back-biased *pn* junctions block the current outside the diffused region. The *proton implantation* method creates regions of high resistivity, thus restricting the current to a narrow path between these regions. The *inner-stripe confinement* technique grows the lasing structure above a channel etched into planar material. Back-biased *pn* junctions restrict the current on both sides of the channel. When the active layer is discontinuous, as in a buried heterostructure, current can be blocked on both sides of the mesa by growing *pn* junctions that are reverse-biased when the device is operating. A laser diode can use more than one current-confining technique.

In a double-heterojunction laser, the highest-order transverse mode that can be excited depends on the waveguide thickness and on the refractive-index differentials at the waveguide boundaries.[1] If the refractive-index differentials are kept at approximately 0.08, then only the fundamental transverse mode will propagate if the active area is thinner than 1 μm.

When designing the width and thickness of the optical cavity, a tradeoff must be made between current density and output beam width. As either the width or the thickness of the active region is increased, a narrowing occurs of the lateral or transverse beam widths, respectively, but at the expense of an increase in the threshold current density. Most positive-index waveguide devices having a lasing spot 3 μm wide by 0.6 μm high. This is significantly greater than the active-layer thickness, since about half the light travels in the confining layers. Such lasers can operate reliably only up to continuous-wave (CW) output powers of 3–5 mW. Here, the transverse and lateral half-power beam widths shown in Fig. 4-18 are about $\theta_\perp \simeq 30$–$50°$ and $\theta_\parallel \simeq 5$–$10°$, respectively.

Although the active layer in a standard double-heterostructure laser is thin enough (1–$3\,\mu$m) to confine electrons and the optical field, the electronic and optical properties remain the same as in the bulk material. This limits the achievable threshold current density, modulation speed, and linewidth of the device. *Quantum-well lasers* overcome these limitations by having an active-layer thickness around 10 nm.[33-35] This changes the electronic and optical properties dramatically, because the dimensionality of the free-electron motion is reduced from three to two dimensions. As shown in Fig. 4-26, the restriction of the carrier motion normal to the active layer results in a quantization of the energy levels. The possible energy-level transitions which lead to photon emission are designated by ΔE_{ij} (see Prob. 4-16). Both single quantum-well (SQW) and multiple quantum-well (MQW) lasers have been fabricated. These structures contain single and multiple active regions, respectively. The layers separating the active regions are called *barrier layers*. The MQW lasers have a better optical-mode confinement, which results in a lower threshold current density. The wavelength of the output light can be changed by adjusting the layer thickness d. For example, in an InGaAs quantum-well laser, the peak output wavelength moves from 1550 nm when $d = 10$ nm to 1500 nm when $d = 8$ nm.

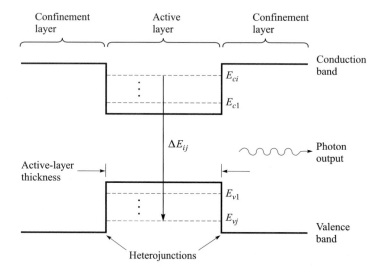

FIGURE 4-26
Energy-band diagram for a quantum layer in a multiple quantum-well (MQW) laser. The parameter ΔE_{ij} represents the allowed energy-level transitions.

4.3.6 Single-Mode Lasers

For high-speed long-distance communications one needs single-mode lasers, which must contain only a single longitudinal mode and a single transverse mode. Consequently, the spectral width of the optical emission is very narrow.

One way of restricting a laser to have only one longitudinal mode is to reduce the length L of the lasing cavity to the point where the frequency separation $\Delta\nu$ of the adjacent modes given in Eq. (4-45) is larger than the laser transition line width; that is, only a single longitudinal mode falls within the gain bandwidth of the device. For example, for a Fabry-Perot cavity, all longitudinal modes have nearly equal losses and are spaced by about 1 nm in a 250-μm-long cavity at 1300 nm. By reducing L from 250 μm to 25 μm, the mode spacing increases from 1 nm to 10 nm. However, these lengths make the device hard to handle, and they are limited to optical output powers of only a few milliwatts.[36]

Alternative devices were thus developed. Among these are vertical-cavity surface-emitting lasers, structures that have a built-in frequency-selective grating, and tunable lasers. Here, we look at the first two structures. Tunable lasers are discussed in Chap. 10 in relation to their use in multiple-wavelength optical links. The special feature of a vertical-cavity surface-emitting laser (abbreviated as VCSEL or VCL)[37-41] is that the light emission is perpendicular to the semiconductor surface, as shown in Fig. 4-27. This feature facilitates the integration of multiple lasers onto a single chip in one- or two-dimensional arrays, which makes them attractive for wavelength-division-multiplexing applications. The active-region volume of these devices is very small, which leads to very low threshold currents ($< 100\,\mu$A). In addition, for an equivalent output power compared to edge-emitting lasers, the modulation bandwidths are much greater, since the higher photon densities reduce radiative lifetimes. The mirror system used in

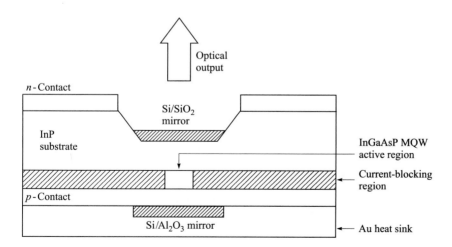

FIGURE 4-27
Basic architecture of a vertical-cavity surface-emitting laser.

VCSELs is of critical importance, since maximum reflectivity is needed for efficient operation.[40] Figure 4-27 shows one mirror system that consists of a semiconductor material, such as Si/SiO_2, as one material and an oxide layer, such as Si/Al_2O_3, as the other material.

Three types of laser configurations using a built-in *frequency-selective reflector* are shown in Fig. 4-28. In each case, the frequency-selective reflector is a corrugated grating which is a passive waveguide layer adjacent to the active region. The optical wave propagates parallel to this grating. The operation of

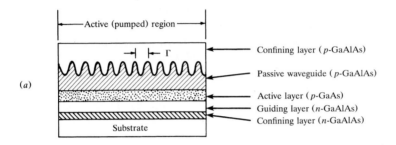

(a)

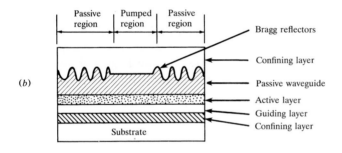

(b)

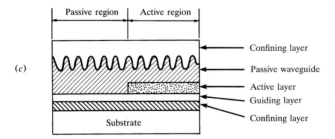

(c)

FIGURE 4-28
Three types of laser structures using built-in frequency-selective resonator gratings: (*a*) distributed-feedback (DFB) laser, (*b*) distributed-Bragg-reflector (DBR) laser, and (*c*) distributed-reflector (DR) laser.

these types of lasers is based on the distributed Bragg phase-grating reflector.[30] A phase grating is essentially a region of periodically varying refractive index that causes two counterpropagating traveling waves to couple. The coupling is at a maximum for wavelengths close to the Bragg wavelength λ_B, which is related to the period Λ of the corrugations by

$$\lambda_B = \frac{2n_e\Lambda}{k} \tag{4-47}$$

where n_e is the effective refractive index of the mode and k is the order of the grating. First-order gratings ($k = 1$) provide the strongest coupling, but sometimes second-order gratings are used since their larger corrugation period makes fabrication easier. Lasers based on this architecture exhibit good single-mode longitudinal operation with low sensitivity to drive-current and temperature variations.

In the *distributed-feedback* (DFB) laser,[29-31] the grating for the wavelength selector is formed over the entire active region. As shown in Fig. 4-29, in an ideal

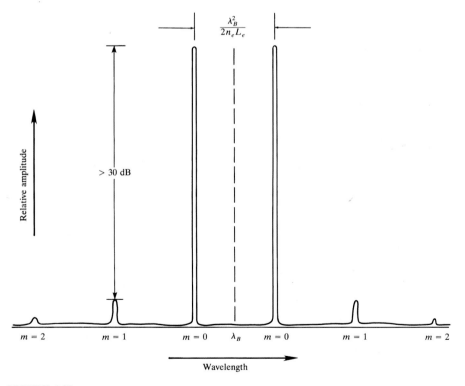

FIGURE 4-29
Output spectrum symmetrically distributed around λ_B in an idealized distributed-feedback (DFB) laser diode.

DFB laser the longitudinal modes are spaced symmetrically around λ_B at wavelengths given by

$$\lambda = \lambda_B \pm \frac{\lambda_B^2}{2n_e L_e}(m + \tfrac{1}{2}) \tag{4-48}$$

where $m = 0, 1, 2, \ldots$ is the mode order and L_e is the effective grating length. The amplitudes of successively higher-order lasing modes are greatly reduced from the zero-order amplitude; for example, the first-order mode ($m = 1$) is usually more than 30 dB down from the zero-order amplitude ($m = 0$).

Theoretically, in a DFB laser that has both ends antireflection-coated, the two zero-order modes on either side of the Bragg wavelength should experience the same lowest threshold gain and would lase simultaneously in an idealized symmetrical structure. However, in practice, the randomness of the cleaving process lifts the degeneracy in the modal gain and results in single-mode operation. This facet asymmetry can be further increased by putting a high-reflection coating on one end and a low-reflection coating on the other; for example, around 2 percent on the front facet and 30 percent on the rear facet. Variations on the DFB design have been the introduction of a $\pi/2$ optical phase shift (i.e., a quarter wavelength) in the corrugation at the center of the optical cavity to make the laser oscillate near the Bragg wavelength, since reflections occur most effectively at this wavelength.[42,43]

For the *distributed-Bragg-reflector* (DBR) laser,[31,44,45] the gratings are located at the ends of the normal active layer of the laser to replace the cleaved end mirrors used in the Fabry-Perot optical resonator (Fig. 4-28b). The *distributed-reflector* laser consists of active and passive distributed reflectors (Fig. 4-28c). This structure improves the lasing properties of conventional DFB and DBR lasers, and has a high efficiency and high output capability.

4.3.7 Modulation of Laser Diodes

The process of imposing information on a light stream is called *modulation*. This can be realized either by directly varying the laser drive current with the information stream to produce a varying optical output power, or by using an external modulator to modify a steady optical power level emitted by the laser. *External modulation* is needed for high-speed systems (> 2.5 Gb/s) to minimize undesirable nonlinear effects such a chirping (see Sec. 12.5). A variety of external modulators are commercially available, either as a separate device or as an integral part of the laser transmitter package.[46] The CD-ROM contains a simulation module that compares direct and external modulation for various bit rates. The basic limitation on the direct modulation rate of laser diodes depends on the spontaneous (radiative) and stimulated carrier lifetimes and the photon lifetime. The *spontaneous lifetime* τ_{sp} is a function of the semiconductor band structure and the carrier concentration. At room temperature the radiative lifetime τ_r is about 1 ns in GaAs-based materials for dopant concentrations on the order of $10^{19}\,\text{cm}^{-3}$. The *stimulated carrier lifetime* τ_{st} depends on the optical density in the lasing cavity and is on the order of 10 ps. The *photon lifetime* τ_{ph} is the average time that the photon resides in the lasing cavity before being lost either by absorption or by emission through the facets. In a Fabry-Perot cavity, the photon lifetime is[1]

$$\tau_{\mathrm{ph}}^{-1} = \frac{c}{n}\left(\bar{\alpha} + \frac{1}{2L}\ln\frac{1}{R_1 R_2}\right) = \frac{c}{n}g_{\mathrm{th}} \tag{4-49}$$

For a typical value of $g_{\mathrm{th}} = 50\,\mathrm{cm}^{-1}$ and a refractive index in the lasing material of $n = 3.5$, the photon lifetime is approximately $\tau_{\mathrm{ph}} = 2\,\mathrm{ps}$. This value sets the upper limit to the direct modulation capability of the laser diode.

A laser diode can readily be pulse-modulated since the photon lifetime is much smaller than the carrier lifetime. If the laser is completely turned off after each pulse, the spontaneous carrier lifetime will limit the modulation rate. This is because, at the onset of a current pulse of amplitude I_p, a period of time t_d given by (see Prob. 4-19)

$$t_d = \tau \ln \frac{I_p}{I_p + (I_B - I_{\mathrm{th}})} \tag{4-50}$$

is needed to achieve the population inversion necessary to produce a gain that is sufficient to overcome the optical losses in the lasing cavity. In Eq. (4-50) the parameter I_B is the bias current and τ is the average lifetime of the carriers in the combination region when the total current $I = I_p + I_B$ is close to I_{th}. From Eq. (4-50), it is clear that the delay time can be eliminated by dc-biasing the diode at the lasing threshold current. Pulse modulation is then carried out by modulating the laser only in the operating region above threshold. In this region, the carrier lifetime is now shortened to the stimulated emission lifetime, so that high modulation rates are possible.

When using a directly modulated laser diode for high-speed transmission systems, the modulation frequency can be no larger than the frequency of the relaxation oscillations of the laser field. The relaxation oscillation depends on both the spontaneous lifetime and the photon lifetime. Theoretically, assuming a linear dependence of the optical gain on carrier density, the relaxation oscillation occurs approximately at[1]

$$f = \frac{1}{2\pi} \frac{1}{(\tau_{\mathrm{sp}}\tau_{\mathrm{ph}})^{1/2}} \left(\frac{I}{I_{\mathrm{th}}} - 1\right)^{1/2} \tag{4-51}$$

Since τ_{sp} is about 1 ns and τ_{ph} is on the order of 2 ps for a 300-μm-long laser, then when the injection current is about twice the threshold current, the maximum modulation frequency is a few gigahertz. An example of a laser that has a relaxation-oscillation peak at 3 GHz is shown in Fig. 4-30.

Analog modulation of laser diodes is carried out by making the drive current above threshold proportional to the baseband information signal. A requirement for this modulation scheme is that a linear relation exist between the light output and the current input. However, signal degradation resulting from non-linearities that are a consequence of the transient response characteristics of laser

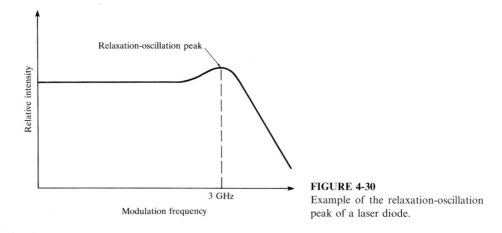

FIGURE 4-30
Example of the relaxation-oscillation
peak of a laser diode.

diodes makes the implementation of analog intensity modulation susceptible to
both intermodulation and cross-modulation effects. This can be alleviated by
using pulse code modulation or through special distortion compensation techniques, as is discussed in more detail in Sec. 4.4.

4.3.8 Temperature Effects

An important factor to consider in the application of laser diodes is the temperature dependence of the threshold current $I_{th}(T)$. This parameter increases with
temperature in all types of semiconductor lasers because of various complex
temperature-dependent factors.[47-49] The complexity of these factors prevents
the formulation of a single equation that holds for all devices and temperature
ranges. However, the temperature variation of I_{th} can be approximated by the
empirical expression

$$I_{th}(T) = I_z \, e^{T/T_0} \tag{4-52}$$

where T_0 is a measure of the relative temperature insensitivity and I_z is a constant.
For a conventional stripe-geometry GaAlAs laser diode, T_0 is typically 120–165°C
in the vicinity of room temperature. An example of a laser diode with $T_0 = 135$°C
and $I_z = 52$ A is shown in Fig. 4-31. The variation in I_{th} with temperature is 0.8
percent/°C, as is shown in Fig. 4-32. Smaller dependences of I_{th} on temperature
have been demonstrated for GaAlAs quantum-well heterostructure lasers. For
these lasers, T_0 can be as high as 437°C. The temperature dependence of I_{th} for
this device is also shown in Fig. 4-32. The threshold variation for this particular
laser type is 0.23 percent/°C. Experimental values[47] of T_0 for 1300-nm InGaAsP
lasers are typically 60–80 K (333–353°C).

For the laser diode shown in Fig. 4-31, the threshold current increase by a
factor of about 1.4 between 20 and 60°C. In addition, the lasing threshold can
change as the laser ages. Consequently, if a constant optical output power level is
to be maintained as the temperature of the laser changes or as the laser ages, it is

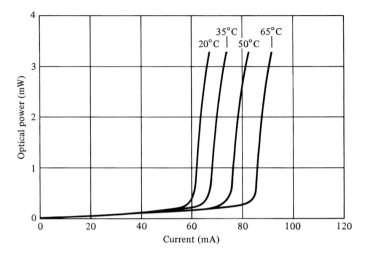

FIGURE 4-31
Temperature-dependent behavior of the optical output power as a function of the bias current for a particular laser diode.

necessary to adjust the dc-bias current level. Possible methods for achieving this automatically are optical feedback and feedforward schemes,[50-53] temperature-matching transistors,[54] and predistortion techniques.[55]

Optical feedback can be carried out by using a photodetector either to sense the variation in optical power emitted from the rear facet of the laser or to tap off and monitor a small portion of the fiber-coupled power emitted from the front facet. The photodetector compares the optical power output with a reference level

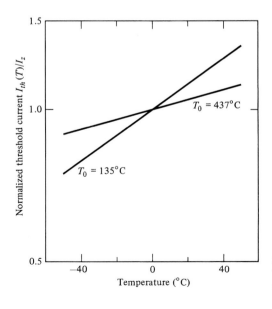

FIGURE 4-32
Variation with temperature of the threshold current I_{th} for two types of laser diodes.

and adjusts the dc-bias current level automatically to maintain a constant peak light output relative to the reference. The photodetector used must have a stable long-term responsivity which remains constant over a wide temperature range. For operation in the 800-to-900-nm region, a silicon *pin* photodiode generally exhibits these characteristics (see Chap. 6).

An example of a feedback-stabilizing circuit[50] that can be used for a digital transmitter is shown in Fig. 4-33. In this scheme, the light emerging from the rear facet of the laser is monitored by a *pin* photodiode. With this circuit, the electric input signal pattern is compared with the optical output level of the laser diode. This effectively prevents the feedback circuit from erroneously raising the bias current level during long sequences of digital zeros or during a period in which there is no input signal on the channel. In this circuit, the dc reference through resistor R_1 sets the bias current at the proper operating point during long sequences of zeros. When this bias current is added to the laser drive current, the desired peak output power from the laser is obtained. The resistor R_2 balances the signal reference current against the *pin* photocurrent for a 50-percent duty ratio at 25°C. As the lasing threshold changes because of aging or temperature variations, the bias current I_B is automatically adjusted to maintain a balance between the data reference and the *pin* photocurrent. A more sophisticated version of this circuit[51] simultaneously and independently controls both the bias current and the modulation current.

Another standard method of stabilizing the optical output of a laser diode is to use a miniature thermoelectric cooler. This device maintains the laser at a constant temperature and thus stabilizes the output level. Normally, a thermoelectric cooler is used in conjunction with a rear-facet detector feedback loop, as is shown in Fig. 4-34.

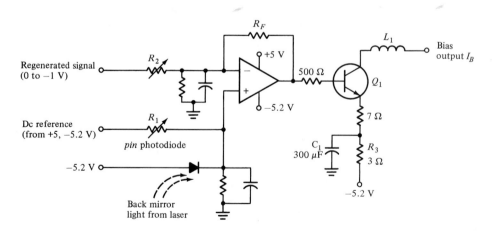

FIGURE 4-33
Example of a bias circuit that provides feedback stabilization of laser output power. (Reproduced with permission from Shumate, Chen, and Dorman,[50] © 1978, The American Telephone and Telegraph Company.)

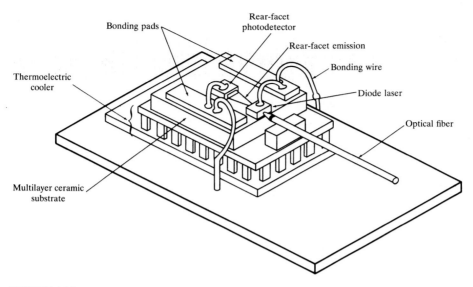

FIGURE 4-34
Construction of a laser diode transmitter using a thermoelectric cooler for temperature stabilization.

4.4 LIGHT SOURCE LINEARITY

High-radiance LEDs and laser diodes are well-suited optical sources for wideband analog applications provided a method is implemented to compensate for any nonlinearities of these devices. In an analog system, the time-varying electric analog signal $s(t)$ is used to modulate directly an optical source about a bias current point I_B, as shown in Fig. 4-35. With no signal input, the optical power output is P_t. When the signal $s(t)$ is applied, the optical output power $P(t)$ is

$$P(t) = P_t[1 + ms(t)] \tag{4-53}$$

Here, m is the *modulation index* (or *modulation depth*) defined as

$$m = \frac{\Delta I}{I'_B} \tag{4-54}$$

where $I'_B = I_B$ for LEDs and $I'_B = I_B - I_{th}$ for laser diodes. The parameter ΔI is the variation in current about the bias point. To prevent distortions in the output signal, the modulation must be confined to the linear region of the curve for optical output versus drive current. Furthermore, if ΔI is greater than I'_B (i.e., m is greater than 100 percent), the lower portion of the signal gets cut off and severe distortion will result. Typical m values for analog applications range from 0.25 to 0.50.

In analog applications, any device nonlinearities will create frequency components in the output signal that were not present in the input signal.[56] Two important nonlinear effects are harmonic and intermodulation distortions. If

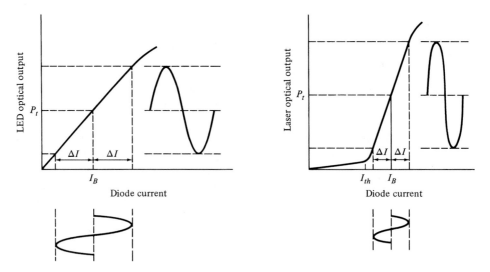

FIGURE 4-35
Bias point and amplitude modulation range for analog applications of LEDs and laser diodes.

the signal input to a nonlinear device is a simple cosine wave $x(t) = A \cos \omega t$, the output will be

$$y(t) = A_0 + A_1 \cos \omega t + A_2 \cos 2\,\omega t + A_3 \cos 3\,\omega t + \cdots \qquad (4\text{-}55)$$

That is, the output signal will consist of a component at the input frequency ω plus spurious components at zero frequency, at the second harmonic frequency 2ω, at the third harmonic frequency 3ω, and so on. This effect is known as *harmonic distortion*. The amount of nth-order distortion in decibels is given by

$$n\text{th-order harmonic distortion} = 20 \log \frac{A_n}{A_1} \qquad (4\text{-}56)$$

To determine *intermodulation distortion*, the modulating signal of a non-linear device is taken to be the sum of two cosine waves $x(t) = A_1 \cos \omega_1 t + A_2 \cos \omega_2 t$. The output signal will then be of the form

$$y(t) = \sum_{m,n} B_{mn} \cos(m\omega_1 + n\omega_2) \qquad (4\text{-}57)$$

where m and $n = 0, \pm 1, \pm 2, \pm 3, \ldots$ This signal includes all the harmonics of ω_1 and ω_2 plus cross-product terms such as $\omega_2 - \omega_1$, $\omega_2 + \omega_1$, $\omega_2 - 2\omega_1$, $\omega_2 + 2\omega_1$, and so on. The sum and difference frequencies give rise to the intermodulation distortion. The sum of the absolute values of the coefficients m and n determines the order of the intermodulation distortion. For example, the second-order inter-modulation products are at $\omega_1 \pm \omega_2$ with amplitude B_{11}, the third-order intermod-ulation products are at $\omega_1 \pm 2\omega_2$ and $2\omega_1 \pm \omega_2$ with amplitudes B_{12} and B_{21}, and so on. (Harmonic distortions are also present wherever either $m \neq 0$ and $n = 0$ or when $m = 0$ and $n \neq 0$. The corresponding amplitudes are B_{m0} and B_{0n}, respectively.) In general, the odd-order intermodulation products having $m = n \pm 1$ (such as $2\omega_1 - \omega_2$, $2\omega_2 - \omega_1$, $3\omega_1 - 2\omega_2$, etc.) are the most troublesome, since they may fall within the bandwidth of the channel. Of these, usually only the third-order terms are important, since the amplitudes of higher-order terms tend to be significantly smaller. If the operating frequency band is less than an octave, all other intermodulation products will fall outside the passband and can be eliminated with appropriate filters in the receiver.

Nonlinear distortions in LEDs are due to effects depending on the carrier injection level, radiative recombination, and other subsidiary mechanisms, as is described in detail by Asatani and Kimura.[57] In certain laser diodes, such as gain-guided devices, there can be nonlinearities in the curve for optical power output versus diode current, as is illustrated in Fig. 4-36. These nonlinearities are a result of inhomogeneities in the active region of the device and also arise from power switching between the dominant lateral modes in the laser. They are generally referred to as "kinks." These kinks are generally not seen in modern laser diodes that use the structures described in Sec. 4.3.5 and 4.3.6. Power saturation (as indicated by a downward curving of the output-versus-current curve) can occur at high output levels because of active-layer heating.

Total harmonic distortions[58-61] in GaAlAs LEDs and laser diodes tend to be in the range of 30–40 dB below the output at the fundamental modulation frequency for modulation depths around 0.5. The second- and third-order harmonic distortions as a function of bias current for several modulation frequencies are shown in Fig. 4-37 for a GaAlAs double-heterojunction LED.[58] The harmonic distortions decrease with increasing bias current but become large at higher modulation frequencies. The intermodulation distortion curves (not shown) follow the same characteristics as those in Fig. 4-37, but are 5–8 dB worse.

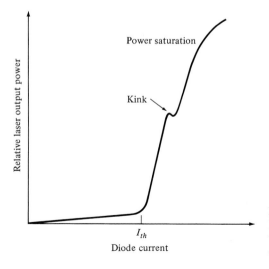

FIGURE 4-36
Example of a kink and power saturation in the curve for optical output power versus drive current of a laser diode.

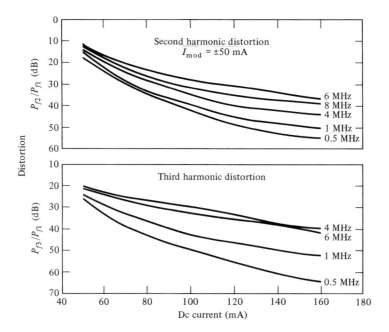

FIGURE 4-37
Second-order and third-order harmonic distortions as a function of bias current in a GaAlAs LED for several modulation frequencies. The distortion is given in terms of the power P_{fn} as the nth harmonic relative to the power P_{f1} at the modulation frequency f_1. (Reproduced with permission from Dawson.[58])

A number of compensation techniques for linearization of optical sources in analog communication systems have been investigated. These methods include circuit techniques such as complementary distortion,[57,61] negative feedback,[62] selective harmonic compensation,[63] and quasi-feedforward compensation,[64] and the use of pulse position modulation (PPM) schemes.[65] One of the most successful circuit design techniques is the quasi-feedforward method, with which a 30- to 40-dB reduction in total harmonic distortion has been achieved.

4.5 MODAL, PARTITION, AND REFLECTION NOISE

Three significant factors associated with the operating characteristics of laser diodes arise in high-speed digital and analog applications. These phenomena are called modal or speckle noise,[66-73] mode-partition noise,[73-77] and reflection noise.[73,78-81] These factors are important because they can introduce receiver output noise, which can be particularly serious for analog systems (see Chaps. 8 and 9).

When light from a coherent laser is launched into a multimode fiber, a number of propagating modes of the fiber are normally excited. As long as these modes retain their relative phase coherence, the radiation pattern seen at the end of the fiber (or at any point along the fiber) takes on the form of a speckle pattern. This is the result of constructive and destructive interference between propagating modes at any given plane. An example of this is shown in Fig. 4-38. The number of speckles in the pattern approximates the number of propagating modes. As the light travels along the fiber, a combination of mode-dependent losses, changes in phase between modes, and fluctuations in the distribution of energy among the various fiber modes will change the modal interference and result in a different speckle pattern. *Modal* or *speckle noise* occurs when any losses that are speckle-pattern-dependent are present in a link. Example of such losses are splices, connectors, microbends, and photodetectors with nonuniform responsivity across the photosensitive area. Noise is generated when the speckle pattern *changes in time* so as to vary the optical power transmitted through the particular loss element. The continually changing speckle pattern that falls on the photodetector thus produces a time-varying noise in the received signal, which degrades receiver performance. Narrowband, high-coherence sources, such as single-mode lasers, result in more modal noise than broadband sources. Incoherent sources, such as LEDs, do not produce modal noise. The use of single-mode fibers eliminates this problem.

Mode-partition noise is associated with intensity fluctuations in the longitudinal modes of a laser diode. This is the dominant noise in single-mode fibers. The output from a laser diode generally can come from more than one longitudinal mode, as shown in Fig. 4-39. The optical output may arise from all of the modes simultaneously, or it may switch from one mode (or group of modes) to another randomly in time. Intensity fluctuations can occur among the various modes in a multimode laser even when the total optical output does not vary as exhibited in Fig. 4-39. Since the output pattern of a laser diode is highly directional, the light from these fluctuating modes can be coupled into a fiber with high coupling efficiency. Each of the longitudinal modes that is coupled into the fiber

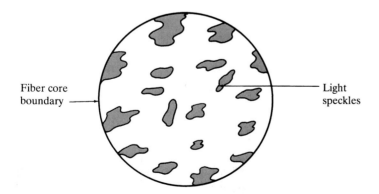

FIGURE 4-38
Example of a speckle pattern that is produced when coherent laser light is launched into a multimode fiber.

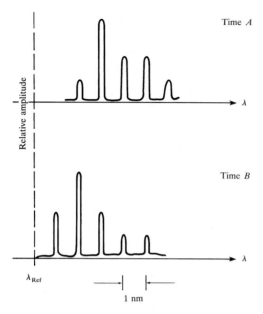

FIGURE 4-39
Time-resolved dynamic spectra of a laser diode. Different modes or groups of modes dominate the optical output at different times. The modes are approximately 1 nm apart.

has a different attenuation and time delay because each is associated with a slightly different wavelength (see Sec. 3.3). Since the power fluctuations among the dominant modes can be quite large, significant variations in signal levels can occur at the receiver in systems with high fiber dispersion.

Reflection noise is associated with laser diode output linearity distortion caused by some of the light output being reflected back into the laser cavity from fiber joints. This reflected power couples with the lasing modes, thereby causing their phases to vary. This produces a periodically modulated noise spectrum that is peaked on the low-frequency side of the intrinsic noise profile. The fundamental frequency of this noise is determined by the roundtrip delay of the light from the laser to the reflecting point and back again. Depending on the roundtrip time, these reflections can create noise peaks in the frequency region where optical fiber data transmission systems operate, even though the lasers themselves are very noise-free at these frequencies. Reflection noise problems can be greatly reduced by using optical isolators between the laser diode and the optical fiber transmission line or by using index-matching fluid in the gaps at fiber-to-fiber joints to eliminate reflections at the fiber–air interfaces.

4.6 RELIABILITY CONSIDERATIONS

The reliability of double-heterojunction LEDs and laser diodes is of importance in that these devices are of greatest interest in optical communications systems. The lifetimes of these sources are affected by both operating conditions and fabrication techniques. Thus, it is important to understand the relationships between light source operation characteristics, degradation mechanisms, and system reliability requirements. A comprehensive review of the reliability of GaAlAs laser diodes has been presented by Ettenberg and Kressel,[82] to which the reader is referred for further details and an extensive list of references. The extension of these results to LEDs is straightforward.[83-87] Reliabilities of InGaAsP LEDs and lasers are given in Refs. 88–91.

Lifetime tests of optical sources are carried out either at room temperature or at elevated temperatures to accelerate the degradation process. A commonly used elevated temperature is 70°C. The two most popular techniques for determining the lifetime of an optical source either maintain constant light output by increasing the bias current automatically or keep the current constant and monitor the optical output level. In the first case, the end of life of the device is assumed to be reached when the source can no longer put out a specified power at the maximum current value for CW (continuous-wave) operation. In the second case, the lifetime is determined by the time taken for the optical output power to decrease by 3 dB.

Degradation of light sources can be divided into three basic categories: internal damage and ohmic contact degradation, which hold for both lasers and LEDs, and damage to the facets of laser diodes.

The limiting factor on LED and laser diode lifetime is internal degradation. This effect arises from the migration of crystal defects into the active region of the light source. These defects decrease the internal quantum efficiency and increase the optical absorption. Fabrication steps that can be taken to minimize internal degradation include the use of substrates with low surface dislocation densities (less than 2×10^3 dislocation/cm^2), keeping work-damaged edges out of the diode current path, and minimizing stresses in the active region (to less than 10^8 dyn/cm^2).

For high-quality sources having lifetimes which follow a slow internal-degradation mode, the optical power P decreases with time according to the exponential relationship

$$P = P_0 \, e^{-t/\tau_m} \tag{4-58}$$

Here, P_0 is the initial optical power at time $t = 0$, and τ_m is a time constant for the degradation process, which is approximately twice the -3-dB mean time to failure. Since the operating lifetime depends on both the current density J and the junction temperature T, internal degradation can be accelerated by increasing either one of these parameters.

The operating lifetime τ_s has been found experimentally to depend on the current density J through the relation

$$\tau_s \propto J^{-n} \tag{4-59}$$

where $1.5 \leq n \leq 2.0$. For example, by doubling the current density, the lifetime decreases by a factor of 3–4. Since the degradation rate of optical sources increases with temperature, an Arrhenius relationship of the form

$$\tau_s = K \, e^{E_A/k_B T} \tag{4-60}$$

has been sought. Here, E_A is an activation energy characterizing the lifetime τ_s, k_B is Boltzmann's constant, T is the absolute temperature at which τ_s was evaluated, and K is a constant. The problem in establishing such an expression is that several competing factors are likely to contribute to the degradation, thereby making it difficult to estimate the activation energy E_A. Activation energies for laser degradation reported in the literature have ranged from 0.3 to 1.0 eV. For practical calculations, a value of 0.7 eV is generally used.

Equations (4-59) and (4-60) indicate that, to increase the light source lifetime, it is advantageous to operate these devices at as low a current and temperature as is practicable. Examples[85] of the luminescent output of InGaAsP LEDs as a function of time for different temperatures are shown in Fig. 4-40. At temperatures below 120°C the output power remains almost constant over the entire measured 15,000-h (1.7-year) operating time. At higher temperatures the power output drops as a function of time. For example, at 230°C the optical power has dropped to one-half its initial value (a 3-dB decrease) after approximately 3000 h (4.1 months) of operation. The activation energy of these lasers is about 1.0 eV.

A second fabrication-related degradation mechanism is ohmic contact deterioration. In LEDs and laser diodes the thermal resistance of the contact between the light source chip and the device heat sink occasionally increases with time. This effect is a function of the solder used to bond the chip to the heat sink, the current density through the contact, and the contact temperature. An increase in the thermal resistance results in a rise in the junction temperature for a fixed operating current. This, in turn, leads to a decrease in the optical output

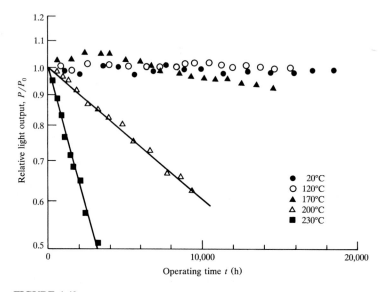

FIGURE 4-40

Normalized output power as a function of operating time for five ambient temperatures. P_0 is the initial optical output power. (Reproduced with permission from Yamakoshi et al.[83])

power. However, careful designs and implementations of high-quality bonding procedures have minimized effects resulting from contact degradation.

Facet damage is a degradation problem that exists for laser diodes. This degradation reduces the laser mirror reflectivity and increases the nonradiative carrier recombination at the laser facets. The two types of facet damage that can occur are generally referred to as *catastrophic facet degradation* and *facet erosion.* Catastrophic facet degradation is mechanical damage of the facets that may arise after short operating times of laser diodes at high optical power densities. This damage tends to reduce greatly the facet reflectivity, thereby increasing the threshold current and decreasing the external quantum efficiency. The catastrophic facet degradation has been observed to be a function of the optical power density and the pulse length.

Facet erosion is a gradual degradation occurring over a longer period of time than catastrophic facet damage. The decrease in mirror reflectivity and the increase in nonradiative recombination at the facets owing to facet erosion lower the internal quantum efficiency of the laser and increase the threshold current. In GaAlAs lasers, facet erosion arises from oxidation of the mirror surface. It is speculated that the oxidation process is stimulated by the optical radiation emitted from the laser. Facet erosion is minimized by depositing a half-wavelength-thick Al_2O_3 film on the facet. This type of coating acts as a moisture barrier and does not affect the mirror reflectivity or the lasing threshold current.

A comparison[82] of two definitions of failure for laser diodes operating at 70°C is shown in Fig. 4-41. The lower trace shows the time required for the laser output to drop to one-half its initial value when a constant current passes through the device. This is the "3-dB life."

The "end-of-life" failure is given by the top trace in Fig. 4-41. This condition is defined as the time at which the device can no longer emit a fixed power level (1.25 mW in this case) at the 70°C heat-sink temperature. The mean operating times (time for 50 percent of the lasers to fail) are 3800 h and 1900 h for the end-of-life and 3-dB-life conditions, respectively. The right-hand ordinate of Fig. 4-41 gives an estimate of the operating time at 22°C, assuming an activation energy of 0.7 eV.

PROBLEMS

4-1. Measurements show that the band-gap energy E_g for GaAs varies with temperature according to the empirical formula

$$E_g(T) \approx 1.55 - 4.3 \times 10^{-4}T$$

where E_g is given in electron volts (eV).

(a) Using this expression, show that the temperature dependence of the intrinsic electron concentration n_i is

$$n_i = 5 \times 10^{15} T^{3/2} e^{-8991/T}$$

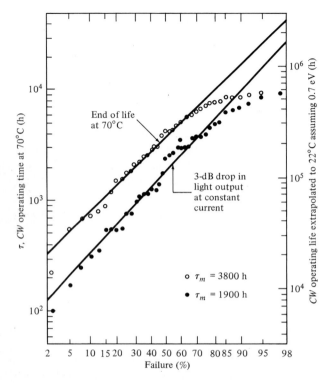

FIGURE 4-41
Time-to-failure plot on log-nor-mal coordinates for 40 low-threshold (≈ 50 mA) oxide-defined stripe lasers at a 70°C heat-sink temperature. τ_m is the time it took for 50 percent of the lasers to fail for the two types of failure mechanisms. (Reproduced with permission from Ettenberg and Kressel,[82] © 1980, IEEE.)

(b) Using a computer, plot the values of n_i as a function of temperature over the range $273\ \text{K} \le T \le 373\ \text{K}(0°\text{C} \le T \le 100°\text{C})$.

4-2. Repeat the steps given in Example 4-2 for a p-type semiconductor. In particular, show that when the net acceptor concentration is much greater than n_i, we have $p_p = N_A$ and $n_p = n_i^2/N_A$.

4-3. An engineer has two $\text{Ga}_{1-x}\text{Al}_x\text{As}$ LEDs: one has a band-gap energy of 1.540 eV and the other has $x = 0.015$.

(a) Find the aluminum mole fraction x and the emission wavelength for the first LED.

(b) Find the band-gap energy and the emission wavelength of the other LED.

4-4. The lattice spacing of $\text{In}_{1-x}\text{Ga}_x\text{As}_y\text{P}_{1-y}$ has been shown to obey Vegard's law.[92] This states that for quaternary alloys of the form $\text{A}_{1-x}\text{B}_x\text{C}_y\text{D}_{1-y}$, where A and B are group III elements (e.g., Al, In, and Ga) and C and D are group V elements (e.g., As, P, and Sb), the lattice spacing $a(x, y)$ of the quaternary alloy can be approximated by

$$a(x, y) = xya(\text{BC}) + x(1 - y)a(\text{BD}) + (1 - x)ya(\text{AC})$$
$$+ (1 - x)(1 - y)a(\text{AD})$$

where the $a(\text{IJ})$ are the lattice spacings of the binary compounds IJ.

(a) Show that for $\text{In}_{1-x}\text{Ga}_x\text{As}_y\text{P}_{1-y}$ with

$$a(\text{GaAs}) = 5.6536 \,\text{Å}$$

$$a(\text{GaP}) = 5.4512 \,\text{Å}$$

$$a(\text{InAs}) = 6.0590 \,\text{Å}$$

$$a(\text{InP}) = 5.8696 \,\text{Å}$$

the quaternary lattice spacing becomes

$$a(x, y) = 0.1894y - 0.4184x + 0.0130xy + 5.8696 \,\text{Å}$$

(b) For quaternary alloys that are lattice-matched to InP, the relation between x and y can be determined by letting $a(x, y) = a(\text{InP})$. Show that since $0 \leq x \leq 0.47$, the resulting expression can be approximated by $y \simeq 2.20x$.

(c) A simple empirical relation that gives the band-gap energy in terms of x and y is[92]

$$E_g(x, y) = 1.35 + 0.668x - 1.17y + 0.758x^2 + 0.18y^2$$

$$- 0.069xy - 0.322x^2y + 0.33xy^2 \,\text{eV}$$

Find the band-gap energy and the peak emission wavelength of $\text{In}_{0.74}\text{Ga}_{0.26}\text{As}_{0.56}\text{P}_{0.44}$.

4-5. Using the expression $E = hc/\lambda$, show why the FWHM power spectral width of LEDs becomes wider at longer wavelengths.

4-6. A double-heterojunction InGaAsP LED emitting at a peak wavelength of 1310 nm has radiative and nonradiative recombination times of 25 and 90 ns, respectively. The drive current is 35 mA.

(a) Find the internal quantum efficiency and the internal power level.

(b) If the refractive index of the light source material is $n = 3.5$, find the power emitted from the device.

4-7. Assume the injected minority carrier lifetime of an LED is 5 ns and that the device has an optical output of 0.30 mW when a constant dc drive current is applied. Plot the optical output power when the LED is modulated at frequencies ranging from 20 to 100 MHz. Note what happens to the LED output power at higher modulation frequencies.

4-8. Consider an LED having a minority carrier lifetime of 5 ns. Find the 3-dB optical bandwidth and the 3-dB electrical bandwidth.

4-9. (a) A GaAlAs laser diode has a 500-μm cavity length which has an effective absorption coefficient of $10 \,\text{cm}^{-1}$. For uncoated facets the reflectivities are 0.32 at each end. What is the optical gain at the lasing threshold?

(b) If one end of the laser is coated with a dielectric reflector so that its reflectivity is now 90 percent, what is the optical gain at the lasing threshold?

(c) If the internal quantum efficiency is 0.65, what is the external quantum efficiency in cases (a) and (b)?

4-10. Find the external quantum efficiency for a $\text{Ga}_{1-x}\text{Al}_x\text{As}$ laser diode (with $x = 0.03$) which has an optical-power-versus-drive-current relationship of 0.5 mW/mA (e.g., as shown in Fig. 4-31).

4-11. Approximate expressions for the transverse and lateral optical-field confinement factors Γ_T and Γ_L, respectively, in a Fabry-Perot lasing cavity are

$$\Gamma_T = \frac{D^2}{2 + D^2} \quad \text{with} \quad D = \frac{2\pi d}{\lambda}\left(n_1^2 - n_2^2\right)^{1/2}$$

and

$$\Gamma_L = \frac{W^2}{2 + W^2} \quad \text{with} \quad W = \frac{2\pi w}{\lambda}\left(n_{\text{eff}}^2 - n_2^2\right)^{1/2}$$

where

$$n_{\text{eff}}^2 = n_2^2 + \Gamma_T\left(n_1^2 - n_2^2\right)$$

Here, w and d are the width and thickness, respectively, of the active layer, and n_1 and n_2 are the refractive indices inside and outside the cavity, respectively.

(a) Consider a 1300-nm InGaAsP laser diode in which the active region is $0.1\,\mu$m thick, $1.0\,\mu$m wide, and $250\,\mu$m long with refractive indices $n_1 = 3.55$ and $n_2 = 3.20$. What are the transverse and lateral optical-field confinement factors?

(b) Given that the total confinement factor is $\Gamma = \Gamma_T\Gamma_L$, what is the gain threshold if the effective absorption coefficient is $\bar{\alpha} = 30\,\text{cm}^{-1}$ and the facet reflectivities are $R_1 = R_2 = 0.31$?

4-12. A GaAs laser emitting at 800 nm has a 400-μm cavity length with a refractive index $n = 3.6$. If the gain g exceeds the total loss α_t throughout the range $750\,\text{nm} < \lambda < 850\,\text{nm}$, how many modes will exist in the laser?

4-13. A laser emitting at $\lambda_0 = 850$ nm has a gain-spectral width of $\sigma = 32$ nm and a peak gain of $g(0) = 50\,\text{cm}^{-1}$. Plot $g(\lambda)$ from Eq. 4-41. If $\alpha_t = 32.2\,\text{cm}^{-1}$, show the region where lasing takes place. If the laser is $400\,\mu$m long and $n = 3.6$, how many modes will be excited in this laser?

4-14. The derivation of Eq. (4-46) assumes that the refractive index n is independent of wavelength.

(a) Show that when n depends on λ, we have

$$\Delta\lambda = \frac{\lambda^2}{2L(n - \lambda dn/d\lambda)}$$

(b) If the group refractive index $(n - \lambda dn/d\lambda)$ is 4.5 for GaAs at 850 nm, what is the mode spacing for a 400-μm-long laser.

4-15. For laser structures that have strong carrier confinement, the threshold current density for stimulated emission J_{th} can to a good approximation be related to the lasing-threshold optical gain g_{th} by $g_{\text{th}} = \beta J_{\text{th}}$, where β is a constant that depends on the specific device construction. Consider a GaAs laser with an optical cavity of length $250\,\mu$m and width $100\,\mu$m. At the normal operating temperature, the gain factor $\beta = 21 \times 10^{-3}\,\text{A/cm}^3$ and the effective absorption coefficient $\bar{\alpha} = 10\,\text{cm}^{-1}$.

(a) If the refractive index is 3.6, find the threshold current density and the threshold current I_{th}. Assume the laser end faces are uncoated and the current is restricted to the optical cavity.

(b) What is the threshold current if the laser cavity width is reduced to $10\,\mu$m?

4-16. From quantum mechanics, the energy levels for electrons and holes in the quantum-well laser structure shown in Fig. 4-26 are given by

$$E_{ci} = E_c + \frac{h^2}{8d^2} \frac{i^2}{m_e} \qquad \text{with } i = 1, 2, 3, \ldots \text{for electrons}$$

and

$$E_{vi} = E_v - \frac{h^2}{8d^2} \frac{j^2}{m_h} \qquad \text{with } j = 1, 2, 3, \ldots \text{for holes}$$

where E_c and E_v are the conduction- and valence-band energies (see Fig. 4-1), d is the active layer thickness, h is Planck's constant, and m_e and m_h are the electron and hole masses as defined in Example 4-1. The possible energy-level transitions which lead to photon emission are given by

$$\Delta E_{ij} = E_{ci} - E_{vj} = E_g + \frac{h^2}{8d^2} \left(\frac{i^2}{m_e} + \frac{j^2}{m_v} \right)$$

If $E_g = 1.43$ eV for GaAs, what is the emission wavelength between the $i = j = 1$ states if the active layer thickness is $d = 5$ nm?

4-17. In a multiple quantum-well laser the temperature dependence of the differential or external quantum efficiency can be described by[48]

$$\eta_{ext}(T) = \eta_i(T) \frac{\alpha_{end}}{N_w [\alpha_w + \gamma(T - T_{th})] + \alpha_{end}}$$

where $\eta_i(T)$ is the internal quantum efficiency, α_{end} is the mirror loss of the lasing cavity as given in Eq. (4-28), N_w is the number of quantum wells, T_{th} is the threshold temperature, α_w is the internal loss of the wells at $T = T_{th}$, and γ is a temperature-dependent internal-loss parameter. Consider a six-well, 350-μm-long MQW laser having the following characteristics: $\alpha_w = 1.25 \, \text{cm}^{-1}$, $\gamma = 0.025 \, \text{cm}^{-1}/\text{K}$, and $T_{th} = 303$ K. The lasing cavity has a standard uncoated facet on the front ($R_1 = 0.31$) and a high-reflection coating on the near facet ($R_2 = 0.96$).

(a) Assuming that the internal quantum efficiency is constant, use a computer to plot the external quantum efficiency as a function of temperature over the range $303 \, \text{K} \leq T \leq 375 \, \text{K}$. Let $\eta_{ext}(T) = 0.8$ at $T = 303$ K.

(b) Given that the optical output power at $T = 303$ K is 30 mW at a drive current of $I_d = 50$ mA, plot the power output as a function of temperature over the range $303 \, \text{K} \leq T \leq 375 \, \text{K}$ at this fixed bias current.

4-18. A distributed-feedback laser has a Bragg wavelength of 1570 nm, a second-order grating with $\Lambda = 460$ nm, and a 300-μm cavity length. Assuming a perfectly symmetrical DFB laser, find the zeroth-, first-, and second-order lasing wavelengths to a tenth of a nanometer. Draw a relative amplitude-versus-wavelength plot.

4-19. When a current pulse is applied to a laser diode, the injected carrier pair density n within the recombination region of width d changes with time according to the relationship

$$\frac{\partial n}{\partial t} = \frac{J}{qd} - \frac{n}{\tau}$$

(a) Assume τ is the average carrier lifetime in the recombination region when the injected carrier pair density is n_{th} near the threshold current density J_{th}. That is, in the steady state we have $\partial n / \partial t = 0$, so that

$$n_{th} = \frac{J_{th}\tau}{qd}$$

If a current pulse of amplitude I_p is applied to an unbiased laser diode, show that the time needed for the onset of stimulated emission is

$$t_d = \tau \ln \frac{I_p}{I_p - I_{th}}$$

Assume the drive current $I = JA$, where J is the current density and A is the area of the active region.

(b) If the laser is now prebiased to a current density $J_B = I_B/A$, so that the initial excess carrier pair density is $n_B = J_B\tau/qd$, then the current density in the active region during a current pulse I_p is $J = J_B + J_p$. Show that in this case Eq. (4-50) results.

4-20. Assume we have an LED that operates with a 5-V bias supply and which we want to drive with a 50-mA peak current. Design the following simple drive circuits:

(a) A common-emitter transistor configuration.

(b) A low-speed driver using a transistor–transistor logic (TTL) gate (e.g., a 7437 buffer).

(c) A high-speed emitter-coupled driver using a commercial emitter-coupled logic (ECL) device (e.g., a 10210 circuit).

4-21. When designing a driver for a laser diode, one must take into account the temperature dependence of the threshold current as noted in Sec. 4.3.8. Design a laser diode transmitter that has a low-speed rear-facet-detection bias-stabilization circuit and a high-speed drive circuit. How would a thermoelectric cooler be incorporated into this design?

4-22. A laser diode has a maximum average output of 1 mW (0 dBm). The laser is to be amplitude-modulated with a signal $x(t)$ that has a dc component of 0.2 and a periodic component of ± 2.56. If the current-input to optical-output relationship is $P(t) = i(t)/10$, find the values of I_0 and m if the modulating current is $i(t) = I_0[1 + mx(t)]$.

4-23. Consider the following Taylor series expansion of the optical-power-versus-drive-current relationship of an optical source about a given bias point:

$$y(t) = a_1 x(t) + a_2 x^2(t) + a_3 x^3(t) + a_4 x^4(t)$$

Let the modulating signal $x(t)$ be the sum of two sinusoidal tones at frequencies ω_1 and ω_2 given by

$$x(t) = b_1 \cos \omega_1 t + b_2 \cos \omega_2 t$$

(a) Find the second-, third-, and fourth-order intermodulation distortion coefficients B_{mn} (where m and $n = \pm 1, \pm 2, \pm 3$, and ± 4) in terms of b_1, b_2, and the a_i.

(b) Find the second-, third-, and fourth-order harmonic distortion coefficients A_2, A_3, and A_4 in terms of b_1, b_2, and the a_i.

4-24. An optical source is selected from a batch characterized as having lifetimes that follow a slow internal degradation mode. The -3-dB mean time to failure of these devices at room temperature is specified as 5×10^4 h. If the device emits 1 mW at room temperature, what is the expected optical output power after 1 month of operation, after 1 year, and after 5 years?

4-25. A group of optical sources is found to have operating lifetimes of 4×10^4 h at 60°C and 6500 h at 90°C., What is the expected lifetime at 20°C if the device lifetime follows an Arrhenius-type relationship?

REFERENCES

1. H. Kressel and J. K. Butler, *Semiconductor Lasers and Heterojunction LEDs*, Academic, New York, 1977.
2. I. Fukuda, *Optical Semiconductor Devices*, Wiley, New York, 1999.
3. L. A. Coldren and S. W. Corzine, *Diode Lasers and Photonic Integrated Circuits*, Wiley, New York, 1995.
4. N. G. Einspruch and W. R. Frensley, *Heterostructures and Quantum Devices*, Academic, New York, 1994.
5. J. W. Goodman, *Physics of Optoelectronic Devices*, Wiley, New York, 1995.
6. P. Vasil'ev, *Ultrafast Diode Lasers: Fundamentals and Applications*, Artech House, Boston, 1995.
7. (*a*) T. P. Lee, C. A. Burrus, Jr., and R. H. Saul, "Light emitting diodes for telecommunications," in S. E. Miller and I. P. Kaminow, eds., *Optical Fiber Telecommunications—II*, Academic, New York, 1988.
   (*b*) J. E. Bowers and M. A. Pollack, "Semiconductor lasers for telecommunications," in S. E. Miller and I. P. Kaminow, eds., *Optical Fiber Telecommunications—II*, Academic, New York, 1988.
8. (*a*) D. Botez, "Laser diodes are power-packed," *IEEE Spectrum*, vol. 22, pp. 43–53, June 1985.
   (*b*) D. Botez, "Recent developments in high-power InGaAsP lasers," *Laser Focus*, vol. 23, pp. 69–79, Mar. 1987.
9. M. Ohtsu, "Tutorial review: Frequency stabilization in semiconductor lasers," *Opt. Quantum Electron.*, vol. 20, pp. 283–300, July 1988.
10. T. P. Lee, "Recent advances in long-wavelength semiconductor lasers for optical fiber communication," *Proc. IEEE*, vol. 79, pp. 253–276, Mar. 1991.
11. (*a*) J. Hecht, "Diode-laser performance rises as structures shrink," *Laser Focus World*, vol. 28, pp. 127–143, May 1992.
    (*b*) J. Hecht, "Long-wavelength diode lasers are tailored for fiber optics," *Laser Focus World*, vol. 28, pp. 79–89, Aug. 1992.
12. (*a*) Special Issues on "Semiconductor Lasers," *IEEE J. Sel. Topics Quantum Electron.*, vol. 1, June 1995; vol. 3, Apr. 1997.
    (*b*) Special Issue on "Semiconductor Optoelectronics," *IEE Proc.—Optoelectron.*, vol. 145, Feb. 1998.
13. G. J. Pendock and D. D. Sampson, "Transmission performance of high bit rate spectrum-sliced WDM systems," *J. Lightwave Tech.*, vol. 14, pp. 2141–2148, Oct. 1996.
14. V. Arya and I. Jacobs, "Optical preamplifier receiver for spectrum-sliced WDM," *J. Lightwave Tech.*, vol. 15, pp. 576–583, Apr. 1997.
15. R. D. Feldman, "Crosstalk and loss in WDM systems employing spectral slicing," *J. Lightwave Tech.*, vol. 15, pp. 1823–1831, Nov. 1997.
16. (*a*) E. Yang, *Microelectronic Devices*, McGraw-Hill, New York, 1989.
    (*b*) M. Zambuto, *Semiconductor Devices*, McGraw-Hill, New York, 1989.
17. D. A. Neamen, *Semiconductor Physics and Devices: Basic Principles*, McGraw-Hill, New York, 2nd ed., 1997.
18. (*a*) S. M. Sze, *Physics of Semiconductor Devices*, Wiley, New York, 1981.
    (*b*) S. M. Sze, *Modern Semiconductor Device Physics*, Wiley, New York, 1988.
19. J. L. Vossen, and W. Kern, *Thin Film Processes II*, Academic, New York, 1991.
20. K. Iga, *Fundamentals of Laser Optics*, Plenum, New York, 1994.
21. (*a*) C. A. Burrus and B. I. Miller, "Small-area double heterostructure AlGaAs electroluminescent diode sources for optical fiber transmission lines," *Opt. Commun.*, vol. 4, pp. 307–309, Dec. 1971.

(*b*) J. P. Wittke, M. Ettenberg, and H. Kressel, "High radiance LED for single fiber optical links," *RCA Rev.*, vol. 37, pp. 159–183, June 1976.

22. A. W. Snyder and J. D. Love, *Optical Waveguide Theory*, Chapman & Hall, New York, 1983, p. 674.

23. T. P. Lee and A. J. Dentai, "Power and modulation bandwidth of GaAs–AlGaAs high radiance LEDs for optical communication systems," *IEEE J. Quantum Electron.*, vol. QE-14, pp. 150–159, Mar. 1978.

24. H. Namizaki, M. Nagano, and S. Nakahara, "Frequency response of GaAlAs light emitting diodes," *IEEE Trans. Electron. Devices*, vol. ED-21, pp. 688–691, 1974.

25. Y. S. Lin and D. A. Smith, "The frequency response of an amplitude modulated GaAs luminescent diode," *Proc. IEEE*, vol. 63, pp. 542–544, Mar. 1975.

26. T. P. Lee, "Effects of junction capacitance on the rise time of LEDs and the turn-on delay of injection lasers," *Bell Sys. Tech. J.*, vol. 54, pp. 53–68, Jan. 1975.

27. I. Hino and K. Iwamoto, "LED pulse response analysis," *IEEE Trans. Electron. Devices*, vol. ED-26, pp. 1238–1242, Aug. 1979.

28. G. Morthier and P. Vankwikelberge, *Handbook of Distributed Feedback Lasers*, Artech House, Boston, 1997.

29. K. Kobayashi and I. Mito, "Single frequency and tunable laser diodes," *J. Lightwave Tech.*, vol. 6, pp. 1623–1633, Nov. 1988.

30. H. Kogelnik and C. V. Shank, "Coupled-wave theory of distributed feedback lasers," *J. Appl. Phys.*, vol.43, pp. 2327–2335, May 1972.

31. H. Ghafouri-Shiraz and B. S. K. Lo, *Distributed Feedback Laser Diodes: Principles and Physical Modeling*, Wiley, New York, 1995.

32. M. Yamada, "Transverse and longitudinal mode control in semiconductor injection lasers," *IEEE J. Quantum Electron.*, vol. 19, pp. 1365–1380, Sept. 1983.

33. W. T. Tsang, "Quantum confinement with heterostructure semiconductor lasers," in R. K. Willardson and A. C. Beer, eds. (R. Dingle vol. ed.), *Semiconductors and Semimetals*, Academic, New York, 1987, chap. 4.

34. P. S. Zory, Jr., ed., *Quantum Well Lasers*, Academic, New York, 1993.

35. C.-Y. Tsai, F.-P. Shih, T.-L. Sung, T.-Y. Wu, C.-H. Chen, and C.-Y. Tsai, "A small-signal analysis of the modulation response of high-speed quantum-well lasers," *IEEE J. Quantum Electron.*, vol. 33, pp. 2084–2096, Nov. 1997.

36. T. P. Lee, C. A. Burrus, R. A. Linke, and R. J. Nelson, "Short-cavity, single-frequency InGaAsP buried heterostructure lasers," *Electron. Lett.*, vol. 19, pp. 82–84, Feb. 1983.

37. S. Kinoshita and K. Iga, "Circular buried heterostructure GaAlAs/GaAs surface-emitting lasers," *IEEE J. Quantum Electron.*, vol. 23, pp. 882–888, June 1987.

38. T. E. Sale, *Vertical Cavity Surface Emitting Laser*, Wiley, New York, 1995.

39. G. R. Hadley, K. L. Lear, M. E. Warren, K. D. Choquette, J. W. Scott, and S. W. Corzine, "Comprehensive numerical modeling of vertical-cavity surface-emitting lasers," *IEEE J. Quantum Electron.*, vol. 32, pp. 607–615, Apr. 1996.

40. N. M, Margalit, S. Z. Zhang, and J. E. Bowers, "Vertical cavity lasers for telecom applications," *IEEE Commun. Mag.*, vol. 35, pp. 164–170, May 1997.

41. W. W. Chow, K. D. Choquette, M. H. Crawford, K. L. Lear, and G. R. Hadley, "Design, fabrication, and performance of infrared and visible vertical-cavity surface-emitting lasers," *IEEE J. Quantum Electron.*, vol. 33, pp. 1810–1824, Oct. 1997.

42. S. Akiba, M. Usami, and K. Utaka, "1.5-μm $\lambda/4$ shifted InGaAsP DFB lasers," *J. Lightwave Tech.*, vol. 5, pp. 1564–1573, Nov. 1987.

43. J. E. A. Whiteaway, G. H. B. Thompson, A. J. Collar, and C. J. Armistead, "The design and assessment of $\lambda/4$ phase-shifted DFB laser structures," *IEEE J. Quantum Electron.*, vol. 25, pp. 1261–1279, June 1989.

44. Y. Suematsu, S. Arai, and K. Kinoshita, "Dynamic single-mode semiconductor lasers with a distributed reflector," *J. Lightwave Tech.*, vol. 1, pp. 161–176, Mar. 1983.

45. G. M. Smith, J. S. Hughes, R. M. Lammert, M. L. Osowski, G. C. Papen, J. T. Verdeyen, and J. J. Coleman, "Very narrow linewidth asymmetric cladding InGaAs–GaAs ridge waveguide distributed Bragg reflector lasers," *IEEE Photonics Tech. Lett.*, vol. 8, pp. 476–478, Apr. 1996.

46. I. Croston and R. Harley, "Modulators provide key to WDM," *Lightwave*, vol. 14, pp. 50–53, June 1997.

47. G. P. Agrawal and N. K. Dutta, *Semiconductor Lasers*, Van Nostrand Reinhold, New York, 2nd ed., 1993.

48. S. Seki and K. Yokoyama, "Power penalty in 1.3-μm InP-based strained-layer multiple-quantum-well lasers at elevated temperatures," *IEEE Photonics Tech. Lett.*, vol. 9, pp. 1205–1207, Sept. 1997.

49. G. H. B. Thompson, "Temperature dependence of threshold current in GaInAsP DH lasers," *IEE Proc.*, vol. 128, pp. 37–43, Apr. 1981.

50. P. W. Shumate, Jr., F. S. Chen, and P. W. Dorman, "GaAlAs transmitter for lightwave transmission," *Bell Sys. Tech., J.*, vol. 57, pp. 1823–1836, July/Aug. 1978.

51. F. S. Chen, "Simultaneous feedback control of bias and modulation currents for injection lasers," *Electron. Lett.*, vol. 16, pp. 7–8, Jan. 1980.

52. L.-S. Fock, A. Kwan, and R. S. Tucker, "Reduction of semiconductor laser intensity noise by feedforward compensation: Experiment and theory," *J. Lightwave Tech.*, vol. 10, pp. 1919–1925, Dec. 1992.

53. A. V. Naumenko, N. A. Loike, S. I. Turovets, P. S. Spencer, and K. A. Shore, "Bias current impulsive feedback control of nonlinear dynamics in external cavity laser diodes," *Electron. Lett.*, vol. 34, pp. 181–182, Jan. 22, 1998.

54. M. Ettenberg, D. R. Patterson, and E. J. Denlinger, "A temperature-compensated laser module for optical communications," *RCA Rev.*, vol. 40, pp. 103–114, June 1979.

55. H.-T. Lin and Y.-H. Kao, "A predistortion technique for DFB laser diodes in lightwave CATV transmission," *IEICE Trans. Commun.*, vol. E79-B, pp. 1671–1676, Nov. 1996.

56. (a) A. B. Carlson, *Communication Systems*, McGraw-Hill, New York, 3rd ed., 1986.
    (b) T. T. Ha, *Solid-State Microwave Amplifier Design*, Wiley, New York, 1981, chap. 6.
    (c) H. Taub and D. L. Shilling, *Principles of Communication Systems*, McGraw-Hill, New York, 2nd ed., 1986.

57. K. Asatani and T. Kimura, "Analysis of LED nonlinear distortions," *IEEE Trans. Electron. Devices*, vol. 25, pp. 199–207, Feb. 1978; "Linearization of LED nonlinearity by predistortions," *ibid.*, pp. 207–212.

58. R. W. Dawson, "Frequency and bias dependence of video distortion in Burrus-type homostructure and heterostructure LEDs," *IEEE Trans. Electron. Devices*, vol. ED-25, pp. 550–551, May 1978.

59. F. D. King, J. Straus, O. I. Szentesi, and A. J. Springthorpe, "High-radiance long-lived LEDs for analogue signalling," *Proc. IEE*, vol. 123, pp. 619–622, June 1976.

60. T. Ozeki and E. H. Hara, "Measurement of nonlinear distortion in light emitting diodes," *Electron. Lett.*, vol. 12, pp. 78–80, Feb. 1976.

61. K. Asatani, "Nonlinearity and its compensation of semiconductor laser diodes for analog intensity modulation systems," *IEEE Trans. Commun.*, vol. COM-28, pp. 297–300, Feb. 1980.

62. (a) J. Straus, "Linearized transmitters for analog fiber links," *Laser Focus*, vol. 14, pp. 54–61, Oct. 1978.
    (b) M. Ohtsu, M. Murata, and M. Kourogi, "FM noise reduction and subkilohertz linewidth of an AlGaAs laser by negative electrical feedback," *IEEE J. Quantum Electron*, vol. 26, pp. 231–241, Feb. 1990.

63. J. Straus, A. J. Springthorpe, and O. I. Szentesi, "Phase shift modulation technique for the linearization of analog transmitters," *Electron. Lett.*, vol. 13, pp. 149–151, Mar. 1977.

64. (a) R. E. Patterson, J. Straus, G. Blenman, and T. Witkowicz, "Linearization of multichannel analog optical transmitters by quasi-feedforward compensation technique," *IEEE Trans. Commun.*, vol. COM-27, pp. 582–588, Mar. 1979.
    (b) J. Straus and O. I. Szentesi, "Linearization of optical transmitters by a quasi-feedforward compensation technique," *Electron. Lett.*, vol. 13, pp. 158–159, Mar. 1977.

65. D. Kato, "High-quality broadband optical communication by TDM-PAM: Nonlinearity in laser diodes," *IEEE J. Quantum Electron.*, vol. QE-14, pp. 343–346, May 1978.

66. K. O. Hill, Y. Tremblay, and B. S. Kawasaki, "Modal noise in multimode fiber links: Theory and experiment," *Opt. Lett.*, vol. 5, pp. 270–272, June 1980.

67. K. Sato and K. Asatani, "Speckle noise reduction in fiber optic analog video transmission using semiconductor laser diodes," *IEEE Trans. Commun.*, vol. COM-29, pp. 1017–1024, July 1981.

68. A. R. Michelson and A. Weierholt, "Modal-noise limited signal-to-noise ratios in multimode optical fibers," *Appl. Opt.*, vol. 22, pp. 3084–3089, Oct. 1983.

69. K. Petermann, "Nonlinear distortions and noise in optical communication systems due to fiber connectors," *IEEE J. Quantum Electron.*, vol. QE-16, pp. 761–770, July 1980.

70. T. Kanada, "Evaluation of modal noise in multimode fiber-optic systems," *J. Lightwave Tech.*, vol. LT-2, pp. 11–18, Feb. 1984.

71. P. E. Couch and R. E. Epworth, "Reproducible modal-noise measurements in system design and analysis," *J. Lightwave Tech.*, vol. LT-1, pp. 591–595, Dec. 1983.

72. F. M. Sears, I. A. White, R. B. Kummer, and F. T. Stone, "Probability of modal noise in single-mode lightguide systems," *J. Lightwave Tech.*, vol. LT-4, pp. 652–655, June 1986.

73. K. Petermann and G. Arnold, "Noise and distortion characteristics of semiconductor lasers in optical fiber communication systems," *IEEE J. Quantum Electron.*, vol. QE-18, pp. 543–554, Apr. 1982.

74. N. H. Jensen, H. Olesen, and K. E. Stubkjaer, "Partition noise in semiconductor lasers under CW and pulsed operation," *IEEE J. Quantum Electron.*, vol. QE-18, pp. 71–80, Jan. 1987.

75. M. Ohtsu and Y. Teramachi, "Analyses of mode partition and mode hopping in semiconductor lasers," *IEEE J. Quantum Electron.*, vol. 25, pp. 31–38, Jan. 1989.

76. (*a*) C. H. Henry, P. S. Henry, and M. Lax "Partition fluctuations in nearly single longitudinal mode lasers," *J. Lightwave Tech.*, vol. LT-2, pp. 209–216, June 1984.
    (*b*) S. E. Miller, "On the prediction of the mode-partitioning floor in injection lasers with multiple side modes at 2 and 10 Gb/s," *IEEE J. Quantum Electron.*, vol. 26, pp. 242–249, Feb. 1990.

77. E. E. Basch, R. F. Kearns, and T. G. Brown, "The influence of mode partition fluctuations in nearly single-longitudinal-mode lasers on receiver sensitivity," *J. Lightwave Tech.*, vol. LT-4, pp. 516–519, May 1986.

78. O. Hirota and Y. Suematsu, "Noise properties of injection lasers due to reflected waves," *IEEE J. Quantum Electron.*, vol. QE-15, pp. 142–149, Mar. 1979.

79. Y. C. Chen, "Noise characteristics of semiconductor laser diodes coupled to short optical fibers," *Appl. Phys. Lett.*, vol. 37, pp. 587–589, Oct. 1980.

80. G. P. Agrawal, N. A. Olsson, and N. K. Dutta, "Effect of far-end reflections on intensity and phase noise in InGaAsP semiconductor lasers," *Appl. Phys. Lett.*, vol. 45, pp. 597–599, Sept. 1984.

81. W. I. Way and M. M. Choy, "Optical feedback on linearity performance of 1300 nm DFB and multimode lasers under microwave intensity modulation," *J. Lightwave Tech.*, vol. 6, pp. 100–108, Jan. 1988.

82. M. Ettenberg and H. Kressel, "The reliability of (AlGa)As CW laser diodes," *IEEE J. Quantum Electron.*, vol. QE-16, pp. 186–196, Feb. 1980.

83. S. Yamakoshi, O. Hasegawa, H. Hamaguchi, M. Abe, and T. Yamaoka, "Degradation of high-radiance $Ga_{1-x}Al_xAs$ LEDs," *Appl. Phys. Lett.*, vol. 31, pp. 627–629, Nov. 1977.

84. L. R. Dawson, V. G. Keramidas, and C. L. Zipfel, "Reliable, high-speed LEDs for short-haul optical data links," *Bell Sys. Tech. J.*, vol. 59, pp. 161–168, Feb. 1980.

85. S. L. Chuang, A. Ishibashi, S. Kijima, N. Nakajama, M. Ukita, and S. Taniguchi, "Kinetic model for degradation of light-emitting diodes," *IEEE J. Quantum Electron.*, vol. 33, pp. 970–979, June 1997.

86. N. J. Frigo, K. C. Reichmann, and P. P. Iannone, "Thermal characteristics of light-emitting diodes and their effect on passive optical networks," *IEEE Photonics Tech. Lett.*, vol. 9, pp. 1164–1166, Aug. 1997.

87. A. K. Dutta, K. Ueda, K. Hara, and K. Kobayashi, "High brightness and reliable AlGaInP-based light-emitting diode for POF data links," *IEEE Photonics Tech. Lett.*, vol. 9, pp. 1567–1569, Dec. 1997.

88. S. Yamakoshi, M. Abe, O. Wada, S. Komiya, and T. Sakurai, "Reliability of high-radiance InGaAsP/InP LEDs operating in the 1.2–1.3 μm wavelength," *IEEE J. Quantum Electron.*, vol. QE-17, pp. 167–173, Feb. 1981.

89. A. R. Goodwin, I. G. A. Davis, R. M. Gibb, and R. H. Murphy, "The design and realization of a high reliability semiconductor laser for single-mode fiber-optical communication links," *J. Lightwave Tech.*, vol. 6, pp. 1424–1434, Sept. 1988.

90. M. Fukuda, O. Fujita, and S. Uehara, "Homogeneous degradation of surface emitting type InGaAsP/InP light emitting diodes," *J. Lightwave Tech.*, vol. 6, pp. 1808–1814, Dec. 1988.

91. M. Fukuda, "Lasers and LED reliability update," *J. Lightwave Tech.*, vol. 6, pp. 1488–1495, Oct. 1988.

92. R. E. Nahory, M. A. Pollack, W. D. Johnston, Jr., and R. L. Barns, "Band gap versus composition and demonstration of Vegard's law for InGaAsP lattice matched to InP," *Appl. Phys. Lett.*, vol. 33, pp. 659–661, Oct. 1978.

CHAPTER
5

POWER
LAUNCHING
AND COUPLING

In implementing an optical fiber link, two of the major system questions are how to launch optical power into a particular fiber from some type of luminescent source and how to couple optical power from one fiber into another. Launching optical power from a source into a fiber entails considerations such as the numerical aperture, core size, refractive-index profile, and core–cladding index difference of the fiber, plus the size, radiance, and angular power distribution of the optical source.

A measure of the amount of optical power emitted from a source that can be coupled into a fiber is usually given by the *coupling efficiency η* defined as

$$\eta = \frac{P_F}{P_S}$$

Here, P_F is the power coupled into the fiber and P_S is the power emitted from the light source. The launching or coupling efficiency depends on the type of fiber that is attached to the source and on the coupling process; for example, whether or not lenses or other coupling improvement schemes are used.

In practice, many source suppliers offer devices with a short length of optical fiber (1 m or less) already attached in an optimum power-coupling configuration. This section of fiber is generally referred to as a *flylead* or a *pigtail*. The power-launching problem for these pigtailed sources thus reduces to a simpler one of coupling optical power from one fiber into another. The effects to be considered in this case include fiber misalignments; different core sizes, numerical apertures, and core refractive-index profiles; plus the need for clean and smooth fiber end faces that are perpendicular to the fiber axis.

Care must also be exercised when measuring the coupling efficiency between the fiber flylead and the cabled fiber, since the source can launch a significant amount of optical power into the cladding of the flylead. Although this power may be present at the end of the short flylead, it will not be coupled into the core of the following fiber. A true measurement of the power available from the flylead for coupling into a fiber can be determined only by stripping off the cladding modes before measuring the output optical power.

5.1   SOURCE-TO-FIBER POWER LAUNCHING

A convenient and useful measure of the optical output of a luminescent source is its radiance (or brightness) B at a given diode drive current. *Radiance* is the optical power radiated into a unit solid angle per unit emitting surface area and is generally specified in terms of watts per square centimeter per steradian. Since the optical power that can be coupled into a fiber depends on the radiance (i.e., on the spatial distribution of the optical power), the radiance of an optical source rather than the total output power is the important parameter when considering source-to-fiber coupling efficiencies.

5.1.1   Source Output Pattern

To determine the optical power-accepting capability of a fiber, the spatial radiation pattern of the source must first be known. This pattern can be fairly complex. Consider Fig. 5-1, which shows a spherical coordinate system characterized by R, θ, and ϕ, with the normal to the emitting surface being the polar axis. The radiance may be a function of both θ and ϕ, and can also vary from point to point on the emitting surface. A reasonable assumption for simplicity of analysis is to take the emission to be uniform across the source area.

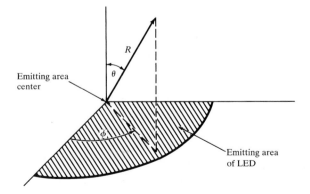

FIGURE 5-1
Spherical coordinate system for characterizing the emission pattern from an optical source.

Surface-emitting LEDs are characterized by their lambertian output pattern, which means the source is equally bright when viewed from any direction. The power delivered at an angle θ, measured relative to a normal to the emitting surface, varies as $\cos\theta$ because the projected area of the emitting surface varies as $\cos\theta$ with viewing direction. The emission pattern for a lambertian source thus follows the relationship

$$B(\theta, \phi) = B_0 \cos\theta \qquad (5\text{-}1)$$

where B_0 is the radiance along the normal to the radiating surface. The radiance pattern for this source is shown in Fig. 5-2.

Edge-emitting LEDs and laser diodes have a more complex emission pattern. These devices have different radiances $B(\theta, 0°)$ and $B(\theta, 90°)$ in the planes parallel and normal, respectively, to the emitting-junction plane of the device. These radiances can be approximated by the general form[1]

$$\frac{1}{B(\theta, \phi)} = \frac{\sin^2\phi}{B_0 \cos^T\theta} + \frac{\cos^2\phi}{B_0 \cos^L\theta} \qquad (5\text{-}2)$$

The integers T and L are the transverse and lateral power distribution coefficients, respectively. In general, for edge emitters, $L = 1$ (which is a lambertian distribution with a 120° half-power beam width) and T is significantly larger. For laser diodes, L can take on values over 100.

Example 5-1. Figure 5-2 compares a lambertian pattern with a laser diode that has a lateral ($\phi = 0°$) half-power beam width of $2\theta = 10°$. In this case, from Eq. (5-2), we have

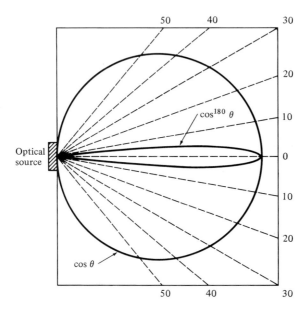

FIGURE 5-2
Radiance patterns for a lambertian source and the lateral output of a highly directional laser diode. Both sources have B_0 normalized to unity.

$$B(\theta = 5°, \phi = 0°) = B_0(\cos 5°)^L = \tfrac{1}{2}B_0$$

Solving for L, we have

$$L = \frac{\log 0.5}{\log(\cos 5°)} = \frac{\log 0.5}{\log 0.9962} = 182$$

The much narrower output beam from a laser diode allows significantly more light to be coupled into an optical fiber.

5.1.2 Power-Coupling Calculation

To calculate the maximum optical power coupled into a fiber, consider first the case shown in Fig. 5-3 for a symmetric source of brightness $B(A_s, \Omega_s)$, where A_s and Ω_s are the area and solid emission angle of the source, respectively. Here, the fiber end face is centered over the emitting surface of the source and is positioned as close to it as possible. The coupled power can be found using the relationship

$$
\begin{aligned}
P &= \int_{A_f} dA_s \int_{\Omega_f} d\Omega_s \, B(A_s, \Omega_s) \\
&= \int_0^{r_m} \int_0^{2\pi} \left[\int_0^{2\pi} \int_0^{\theta_{0,\max}} B(\theta, \phi) \sin\theta \, d\theta \, d\phi \right] d\theta_s \, r \, dr
\end{aligned}
\tag{5-3}
$$

where the area and solid acceptance angle of the fiber define the limits of the integrals. In this expression, first the radiance $B(\theta, \phi)$ from an individual radiating point source on the emitting surface is integrated over the solid acceptance angle of the fiber. This is shown by the expression in square brackets, where $\theta_{0,\max}$ is the maximum acceptance angle of the fiber, which is related to the numerical aperture NA through Eq. (2-23). The total coupled power is then determined by summing up the contributions from each individual emitting-point source of incremental area $d\theta_s \, r \, dr$; that is, integrating over the emitting area. For simplicity, here the

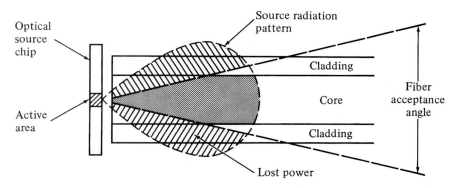

FIGURE 5-3

Schematic diagram of an optical source coupled to an optical fiber. Light outside of the acceptance angle is lost.

emitting surface is taken as being circular. If the source radius r_s is less than the fiber-core radius a, then the upper integration limit $r_m = r_s$; for source areas larger than the fiber-core area, $r_m = a$.

As an example, assume a surface-emitting LED of radius r_s less than the fiber-core radius a. Since this is a lambertian emitter, Eq. (5-1) applies and Eq. (5-3) becomes

$$
P = \int_0^{r_s} \int_0^{2\pi} \left(2\pi B_0 \int_0^{\theta_{0,\max}} \cos\theta \sin\theta \, d\theta \right) d\theta_s \, r \, dr
$$

$$
= \pi B_0 \int_0^{r_s} \int_0^{2\pi} \sin^2\theta_{0,\max} \, d\theta_s \, r \, dr
$$

$$
= \pi B_0 \int_0^{r_s} \int_0^{2\pi} \mathrm{NA}^2 d\theta_s \, r \, dr \tag{5-4}
$$

where the numerical aperture NA is defined by Eq. (2-23). For step-index fibers the numerical aperture is independent of the positions θ_s and r on the fiber end face, so that Eq. (5-4) becomes (for $r_s < a$)

$$
P_{\mathrm{LED,step}} = \pi^2 r_s^2 B_0 (\mathrm{NA})^2 \simeq 2\pi^2 r_s^2 B_0 n_1^2 \Delta \tag{5-5}
$$

Consider now the total optical power P_s that is emitted from the source of area A_s into a hemisphere (2π sr). This is given by

$$
P_s = A_s \int_0^{2\pi} \int_0^{\pi/2} B(\theta, \phi) \sin\theta \, d\theta \, d\phi
$$

$$
= \pi r_s^2 \, 2\pi B_0 \int_0^{\pi/2} \cos\theta \sin\theta \, d\theta
$$

$$
= \pi^2 r_s^2 B_0 \tag{5-6}
$$

Equation (5-5) can, therefore, be expressed in terms of P_s:

$$
P_{\mathrm{LED,step}} = P_s (\mathrm{NA})^2 \qquad \text{for} \qquad r_s \le a \tag{5-7}
$$

When the radius of the emitting area is larger than the radius a of the fiber-core area, Eq. (5-7) becomes

$$
P_{\mathrm{LED,step}} = \left(\frac{a}{r_s} \right)^2 P_s (\mathrm{NA})^2 \qquad \text{for} \qquad r_s > a \tag{5-8}
$$

Example 5-2. Consider an LED that has a circular emitting area of radius $35\,\mu\mathrm{m}$ and a lambertian emission pattern with $150\ \mathrm{W/(cm^2 \cdot sr)}$ axial radiance at a given drive current. Let us compare the optical powers coupled into two step-index fibers, one of which has a core radius of $25\,\mu\mathrm{m}$ with $\mathrm{NA} = 0.20$ and the other which has a core radius of $50\,\mu\mathrm{m}$ with $\mathrm{NA} = 0.20$. For the larger core fiber, we use Eqs. (5-6) and (5-7) to get

$$P_{\text{LED,step}} = P_s(\text{NA})^2 = \pi^2 r_s^2 B_0 (\text{NA})^2$$

$$= \pi^2 (0.0035\,\text{cm})^2 [150\,\text{W}/(\text{cm}^2 \cdot \text{sr})](0.20)^2 = 0.725\,\text{mW}$$

For the case when the fiber end-face area is smaller than the emitting surface area, we use Eq. (5-8). Thus, the coupled power is less than the above case by the ratio of the radii squared:

$$P_{\text{LED,step}} = \left(\frac{25\,\mu\text{m}}{35\,\mu\text{m}}\right)^2 P_s(\text{NA})^2 = \left(\frac{25\,\mu\text{m}}{35\,\mu\text{m}}\right)^2 (0.725\,\text{mW}) = 0.37\,\text{mW}$$

In the case of a graded-index fiber, the numerical aperture depends on the distance r from the fiber axis through the relationship defined by Eq. (2-80). Thus, using Eqs. (2-80a) and (2-80b), the power coupled from a surface-emitting LED into a graded-index fiber becomes (for $r_s < a$)

$$P_{\text{LED,graded}} = 2\pi^2 B_0 \int_0^{r_s} [n^2(r) - n_2^2]r\,dr$$

$$= 2\pi^2 r_s^2 B_0 n_1^2 \Delta \left[1 - \frac{2}{\alpha + 2}\left(\frac{r_s}{a}\right)^\alpha\right]$$

$$= 2P_s n_1^2 \Delta \left[1 - \frac{2}{\alpha + 2}\left(\frac{r_s}{a}\right)^\alpha\right] \tag{5-9}$$

where the last expression was obtained from Eq. (5-6).

For computer-based analyses, a Fourier technique can be used in place of a numerical integration of the above expressions to rapidly calculate the optical power coupled from an LED into a large-core fiber.[2] Furthermore, the foregoing analyses assumed perfect coupling conditions between the source and the fiber. This can be achieved only if the refractive index of the medium separating the source and the fiber end matches the index n_1 of the fiber core. If the refractive index n of this medium is different from n_1, then, for perpendicular fiber end faces, the power coupled into the fiber reduces by the factor

$$R = \left(\frac{n_1 - n}{n_1 + n}\right)^2 \tag{5-10}$$

where R is the *Fresnel reflection* or the *reflectivity* at the fiber-core end face. The ratio $r = (n_1 - n)/(n_1 + n)$, which is known as the *reflection coefficient*, relates the amplitude of the reflected wave to the amplitude of the incident wave.

Example 5-3. A GaAs optical source with a refractive index of 3.6 is coupled to a silica fiber that has a refractive index of 1.48. If the fiber end and the source are in close physical contact, then, from Eq. (5-10), the Fresnel reflection at the interface is

$$R = \left(\frac{n_1 - n}{n_1 + n}\right)^2 = \left(\frac{3.60 - 1.48}{3.60 + 1.48}\right)^2 = 0.174$$

This value of R corresponds to a reflection of 17.4 percent of the emitted optical power back into the source. Given that

$$P_{coupled} = (1 - R)P_{emitted}$$

the power loss L in decibels is found from

$$L = -10 \log\left(\frac{P_{coupled}}{P_{emitted}}\right) = -10 \log(1 - R) = -10 \log(0.826) = 0.83 \, \text{dB}$$

This number can be reduced by having an index-matching material between the source and the fiber end.

The calculation of power coupling for nonlambertian emitters following a cylindrical $\cos^m \theta$ distribution is left as an exercise. The power launched into a fiber from an edge-emitting LED that has a noncylindrical distribution is rather complex. An example of this has been given by Marcuse,[3] to which the reader is referred for details. Section 5.4 presents a simplified analysis of this in the discussion on coupling LEDs to single-mode fibers.

5.1.3   Power Launching versus Wavelength

It is of interest to note that the optical power launched into a fiber does not depend on the wavelength of the source but only on its brightness; that is, its radiance. Let us explore this a little further. We saw in Eq. (2-81) that the number of modes that can propagate in a graded-index fiber of core size a and index profile α is

$$M = \frac{\alpha}{\alpha + 2}\left(\frac{2\pi a n_1}{\lambda}\right)^2 \Delta \tag{5-11}$$

Thus, for example, twice as many modes propagate in a given fiber at 900 nm than at 1300 nm.

The radiated power per mode, P_s/M, from a source at a particular wavelength is given by the radiance multiplied by the square of the nominal source wavelength,[4]

$$\frac{P_s}{M} = B_0 \lambda^2 \tag{5-12}$$

Thus, twice as much power is launched into a given mode at 1300 nm than at 900 nm. Hence, two identically sized sources operating at different wavelengths but having identical radiances will launch equal amounts of optical power into the same fiber.

5.1.4   Equilibrium Numerical Aperture

As we noted earlier, a light source is often supplied with a short (1- to 2-m) fiber flylead attached to it in order to facilitate coupling the source to a system fiber. To achieve a low coupling loss, this flylead should be connected to a system fiber that has a nominally identical NA and core diameter. A certain amount of optical power (ranging from 0.1 to 1 dB) is lost at this junction, the exact loss depending on the connecting mechanism; this is discussed in Sec. 5.3. In addition to the coupling loss, an excess power loss will occur in the first few tens of meters of the system fiber. This excess loss is a result of nonpropagating modes scattering out of the fiber as the launched modes come to an equilibrium condition (see Sec. 3.4). This is of particular importance for surface-emitting LEDs, which tend to launch power into all modes of the fiber. Fiber-coupled lasers are less prone to this effect since they tend to excite fewer nonpropagating fiber modes.

The excess power loss must be analyzed carefully in any system design, since it can be significantly higher for some types of fibers than for others.[5] An example of the excess power loss is shown in Fig. 5-4 in terms of the fiber numerical aperture. At the input end of the fiber, the light acceptance is described in terms of the launch numerical aperture NA_{in}. If the light-emitting area of the LED is less than the cross-sectional area of the fiber core, then, at this point, the power coupled into the fiber is given by Eq. (5-7), where $NA = NA_{in}$.

However, when the optical power is measured in long fiber lengths after the launched modes have come to equilibrium (which is often taken to occur at 50 m), the effect of the equilibrium numerical aperture NA_{eq} becomes apparent. At this point, the optical power in the fiber scales as

$$P_{eq} = P_{50} \left(\frac{NA_{eq}}{NA_{in}} \right)^2 \tag{5-13}$$

where P_{50} is the power expected in the fiber at the 50-m point based on the launch NA. The degree of mode coupling occurring in a fiber is primarily a function of the core–cladding index difference. It can thus vary significantly among different fiber types. Since most optical fibers attain 80–90 percent of their equilibrium NA after about 50 m, it is the value of NA_{eq} that is important when calculating launched optical power in telecommunication systems.

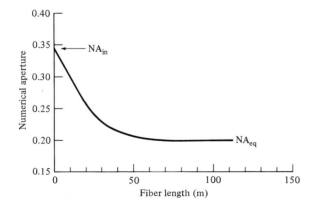

FIGURE 5-4
Example of the change in numerical aperture as a function of fiber length.

5.2 LENSING SCHEMES FOR COUPLING IMPROVEMENT

The optical power-launching analysis given in Sec. 5.1 is based on centering a flat fiber end face directly over the light source as close to it as possible. If the source-emitting area is larger than the fiber-core area, then the resulting optical power coupled into the fiber is the maximum that can be achieved. This is a result of fundamental energy and radiance conservation principles[6] (also known as the *law of brightness*). However, if the emitting area of the source is smaller than the core area, a miniature lens may be placed between the source and the fiber to improve the power-coupling efficiency.

The function of the microlens is to magnify the emitting area of the source to match exactly the core area of the fiber end face. If the emitting area is increased by a magnification factor M, the solid angle within which optical power is coupled to the fiber from the LED is increased by the same factor.

Several possible lensing schemes[1,7-12] are shown in Fig. 5-5. These include a rounded-end fiber, a small glass sphere (nonimaging microsphere) in contact with both the fiber and the source, a larger spherical lens used to image the source on the core area of the fiber end, a cylindrical lens generally formed from a short section of fiber, a system consisting of a spherical-surfaced LED and a spherical-ended fiber, and a taper-ended fiber.

Although these techniques can improve the source-to-fiber coupling efficiency, they also create additional complexities. One problem is that the lens size is similar to the source and fiber-core dimensions, which introduces fabrication and handling difficulties. In the case of the taper-ended fiber, the mechanical alignment must be carried out with greater precision since the coupling efficiency

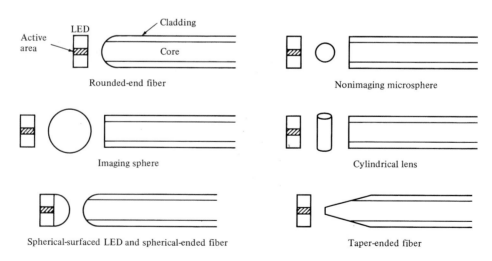

FIGURE 5-5
Examples of possible lensing schemes used to improve optical source-to-fiber coupling efficiency.

becomes a more sharply peaked function of the spatial alignment. However, alignment tolerances are increased for other types of lensing systems.

5.2.1 Nonimaging Microsphere

One of the most efficient lensing methods is the use of a nonimaging microsphere. Let us first examine its use for a surface emitter, as shown in Fig. 5-6. We first make the following practical assumptions: the spherical lens has a refractive index of about 2.0, the outside medium is air ($n = 1.0$), and the emitting area is circular. To collimate the output from the LED, the emitting surface should be located at the focal point of the lens. The focal point can be found from the gaussian lens formula[13]

$$\frac{n}{s} + \frac{n'}{q} = \frac{n' - n}{r} \tag{5-14}$$

where s and q are the object and image distances, respectively, as measured from the lens surface, n is the refractive index of the lens, n' is the refractive index of the outside medium, and r is the radius of curvature of the lens surface.

The following sign conventions are used with Eq. (5-14):

1. Light travels from left to right.
2. Object distances are measured as positive to the left of a vertex and negative to the right.
3. Image distances are measured as positive to the right of a vertex and negative to the left.
4. All convex surfaces encountered by the light have a positive radius of curvature, and concave surfaces have a negative radius.

With the use of these conventions, we shall now find the focal point for the right-hand surface of the lens shown in Fig. 5-6. To find the focal point, we set $q = \infty$ and solve for s in Eq. (5-14), where s is measured from point B. With $n = 2.0$, $n' = 1.0$, $q = \infty$, and $r = -R_L$, Eq. (5-14) yields

$$s = f = 2R_L$$

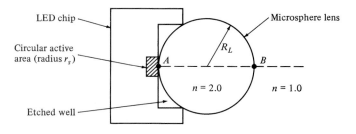

FIGURE 5-6
Schematic diagram of an LED emitter with a microsphere lens.

Thus, the focal point is located on the lens surface at point A. (This, of course, changes if the refractive index of the sphere is not equal to 2.0.)

Placing the LED close to the lens surface thus results in a magnification M of the emitting area. This is given by the ratio of the cross-sectional area of the lens to that of the emitting area:

$$M = \frac{\pi R_L^2}{\pi r_s^2} = \left(\frac{R_L}{r_s}\right)^2 \tag{5-15}$$

Using Eq. (5-4) we can show that, with the lens, the optical power P_L that can be coupled into a full aperture angle 2θ is given by

$$P_L = P_s \left(\frac{R_L}{r_s}\right)^2 \sin^2\theta \tag{5-16}$$

where P_s is the total output power from the LED without the lens.

The theoretical coupling efficiency that can be achieved is based on energy and radiance conservation principles.[14] This efficiency is usually determined by the size of the fiber. For a fiber of radius a and numerical aperture NA, the maximum coupling efficiency η_{max} is given by

$$\eta_{max} = \begin{cases} \left(\frac{a}{r_s}\right)^2 (\text{NA})^2 & \text{for} & \frac{r_s}{a} > \text{NA} \\ 1 & \text{for} & \frac{r_s}{a} \leq \text{NA} \end{cases} \tag{5-17}$$

Thus, when the radius of the emitting area is larger than the fiber radius, no improvement in coupling efficiency is possible with a lens. In this case, the best coupling efficiency is achieved by a direct-butt method.

Based on Eq. (5-17), the theoretical coupling efficiency as a function of the emitting diameter is shown in Fig. 5-7 for a fiber with a numerical aperture of 0.20 and 50-μm core diameter.

5.2.2   Laser Diode-to-Fiber Coupling

As we noted in Chap. 4, edge-emitting laser diodes have an emission pattern that nominally has a full width at half-maximum (FWHM) of 30–50° in the plane perpendicular to the active-area junction and an FWHM of 5–10° in the plane parallel to the junction. Since the angular output distribution of the laser is greater than the fiber acceptance angle, and since the laser emitting area is much smaller than the fiber core, spherical or cylindrical lenses[10,11,15] or optical fiber tapers[16-19] can also be used to improve the coupling efficiency between edge-emitting laser diodes and optical fibers. This also works well for vertical-cavity surface-emitting lasers (VCSELs). Here, coupling efficiencies to multimode fibers of 35 percent result for mass-produced connections of laser arrays to parallel optical fibers, and efficiencies of better than 90 percent are possible by direct (lensless) coupling from a single VCSEL source to a multimode fiber.[20]

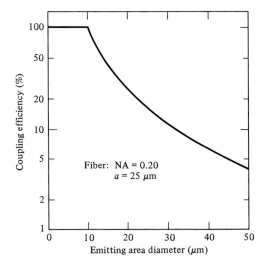

FIGURE 5-7
Theoretical coupling efficiency for a surface-emitting LED as a function of the emitting diameter. Coupling is to a fiber with NA = 0.20 and core radius $a = 25\,\mu$m.

The use of homogeneous glass microsphere lenses has been tested in a series of several hundred laser diode assemblies by Khoe and Kyut.[15] Spherical glass lenses with a refractive index of 1.9 and diameters ranging between 50 and 60 μm were epoxied to the ends of 50-μm core-diameter graded-index fibers having a numerical aperture of 0.2. The measured FWHM values of the laser output beams were as follows:

1. Between 3 and 9 μm for the near-field parallel to the junction.
2. Between 30 and 60° for the field perpendicular to the junction.
3. Between 15 and 55° for the field parallel to the junction.

Coupling efficiencies in these experiments ranged between 50 and 80 percent.

5.3 FIBER-TO-FIBER JOINTS

A significant factor in any fiber optic system installation is the requirement to interconnect fibers in a low-loss manner. These interconnections occur at the optical source, at the photodetector, at intermediate points within a cable where two fibers are joined, and at intermediate points in a link where two cables are connected. The particular technique selected for joining the fibers depends on whether a permanent bond or an easily demountable connection is desired. A permanent bond is generally referred to as a *splice*, whereas a demountable joint is known as a *connector*.

Every joining technique is subject to certain conditions which can cause various amounts of optical power loss at the joint. These losses depend on parameters such as the input power distribution to the joint, the length of the fiber between the optical source and the joint, the geometrical and waveguide characteristics of the two fiber ends at the joint, and the fiber end-face qualities.

The optical power that can be coupled from one fiber to another is limited by the number of modes that can propagate in each fiber. For example, if a fiber in which 500 modes can propagate is connected to a fiber in which only 400 modes can propagate, then, at most, 80 percent of the optical power from the first fiber can be coupled into the second fiber (if we assume that all modes are equally excited). For a graded-index fiber with a core radius a and a cladding index n_2, and with $k = 2\pi/\lambda$, the total number of modes can be found from the expression (the derivation of this is complex)[6]

$$M = k^2 \int_0^a [n^2(r) - n_2^2] r \, dr \qquad (5\text{-}18)$$

where $n(r)$ defines the variation in the refractive-index profile of the core. This can be related to a general local numerical aperture $\text{NA}(r)$ through Eq. (2-80) to yield

$$M = k^2 \int_0^a \text{NA}^2(r) r \, dr$$

$$= k^2 \text{NA}^2(0) \int_0^a \left[1 - \left(\frac{r}{a}\right)^\alpha\right] r \, dr \qquad (5\text{-}19)$$

In general, any two fibers that are to be joined will have varying degrees of differences in their radii a, axial numerical apertures $\text{NA}(0)$, and index profiles α. Thus, the fraction of energy coupled from one fiber to another is proportional to the common mode volume M_{comm} (if a uniform distribution of energy over the modes is assumed). The fiber-to-fiber coupling efficiency η_F is given by

$$\eta_F = \frac{M_{\text{comm}}}{M_E} \qquad (5\text{-}20)$$

where M_E is the number of modes in the *emitting fiber* (the fiber which launches power into the next fiber).

The fiber-to-fiber coupling loss L_F is given in terms of η_F as

$$L_F = -10 \log \eta_F \qquad (5\text{-}21)$$

An analytical estimate of the optical power loss at a joint between multimode fibers is difficult to make, since the loss depends on the power distribution among the modes in the fiber.[21-23] For example, consider first the case where all modes in a fiber are equally excited, as shown in Fig. 5-8a. The emerging optical beam thus fills the entire exit numerical aperture of this emitting fiber. Suppose now that a second identical fiber, which we shall call the *receiving fiber*, is to be joined to the emitting fiber. For the receiving fiber to accept all the optical power emitted by the first fiber, there must be perfect mechanical alignment between the two optical waveguides, and their geometric and waveguide characteristics must match precisely.

On the other hand, if steady-state modal equilibrium has been established in the emitting fiber, most of the energy is concentrated in the lower-order fiber

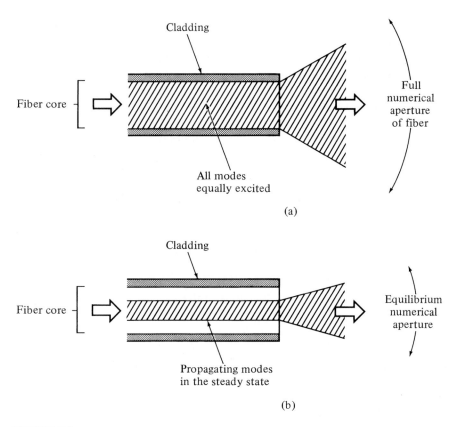

FIGURE 5-8
Different modal distributions of the optical beam emerging from a fiber lead to different degrees of coupling loss. (*a*) When all modes are equally excited, the output beam fills the entire output NA; (*b*) for a steady-state modal distribution, only the equilibrium NA is filled by the output beam.

modes. This means that the optical power is concentrated near the center of the fiber core, as shown in Fig. 5-8*b*. The optical power emerging from the fiber then fills only the equilibrium numerical aperture (see Fig. 5-4). In this case, since the input NA of the receiving fiber is larger than the equilibrium NA of the emitting fiber, slight mechanical misalignments of the two joined fibers and small variations in their geometric characteristics do not contribute significantly to joint loss.

Steady-state modal equilibrium is generally established in long fiber lengths (see Chap. 3). Thus, when estimating joint losses between long fibers, calculations based on a uniform modal power distribution tend to lead to results which may be too pessimistic. However, if a steady-state equilibrium modal distribution is assumed, the estimate may be too optimistic, since mechanical misalignments and fiber-to-fiber variations in characteristics cause a redistribution of power among the modes in the second fiber. As the power propagates along the second fiber, an additional loss will thus occur when a steady-state distribution is again established.

An exact calculation of coupling loss between different optical fibers, which takes into account nonuniform distribution of power among the modes and propagation effects in the second fiber, is lengthy and involved.[24] Here, we shall therefore make the assumption that all modes in the fiber are equally excited. Although this give a somewhat pessimistic prediction of joint loss, it will allow an estimate of the relative effects of losses resulting from mechanical misalignments, geometrical mismatches, and variations in the waveguide properties between two joined fibers.

5.3.1   Mechanical Misalignment

Mechanical alignment is a major problem when joining two fibers, owing to their microscopic size.[25-29] A standard multimode graded-index fiber core is 50–100 μm in diameter, which is roughly the thickness of a human hair, whereas single-mode fibers have diameters on the order of 9 μm. Radiation losses result from mechanical misalignments because the radiation cone of the emitting fiber does not match the acceptance cone of the receiving fiber. The magnitude of the radiation loss depends on the degree of misalignment. The three fundamental types of misalignment between fibers are shown in Fig. 5-9.

Longitudinal separation occurs when the fibers have the same axis but have a gap s between their end faces. *Angular misalignment* results when the two axes form an angle so that the fiber end faces are no longer parallel. *Axial displacement* (which is also often called *lateral displacement*) results when the axes of the two fibers are separated by a distance d.

The most common misalignment occurring in practice, which also causes the greatest power loss, is axial displacement. This axial offset reduces the overlap area of the two fiber-core end faces, as illustrated in Fig. 5-10, and consequently, reduces the amount of optical power that can be coupled from one fiber into the other.

To illustrate the effects of axial misalignment, let us first consider the simple case of two identical step-index fibers of radii a. Suppose that their axes are offset by a separation d at the common junction, as is shown in Fig. 5-10, and assume there is a uniform modal power distribution in the emitting fiber. Since the numerical aperture is constant across the end faces of the two fibers, the optical

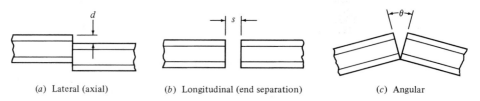

(a) Lateral (axial)   (b) Longitudinal (end separation)   (c) Angular

FIGURE 5-9
Three types of mechanical misalignments that can occur between two joined fibers.

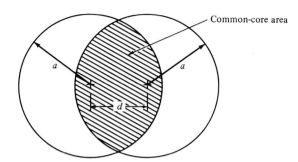

Common-core area

FIGURE 5-10
Axial offset reduces the common-core
area of the two fiber end faces.

power coupled from one fiber to another is simply proportional to the common
area A_{comm} of the two fiber cores. It is straightforward to show that this is (see
Prob. 5-9)

$$A_{\text{comm}} = 2a^2 \arccos \frac{d}{2a} - d\left(a^2 - \frac{d^2}{4}\right)^{1/2} \qquad (5\text{-}22)$$

For the step-index fiber, the coupling efficiency is simply the ratio of the
common-core area of the core end-face area,

$$\eta_{F,\text{step}} = \frac{A_{\text{comm}}}{\pi a^2} = \frac{2}{\pi} \arccos \frac{d}{2a} - \frac{d}{\pi a}\left[1 - \left(\frac{d}{2a}\right)^2\right]^{1/2} \qquad (5\text{-}23)$$

The calculation of power coupled from one graded-index fiber into another
identical one is more complex, since the numerical aperture varies across the fiber
end face. Because of this, the total power coupled into the receiving fiber at a
given point in the common-core area is limited by the numerical aperture of the
transmitting or receiving fiber, depending on which is smaller at that point.

If the end face of a graded-index fiber is uniformly illuminated, the optical
power accepted by the core will be that power which falls within the numerical
aperture of the fiber. The optical power density $p(r)$ at a point r on the fiber end is
proportional to the square of the local numerical aperture $NA(r)$ at that point:[30]

$$p(r) = p(0)\frac{NA^2(r)}{NA^2(0)} \qquad (5\text{-}24)$$

where $NA(r)$ and $NA(0)$ are defined by Eqs. (2-80a) and (2-80b), respectively. The
parameter $p(0)$ is the power density at the core axis, which is related to the total
power P in the fiber by

$$P = \int_0^{2\pi} \int_0^a p(r) r \, dr \, d\theta \qquad (5\text{-}25)$$

For an arbitrary index profile, the double integral in Eq. (5-25) must be evaluated
numerically. However, an analytic expression can be found by using a fiber with a

parabolic index profile ($\alpha = 2.0$). Using Eq. (2-80), the power density expression at a point r given by Eq. (5-24) becomes

$$p(r) = p(0)\left[1 - \left(\frac{r}{a}\right)^2\right] \tag{5-26}$$

Using Eqs. (5-25) and (5-26), the relationship between the axial power density $p(0)$ and the total power P in the emitting fiber is

$$P = \frac{\pi a^2}{2} p(0) \tag{5-27}$$

Let us now calculate the power transmitted across the butt joint of the two parabolic graded-index fibers with an axial offset d, as shown in Fig. 5-11. The overlap region must be considered separately for the areas A_1 and A_2. In area A_1 the numerical aperture is limited by that of the emitting fiber, whereas in area A_2 the numerical aperture of the receiving fiber is smaller than that of the emitting fiber. The vertical dashed line separating the two areas is the locus of points where the numerical apertures are equal.

To determine the power coupled into the receiving fiber, the power density given by Eq. (5-26) is integrated separately over areas A_1 and A_2. Since the numerical aperture of the emitting fiber is smaller than that of the receiving fiber in area A_1, all of the power emitted in this region will be accepted by the receiving fiber. The received power P_1 in area A_1 is thus

$$P_1 = 2 \int_0^{\theta_1} \int_{r_1}^{a} p(r)r \, dr \, d\theta$$

$$= 2p(0) \int_0^{\theta_1} \int_{r_1}^{a} \left[1 - \left(\frac{r}{a}\right)^2\right] r \, dr \, d\theta \tag{5-28}$$

where the limits of integration, shown in Fig. 5-12, are

$$r_1 = \frac{d}{2\cos\theta}$$

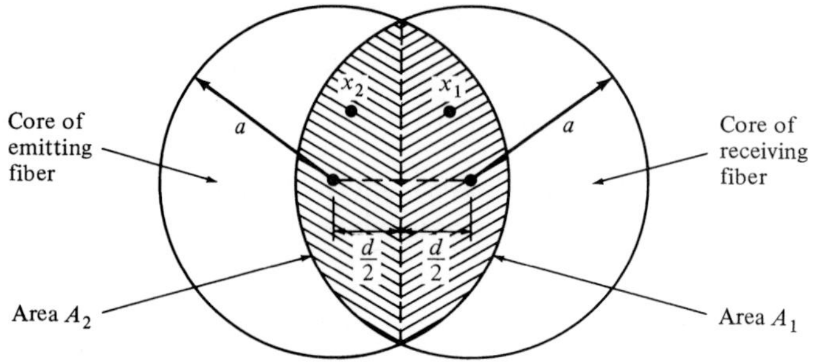

FIGURE 5-11
Core overlap region for two identical parabolic graded-index fibers with an axial separation d. Points x_1 and x_2 are arbitrary points of symmetry in areas A_1 and A_2.

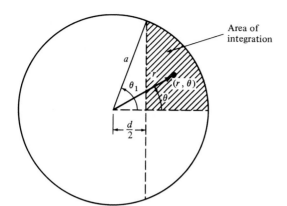

Area of integration

FIGURE 5-12
Area and limits of integration for the common-core area of two parabolic graded-index fibers.

and

$$\theta_1 = \arccos \frac{d}{2a}$$

Carrying out the integration yields

$$P_1 = \frac{a^2}{2} p(0) \left\{ \arccos \frac{d}{2a} - \left[1 - \left(\frac{d}{2a} \right)^2 \right]^{1/2} \frac{d}{6a} \left(5 - \frac{d^2}{2a^2} \right) \right\} \qquad (5\text{-}29)$$

where $p(0)$ is given by Eq. (5-27). The derivation of Eq. (5-29) is left as an exercise.

In area A_2 the emitting fiber has a larger numerical aperture than the receiving fiber. This means that the receiving fiber will accept only that fraction of the emitted optical power that falls within its own numerical aperture. This power can be found easily from symmetry considerations.[31] The numerical aperture of the receiving fiber at a point x_2 in area A_2 is the same as the numerical aperture of the emitting fiber at the symmetrical point x_1 in area A_1. Thus, the optical power accepted by the receiving fiber at any point x_2 in area A_2 is equal to that emitted from the symmetrical point x_1 in area A_1. The total power P_2 coupled across area A_2 is thus equal to the power P_1 coupled across area A_1. Combining these results, we have that the total power P_T accepted by the receiving fiber is

$$P_T = 2P_1$$

$$= \frac{2}{\pi} P \left\{ \arccos \frac{d}{2a} - \left[1 - \left(\frac{d}{2a} \right)^2 \right]^{1/2} \frac{d}{6a} \left(5 - \frac{d^2}{2a^2} \right) \right\} \qquad (5\text{-}30)$$

When the axial misalignment d is small compared with the core radius a, Eq. (5-30) can be approximated by

$$P_T \simeq P \left(1 - \frac{8d}{3\pi a} \right) \qquad (5\text{-}31)$$

This is accurate to within 1 percent for $d/a < 0.4$. The coupling loss for the offsets given by Eqs. (5-30) and (5-31) is

$$L_F = -10 \log \eta_F = -10 \log \frac{P_T}{P} \qquad (5\text{-}32)$$

The effect of separating the two fiber ends longitudinally by a gap s is shown in Fig. 5-13. Not all the higher-mode optical power emitted in the ring of width x will be intercepted by the receiving fiber. It is straightforward to show that, for a step-index fiber, the loss occurring in this case is

$$L_F = -10 \log \left(\frac{a}{a + s \tan \theta_c} \right)^2 \qquad (5\text{-}33)$$

where θ_c is the critical acceptance angle of the fiber.

When the axes of two joined fiber are angularly misaligned at the joint, the optical power that leaves the emitting fiber outside of the solid acceptance angle of the receiving fiber will be lost. For two step-index fibers that have an angular misalignment θ, the optical power loss at the joint has been shown to be[32,33]

$$L_F = -10 \log \left(\cos \theta \left\{ \frac{1}{2} - \frac{1}{\pi} p(1 - p^2)^{1/2} - \frac{1}{\pi} \arcsin p \right. \right.$$

$$\left. \left. - q \left[\frac{1}{\pi} y(1 - y^2)^{1/2} + \frac{1}{\pi} \arcsin y + \frac{1}{2} \right] \right\} \right) \qquad (5\text{-}34)$$

where

$$p = \frac{\cos \theta_c (1 - \cos \theta)}{\sin \theta_c \sin \theta}$$

$$q = \frac{\cos^3 \theta_c}{(\cos^2 \theta_c - \sin^2 \theta)^{3/2}}$$

$$y = \frac{\cos^2 \theta_c (1 - \cos \theta) - \sin^2 \theta}{\sin \theta_c \cos \theta_c \sin \theta}$$

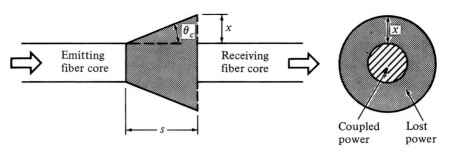

FIGURE 5-13
Loss effect when the fiber ends are separated longitudinally by a gap s.

The derivation of Eq. (5-34) again assumes that all modes are uniformly excited.

An experimental comparison[27] of the losses induced by the three types of mechanical misalignments is shown in Fig. 5-14. The measurements were based on two independent experiments using LED sources and graded-index fibers. The core diameters were 50 and 55 μm for the first and second experiments, respectively. A 1.83-m-long fiber was used in the first test and 20-m length in the second. In either case, the power output from the fibers was first optimized. The fibers were then cut at the center, so that the mechanical misalignment loss measurements were carried out on identical fibers. The axial offset and longitudinal separation losses are plotted as functions of misalignment normalized to the core radius. A normalized angular misalignment of 0.1 corresponds to a 1° angular offset.

Example 5-4. Suppose two graded index fibers are misaligned with an axial offset of $d = 0.3a$. From Eq. (5-30), we have that the fraction of optical power coupled from the first fiber into the second fiber is

$$\frac{P_T}{P} = \frac{2}{\pi}\left\{\arccos(0.15) - [1 - (0.15)^2]^{1/2}\left(\frac{0.15}{3}\right)\left[5 - \frac{(0.3)^2}{2}\right]\right\} = 0.748$$

or, in decibels,

$$10\log\frac{P_T}{P} = -1.27\,\text{dB}$$

which compares with the experimental value shown in Fig. 5-14.

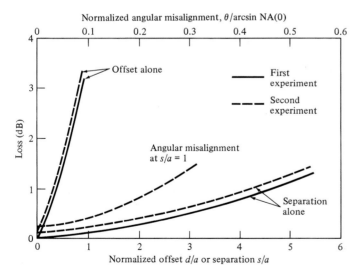

FIGURE 5-14
Experimental comparison of loss (in dB) as a function of mechanical misalignments. (Reproduced with permission from Chu and McCormick,[27] © 1978, AT&T.)

Figure 5-14 shows that, of the three mechanical misalignments, the dominant loss arises from lateral displacement. In practice, angular misalignments of less than $1°$ are readily achievable in splices and connectors. From the experimental data shown in Fig. 5-14, these misalignments result in losses of less than 0.5 dB.

For splices, the separation losses are normally negligible, since the fibers should be in relatively close contact. In most connectors, the fiber ends are intentionally separated by a small gap. This prevents them from rubbing against each other and becoming damaged during connector engagement. Typical gaps in these applications range from 0.025 to 0.10 mm, which results in losses of less than 0.8 dB for a 50-μm-diameter fiber.

5.3.2 Fiber-Related Losses

In addition to mechanical misalignments, differences in the geometrical and waveguide characteristics of any two waveguides being joined can have a profound effect on fiber-to-fiber coupling loss. These include variations in core diameter, core-area ellipticity, numerical aperture, refractive-index profile, and core–cladding concentricity of each fiber. Since these are manufacturer-related variations, the user generally has little control over them. Theoretical and experimental studies[24,34-39] of the effects of these variations have shown that, for a given percentage mismatch, differences in core radii and numerical apertures have a significantly larger effect on joint loss than mismatches in the refractive-index profile or core ellipticity.

The joint losses resulting from core diameter, numerical aperture, and core refractive-index-profile mismatches can easily be found from Eqs. (5-19) and (5-20). For simplicity, let the subscripts E and R refer to the emitting and receiving fibers, respectively. If the radii a_E and a_R are not equal but the axial numerical apertures and the index profiles are equal [$NA_E(0) = NA_R(0)$ and $\alpha_E = \alpha_R$], then the coupling loss is

$$L_F(a) = \begin{cases} -10 \log \left(\dfrac{a_R}{a_E} \right)^2 & \text{for} \quad a_R < a_E \\ 0 & \text{for} \quad a_R \geq a_E \end{cases} \tag{5-35}$$

If the radii and the index profiles of the two coupled fibers are identical but their axial numerical apertures are different, then

$$L_F(NA) = \begin{cases} -10 \log \left[\dfrac{NA_R(0)}{NA_E(0)} \right]^2 & \text{for} \quad NA_R(0) < NA_E(0) \\ 0 & \text{for} \quad NA_R(0) \geq NA_E(0) \end{cases} \tag{5-36}$$

Finally, if the radii and the axial numerical apertures are the same but the core refractive-index profiles differ in two joined fibers, then the coupling loss is

$$L_F(\alpha) = \begin{cases} -10 \log \dfrac{\alpha_R(\alpha_E + 2)}{\alpha_E(\alpha_R + 2)} & \text{for} \quad \alpha_R < \alpha_E \\ 0 & \text{for} \quad \alpha_R \geq \alpha_E \end{cases} \qquad (5\text{-}37)$$

This results because for $\alpha_R < \alpha_E$ the number of modes that can be supported by the receiving fiber is less than the number of modes in the emitting fiber. If $\alpha_R > \alpha_E$, then all modes in the emitting fiber can be captured by the receiving fiber. The derivations of Eqs. (5-35) to (5-37) are left as an exercise (see Probs. 5-13 through 5-15).

5.3.3   Fiber End-Face Preparation

One of the first steps that must be followed before fibers are connected or spliced to each other is to prepare the fiber end faces properly. In order not to have light deflected or scattered at the joint, the fiber ends must be flat, perpendicular to the fiber axis, and smooth. End-preparation techniques that have been extensively used include sawing, grinding and polishing, and controlled fracture.

Conventional grinding and polishing techniques can produce a very smooth surface that is perpendicular to the fiber axis. However, this method is quite time-consuming and requires a fair amount of operator skill. Although it is often implemented in a controlled environment such as a laboratory or a factory, it is not readily adaptable for field use. The procedure employed in the grinding and polishing technique is to use successively finer abrasives to polish the fiber end face. The end face is polished with each successive abrasive until the scratches created by the previous abrasive material are replaced by the finer scratches of the present abrasive. The number of abrasives used depends on the degree of smoothness that is desired.

Controlled-fracture techniques are based on score-and-break methods for cleaving fibers. In this operation, the fiber to be cleaved is first scratched to create a stress concentration at the surface. The fiber is then bent over a curved form while tension is simultaneously applied, as shown in Fig. 5-15. This action produces a stress distribution across the fiber. The maximum stress occurs at the scratch point so that a crack starts to propagate through the fiber.

One can produce a highly smooth and perpendicular end face in this way. A number of different tools based on the controlled-fracture technique have been developed and are being used both in the field and in factory environments. However, the controlled-fracture method requires careful control of the curvature of the fiber and of the amount of tension applied. If the stress distribution across the crack is not properly controlled, the fracture propagating across the fiber can fork into several cracks. This forking produces defects such as a lip or a hackled portion on the fiber end, as shown in Fig. 5-16. The EIA Fiber Optic Test Procedures (FOTP) 57 and 179 define these and other common end-face defects as follows:[40,41]

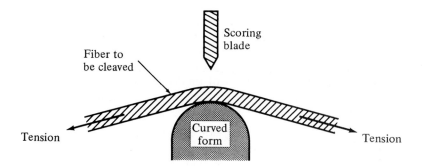

FIGURE 5-15
Controlled-fracture procedure for fiber end preparation.

Lip. This is a sharp protrusion from the edge of a cleaved fiber that prevents the cores from coming in close contact. Excessive lip height can cause fiber damage.

Rolloff. This rounding-off of the edge of a fiber is the opposite condition to lipping. It is also known as *breakover* and can cause high insertion or splice loss.

Chip. A chip is a localized fracture or break at the end of a cleaved fiber.

Hackle. Figure 5-16 shows this as severe irregularities across a fiber end face.

Mist. This is similar to hackle but much less severe.

Spiral or step. These are abrupt changes in the end-face surface topology.

Shattering. This is the result of an uncontrolled fracture and has no definable cleavage or surface characteristics.

5.4   LED COUPLING TO SINGLE-MODE FIBERS

In the early years of optical fiber applications, LEDs were traditionally considered only for multimode-fiber systems. However, around 1985, researchers recognized that edge-emitting LEDs can launch sufficient optical power into a single-mode fiber for transmission at data rates up to 560 Mb/s over several kilometers.[42-49] The interest in this arose because of the cost and reliability advantages of LEDs

FIGURE 5-16
Two examples of improperly cleaved fiber ends.

over laser diodes. Edge-emitting LEDs are used for these applications since they have a laserlike output pattern in the direction perpendicular to the junction plane.

To rigorously evaluate the coupling between an LED and a single-mode fiber we need to use the formalism of electromagnetic theory rather than geometrical optics, because of the monomode nature of the fiber. However, coupling analyses of the output from an edge-emitting LED to a single-mode fiber can be carried out wherein the results of electromagnetic theory are interpreted from a geometrical point of view,[45-47] which involves defining a numerical aperture for the single-mode fiber. Agreement with experimental measurements and with a more exact theory are quite good.[46-49]

Here, we will use the analysis of Reith and Shumate[46] to look at the following two cases: (1) direct coupling of an LED into a single-mode fiber, and (2) coupling into a single-mode fiber from a multimode flylead attached to the LED. In general, edge-emitting LEDs have gaussian near-field output profiles with $1/e^2$ full widths of approximately 0.9 and 22 μm in the directions perpendicular and parallel to the junction plane, respectively. The far-field patterns vary approximately as $\cos^7 \theta$ in the perpendicular direction and as $\cos \theta$ (lambertian) in the parallel direction.

For a source with a circularly asymmetric radiance $B(A_S, \Omega_S)$, Eq. (5-3) is, in general, not separable into contributions from the perpendicular and parallel directions. However, we can approximate the independent contributions by evaluating Eq. (5-3) as if each component were a circularly symmetric source, and then taking the geometric mean to find the total coupling efficiency. Calling these the x (parallel) and y (perpendicular) directions, and letting τ_x and τ_y be the x and y power transmissivities (directional coupling efficiencies), respectively, we can find the maximum LED-to-fiber coupling efficiency η from the relation

$$\eta = \frac{P_{in}}{P_s} = \tau_x \tau_y \tag{5-38}$$

where P_{in} is the optical power launched into the fiber and P_s is the total source output power.

Using a small-angle approximation, we first integrate over the effective solid acceptance angle of the fiber to get πNA_{SM}^2, where the geometrical-optics-based fiber numerical aperture $NA_{SM} = 0.11$. Assuming a gaussian output for the source, then for butt coupling of the LED to the single-mode fiber of radius a, the coupling efficiency in the y direction is

$$\tau_y = \left(\frac{P_{in;y}}{P_s} \right)^{1/2}$$

$$= \left[\frac{\int_0^{2\pi} \int_0^a B_0 \, e^{-2r^2/\omega_y^2} r \, dr \, d\theta_s \pi NA_{SM}^2}{\int_0^{2\pi} \int_0^\infty B_0 \, e^{-2r^2/\omega_y^2} y \, dy \, d\theta_s \int_0^{2\pi} \int_0^{\pi/2} \cos^7 \theta \sin \theta \, d\theta \, d\phi} \right]^{1/2} \tag{5-39}$$

where $P_{in,y}$ is the optical power coupled into the fiber from the y-direction source output, which has a $1/e^2$ LED intensity radius ω_y. One can write a similar set of integrals for τ_x. Letting $a = 4.5\,\mu m$, $\omega_x = 10.8\,\mu m$, and $\omega_y = 0.47\,\mu m$, Reith and Shumate calculated $\tau_x = -12.2$ dB and $\tau_y = -6.6$ dB to yield a total coupling efficiency $\eta = -18.8$ dB. Thus, for example, if the LED emits $200\,\mu W$ (-7 dBm), then $2.6\,\mu W$ (-25.8 dBm) gets coupled into the single-mode fiber.

When a 1- to 2-m multimode-fiber flylead is attached to an edge-emitting LED, the near-field profile of the multimode fiber has the same asymmetry as the LED. In this case, one can assume that the multimode-fiber optical output is a simple gaussian with different beam widths along the x and y directions. Using a similar coupling analysis with effective beam widths of $\omega_x = 19.6\,\mu m$ and $\omega_y = 10.0\,\mu m$, the directional coupling efficiencies are $\tau_x = -7.8$ dB and $\tau_y = -5.2$ dB, yielding a total LED-to-fiber coupling efficiency $\eta = -13.0$ dB.

5.5 FIBER SPLICING

A *fiber splice* is a permanent or semipermanent joint between two fibers. These are typically used to create long optical links or in situations where frequent connection and disconnection are not needed. In making and evaluating such splices, one must take into account the geometrical differences in the two fibers, fiber misalignments at the joint, and the mechanical strength of the splice. This section first addresses general splicing methods and then examines the factors contributing to loss when splicing single-mode fibers.

5.5.1 Splicing Techniques

Fiber splicing techniques include the fusion splice,[50-53] the V-groove mechanical splice,[54-57] and the elastic-tube splice.[58,59] The first technique yields a permanent joint, whereas the other two types of splices can be disassembled if necessary.

Fusion splices are made by thermally bonding together prepared fiber ends, as pictured in Fig. 5-17. In this method, the fiber ends are first prealigned and

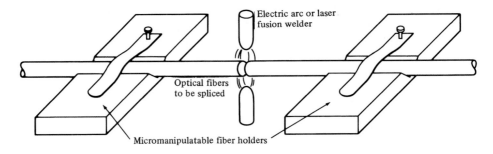

FIGURE 5-17
Fusion splicing of optical fibers.

butted together. This is done either in a grooved fiber holder or under a microscope with micromanipulators. The butt joint is then heated with an electric arc or a laser pulse so that the fiber ends are momentarily melted and hence bonded together. This technique can produce very low splice losses (typically averaging less than 0.06 dB). However, care must be exercised in this technique, since surface damage due to handling, surface defect growth created during heating, and residual stresses induced near the joint as a result of changes in chemical composition arising from the material melting can produce a weak splice.[60,61]

In the V-groove splice technique, the prepared fiber ends are first butted together in a V-shaped groove, as shown in Fig. 5-18. They are then bonded together with an adhesive or are held in place by means of a cover plate. The V-shaped channel can be either a grooved silicon, plastic, ceramic, or metal substrate. The splice loss in this method depends strongly on the fiber size (outside dimensions and core-diameter variations) and eccentricity (the position of the core relative to the center of the fiber).

The elastic-tube splice shown cross-sectionally in Fig. 5-19 is a unique device that automatically performs lateral, longitudinal, and angular alignment. It splices multimode fibers to give losses in the same range as commercial fusion splices, but much less equipment and skill are needed. The splice mechanism is basically a tube made of an elastic material. The central hole diameter is slightly smaller than that of the fiber to be spliced and is tapered on each end for easy fiber insertion. When a fiber is inserted, it expands the hole diameter so that the elastic material exerts a symmetrical force on the fiber. This symmetry feature allows an accurate and automatic alignment of the axes of the two fibers to be joined. A wide range of fiber diameters can be inserted into the elastic tube. Thus, the fibers to be spliced do not have to be equal in diameter, since each fiber moves into position independently relative to the tube axis.

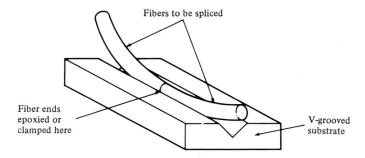

FIGURE 5-18
V-groove optical fiber splicing technique.

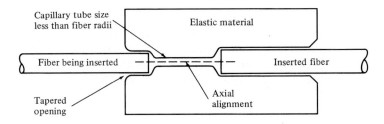

FIGURE 5-19
Schematic of an elastic-tube splice.

5.5.2 Splicing Single-Mode Fibers

As is the case in multimode fibers, in single-mode fibers the lateral (axial) offset loss presents the most serious misalignment. This loss depends on the shape of the propagating mode. For gaussian-shaped beams the loss between identical fibers is[62]

$$L_{SM;lat} = -10 \log\left\{ \exp\left[-\left(\frac{d}{W}\right)^2 \right] \right\} \qquad (5\text{-}40)$$

where the spot size W is the mode-field radius defined in Eq. (2-74), and d is the lateral displacement shown in Fig. 5-9. Since the spot size is only a few micrometers in single-mode fibers, low-loss coupling requires a very high degree of mechanical precision in the axial dimension.

Example 5-5. A single-mode fiber has a normalized frequency $V = 2.40$, a core refractive index $n_1 = 1.47$, a cladding refractive index $n_2 = 1.465$, and a core diameter $2a = 9\,\mu$m. Let us find the insertion losses of a fiber joint having a lateral offset of $1\,\mu$m.

First, using the expression for the mode-field diameter from Prob. 2-24, we have

$$W_0 = a(0.65 + 1.619V^{-3/2} + 2.879V^{-6})$$
$$= 4.5[0.65 + 1.619(2.40)^{-3/2} + 2.879(2.40)^{-6}] = 4.95\,\mu\text{m}$$

Then, from Eq. (5-40), we have

$$L_{SM,lat} = -10\log\{\exp[-(1/4.95)^2]\} = 0.18\,\text{dB}$$

For angular misalignment in single-mode fibers, the loss at a wavelength λ is[62]

$$L_{SM;ang} = -10 \log\left\{ \exp\left[-\left(\frac{\pi n_2 W \theta}{\lambda}\right)^2 \right] \right\} \qquad (5\text{-}41)$$

where n_2 is the refractive index of the cladding, θ is the angular misalignment in radians shown in Fig. 5-9, and W is the mode-field radius.

Example 5-6. Consider the single-mode fiber described in Example 5-5. Let us find the loss at a joint having an angular misalignment of $1°$ at a 1300-nm wavelength. From Eq. (5-41), we have

$$L_{SM;ang} = -10 \log\left\{ \exp\left[-\left(\frac{\pi(1.465)(4.95)(0.0175)}{1.3} \right)^2 \right] \right\} = 0.41 \text{ dB}$$

For a gap s with a material of index n_3, and letting $G = s/kW^2$, the gap loss for identical single-mode fiber splices is

$$L_{SM;gap} = -10 \log \frac{64n_1^2 n_3^2}{(n_1 + n_3)^4 (G^2 + 4)} \tag{5-42}$$

See Eq. (5-43) for a more general equation for dissimilar fibers.

5.6 OPTICAL FIBER CONNECTORS

A wide variety of optical fiber connectors has evolved for numerous different applications. Their uses range from simple single-channel fiber-to-fiber connectors in a benign location to multichannel connectors used in harsh military field environments. Some of the principal requirements of a good connector design are as follows:

1. *Low coupling losses.* The connector assembly must maintain stringent alignment tolerances to assure low mating losses. These low losses must not change significantly during operation or after numerous connects and disconnects.

2. *Interchangeability.* Connectors of the same type must be compatible from one manufacturer to another.

3. *Ease of assembly.* A service technician should readily be able to install the connector in a field environment; that is, in a location other than the connector factory. The connector loss should also be fairly insensitive to the assembly skill of the technician.

4. *Low environmental sensitivity.* Conditions such as temperature, dust, and moisture should have a small effect on connector-loss variations.

5. *Low-cost and reliable construction.* The connector must have a precision suitable to the application, but its cost must not be a major factor in the fiber system.

6. *Ease of connection.* Generally, one should be able to mate and demate the connector, simply, by hand.

5.6.1 Connector Types

Connectors are available in screw-on, bayonet-mount, and push–pull config-urations.[55,63-74] These include both single-channel and multichannel assemblies for cable-to-cable and for cable-to-circuit card connections. The basic coupling mechanisms used in these connectors belong to either the *butt-joint* or the *expanded-beam* classes.

Butt-joint connectors employ a metal, ceramic, or molded-plastic ferrule for each fiber and a precision sleeve into which the ferrule fit. The fiber is epoxied into a precision hole which has been drilled into the ferrule. The mechanical challenges of ferrule connectors include maintaining both the dimensions of the hole diam-eter and its position relative to the ferrule outer surface.

Figure 5-20 shows two popular butt-joint alignment designs used in both multimode and single-mode fiber systems. These are the *straight-sleeve* and the *tapered-sleeve* (or *biconical*) mechanisms. In the straight-sleeve connector, the length of the sleeve and a guide ring on the ferrules determine the end separation of the fibers. The biconical connector uses a tapered sleeve to accept and guide tapered ferrules. Again, the sleeve length and the guide rings maintain a given fiber-end separation.

An expanded-beam connector, illustrated in Fig. 5-21, employs lenses on the ends of the fibers. These lenses either collimate the light emerging from the trans-mitting fiber, or focus the expanded beam onto the core of the receiving fiber. The fiber-to-lens distance is equal to the focal length of the lens. The advantage of this scheme is that, since the beam is collimated, separation of the fiber ends may take place within the connector. Thus, the connector is less dependent on lateral align-ments. In addition, optical processing elements, such as beam splitters and switches, can easily be inserted into the expanded beam between the fiber ends.

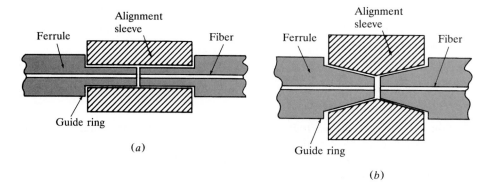

(a)

(b)

FIGURE 5-20
Examples of two popular alignment schemes used in fiber optic connectors: (*a*) straight sleeve, (*b*) tapered sleeve.

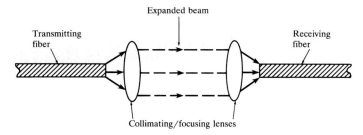

Expanded beam

Transmitting fiber

Receiving fiber

Collimating/focusing lenses

FIGURE 5-21
Schematic representation of an expanded-beam fiber optic connector.

5.6.2 Single-Mode Fiber Connectors

Because of the wide use of single-mode fiber optic links and because of the greater alignment precision required for these systems, this section addresses single-mode connector coupling losses. Based on the gaussian-beam model of single-mode fiber fields,[63] Nemota and Makimoto[75] derived the following coupling loss (in decibels) between single-mode fibers that have unequal mode-field diameters (which is an intrinsic factor) and lateral, longitudinal, and angular offsets plus reflections (which are all extrinsic factors):

$$L_{\mathrm{SM;ff}} = -10 \log \left[\frac{16 n_1^2 n_3^2}{(n_1 + n_3)^4} \frac{4\sigma}{q} \exp\left(-\frac{\rho u}{q} \right) \right] \tag{5-43}$$

where
$$\rho = (kW_1)^2$$
$$q = G^2 + (\sigma + 1)^2$$
$$u = (\sigma + 1)F^2 + 2\sigma FG \sin\theta + \sigma(G^2 + \sigma + 1)\sin^2\theta$$
$$F = \frac{d}{kW_1^2}$$
$$G = \frac{s}{kW_1^2}$$
$$\sigma = (W_2/W_1)^2$$
$$k = 2\pi n_3/\lambda$$
n_1 = core refractive index of fibers
n_3 = refractive index of medium between fibers
λ = wavelength of source
d = lateral offset
s = longitudinal offset
θ = angular misalignment
W_1 = 1/e mode-field radius of transmitting fiber
W_2 = 1/e mode-field radius of receiving fiber

This general equation gives very good correlation with experimental investigations.[64]

5.6.3   Connector Return Loss

A connection point in an optical link can be categorized into four interface types. These consist of either a perpendicular or an angled end-face on the fiber, and either a direct physical contact between the fibers or a contact employing an index-matching material. Each of these methods has a basic application for which it is best suited. The physical-contact type connectors without index-matching material are traditionally used in situations where frequent reconnections are required, such as within a building or on localized premises. Index-matching connectors are standardly employed in outside cable plants where the reconnections are infrequent, but need to have a low loss.

This section gives some details on index-matched and direct physical contacts, and briefly discusses angled interfaces. In each case, these connections require high return losses (low reflection levels) and low insertion loses (high optical-signal throughput levels). The low reflectance levels are desired since optical reflections provide a source of unwanted feedback into the laser cavity. This can affect the optical frequency response, the linewidth, and the internal noise of the laser, which results in degradation of system performance.

Figure 5-22 shows a model of an index-matched connection with perpendicular fiber end faces. In this figure and in the following analyses, offsets and angular misalignments are not taken into account. The connection model shows that the fiber end faces have a thin surface layer of thickness h having a high refractive index n_2 relative to the core index, which is a result of fiber polishing. The fiber core has an index n_0, and the gap width d between the end faces is filled with index-matching material having a refractive index n_1. The return loss RL_{IM} in decibels for the index-matched gap region is given by[76]

$$\text{RL}_{IM} = -10 \log\left\{ 2R\left[1 - \cos\left(\frac{4\pi n_1 d}{\lambda}\right) \right] \right\} \tag{5-44}$$

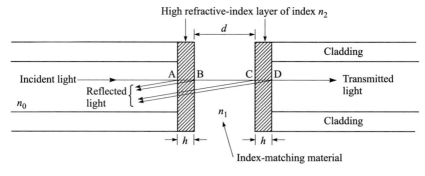

FIGURE 5-22
Model of an index-matched connection with perpendicular fiber end faces.

where

$$R = \frac{r_1^2 + r_2^2 + 2r_1 r_2 \cos \delta}{1 + r_1^2 r_2^2 + 2r_1 r_2 \cos \delta} \tag{5-45}$$

is the reflectivity at a single material-coated end face, and

$$r_1 = \frac{n_0 - n_2}{n_0 + n_2} \quad \text{and} \quad r_2 = \frac{n_2 - n_1}{n_2 + n_1} \tag{5-46}$$

are the reflection coefficients through the core from the high-index layer and through the high-index layer from the core, respectively. The parameter $\delta = (4\pi/\lambda)n_2 h$ is the phase difference in the high-index layer. The factor 2 in Eq. (5-44) accounts for reflections at both fiber end faces. The value of n_2 of the glass surface layer varies from 1.46 to 1.60, and the thickness h ranges from 0 to 0.15 μm.

When the perpendicular end faces are in direct physical contact, the return loss RL_{PC} in decibels is given by[76]

$$\mathrm{RL}_{PC} = -10\log\left\{2R_2\left[1 - \cos\left(\frac{4\pi n_2}{\lambda} 2h\right)\right]\right\} \tag{5-47}$$

where

$$R_2 = \left(\frac{n_0 - n_2}{n_0 + n_2}\right)^2 \tag{5-48}$$

Here, R_2 is the reflectivity at the discontinuity between the refractive indices of the fiber core and the high-index surface layer. In this case, the return loss at a given wavelength depends on the value of the refractive index n_2 and the thickness h of the surface layer.

Connections with angled end-faces are used in applications where an ultra-low reflection is required. Figure 5-23 shows a cross-sectional view of such a connection with a small gap of width d separating the fiber ends. The fiber

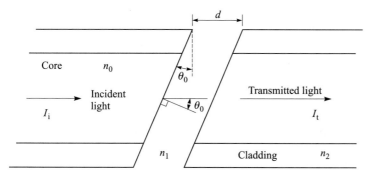

FIGURE 5-23
Connection with angled end faces having a small gap of width d separating the fiber ends.

core has an index n_0, and the material in the gap has a refractive index n_1. The end faces are polished at an angle θ_0 with respect to the plane perpendicular to the fiber axis. This angle is typically $8°$. If I_i and I_t are the incident and throughput optical power intensities, respectively, then the transmitted efficiency T through the connector is[77]

$$T = \frac{I_t}{I_i} = \frac{(1-R)^2}{(1-R)^2 + 4R\sin^2(\beta/2)} \qquad (5\text{-}49)$$

where

$$\frac{\sin\theta_0}{\sin\theta} = \frac{n}{n_0}, \qquad \beta = \frac{4\pi n_1 d \cos\theta}{\lambda}, \qquad \text{and} \qquad R = \left(\frac{n_0 - n_1}{n_0 + n_1}\right)^2$$

The insertion loss for this type of connector with an $8°$ angle will vary from 0 dB for no gap to 0.6 dB for an air gap of width $d = 1.0\,\mu\text{m}$. Note that when an index-matching material is used so that $n_0 = n_1$, then $R = 0$ and $T = 1$. When $n_0 \neq n_1$, the transmitted efficiency (and hence the connector loss) has an oscillatory behavior as a function of the wavelength and the end-face angle.

PROBLEMS

5-1. Analogous to Fig. 5-2, use a computer to plot and compare the emission patterns from a lambertian source and a source with an emission pattern given by $B(\theta) = B_0 \cos^3 \theta$. Assume both sources have the same peak radiance B_0, which is normalized to unity in each case.

5-2. Consider light sources where the emission pattern is given by $B(\theta) = B_0 \cos^m \theta$. Use a computer to plot $B(\theta)$ as a function of m in the range $1 \leq m \leq 20$ at viewing angles of $10°$, $20°$, and $45°$. Assume all sources have the same peak radiance B_0.

5-3. A laser diode has lateral ($\phi = 0°$) and transverse ($\phi = 90°$) half-power beam widths of $2\theta = 60°$ and $30°$, respectively. What are the transverse and lateral power distribution coefficients for this device?

5-4. An LED with a circular emitting area of radius $20\,\mu\text{m}$ has a lambertian emission pattern with a $100\text{-W}/(\text{cm}^2 \cdot \text{sr})$ axial radiance at a 100-mA drive current. How much optical power can be coupled into a step-index fiber having a $100\text{-}\mu\text{m}$ core diameter and $\text{NA} = 0.22$? How much optical power can be coupled from this source into a $50\text{-}\mu\text{m}$ core-diameter graded-index fiber having $\alpha = 2.0$, $n_1 = 1.48$, and $\Delta = 0.01$?

5-5. A GaAs optical source that has a refractive index of 3.600 is closely coupled to a step-index fiber which has a core refractive index of 1.465. If the source size is smaller than the fiber core, and the small gap between the source and the fiber is filled with a gel that has a refractive index of 1.305, what is the power loss in decibels from the source into the fiber?

5-6. Use Eq. (5-3) to derive an expression for the power coupled into a step-index fiber from an LED that has a radiant distribution given by

$$B(\theta) = B_0 \cos^m \theta$$

5-7. On the same graph, plot the maximum coupling efficiencies as a function of the source radius r_s for the following fibers:
    (a) Core radius of 25 μm and NA = 0.16.
    (b) Core radius of 50 μm and NA = 0.20.
Let r_s range from 0 to 50 μm. In what regions can a lens improve the coupling efficiency?

5-8. The end faces of two optical fibers with core refractive indices of 1.485 are perfectly aligned and have a small gap between them. If this gap is filled with a gel that has a refractive index of 1.305, find the optical power in decibels reflected at one interface of this joint. If the gap is very small, what is the power loss in decibels through the joint when no index-matching material is used? Note that $n = 1.0$ for air.

5-9. Verify that Eq. (5-22) gives the common-core area of the two axially misaligned step-index fibers shown in Fig. 5-10. If $d = 0.1a$, what is the coupling efficiency?

5-10. Consider the three fibers having the properties listed in Table P5-10. Use Eq. (5-23) to complete this table for connector losses (in decibels) due to the indicated axial misalignments.

TABLE P5-10

Fiber size: core diameter (μm)/clad diameter (μm)	Coupling loss (dB) for given axial misalignment (μm)			
	1	3	5	10
50/125			0.590	
62.5/125				
100/140				

5-11. Show that when the axial misalignment of d is small compared with the core radius a, Eq. (5-30) can be approximated by Eq. (5-31). Compare Eqs. (5-30) and (5-31) in terms of P_T/P as a function of d/a over the range $0 \le d/a \le 0.4$.

5-12. Consider an optical fiber that has a core refractive index $n_1 = 1.48$ and a numerical aperture NA = 0.20. Using Eqs. (5-32), (5-33), and (5-34), plot the three mechanical misalignment losses in decibels over the following ranges:
    (a) $0 \le d/a \le 1.0$
    (b) $0 \le s/a \le 3.0$
    (c) $0 \le \theta \le 10°$

5-13. Using Eqs. (5-19) and (5-20), show that Eq. (5-35) gives the coupling loss for two fibers with unequal core radii. Plot the coupling loss in decibels as a function of a_R/a_E for $0.5 \le a_R/a_E \le 1.0$.

5-14. Using Eqs. (5-19) and (5-20), show that Eq. (5-36) gives the coupling loss for two fibers with unequal axial numerical apertures. Plot this coupling loss in decibels as a function of $NA_R(0)/NA_E(0)$ over the range $0.5 \le NA_R(0)/NA_E(0) \le 1.0$.

5-15. Show that Eq. (5-37) gives the coupling loss for two fibers with different core refractive-index profiles. Plot this coupling loss in decibels as a function of α_R/α_E over the range $0.75 \le \alpha_R/\alpha_E \le 1.0$. Take $\alpha_E = 2.0$.

5-16. Consider two multimode graded-index fibers that have the characteristics given in Table P5-16. If these two fibers are perfectly aligned with no gap between them, calculate the splice losses and the coupling efficiencies for the following cases:

TABLE P5-16

Parameter	Fiber 1	Fiber 2
Core index n_1	1.46	1.48
Index difference Δ	0.010	0.015
Core radius a	50 μm	62.5 μm
Profile factor α	2.00	1.80

(a) Light going from fiber 1 to fiber 2.

(b) Light going from fiber 2 to fiber 1.

5-17. Consider two identical single-mode optical fibers that have a core refractive index $n_1 = 1.48$ and a mode-field radius $W = 5\,\mu$m at 1300 nm. Assume the material between the fiber ends is air with an index of 1.0. Using Eq. (5-43), plot the following connector losses in decibels (in each case, vary only one alignment parameter, keeping the other two mechanical misalignments fixed at zero):

(a) Lateral offset over the range $0 \le d \le 4\,\mu$m.

(b) Longitudinal offset over the range $0 \le s \le 40\,\mu$m.

(c) Angular misalignment over the range $0 \le \theta \le 2°$.

5-18. Assuming that a single-mode connector has no losses due to extrinsic factors, show that a 10 percent mismatch in mode-field diameters yields a loss of 0.05 dB.

5-19. Consider two fibers that have core refractive indices $n_0 = 1.463$. Assume these fibers are separated by a gap $d = 0.22\,\mu$m, which is filled with a material that has a refractive index $n_1 = 1.467$. Use Eq. (5-44) to plot the return loss as a function of the high-index-layer thickness h over the range $0 \le h \le 0.15\,\mu$m for values of n_2 equal to 1.467, 1.500, and 1.600.

5-20. Consider a connector in which the fibers have angled end-faces and core refractive indices $n_0 = 1.470$, as shown in Fig. 5-23. Assume the gap $d = 1\,\mu$m and the face angle $\theta = 8°$. The connector experiences no loss when the gap is filled with an index-matching material that has $n_1 = 1.470$. Thus, use the relationship

$$L(\lambda) = 10 \log \left[\frac{T(\lambda,\ n_1 = 1.470,\ \theta = 8°)}{T(\lambda,\ n_1 = 1.00,\ \theta = 8°)} \right]$$

to plot the throughput loss with an air gap ($n_1 = 1.0$ for air) as a function of wavelength for 700 nm $\le \lambda \le$ 1800 nm.

REFERENCES

1. (a) Y. Uematsu, T. Ozeki, and Y. Unno, "Efficient power coupling between an MH LED and a taper-ended multimode fiber," *IEEE J. Quantum Electron.*, vol. 15, pp. 86–92, Feb. 1979.

   (b) H. Kuwahara, M. Sasaki, and N. Tokoyo, "Efficient coupling from semiconductor lasers into single-mode fibers with tapered hemispherical ends," *Appl. Opt.*, vol. 19, pp. 2578–2583, Aug. 1980.

2. S.-C. Wang and M. A. Ingram, "A novel Fourier technique for calculating fiber-to-LED coupling efficiency with lateral and longitudinal misalignments," *J. Lightwave Tech.*, vol. 14, pp. 2407–2413, Oct. 1996.

3. D. Marcuse, "Excitation of parabolic-index fibers with incoherent sources," *Bell Sys. Tech. J.*, vol. 54, pp. 1507–1530, Nov. 1975; "LED fundamentals: Comparison of front and edge-emitting diodes," *IEEE J. Quantum Electron.*, vol. 13, pp. 819–827, Oct. 1977.

4. A. Yariv, *Quantum Electronics*, Wiley, New York, 3rd ed., 1988.

5. TIA/EIA-455-54B, *Mode Scrambler Requirements for Overfilled Launching Conditions to Multimode Fibers*, Aug. 1998.

6. M. Born and E. Wolf, *Principles of Optics*, Pergamon, Oxford, 6th ed., 1980.

7. J. G. Ackenhusen, "Microlenses to improve LED-to-fiber coupling and alignment tolerance," *Appl. Opt.*, vol. 18, pp. 3694–3699, Sept. 1979.

8. A. Nicia, "Lens coupling in fiber-optic devices: Efficiency limits," *Appl. Opt.*, vol. 20, pp. 3136–3145, Sept. 1981.

9. L. A. Wang and C. D. Su, "Tolerance analysis of aligning an astigmatic laser diode with a single-mode fiber," *J. Lightwave Tech.*, vol. 14, pp. 2757–2762, Dec. 1996.

10. C. A. Edwards, H. M. Presby, and C. Dragone, "Ideal microlenses for laser to fiber coupling," *J. Lightwave Tech.*, vol. 11, pp. 252–257, Feb. 1993.

11. K. Shiraishi, N. Oyama, K. Matsumura, I. Ohishi, and S. Suga, "A fiber lens with a long working distance for integrated coupling between laser diodes and single-mode fibers," *J. Lightwave Tech.*, vol. 13, pp. 1736–1744, Aug. 1995.

12. Z. L. Liau, D. Z. Tsang, and J. N. Walpole, "Simple compact diode-laser/microlens packaging," *IEEE J. Quantum Electron.*, vol. 33, pp. 457–461, Mar. 1997.

13. See any general physics or introductory optics book; for example:
    (a) F. A. Jenkins and H. E. White, *Fundamentals of Optics*, McGraw-Hill, New York, 4th ed., 1976.
    (b) E. Hecht and A. Zajac, *Optics*, Addison-Wesley, Reading MA, 2nd ed., 1987.
    (c) R. Ditteon, *Modern Geometrical Optics*, Wiley, New York, 1997.

14. M. C. Hudson, "Calculation of the maximum optical coupling efficiency into multimode optical waveguides," *Appl. Opt.*, vol. 13, pp. 1029–1033, May 1974.

15. G. K. Khoe and G. Kuyt, "Realistic efficiency of coupling light from GaAs laser diodes into parabolic-index optical fibers," *Electron. Lett.*, vol. 14, pp. 667–669, Sept. 28, 1978.

16. H. M. Presby, N. Amitay, R. Scotti, and A. F. Benner, "Laser-to-fiber coupling via optical fiber up-tapers," *J. Lightwave Tech.*, vol. 7, pp. 274–277, Aug. 1989.

17. V. Vusirikala, S. S. Saini, R. E. Bartolo, R. Whaley, S. Agarwala, M. Dagenais, F. G. Johnson, and D. Stone, "High butt-coupling efficiency to single-mode fibers using a 1.55-μm InGaAsP laser integrated with a tapered ridge mode transformer," *IEEE Photonics Tech. Lett.*, vol. 9, pp. 1472–1474, Nov. 1997.

18. I. Moerman, P. P. Van Daele, and P. M. Demeester, "A review of fabrication technologies for the monolithic integration of tapers with III–V semiconductor devices," *IEEE J. Sel. Topics Quantum Electron.*, vol. 3, pp. 1308–1320, Dec. 1997.

19. A. Safaai-Jazi and V. Suppanitchakij, "A tapered graded-index lens: Analysis of transmission properties and applications in fiber-optic communication systems," *IEEE J. Quantum Electron.*, vol. 33, pp. 2159–2166, Dec. 1997.

20. (a) K. Matsuda, T. Yoshida, Y. Kobayashi, and T. Chino, "A surface-emitting laser array with backside guiding holes for passive alignment to parallel optical fibers," *IEEE Photonics Tech. Lett.*, vol. 8, pp. 494–496, Apr. 1996.
    (b) J. Heinrich, E. Zeeb, and K. J. Ebeling, "Butt-coupling efficiency of VCSELs into multimode fibers," *IEEE Photonics Tech. Lett.*, vol. 9, pp. 1555–1557, Dec. 1997.

21. D. H. Rice and G. E. Keiser, "Short-haul fiber-optic link connector loss," *Int. Wire & Cable Symp. Proc.*, Nov. 13–15, 1984, Reno, NV, pp. 190–192.

22. Y. Daido, E. Miyauchi, and T. Iwama, "Measuring fiber connection loss using steady-state power distribution: A method," *Appl. Opt.*, vol. 20, pp. 451–456, Feb. 1981.

23. M. J. Hackert, "Evolution of power distributions in fiber optic systems: Development of a measurement strategy," *Fiber & Integrated Optics*, vol. 8, pp. 163–167, 1989.

24. P. DiVita and U. Rossi, "Realistic evaluation of coupling loss between different optical fibers," *J. Opt. Commun.*, vol. 1, pp. 26–32, Sept. 1980; "Evaluation of splice losses induced by mismatch in fiber parameters," *Opt. Quantum Electron.*, vol. 13, pp. 91–94, Jan. 1981.

25. M. J. Adams, D. N. Payne, and F. M. E. Sladen, "Splicing tolerances in graded index fibers," *Appl. Phys. Lett.*, vol. 28, pp. 524–526, May 1976.

26. D. Gloge, "Offset and tilt loss in optical fiber splices," *Bell Sys. Tech. J.,* vol. 55, pp. 905–916, Sept. 1976.

27. T. C. Chu and A. R. McCormick, "Measurement of loss due to offset, end separation and angular misalignment in graded index fibers excited by an incoherent source," *Bell Sys. Tech. J.,* vol. 57, pp. 595–602, Mar. 1978.

28. P. DiVita and U. Rossi, "Theory of power coupling between multimode optical fibers," *Opt. Quantum Electron.,* vol. 10, pp. 107–117, Jan. 1978.

29. C. M. Miller, "Transmission vs. transverse offset for parabolic-profile fiber splices with unequal core diameters," *Bell Sys. Tech. J.,* vol. 55, pp. 917–927, Sept. 1976.

30. D. Gloge and E. A. J. Marcatili, "Multimode theory of graded-core fibers," *Bell Sys. Tech. J.,* vol. 52, pp. 1563–1578, Nov. 1973.

31. H. G. Unger, *Planar Optical Waveguides and Fibres,* Clarendon, Oxford, 1977.

32. F. L. Thiel and R. M. Hawk, "Optical waveguide cable connection," *Appl. Opt.,* vol. 15, pp. 2785–2791, Nov. 1976.

33. F. L. Thiel and D. H. Davis, "Contributions of optical-waveguide manufacturing variations to joint loss," *Electron, Lett.,* vol. 12, pp. 340–341, June 1976.

34. S. C. Mettler, "A general characterization of splice loss for multimode optical fibers," *Bell Sys. Tech. J.,* vol. 58, pp. 2163–2182, Dec. 1979.

35. D. J. Bond and P. Hensel, "The effects on joint losses of tolerances in some geometrical parameters of optical fibres," *Opt. Quantum Electron.,* vol. 13, pp. 11–18, Jan. 1981.

36. S. C. Mettler and C. M. Miller, "Optical fiber splicing," in S. E. Miller and I. P. Kaminow, eds., *Optical Fiber Telecommunications—II,* Academic, New York, 1988.

37. V. C. Y. So, R. P. Hughes, J. B. Lamont, and P. J. Vella, "Splice loss measurement using local launch and detect," *J. Lightwave Tech.,* vol. LT-5, pp. 1663–1666, Dec. 1987.

38. D. W. Peckham and C. R. Lovelace, "Multimode optical fiber splice loss: Relating system and laboratory measurements," *J. Lightwave Tech.,* vol. LT-5, pp. 1630–1636, Dec. 1987.

39. G. Cancellieri and U. Ravaioli, *Measurements of Optical Fibers and Devices,* Artech House, Dedham, MA, 1984.

40. TIA/EIA-455-57B (FOTP-57B), *Preparation and Examination of Optical Fiber Endface for Testing Purposes,* Feb. 1996.

41. TIA/EIA-455-179 (FOTP-179), *Inspection of Cleaved Fiber Endfaces by Interferometry,* May 1988.

42. D. M. Fye, R. Olshansky, J. LaCourse, W. Powazinik, and R. B. Lauer, "Low-current, 1.3-μm edge-emitting LED for single-mode subscriber loop applications," *Electron. Lett.,* vol. 22, pp. 87–88, Jan. 1986.

43. G. K. Chang, H. P. Leblanc, and P. W. Shumate, "Novel high-speed LED transmitter for single-mode fiber and wideband loop transmission systems," *Electron. Lett.,* vol. 23, pp. 1338–1340, Dec. 3, 1987.

44. T. Tsubota, Y. Kashima, H. Takano, and Y. Hirose, "InGaAsP/InP long-wavelength high-efficiency edge-emitting LED for single-mode fiber optic communication," *Fiber Integr. Optics.,* vol. 7, no. 4, pp. 353–360, 1988.

45. D. N. Christodoulides, L. A. Reith, and M. A. Saifi, "Coupling efficiency and sensitivity of an LED to a single-mode fiber," *Electron. Lett.,* vol. 22, pp. 1110–1111, Oct. 1986.

46. L. A. Reith and P. A. Shumate, "Coupling sensitivity of an edge-emitting LED to a single-mode fiber," *J. Lightwave Tech.,* vol. LT-5, pp. 29–34, Jan. 1987.

47. B. Hillerich, "New analysis of LED to a single-mode fiber coupling," *Electron. Lett.,* vol. 22, pp. 1176–1177, Oct. 1986; "Efficiency and alignment tolerances of LED to a single-mode fiber coupling—theory and experiment," *Opt. Quantum Electron.,* vol. 19, no. 4, pp. 209–222, July 1987.

48. W. van Etten, "Coupling of LED light into a single-mode fiber," *J. Opt. Commun.,* vol. 9, no. 3, pp. 100–101, Sept. 1988.

49. D. N. Christodoulides, L. A. Reith, and M. A. Saifi, "Theory of LED coupling to single-mode fibers," *J. Lightwave Tech.,* vol. LT-5, pp. 1623–1629, Nov. 1987.

50. J. T. Krause, C. R. Kurkjian, and U. C. Paek, "Strength of fusion splices for fiber lightguides," *Electron. Lett.,* vol. 17, pp. 232–233, Mar. 1981.

51. T. Yamada, Y. Ohsato, M. Yoshinuma, T. Tanaka, and K.-I. Itoh, "Arc fusion splicer with profile alignment system for high-strength low-loss optical submarine cable," *J. Lightwave Tech.*, vol. LT-4, pp. 1204–1210, Aug. 1986.

52. M. Fujise, Y. Iwamoto, and S. Takei, "Self core-alignment arc-fusion splicer based on a simple local monitoring method," *J. Lightwave Tech.*, vol. LT-4, pp. 1211–1218, Aug. 1986.

53. G. D. Khoe, J. A. Luijendijk, and L. J. C. Vroomen, "Arc-welded monomode fiber splices made with the aid of local injection and detection of blue light," *J. Lightwave Tech.*, vol. LT-4, pp. 1219–1222, Aug. 1986.

54. E. E. Basch, R. A. Beaudette, and H. A. Carnes, "Optical transmission for interoffice trunks," *IEEE Trans. Commun.*, vol. COM-26, pp. 1007–1014, July 1978.

55. C. M. Miller, S. C. Mettler, and I. A. White, *Optical Fiber Splices and Connectors*, Marcel Dekker, New York, 1986.

56. D. B. Keck, A. J. Morrow, D. A. Nolan, and D. A. Thompson, "Passive components in the subscriber loop," *J. Lightwave Tech.*, vol. 7, pp. 1623–1633, Nov. 1989.

57. R. A. Patterson, "A new low-cost high-performance mechanical optical fiber splicing system for construction and restoration in the subscriber loop," *J. Lightwave Tech.*, vol. 7, pp. 1682–1688, Nov. 1989.

58. P. Melman and W. J. Carlsen, "Elastic-tube splice performance with single-mode and multimode fibers," *Electron. Lett.*, vol. 18, no. 8, pp. 320–321, Apr. 1982.

59. D. M. Knecht, W. J. Carlsen, and P. Melman, "Fiber-optic field splice," *Proc. SPIE Intl. Soc. Opt. Eng., Tech. Symp.*, Los Angeles, vol. 326, pp. 57–60, Jan. 25–29, 1982.

60. J. T. Krause, W. A. Reed, and K. L. Walker, "Splice loss of single-mode fiber as related to fusion time, temperature, and index profile alteration," *J. Lightwave Tech.*, vol. LT-4, pp. 837–840, July 1986.

61. E. Serafini, "Statistical approach to the optimization of optical fiber fusion splicing in the field," *J. Lightwave Tech.*, vol. 7, pp. 431–435, Feb. 1989.

62. D. Marcuse, D. Gloge, and E. A. J. Marcatili, "Guiding properties of fibers," in S. E. Miller and A. G. Chynoweth, eds., *Optical Fiber Telecommunications*, Academic New York, 1979.

63. D. Marcuse, "Loss analysis of single-mode splices," *Bell Sys. Tech. J.*, vol. 56, pp. 703–718, May 1977.

64. W. C. Young and D. R. Frey, "Fiber connectors," in S. E. Miller and I. P. Kaminow, eds., *Optical Fiber Telecommunications—II*, Academic, New York, 1988.

65. T. Ormand, "Fiber optic connectors come of age," *EDN*, vol. 35, pp. 89–96, Feb. 15, 1990.

66. K. Fleck, "Fiber-optic interconnect sales reach one billion," *Lightwave*, vol. 12, pp. 31–33, July 1997.

67. P. McGlaughlin, "Quick-connect fiber-optic connectors," *Lightwave,* vol. 12, pp. 86–88, Sept. 1997.

68. S. Iwano, R. Nagase, K. Kanayama, E. Sugita, K. Yasuda, and Y. Ando, "Compact and self-retentive multi-ferrule optical backplane connector," *J. Lightwave Tech.*, vol. 10, pp. 1356–1362, Oct. 1992.

69. T. Shintaku, E. Sugita, and R. Nagase, "Highly stable physical-contact optical fiber connectors with spherical convex ends," *J. Lightwave Tech.*, vol. 11, pp. 241–248, Feb. 1993.

70. M. Takaya, M. Kihara, and S. Nagasawa, "Design and performance of a multifiber backplane type connector," *IEEE Photonics Tech. Lett.*, vol. 8, pp. 655–657, May 1996.

71. M. Takahashi, "Improved design of APC optical connectors with slanted angle of 12° for dispersion shifted optical fiber," *J. Lightwave Tech.*, vol. 16, pp. 567–572, Apr. 1998.

72. Y. Takeuchi, S. Mitachi, and R. Nagase, "High-strength glass-ceramic ferrule for SC-type single-mode optical fiber connector," *IEEE Photonics Tech. Lett.*, vol. 9, pp. 1502–1504, Nov. 1997.

73. M. Ramos, I. Verrier, J. P. Goure, and P. Mottier, "Efficient ball lens coupling between a single-mode optical fiber and a silica microguide," *J. Opt. Commun.*, vol. 16, pp. 179–185, Oct. 1995.

74. G. D. Landry and T. A. Maldonado, "Ray tracing through two ball uniaxial sapphire lens system in a single-mode fiber-to-fiber coupler," *J. Lightwave Tech.*, vol. 14, pp. 509–512, Mar. 1996.

75. S. Nemota and T. Makimoto, "Analysis of splice loss in single-mode fibers using a gaussian field approximation," *Opt. Quantum Electron.*, vol. 11, no. 5, pp. 447–457, Sept. 1979.

76. M. Kihara, S. Nagasawa, and T. Tanifuji, "Return loss characteristics of optical fiber connectors," *J. Lightwave Tech.*, vol. 14, pp. 1986–1991, Sept. 1996.

77. M. Kihara, S. Nagasawa, and T. Tanifuji, "Design and performance of an angled physical contact type multifiber connector," *J. Lightwave Tech.*, vol. 14, pp. 542–548, Apr. 1996.

CHAPTER
6

PHOTODETECTORS

At the output end of an optical transmission line, there must be a receiving device which interprets the information contained in the optical signal. The first element of this receiver is a photodetector. The photodetector senses the luminescent power falling upon it and converts the variation of this optical power into a correspondingly varying electric current. Since the optical signal is generally weakened and distorted when it emerges from the end of the fiber, the photodetector must meet very high performance requirements. Among the foremost of these requirements are a high response or sensitivity in the emission wavelength range of the optical source being used, a minimum addition of noise to the system, and a fast response speed or sufficient bandwidth to handle the desired data rate. The photodetector should also be insensitive to variations in temperature, be compatible with the physical dimensions of the optical fiber, have a reasonable cost in relation to the other components of the system, and have a long operating life.

Several different types of photodetectors are in existence. Among these are photomultipliers,[1-3] pyroelectric detectors,[4] and semiconductor-based photoconductors, phototransistors, and photodiodes.[5] However, many of these detectors do not meet one or more of the foregoing requirements. Photomultipliers consisting of a photocathode and an electron multiplier packaged in a vacuum tube are capable of very high gain and very low noise. Unfortunately, their large size and high voltage requirements make them unsuitable for optical fiber systems. Pyroelectric photodetectors involve the conversion of photons to heat. Photon absorption results in a temperature change of the detector material. This gives rise to a variation in the dielectric constant which is usually measured as a capacitance change. The response of this detector is quite flat over a broad spectral band, but

243

its speed is limited by the detector cooling rate after it has been excited. Its principal use is for detecting high-speed laser pulses, and it is not well suited for optical fiber systems.

Of the semiconductor-based photodetectors, the photodiode is used almost exclusively for fiber optic systems because of its small size, suitable material, high sensitivity, and fast response time. The two types of photodiodes used are the *pin* photodetector and the avalanche photodiode (APD). Detailed reviews of these photodiodes have been presented in the literature.[3,5-10] We shall examine the fundamental characteristics of these two device types in the following sections. In describing these components, we shall make use of the elementary principles of semiconductor device physics given in Sec. 4.1. Basic discussions of photodetection processes can be found in various texts.[11,12]

6.1 PHYSICAL PRINCIPLES OF PHOTODIODES

6.1.1 The *pin* Photodetector

The most common semiconductor photodetector is the *pin* photodiode, shown schematically in Fig. 6-1. The device structure consists of *p* and *n* regions separated by a very lightly *n*-doped intrinsic (*i*) region. In normal operation a sufficiently large reverse-bias voltage is applied across the device so that the intrinsic region is fully depleted of carriers. That is, the intrinsic *n* and *p* carrier concentrations are negligibly small in comparison with the impurity concentration in this region.

When an incident photon has an energy greater than or equal to the bandgap energy of the semiconductor material, the photon can give up its energy and excite an electron from the valance band to the conduction band. This process generates free electron–hole pairs, which are known as *photocarriers* since they are photon-generated charge carriers, as is shown in Fig. 6-2. The photodetector is normally designed so that these carriers are generated mainly in the depletion region (the depleted intrinsic region) where most of the incident light is absorbed. The high electric field present in the depletion region causes the carriers to separate and be collected across the reverse-biased junction. This gives rise to a current

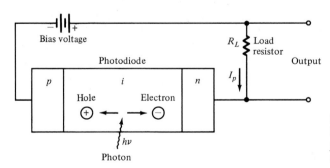

FIGURE 6-1
Schematic representation of a *pin* photodiode circuit with an applied reverse bias.

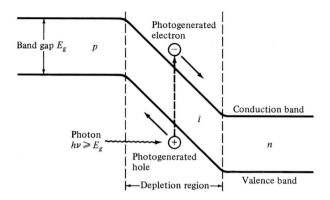

Band gap E_g p

Photogenerated electron

Photon $h\nu \geqslant E_g$

Photogenerated hole

i

Conduction band

n

Valence band

←—Depletion region—→

FIGURE 6-2
Simple energy-band diagram for a *pin* photodiode. Photons with an energy greater than or equal to the band-gap energy E_g can generate free electron–hole pairs which act as photo-current carriers.

flow in an external circuit, with one electron flowing for every carrier pair generated. This current flow is known as the *photocurrent*.

As the charge carriers flow through the material, some electron–hole pairs will recombine and hence disappear. On the average, the charge carriers move a distance L_n or L_p for electrons and holes, respectively. This distance is known as the *diffusion length*. The time it takes for an electron or hole to recombine is known as the *carrier lifetime* and is represented by τ_n and τ_p, respectively. The lifetimes and the diffusion lengths are related by the expressions

$$L_n = (D_n\tau_n)^{1/2} \quad \text{and} \quad L_p = (D_p\tau_p)^{1/2}$$

where D_n and D_p are the electron and hole diffusion coefficients (or constants), respectively, which are expressed in units of centimeters squared per second.

Optical radiation is absorbed in the semiconductor material according to the exponential law

$$P(x) = P_0(1 - e^{-\alpha_s(\lambda)x}) \tag{6-1}$$

Here, $\alpha_s(\lambda)$ is the *absorption coefficient* at a wavelength λ, P_0 is the incident optical power level, and $P(x)$ is the optical power absorbed in a distance x.

The dependence of the optical absorption coefficient on wavelength is shown in Fig. 6-3 for several photodiode materials.[13] As the curves clearly show, α_s depends strongly on the wavelength. Thus, a particular semiconductor material can be used only over a limited wavelength range. The upper wavelength cutoff λ_c is determined by the band-gap energy E_g of the material. If E_g is expressed in units of electron volts (eV), then λ_c is given in units of micrometers (μm) by

$$\lambda_c(\mu m) = \frac{hc}{E_g} = \frac{1.24}{E_g \text{ (eV)}} \tag{6-2}$$

The cutoff wavelength is about $1.06\,\mu$m for Si and $1.6\,\mu$m for Ge. For longer wavelengths, the photon energy is not sufficient to excite an electron from the valence to the conduction band.

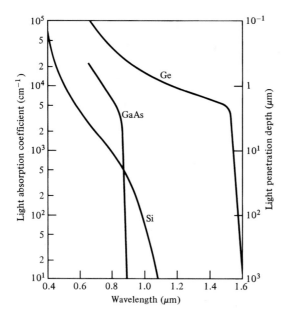

FIGURE 6-3

Optical absorption coefficient as a function of wavelength for silicon, germanium, and gallium arsenide. (Reproduced with permission from Miller, Marcatili, and Li,[13] © 1973, IEEE.)

Example 6-1. A photodiode is constructed of GaAs, which has a band-gap energy of 1.43 eV at 300 K. From Eq. (6-2), the long-wavelength cutoff is

$$\lambda_c = \frac{hc}{E_g} = \frac{(6.625 \times 10^{-34}\ \text{J} \cdot \text{s})(3 \times 10^8\ \text{m/s})}{(1.43\ \text{eV})(1.6 \times 10^{-19}\ \text{J/ev})} = 869\ \text{nm}$$

This GaAs photodiode will not operate for photons of wavelength greater than 869 nm.

At the lower-wavelength end, the photoresponse cuts off as a result of the very large values of α_s at the shorter wavelengths. In this case, the photons are absorbed very close to the photodetector surface, where the recombination time of the generated electron–hole pairs is very short. The generated carriers thus recombine before they can be collected by the photodetector circuitry.

If the depletion region has a width w, then, from Eq. (6-1), the total power absorbed in the distance w is

$$P(w) = P_0(1 - e^{-\alpha_s w}) \tag{6-3}$$

If we take into account a reflectivity R_f at the entrance face of the photodiode, then the primary photocurrent I_p resulting from the power absorption of Eq. (6-3) is given by

$$I_p = \frac{q}{h\nu} P_0(1 - e^{-\alpha_s w})(1 - R_f) \tag{6-4}$$

where P_0 is the optical power incident on the photodetector, q is the electron charge, and $h\nu$ is the photon energy.

Two important characteristics of a photodetector are its quantum efficiency and its response speed. These parameters depend on the material band gap, the operating wavelength, and the doping and thickness of the p, i, and n regions of the device. The *quantum efficiency* η is the number of the electron–hole carrier pairs generated per incident photon of energy $h\nu$ and is given by

$$\eta = \frac{\text{number of electron-hole pairs generated}}{\text{number of incident photons}} = \frac{I_p/q}{P_0/h\nu} \tag{6-5}$$

Here, I_p is the average photocurrent generated by a steady-state average optical power P_0 incident on the photodetector.

> **Example 6-2.** In a 100-ns pulse, 6×10^6 photons at a wavelength of 1300 nm fall on an InGaAs photodetector. On the average, 5.4×10^6 electron–hole (e–h) pairs are generated. The quantum efficiency is found from Eq. (6-5) as
>
> $$\eta = \frac{\text{number of e-h pairs generated}}{\text{number of incident photons}} = \frac{5.4 \times 10^6}{6 \times 10^6} = 0.90$$
>
> Thus, the quantum efficiency at 1300 nm is 90 percent.

In a practical photodiode, 100 photons will create between 30 and 95 electron–hole pairs, thus giving a detector quantum efficiency ranging from 30 to 95 percent. To achieve a high quantum efficiency, the depletion layer must be thick enough to permit a large fraction of the incident light to be absorbed. However, the thicker the depletion layer, the longer it takes for the photogenerated carriers to drift across the reverse-biased junction. Since the carrier drift time determines the response speed of the photodiode, a compromise has to be made between response speed and quantum efficiency. We shall discuss this further in Sec. 6.3

The performance of a photodiode is often characterized by the *responsivity* $\mathscr{R}$. This is related to the quantum efficiency by

$$\mathscr{R} = \frac{I_p}{P_0} = \frac{\eta q}{h\nu} \tag{6-6}$$

This parameter is quite useful, since it specifies the photocurrent generated per unit optical power. Typical *pin* photodiode responsivities as a function of wavelength are shown in Fig. 6-4. Representative values are 0.65 A/W for silicon at 900 nm and 0.45 A/W for germanium at 1.3 μm. For InGaAs, typical values are 0.9 A/W at 1.3 μm and 1.0 A/W at 1.55 μm.

> **Example 6.3.** Photons of energy 1.53×10^{-19} J are incident on a photodiode which has a responsivity of 0.65 A/W. If the optical power level is 10 μW, then from Eq. (6-6) the photocurrent generated is
>
> $$I_p = \mathscr{R}P_0 = (0.65 \, \text{A/W})(10 \, \mu\text{W}) = 6.5 \, \mu\text{A}$$

In most photodiodes the quantum efficiency is independent of the power level falling on the detector at a given photon energy. Thus, the responsivity is a linear function of the optical power. That is, the photocurrent I_p is directly

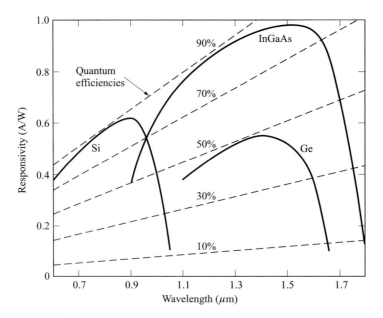

FIGURE 6-4
Comparison of the responsivity and quantum efficiency as a function of wavelength for *pin* photo-diodes constructed of different materials.

proportional to the optical power P_0 incident upon the photodetector, so that the responsivity $\mathscr{R}$ is constant at a given wavelength (a given value of hv). Note, however, that the quantum efficiency is not a constant at all wavelengths, since it varies according to the photon energy. Consequently, the responsivity is a function of the wavelength and of the photodiode material (since different materials have different band-gap energies). For a given material, as the wavelength of the incident photon becomes longer, the photon energy becomes less than that required to excite an electron from the valence band to the conduction band. The responsivity thus falls off rapidly beyond the cutoff wavelength, as can be seen in Fig. 6-4.

Example 6-4. As shown in Fig. 6-4, for the wavelength range 1300 nm $< \lambda <$ 1600 nm, the quantum efficiency for InGaAs is around 90 percent. Thus, in this wavelength range the responsivity is

$$\mathscr{R} = \frac{\eta q}{hv} = \frac{\eta q \lambda}{hc} = \frac{(0.90)(1.6 \times 10^{-19} \text{ C})\lambda}{(6.625 \times 10^{-34} \text{ J} \cdot \text{s})(3 \times 10^8 \text{ m/s})} = 7.25 \times 10^5 \lambda$$

For example, at 1300 nm we have

$$\mathscr{R} = [7.25 \times 10^5 (\text{A/W})/\text{m}](1.30 \times 10^{-6} \text{ m}) = 0.92 \text{ A/W}$$

At wavelengths higher than 1600 nm, the photon energy is not sufficient to excite an electron from the valence band to the conduction band. For example, $In_{0.53}Ga_{0.47}As$ has an energy gap $E_g = 0.73$ eV, so that from Eq. (6-2) the cutoff wavelength is

$$\lambda_c = \frac{1.24}{E_g} = \frac{1.24}{0.73} = 1.7\,\mu m$$

At wavelengths less than 1100 nm, the photons are absorbed very close to the photodetector surface, where the recombination rate of the generated electron–hole pairs is very short. The responsivity thus decreases rapidly for smaller wavelengths, since many of the generated carriers do not contribute to the photocurrent.

6.1.2 Avalanche Photodiodes

Avalanche photodiodes (APDs) internally multiply the primary signal photocurrent before it enters the input circuitry of the following amplifier. This increases receiver sensitivity, since the photocurrent is multiplied before encountering the thermal noise associated with the receiver circuit. In order for carrier multiplication to take place, the photogenerated carriers must traverse a region where a very high electric field is present. In this high-field region, a photogenerated electron or hole can gain enough energy so that it ionizes bound electrons in the valence band upon colliding with them. This carrier multiplication mechanism is known as *impact ionization*. The newly created carriers are also accelerated by the high electric field, thus gaining enough energy to cause further impact ionization. This phenomenon is the *avalanche effect*. Below the diode breakdown voltage a finite total number of carriers are created, whereas above breakdown the number can be infinite.

A commonly used structure for achieving carrier multiplication with very little excess noise is the *reach-through* construction[7,13-16] shown in Fig. 6-5. The reach-through avalanche photodiode (RAPD) is composed of a high-resistivity *p*-type material deposited as an epitaxial layer on a p^+ (heavily doped *p*-type) substrate. A *p*-type diffusion or ion implant is then made in the high-resistivity material, followed by the construction of an n^+ (heavily doped *n*-type) layer. For silicon, the dopants used to form these layers are normally boron and phosphorus, respectively. This configuration is referred to as $p^+\pi p n^+$ *reach-through* structure. The π layer is basically an intrinsic material that inadvertently has some *p* doping because of imperfect purification. Section 6.5 describes more complex structures used for InGaAs APDs.

The term "reach-through" arises from the photodiode operation. When a low reverse-bias voltage is applied, most of the potential drop is across the pn^+ junction. The depletion layer widens with increasing bias until a certain voltage is reached at which the peak electric field at the pn^+ junction is about 5–10 percent below that needed to cause avalanche breakdown. At this point, the depletion layer just "reaches through" to the nearly intrinsic π region.

In normal usage, the RAPD is operated in the fully depleted mode. Light enters the device through the p^+ region and is absorbed in the π material, which

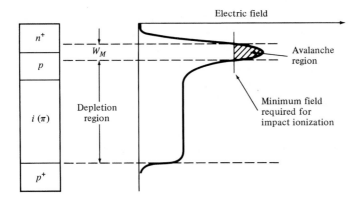

FIGURE 6-5
Reach-through avalanche photodiode structure and the electric fields in the depletion and multiplication regions.

acts as the collection region for the photogenerated carriers. Upon being absorbed, the photon gives up its energy, thereby creating electron–hole pairs, which are then separated by the electric field in the π region. The photogenerated electrons drift through the π region in the pn^+ junction, where a high electric field exists. It is in this high-field region that carrier multiplication takes place.

The average number of electron–hole pairs created by a carrier per unit distance traveled is called the *ionization rate*. Most materials exhibit different *electron ionization rates* α and *hole ionization rates* β. Experimentally obtained values of α and β for five different semiconductor materials are shown in Fig. 6-6. The ratio $k = \beta/\alpha$ of the two ionization rates is a measure of the photodetector performance. As we shall see in Sec. 6.4, avalanche photodiodes constructed of materials in which one type of carrier largely dominates impact ionization exhibit low noise and large gain-bandwidth products. Of all the materials shown in Fig. 6-6, only silicon has a significant difference between electron and hole ionization rates.[16–34]

The multiplication M for all carriers generated in the photodiode is defined by

$$M = \frac{I_M}{I_p} \tag{6-7}$$

where I_M is the average value of the total multiplied output current and I_p is the primary unmultiplied photocurrent defined in Eq. (6-4). In practice, the avalanche mechanism is a statistical process, since not every carrier pair generated in the diode experiences the same multiplication. Thus, the measured value of M is expressed as an average quantity.

Example 6-5. A given silicon avalanche photodiode has a quantum efficiency of 65 percent at a wavelength of 900 nm. Suppose $0.5\,\mu$W of optical power produces a

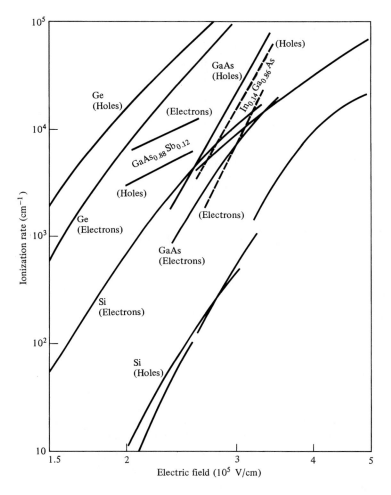

FIGURE 6-6
Carrier ionization rates obtained experimentally for silicon, germanium, gallium arsenide, gallium arsenide antimonide, and indium gallium arsenide. (Reproduced with permission from Melchior.[2])

multiplied photocurrent of $10\,\mu A$. Let us find the multiplication M. From Eq. (6-6), the primary photocurrent is

$$I_p = \mathscr{R}P_0 = \frac{\eta q}{h\nu}P_0 = \frac{\eta q \lambda}{hc}P_0$$

$$= \frac{(0.65)(1.6 \times 10^{-19}\,\text{C})(9 \times 10^{-7}\,\text{m})}{(6.625 \times 10^{-34}\,\text{J} \cdot \text{s})(3 \times 10^8\,\text{m/s})}\,5 \times 10^{-7}\,\text{W} = 0.235\,\mu A$$

From Eq. (6-7), the multiplication is

$$M = \frac{I_M}{I_p} = \frac{10\,\mu A}{0.235\,\mu A} = 43$$

Thus, the primary photocurrent is multiplied by a factor of 43.

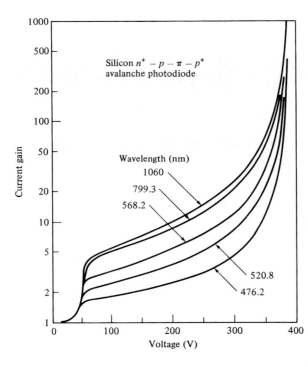

FIGURE 6-7

Typical room-temperature current gains of a silicon reach-through avalanche photodiode for different wavelengths as a function of bias voltage. (Reproduced with permission from Melchior, Hartman, Schinke, and Seidel,[15] © 1978, AT&T.)

Typical current gains for different wavelengths[15] as a function of bias voltage for a silicon reach-through avalanche photodiode are shown in Fig. 6-7. The dependence of the gain on the excitation wavelength is attributable to mixed initiation of the avalanche process by electrons and holes when most of the light is absorbed in the n^+p region close to the detector surface. This is especially noticeable at short wavelengths, where a larger portion of the optical power is absorbed close to the surface than at longer wavelengths. In silicon, since the ionization coefficient for holes is smaller than that for electrons, the total current gain is reduced at the short wavelengths.

Analogous to the *pin* photodiode, the performance of an APD is characterized by its responsivity $\mathscr{R}_{\text{APD}}$, which is given by

$$\mathscr{R}_{\text{APD}} = \frac{\eta q}{h\nu} M = \mathscr{R}_0 M \qquad (6\text{-}8)$$

where $\mathscr{R}_0$ is the unity gain responsivity.

6.2   PHOTODETECTOR NOISE

In fiber optic communication systems, the photodiode is generally required to detect very weak optical signals. Detection of the weakest possible optical signals requires that the photodetector and its following amplification circuitry be optimized so that a given signal-to-noise ratio is maintained. The power signal-to-noise ratio S/N at the output of an optical receiver is defined by

$$\frac{S}{N} = \frac{\text{signal power from photocurrent}}{\text{photodetector noise power} + \text{amplifier noise power}} \tag{6-9}$$

The noise sources in the receiver arise from the photodetector noises resulting from the statistical nature of the photon-to-electron conversion process and the thermal noises associated with the amplifier circuitry.

To achieve a high signal-to-noise ratio, the following conditions should be met:

1. The photodetector must have a high quantum efficiency to generate a large signal power.
2. The photodetector and amplifier noises should be kept as low as possible.

From most applications, it is the noise currents which determine the minimum optical power level that can be detected, since the photodiode quantum efficiency is normally close to its maximum possible value.

The sensitivity of a photodetector in an optical fiber communication system is describable in terms of the *minimum detectable optical power*. This is the optical power necessary to produce a photocurrent of the same magnitude as the root mean square (rms) of the total noise current, or equivalently, a signal-to-noise ratio of 1. A thorough understanding of the source, characteristics, and interrelationships of the various noises in a photodetector is therefore necessary to make a reliable design and to evaluate optical receivers.

6.2.1 Noise Sources

To see the interrelationship of the different types of noises affecting the signal-to-noise ratio, let us examine the simple receiver model and its equivalent circuit shown in Fig. 6-8. The photodiode has a small series resistance R_s, a total capacitance C_d consisting of junction and packaging capacitances, and a bias (or load) resistor R_L. The amplifier following the photodiode has an input capacitance C_a

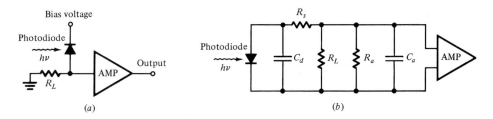

FIGURE 6-8
(*a*) Simple model of a photodetector receiver, and (*b*) its equivalent circuit.

and a resistance R_a. For practical purposes, R_s is much smaller than the load resistance R_L and can be neglected.

If a modulated signal of optical power $P(t)$ falls on the detector, the primary photocurrent $i_{ph}(t)$ generated is

$$i_{ph}(t) = \frac{\eta q}{h\nu} P(t) \qquad (6\text{-}10)$$

The primary current consists of a dc value I_p, which is the average photocurrent due to the signal power, and a signal component $i_p(t)$. For *pin* photodiodes the mean-square signal current $\langle i_s^2 \rangle$ is

$$\langle i_s^2 \rangle = \sigma_{s,pin}^2 = \langle i_p^2(t) \rangle \qquad (6\text{-}11a)$$

where σ is the variance. For avalanche photodetectors,

$$\langle i_s^2 \rangle = \sigma_{s,\text{APD}}^2 = \langle i_p^2(t) \rangle M^2 \qquad (6\text{-}11b)$$

where M is the average of the statistically varying avalanche gain as defined in Eq. (6-7). For a sinusoidally varying input signal of modulation index m, the signal component $\langle i_p^2 \rangle$ is of the form (see Prob. 6-5)

$$\langle i_p^2(t) \rangle = \sigma_p^2 = \frac{m^2}{2} I_p^2 \qquad (6\text{-}12)$$

where m is defined in Eq. (4-54).

The principal noises associated with photodetectors that have no internal gain are quantum noise, dark-current noise generated in the bulk material of the photodiode, and surface leakage current noise. The *quantum* or *shot noise* arises from the statistical nature of the production and collection of photoelectrons when an optical signal is incident on a photodetector. It has been demonstrated[35] that these statistics follow a Poisson process. Since the fluctuations in the number of photocarriers created from the photoelectric effect are a fundamental property of the photodetection process, they set the lower limit on the receiver sensitivity when all other conditions are optimized. The quantum noise current has a mean-square value in a bandwidth B which is proportional to the average value of the photocurrent I_p:

$$\langle i_Q^2 \rangle = \sigma_Q^2 = 2q I_p B M^2 F(M) \qquad (6\text{-}13)$$

where $F(M)$ is a noise figure associated with the random nature of the avalanche process. From experimental results, it has been found that to a reasonable approximation $F(M) \simeq M^x$, where x (with $0 \le x \le 1.0$) depends on the material. This is discussed in more detail in Sec. 6.4. For *pin* photodiodes, M and $F(M)$ are unity.

The photodiode dark current is the current that continues to flow through the bias circuit of the device when no light is incident on the photodiode. This is a combination of bulk and surface currents. The *bulk dark current* i_{DB} arises from electrons and/or holes which are thermally generated in the *pn* junction of the photodiode. In an APD, these liberated carriers also get accelerated by the high electric field present at the *pn* junction, and are therefore multiplied by the avalanche gain mechanism. The mean-square value of this current is given by

$$\langle i_{DB}^2 \rangle = \sigma_{DB}^2 = 2qI_D M^2 F(M)B \tag{6-14}$$

where I_D is the primary (unmultiplied) detector bulk dark current.

The *surface dark current* is also referred to as a *surface leakage current* or simply the leakage current. It is dependent on surface defects, cleanliness, bias voltage, and surface area. An effective way of reducing surface dark current is through the use of a guard ring structure which shunts surface leakage currents away from the load resistor. The mean-square value of the surface dark current is given by

$$\langle i_{DS}^2 \rangle = \sigma_{DS}^2 = 2qI_L B \tag{6-15}$$

where I_L is the surface leakage current. Note that since avalanche multiplication is a bulk effect, the surface dark current is not affected by the avalanche gain.

A comparison[30] of typical dark currents for Si, Ge, GaAs, and $In_x Ga_{1-x}As$ photodiodes is given in Fig. 6-9 as a function of applied voltage normalized to the breakdown voltage V_B. Note that for $In_x Ga_{1-x}As$ photodiodes the dark current increases with the composition x. Under a reverse bias, both dark currents also increases with the area. The surface dark current increases in proportion to the

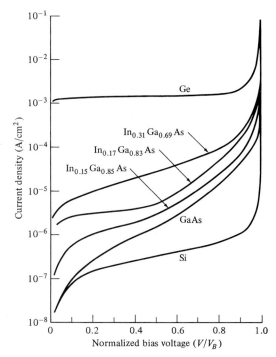

FIGURE 6-9
A comparison of typical dark currents for Si, Ge, GaAs, and InGaAs photodiodes as a function of normalized bias voltage. (Reproduced with permission from Susa, Yamauchi, and Kanbe,[30] © 1980, IEEE.)

square root of the active area, and the bulk dark current is directly proportional to the area.

Since the dark currents and the signal current are uncorrelated, the total mean-square photodetector noise current $\langle i_N^2 \rangle$ can be written as

$$\langle i_N^2 \rangle = \sigma_N^2 = \langle i_Q^2 \rangle + \langle i_{DB}^2 \rangle + \langle i_{DS}^2 \rangle = \sigma_Q^2 + \sigma_{DB}^2 + \sigma_{DS}^2$$

$$= 2q(I_p + I_D)M^2 F(M)B + 2qI_L B \tag{6-16}$$

To simplify the analysis of the receiver circuitry, we shall assume here that the amplifier input impedance is much greater than the load resistance, so that its thermal noise is much smaller than that of R_L. The photodetector load resistor contributes a mean-square thermal (Johnson) noise current

$$\langle i_T^2 \rangle = \sigma_T^2 = \frac{4k_B T}{R_L} B \tag{6-17}$$

where k_B is Boltzmann's constant and T is the absolute temperature. This noise can be reduced by using a load resistor which is large but still consistent with the receiver bandwidth requirements. Further details on this are given in Chap. 7 along with a detailed discussion of the amplifier noise current i_{amp}.

Example 6-6. An InGaAs *pin* photodiode has the following parameters at a wavelength of 1300 nm: $I_D = 4$ nA, $\eta = 0.90$, $R_L = 1000\,\Omega$, and the surface leakage current is negligible. The incident optical power is 300 nW (-35 dBm), and the receiver bandwidth is 20 MHz. Let us find the various noise terms of the receiver.

First, we need to find the primary photocurrent. From Eq. (6-6),

$$I_p = \mathcal{R}P_0 = \frac{\eta q}{h\nu} P_0 = \frac{\eta q \lambda}{hc} P_0$$

$$= \frac{(0.90)(1.6 \times 10^{-19}\,\text{C})(1.3 \times 10^{-6}\,\text{m})}{(6.625 \times 10^{-34}\,\text{J} \cdot \text{s})(3 \times 10^8\,\text{m/s})} 3 \times 10^{-7}\,\text{W} = 0.282\,\mu\text{A}$$

From Eq. (6-13), the mean-square quantum noise current for a *pin* photodiode is

$$\langle I_Q^2 \rangle = 2qI_p B = 2(1.6 \times 10^{-19}\,\text{C})(0.282 \times 10^{-6}\,\text{A})(20 \times 10^6\,\text{Hz})$$

$$= 1.80 \times 10^{-18}\,\text{A}^2$$

or

$$\langle I_Q^2 \rangle^{1/2} = 1.34\,\text{nA}$$

From Eq. (6-14), the mean-square dark current is

$$\langle I_{DB}^2 \rangle = 2qI_D B = 2(1.6 \times 10^{-19}\,\text{C})(4 \times 10^{-9}\,\text{A})(20 \times 10^6\,\text{Hz})$$

$$= 2.56 \times 10^{-20}\,\text{A}^2$$

or

$$\langle I_{DB}^2 \rangle^{1/2} = 0.16\,\text{nA}$$

The mean-square thermal noise current for the receiver is found from Eq. (6-17) as

$$\langle I_T^2 \rangle = \frac{4k_B T}{R_L} B = \frac{4(1.38 \times 10^{-23} \, \text{J/K})(293 \, \text{K})}{1 \, \text{k}\Omega} 20 \times 10^6 \, \text{Hz}$$

$$= 323 \times 10^{-18} \, \text{A}^2$$

or

$$\langle I_T^2 \rangle^{1/2} = 18 \, \text{nA}$$

Thus, for this receiver the rms thermal noise current is about 14 times greater than the rms shot noise current and about 100 times greater than the rms dark current.

6.2.2 Signal-to-Noise Ratio

Substituting Eqs. (6-11), (6-16), and (6-17) into Eq. (6-9) for the signal-to-noise ratio at the input of the amplifier, we have

$$\frac{S}{N} = \frac{\langle i_p^2 \rangle M^2}{2q(I_p + I_D)M^2 F(M)B + 2qI_L B + 4k_B TB/R_L} \tag{6-18}$$

In general, when *pin* photodiodes are used, the dominating noise currents are those of the detector load resistor (the thermal current i_T) and the active elements of the amplifier circuitry (i_{amp}). For avalanche photodiodes, the thermal noise is of lesser importance and the photodetector noises usually dominate.[36]

From Eq. (6-18), it can be seen that the signal power is multiplied by M^2 and the quantum noise plus bulk dark current is multiplied by $M^2 F(M)$. The surface-leakage current is not altered by the avalanche gain mechanism. Since the noise figure $F(M)$ increases with M, there always exists an optimum value of M that maximizes the signal-to-noise ratio. The optimum gain at the maximum signal-to-noise ratio can be found by differentiating Eq. (6-18) with respect to M, setting the result equal to zero, and solving for M. Doing so for a sinusoidally modulated signal, with $m = 1$ and $F(M)$ approximated by M^x, yields

$$M_{\text{opt}}^{x+2} = \frac{2qI_L + 4k_B T/R_L}{xq(I_p + I_D)} \tag{6-19}$$

6.3 DETECTOR RESPONSE TIME

6.3.1 Depletion Layer Photocurrent

To understand the frequency response of photodiodes, let us first consider the schematic representation of a reverse-biased *pin* photodiode shown in Fig. 6-10. Light enters the device through the *p* layer and produces electron–hole pairs as it is absorbed in the semiconductor material. Those electron–hole pairs that are generated in the depletion region or within a diffusion length of it will be separated by the reverse-bias-voltage-induced electric field, thereby leading to a current flow in the external circuit as the carriers drift across the depletion layer.

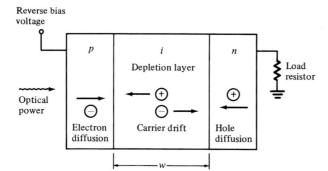

FIGURE 6-10

Schematic representation of a reverse-biased *pin* photodiode.

Under steady-state conditions, the total current density J_{tot} flowing through the reverse-biased depletion layer is[37]

$$J_{tot} = J_{dr} + J_{diff} \tag{6-20}$$

Here, J_{dr} is the drift current density resulting from carriers generated inside the depletion region, and J_{diff} is the diffusion current density arising from the carriers that are produced outside of the depletion layer in the bulk of the semiconductor (i.e., in the n and p regions) and diffuse into the reverse-biased junction. The drift current density can be found from Eq. (6-4):

$$J_{dr} = \frac{I_p}{A} = q\Phi_0(1 - e^{-\alpha_s w}) \tag{6-21}$$

where A is the photodiode area and Φ_0 is the incident photon flux per unit area given by

$$\Phi_0 = \frac{P_0(1 - R_f)}{Ah\nu} \tag{6-22}$$

The surface p layer of a *pin* photodiode is normally very thin. The diffusion current is thus principally determined by hole diffusion from the bulk n region. The hole diffusion in this material can be determined by the one-dimensional diffusion equation[12]

$$D_p \frac{\partial^2 p_n}{\partial x^2} - \frac{p_n - p_{n0}}{\tau_p} + G(x) = 0 \tag{6-23}$$

where D_p is the hole diffusion coefficient, p_n is the hole concentration in the n-type material, τ_p is the excess hole lifetime, p_{n0} is the equilibrium hole density, and $G(x)$ is the electron–hole generation rate given by

$$G(x) = \Phi_0 \alpha_s \, e^{-\alpha_s x} \tag{6-24}$$

From Eq. (6-23), the diffusion current density is found to be (see Prob. 6-10)

$$J_{diff} = q\Phi_0 \frac{\alpha_s L_p}{1 + \alpha_s L_p} e^{-\alpha_s w} + q p_{n0} \frac{D_p}{L_p} \tag{6-25}$$

Substituting Eqs. (6-21) and (6-25) into Eq. (6-20), we have that the total current density through the reverse-biased depletion layer is

$$J_{\text{tot}} = q\Phi_0 \left(1 - \frac{e^{-\alpha_s w}}{1 + \alpha_s L_p} \right) + q p_{n0} \frac{D_p}{L_p} \tag{6-26}$$

The term involving p_{n0} is normally small, so that the total photogenerated current is proportional to the photon flux Φ_0.

6.3.2 Response Time

The response time of a photodiode together with its output circuit (see Fig. 6-8) depends mainly on the following three factors:

1. The transit time of the photocarriers in the depletion region.
2. The diffusion time of the photocarriers generated outside the depletion region.
3. The *RC* time constant of the photodiode and its associated circuit.

The photodiode parameters responsible for these three factors are the absorption coefficient α_s, the depletion region width w, the photodiode junction and package capacitances, the amplifier capacitance, the detector load resistance, the amplifier input resistance, and the photodiode series resistance. The photodiode series resistance is generally only a few ohms and can be neglected in comparison with the large load resistance and the amplifier input resistance.

Let us first look at the transit time of the photocarriers in the depletion region. The response speed of a photodiode is fundamentally limited by the time it takes photogenerated carriers to travel across the depletion region. This transit time t_d depends on the carrier drift velocity v_d and the depletion layer width w, and is given by

$$t_d = \frac{w}{v_d} \tag{6-27}$$

In general, the electric field in the depletion region is large enough so that the carriers have reached their scattering-limited velocity. For silicon, the maximum velocities for electrons and holes are 8.4×10^6 and 4.4×10^6 cm/s, respectively, when the field strength is on the order of 2×10^4 V/cm. A typical high-speed silicon photodiode with a 10-μm depletion layer width thus has a response time limit of about 0.1 ns.

The diffusion processes are slow compared with the drift of carriers in the high-field region. Therefore, to have a high-speed photodiode, the photocarriers should be generated in the depletion region or so close to it that the diffusion times are less than or equal to the carrier drift times. The effect of long diffusion times can be seen by considering the photodiode response time. This response time is described by the rise time and fall time of the detector output when the detector is illuminated by a step input of optical radiation. The rise time τ_r is typically measured from the 10- to the 90-percent points of the leading edge of the output pulse, as is shown in Fig. 6-11. For fully depleted photodiodes the rise time τ_r and fall time τ_f are generally the same. However, they can be different at low bias levels where the photodiode is not fully depleted, since the photon collection time then starts to become a significant contributor to the rise time. In this case, charge carries produced in the depletion region are separated and collected quickly. On the other hand, electron–hole pairs generated in the n and p regions must slowly diffuse to the depletion region before they can be separated and collected. A typical response time of a partially depleted photodiode is shown in Fig. 6-12. The fast carriers allow the device output to rise to 50 percent of its maximum value in approximately 1 ns, but the slow carriers cause a relatively long delay before the output reaches its maximum value.

To achieve a high quantum efficiency, the depletion layer width must be much larger than $1/\alpha_s$ (the inverse of the absorption coefficient), so that most of the light will be absorbed. The response to a rectangular input pulse of a low-capacitance photodiode having $w \gg 1/\alpha_s$ is shown in Fig. 6-13b. The rise and fall times of the photodiode follow the input pulse quite well. If the photodiode capacitance is larger, the response time becomes limited by the RC time constant of the load resistor R_L and the photodiode capacitance. The photodetector response then begins to appear as that shown in Fig. 6-13c.

If the depletion layer is too narrow, any carriers created in the undepleted material would have to diffuse back into the depletion region before they could be collected. Devices with very thin depletion regions thus tend to show distinct slow- and fast-response components, as shown in Fig. 6-13d. The fast component in the rise time is due to carriers generated in the depletion region, whereas the slow component arises from the diffusion of carriers that are created with a distance L_n from the edge of the depletion region. At the end of the optical pulse, the carries in

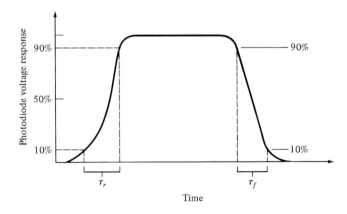

FIGURE 6-11
Photodiode response to an optical input pulse showing the 10- to 90-percent rise time and the 10- to 90-percent fall time.

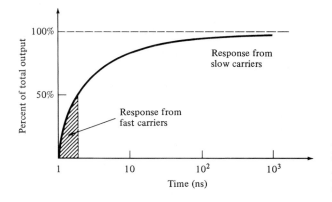

FIGURE 6-12
Typical response time of a photodiode that is not fully depleted.

the depletion region are collected quickly, which results in the fast-detector-response component in the fall time. The diffusion of carriers which are within a distance L_n of the depletion region edge appears as the slowly decaying tail at the end of the pulse. Also, if w is too thin, the junction capacitance will become excessive. The junction capacitance C_j is

$$C_j = \frac{\epsilon_s A}{w} \tag{6-28}$$

where ϵ_s = the permittivity of the semiconductor material = $\epsilon_0 K_s$
   K_s = the semiconductor dielectric constant
   ϵ_0 = 8.8542×10^{-12} F/m is the free-space permittivity
   A = the diffusion layer area

This excessiveness will then give rise to a large RC time constant which limits the detector response time. A reasonable compromise between high-frequency response and high quantum efficiency is found for absorption region thicknesses between $1/\alpha_s$ and $2/\alpha_s$.

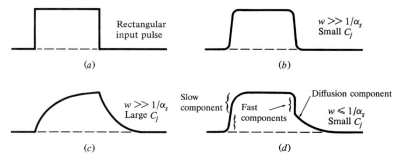

FIGURE 6-13
Photodiode pulse responses under various detector parameters.

If R_T is the combination of the load and amplifier input resistances and C_T is the sum of the photodiode and amplifier capacitances, as shown in Fig. 6-8, the detector behaves approximately like a simple RC low-pass filter with a passband given by

$$B = \frac{1}{2\pi R_T C_T} \tag{6-29}$$

Example 6-7. If the photodiode capacitance is 3 pF, the amplifier capacitance is 4 pF, the load resistor is 1 kΩ, and the amplifier input resistance is 1 MΩ, then $C_T = 7$ pF and $R_T \simeq 1$ kΩ, so that the circuit bandwidth is

$$B = \frac{1}{2\pi R_T C_T} = 23\ \text{MHz} \tag{6-30}$$

If we reduce the photodetector load resistance to 50 Ω, then the circuit bandwidth becomes $B = 455$ MHz.

6.4 AVALANCHE MULTIPLICATION NOISE

As we noted earlier, the avalanche process is statistical in nature, since not every photogenerated carrier pair undergoes the same multiplication.[38-41] The probability distribution of possible gains that any particular electron–hole pair might experience is sufficiently wide so that the mean-square gain is greater than the average gain squared. That is, if m denotes the statistically varying gain, then

$$\langle m^2 \rangle > \langle m \rangle^2 = M^2 \tag{6-31}$$

where the symbols $\langle\ \rangle$ denote an ensemble average and $\langle m \rangle = M$ is the average carrier gain defined in Eq. (6-7). Since the noise created by the avalanche process depends on the mean-square gain $\langle m^2 \rangle$, the noise in an avalanche photodiode can be relatively high. From experimental observations it has been found that, in general, $\langle m^2 \rangle$ can be approximated by

$$\langle m^2 \rangle \simeq M^{2+x} \tag{6-32}$$

where the exponent x varies between 0 and 1.0 depending on the photodiode material and structure.

The ratio of the actual noise generated in an avalanche photodiode to the noise that would exist if all carrier pairs were multiplied by exactly M is called the *excess noise factor* F and is defined by

$$F = \frac{\langle m^2 \rangle}{\langle m \rangle^2} = \frac{\langle m^2 \rangle}{M^2} \tag{6-33}$$

This excess noise factor is a measure of the increase in detector noise resulting from the randomness of the multiplication process. It depends on the ratio of the electron and hole ionization rates and on the carrier multiplication.

The derivation of an expression for F is complex, since the electric field in the avalanche region (of width W_m, as shown in Fig. 6-5) is not uniform, and both holes and electrons produce impact ionization. McIntyre[40] has shown that, for injected electrons and holes, the excess noise factors are

$$F_e = \frac{k_2 - k_1^2}{1 - k_2} M_e + 2\left[1 - \frac{k_1(1 - k_1)}{1 - k_2}\right] - \frac{(1 - k_1)^2}{M_e(1 - k_2)} \tag{6-34}$$

$$F_h = \frac{k_2 - k_1^2}{k_1^2(1 - k_2)} M_h - 2\left[\frac{k_2(1 - k_1)}{k_1^2(1 - k_2)} - 1\right] + \frac{(1 - k_1)^2 k_2}{k_1^2(1 - k_2)M_h} \tag{6-35}$$

where the subscripts e and h refer to electrons and holes, respectively. The weighted ionization rate ratios k_1 and k_2 take into account the nonuniformity of the gain and the carrier ionization rates in the avalanche region. They are given by

$$k_1 = \frac{\int_0^{W_M} \beta(x)M(x)\,dx}{\int_0^{W_M} \alpha(x)M(x)\,dx} \tag{6-36}$$

$$k_2 = \frac{\int_0^{W_M} \beta(x)M^2(x)\,dx}{\int_0^{W_M} \alpha(x)M^2(x)\,dx} \tag{6-37}$$

where $\alpha(x)$ and $\beta(x)$ are the electron and hole ionization rates, respectively.

Normally, to a first approximation k_1 and k_2 do not change much with variations in gain and can be considered as constant and equal. Thus, Eq. (6-34) and (6-35) can be simplified as[7]

$$F_e = M_e\left[1 - (1 - k_{\text{eff}})\left(1 - \frac{1}{M_e}\right)^2\right]$$

$$= k_{\text{eff}} M_e + \left(2 - \frac{1}{M_e}\right)(1 - k_{\text{eff}}) \tag{6-38}$$

for electron injection, and

$$F_h = M_h\left[1 - \left(1 - \frac{1}{k'_{\text{eff}}}\right)\left(1 - \frac{1}{M_h}\right)^2\right]$$

$$= k'_{\text{eff}} M_h - \left(2 - \frac{1}{M_h}\right)(k'_{\text{eff}} - 1) \tag{6-39}$$

for hole injection, where the effective ionization rate ratios are

$$k_{\text{eff}} = \frac{k_2 - k_1^2}{1 - k_2} \simeq k_2$$

$$k'_{\text{eff}} = \frac{k_{\text{eff}}}{k_1^2} \simeq \frac{k_2}{k_1^2} \tag{6-40}$$

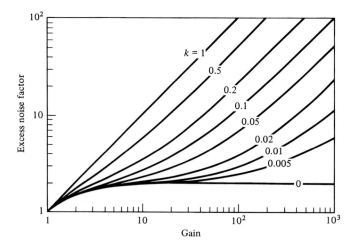

FIGURE 6-14

Variation of the electron excess noise factor F_e as a function of the electron gain for various values of the effective ionization rate ratio k_{eff}. (Reproduced with permission from Webb, McIntyre, and Conradi.[7])

Figure 6-14 shows F_e as a function of the average electron gain M_e for various values of the effective ionization rate ratio k_{eff}. If the ionization rates are equal, the excess noise is at its maximum so that F_e is at its upper limit of M_e. As the ratio β/α decreases from unity, the electron ionization rate starts to be the dominant contributor to impact ionization, and the excess noise factor becomes smaller. If only electrons cause ionization, $\beta = 0$ and F_e reaches its lower limit of 2.

This shows that to keep the excess noise factor at a minimum, it is desirable to have small values of k_{eff}. Referring back to Fig. 6-6, we thus see the superiority of silicon over other materials for making avalanche photodiodes. The effective ionization rate ratio k_{eff} varies between 0.015 and 0.035 for silicon, between 0.3 and 0.5 for indium gallium arsenide, and between 0.6 and 1.0 for germanium.

From the empirical relationship for the mean-square gain given by Eq. (6-32), the excess noise factor can be approximated by

$$F = M^x \tag{6-41}$$

The parameter x takes on values of 0.3 for Si, 0.7 for InGaAs, and 1.0 for Ge avalanche photodiodes.

6.5 STRUCTURES FOR InGaAs APDs

To improve the performance of InGaAs APDs, various complex device architectures have been devised. One widely used structure is the *separate-absorption-and-multiplication* (SAM) APD configuration.[42,43] As Fig. 6-15 shows, this structure

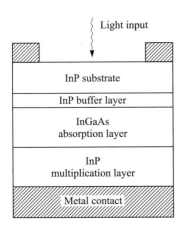

Light input

FIGURE 6-15
Simple diagram of a SAM APD structure (layers are not drawn to scale).

uses different materials in the absorption and multiplication regions, with each region being optimized for a particular function. Here, light enters the APD through the InP substrate. Since this material has a larger energy band-gap, it allows long-wavelength photons to pass through to the InGaAs absorption region where electron–hole pairs are generated. Following this is an InP layer that is used for the multiplication region because high electric fields needed for the gain mechanism can exist in InP without tunneling breakdown. This device structure gets its name SAM as a result of the separation of the absorption and multiplication regions.

Variations on the SAM structure include adding other layers to the device. These include

● Using a grading layer between the absorption and multiplication regions to increase the response time and bandwidth of the device.
● Adding a charge layer that provides better control of the electric field profile.
● Incorporating a resonant cavity that decouples the optical and electrical path lengths to achieve high quantum efficiencies and wide bandwidths simultaneously.

Another popular design for InGaAs APDs is the *superlattice structure*.[28,33] In these devices, the multiplication region is around 250 nm thick and consists of, for example, 13 layers of 9-nm-thick InAlGaAs quantum wells separated by 12-nm-thick InAlAs barrier layers. This structure improves the speed and sensitivity of InGaAs APDs, thereby allowing them to be used for applications such as 10-Gb/s long-distance systems (e.g., SONET OC-192/SDH STM-64 links).

6.6   TEMPERATURE EFFECT ON AVALANCHE GAIN

The gain mechanism of an avalanche photodiode is very temperature-sensitive because of the temperature dependence of the electron and hole ionization rates.[44-46] This temperature dependence is particularly critical at high bias voltages, where small changes in temperature can cause large variations in gain. An example of this is shown in Fig. 6-16 for a silicon avalanche photodiode. For example, if the operating temperature decreases and the applied bias voltage is kept constant, the ionization rates for electrons and holes will increase and so will the avalanche gain.

To maintain a constant gain as the temperature changes, the electric field in the multiplying region of the *pn* junction must also be changed. This requires that the receiver incorporate a compensation circuit which adjusts the applied bias voltage on the photodetector when the temperature changes.

The dependence of gain on temperature has been studied in detail by Conradi.[46] In that work, the gain curves were described by using the explicit temperature dependence of the ionization rates α and β together with a detailed knowledge of the device structure. Although excellent agreement was found between the theoretically computed and the experimentally measured gains, the calculations are rather involved. However, a simple temperature-dependent expression can be obtained from the empirical relationship[47]

$$M = \frac{1}{1 - (V/V_B)^n} \tag{6-42}$$

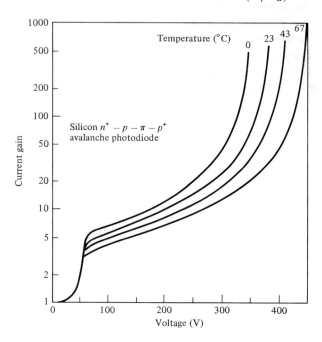

FIGURE 6-16

Example of how the gain mechanism of a silicon avalanche photodiode depends on temperature. The measurements for this device were carried out at 825 nm. (Reproduced with permission from Melchior, Hartman, Schinke, and Seidel,[15] © 1978, AT&T.)

where V_B is the breakdown voltage at which M goes to infinity; the parameter n varies between 2.5 and 7, depending on the material; and $V = V_a - I_M R_M$, with V_a being the reverse-bias voltage applied to the detector, I_M is the multiplied photocurrent, and R_M accounts for the photodiode series resistance and the detector load resistance. Since the breakdown voltage is known to vary with temperatures as[48,49]

$$V_B(T) = V_B(T_0)[1 + a(T - T_0)] \qquad (6\text{-}43)$$

the temperature dependence of the avalanche gain can be approximated by substituting Eq. (6-43) into Eq. (6-42) together with the expression

$$n(T) = n(T_0)[1 + b(T - T_0)] \qquad (6\text{-}44)$$

The constants a and b are positive for reach-through avalanche photodiodes and can be determined from experimental curves of gain versus temperature.

6.7 COMPARISONS OF PHOTODETECTORS

This section summarizes some generic operating characteristics of Si, Ge, and InGaAs photodiodes. Tables 6-1 and 6-2 list the performance values for *pin* and avalanche photodiodes, respectively. The values were derived from various

TABLE 6-1
Generic operating parameters of Si, Ge, and InGaAs *pin* photodiodes

Parameter	Symbol	Unit	Si	Ge	InGaAs
Wavelength range	λ	nm	400–1100	800–1650	1100–1700
Responsivity	$\mathscr{R}$	A/W	0.4–0.6	0.4–0.5	0.75–0.95
Dark current	I_D	nA	1–10	50–500	0.5–2.0
Rise time	τ_r	ns	0.5–1	0.1–0.5	0.05–0.5
Bandwidth	B	GHz	0.3–0.7	0.5–3	1–2
Bias voltage	V_B	V	5	5–10	5

TABLE 6-2
Generic operating parameters of Si, Ge, and InGaAs avalanche photodiodes

Parameter	Symbol	Unit	Si	Ge	InGaAs
Wavelength range	λ	nm	400–1100	800–1650	1100–1700
Avalanche gain	M	—	20–400	50–200	10–40
Dark current	I_D	nA	0.1–1	50–500	10–50 @ $M = 10$
Rise time	τ_r	ns	0.1–2	0.5–0.8	0.1–0.5
Gain · bandwidth	$M \cdot B$	GHz	100–400	2–10	20–250
Bias voltage	V_B	V	150–400	20–40	20–30

vendor data sheets and from performance numbers reported in the literature. They are given as guidelines for comparison purposes. Detailed values on specific devices for particular applications can be obtained from photodetector and receiver module suppliers.

For short-distance applications, Si devices operating around 850 nm provide relatively inexpensive solutions for most links. Longer links usually require operation in the 1300-nm and 1550-nm windows; here, one normally uses InGaAs-based devices.

PROBLEMS

6-1. Consider the absorption coefficient of silicon as a function of wavelength, as shown in Fig. P6-1. Ignoring reflections at the photodiode surface, plot the quantum efficiency for depletion layer widths of 1, 5, 10, 20, and 50 μm over the wavelength range 0.6–1.0 μm.

6-2. If an optical power level P_0 is incident on a photodiode, the electron–hole generation rate $G(x)$ in the photodetector is given by

$$G(x) = \Phi_0 \alpha_s \, e^{-\sigma_s x}$$

Here, Φ_0 is the incident photon flux per unit area given by

$$\Phi_0 = \frac{P_0(1 - R_f)}{Ah\nu}$$

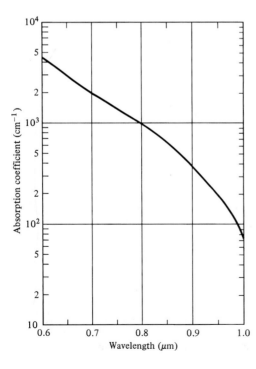

FIGURE P6-1
Absorption coefficient of Si as a function of wavelength.

where A is the detector area. From this, show that the primary photocurrent generated in the depletion region of width w is given by Eq. (6-4).

6-3. Using the data from Fig. P6-1, plot the responsivity over the wavelength range 0.6–1.0 μm for a silicon *pin* photodiode having a 20-μm-thick depletion layer. Assume $R_f = 0$.

6-4. The low-frequency gain M_0 of an avalanche photodiode depends on the carrier ionization rate and on the width of the multiplication region, both of which depend on the applied reverse-bias voltage V_a. This gain can be described by the empirical relationship[47]

$$M_0 = \frac{I_M}{I_p} = \frac{1}{1 - \left(\dfrac{V_a - I_M R_M}{V_B}\right)^n} \tag{P6-4}$$

where V_B is the breakdown voltage at which M_0 goes to infinity ($M_0 \to \infty$), I_M is the total multiplied current, and R_M accounts for the photodiode series resistance and the detector load resistance. The exponential factor n depends on the semiconductor material and its doping profile. Its value varies between about 2.5 and 7.

(a) Show that for applied voltages near the breakdown voltage, at which point $V_B \gg I_M R_M$, Eq. (P6-4) can be approximated by

$$M_0 = \frac{I_M}{I_p} \simeq \frac{V_B}{n(V_B - V_a + I_M R_M)} \simeq \frac{V_B}{n I_M R_M}$$

(b) The maximum value of M_0 occurs when $V_a = V_B$. Show that, at this point,

$$M_{0,\text{max}} = \left(\frac{V_B}{n R_M I_p}\right)^{1/2}$$

6-5. Consider a sinusoidally modulated optical signal $P(t)$ of frequency ω, modulation index m, and average power P_0 given by

$$P(t) = P_0(1 + m\cos\omega t)^2$$

Show that when this optical signal falls on a photodetector, the mean-square signal current $\langle i_s^2 \rangle$ generated consists of a dc (average) component I_p and a signal current i_p given by

$$\langle i_s^2 \rangle = I_p^2 + \langle i_p^2 \rangle = (\mathscr{R}_0 P_0)^2 + \tfrac{1}{2}(m\mathscr{R}_0 P_0)^2$$

where the responsivity $\mathscr{R}_0$ is given by Eq. (6-6).

6-6. An InGaAs *pin* photodiode has the following parameters at 1550 nm: $I_D = 1.0\,\text{nA}$, $\eta = 0.95$, $R_L = 500\,\Omega$, and the surface leakage current is negligible. The incident optical power is 500 nW (-33 dBm) and the receiver bandwidth is 150 MHz. Compare the noise currents given by Eq. (6-14), (6-15), and (6-16).

6-7. Consider an avalanche photodiode receiver that has the following parameters: dark current $I_D = 1$ nA, leakage current $I_L = 1$ nA, quantum efficiency $\eta = 0.85$, gain $M = 100$, excess noise factor $F = M^{1/2}$, load resistor $R_L = 10^4\,\Omega$, and bandwidth $B = 10$ kHz. Suppose a sinusoidally varying 850-nm signal having a modulation index $m = 0.85$ falls on the photodiode, which is at room temperature ($T = 300$ K). To compare the contributions form the various noise terms to the signal-to-noise ratio for this particular set of parameters, plot the following terms in decibels [i.e.,

$10 \log(S/N)$] as a function of the average received optical power P_0. Let P_0 range from -70 to 0 dBm; that is, from 0.1 nW to 1.0 mW:

(a)
$$\left(\frac{S}{N}\right)_Q = \frac{\langle i_s^2 \rangle}{\langle i_Q^2 \rangle}$$

(b)
$$\left(\frac{S}{N}\right)_{DB} = \frac{\langle i_s^2 \rangle}{\langle i_{DB}^2 \rangle}$$

(c)
$$\left(\frac{S}{N}\right)_{DS} = \frac{\langle i_s^2 \rangle}{\langle i_{DS}^2 \rangle}$$

(d)
$$\left(\frac{S}{N}\right)_T = \frac{\langle i_s^2 \rangle}{\langle i_T^2 \rangle}$$

What happens to these curves if either the load resistor, the gain, the dark current, or the bandwidth is changed?

6-8. Suppose an avalanche photodiode has the following parameters: $I_L = 1\,\text{nA}$, $I_D = 1\,\text{nA}$, $\eta = 0.85$, $F = M^{1/2}$, $R_L = 10^3\,\Omega$, and $B = 1\,\text{kHz}$. Consider a sinusoidally varying 850-nm signal, which has a modulation index $m = 0.85$ and an average power level $P_0 = -50$ dBm, to fall on the detector at room temperature. Plot the signal-to-noise ratio as a function of M for gains ranging from 20 to 100. At what value of M does the maximum signal-to-noise ratio occur?

6-9. Derive Eq. (6-19).

6-10. (a) Show that under the boundary conditions

$$p_n = p_{n0} \qquad \text{for} \qquad x = \infty$$

and

$$p_n = 0 \qquad \text{for} \qquad x = w$$

the solution to Eq. (6-23) is given by

$$p_n = p_{n0} - (p_{n0} + Be^{-\alpha_s w})\, e^{(w-x)/L_p} + Be^{-\alpha_s x}$$

where $L_p = (D_p \tau_p)^{1/2}$ is the diffusion length and

$$B = \left(\frac{\Phi_0}{D_p}\right) \frac{\alpha_s L_p^2}{1 - \alpha_s^2 L_p^2}$$

(b) Derive Eq. (6-25) using the relationship

$$J_{\text{diff}} = q D_p \left(\frac{\partial p_n}{\partial x}\right)_{x=w}$$

(c) Verify that J_{tot} is given by Eq. (6-26).

6-11. Consider a modulated photon flux density

$$\Phi = \Phi_0\, e^{j\omega t} \quad \text{photons}/(\text{s} \cdot \text{cm}^2)$$

to fall on a photodetector, where ω is the modulation frequency. The total current through the depletion region generated by this photon flux can be shown to be[37]

$$J_{\text{tot}} = \left(\frac{j\omega\epsilon_s V}{w} + q\Phi_0 \frac{1 - e^{-j\omega t_d}}{j\omega t_d} \right) e^{j\omega t}$$

where ϵ_s is the material permittivity, V is the voltage across the depletion layer, and t_d is the transit time of carriers through the depletion region.

(a) From the short-circuit current density ($V = 0$), find the value of ωt_d at which the photocurrent amplitude is reduced by $\sqrt{2}$.

(b) If the depletion region thickness is assumed to be $1/\alpha_s$, what is the 3-dB modulation frequency in terms of α_s and v_d (the drift velocity)?

6-12. Suppose we have a silicon *pin* photodiode which has a depletion layer width $w = 20\,\mu\text{m}$, an area $A = 0.05\,\text{mm}^2$, and a dielectric constant $K_s = 11.7$. If the photodiode is to operate with a 10-kΩ load resistor at 800 nm, where the absorption coefficient $\alpha_s = 10^3\,\text{cm}^{-1}$, compare the RC time constant and the carrier drift time of this device. Is carrier diffusion time of importance in this photodiode?

6-13. Verify that, when the weighted ionization rate ratios k_1 and k_2 are assumed to be approximately equal, Eqs. (6-34) and (6-35) can be simplified to yield Eqs. (6-38) and (6-39).

6-14. Derive the limits of F_e given by Eq. (6-38) when (a) only electrons cause ionization; (b) the ionization rates α and β are equal.

REFERENCES

1. H. Melchior, "Sensitive high speed photodetectors for the demodulation of visible and near infrared light," *J. Luminescence*, vol. 7, pp. 390–414, 1973.
2. H. Melchior, "Detectors for lightwave communications," *Phys. Today*, vol. 30, pp. 32–39, Nov. 1977.
3. (a) F. Capasso, "Multilayer avalanche photodiodes and solid-state photomultipliers," *Laser Focus/Electro-Optics*, vol. 20, pp. 84–101, July 1984.
   (b) Y. Wang, D. H. Park, and K. F. Brennan, "Theoretical analysis of confined quantum state GaAs/AlGaAs solid-state photomultipliers," *IEEE J. Quantum Electron.*, vol. 26, pp. 285–295, Feb. 1990.
4. E. H. Putley, "The pyro-electric detector," in R. K. Willardson and A. C. Beer, eds., *Semiconductors and Semimetals*, vol. 5, Academic, New York, 1970; "Photoconductive detectors," vol. 12, Academic, New York, 1977.
5. S. R. Forrest, "Optical detectors: Three contenders," *IEEE Spectrum*, vol. 23, pp. 76–84, May 1986.
6. G. E. Stillman and C. M. Wolfe, "Avalanche photodiodes," in R. K. Willardson and A. C. Beer, eds., *Semiconductors and Semimetals*, vol. 12, Academic, New York, 1977.
7. P. P. Webb, R. J. McIntyre, and J. Conradi, "Properties of avalanche photodiodes," *RCA Rev.*, vol. 35, pp. 234–278, June 1974.
8. J. C. Campbell, "Heterojunction photodetectors for optical communications," in N. G. Einspruch and W. R. Frensley, eds., *Heterostructures and Quantum Devices*, Academic, New York, 1994.
9. T. P. Lee and T. Li, "Photodetectors," in S. E. Miller and A. C. Chynoweth, eds., *Optical Fiber Telecommunications*, Academic, New York, 1979.
10. S. R. Forrest, "Optical detectors for lightwave communications," in S. E. Miller and I. P. Kaminow, eds., *Optical Fiber Telecommunications—II*. Academic, New York, 1988.
11. C. Pollock, *Fundamentals of Optoelectronics*, Irwin, Chicago, 1995.
12. S. M. Sze, *Physics of Semiconductor Devices*, Wiley, New York, 2nd ed., 1981, chap. 13.
13. S. E. Miller, E. A. J. Marcatili, and T. Li, "Research toward optical-fiber transmission systems," *Proc. IEEE*, vol. 61, pp. 1703–1751, Dec. 1973.

14. J.-W. Hong, Y.-W. Chen, W.-L. Laih, Y.-K. Fanf, C.-Y. Chang, and C. Gong, "The hydro-genated amorphous silicon reach-through avalanche photodiode," *IEEE J. Quantum Electron.*, vol. 26, pp. 280–284, Feb. 1990.

15. H. Melchior, A. R. Hartman, D. P. Schinke, and T. E. Seidel, "Planar epitaxial silicon avalanche photodiode," *Bell Sys. Tech. J.,* vol. 57, pp. 1791–1807, July/Aug. 1978.

16. J. T. K. Tang and K. B. Letaief, "The use of WMC distribution for performance evaluation of APD optical communication systems," *IEEE Trans. Commun.*, vol. 46, pp. 279–285, Feb. 1998.

17. H. Sudo, Y. Nakano, and G. Iwanea, "Reliability of germanium avalanche photodiodes for optical transmission systems," *IEEE Trans. Electron. Devices*, vol. 33, pp. 98–103, Jan. 1986.

18. T. P. Lee, C. A. Burrus, A. G. Dentai, A. A. Ballman, and W. A. Bonner, "High avalanche gain in small-area InP photodiodes," *Appl. Phys. Lett.,* vol 35, pp. 511–513, Oct. 1979.

19. R. Yeats and S. H. Chiao, "Long-wavelength InGaAsP avalanche photodiodes," *Appl. Phys. Lett.*, vol. 34, pp. 581–583, May 1979; "Leakage current in InGaAsP avalanche photodiodes," *ibid.*, vol. 36, pp. 160–170, Jan. 1980.

20. T. Shirai, S. Yamasaki, F. Osaka, K. Nakajima, and T. Kaneda, "Multiplication noise in planar InP/InGaAsP heterostructure avalanche photodiodes," *Appl. Phys. Lett.*, vol. 40, pp. 532–533, Mar. 1982.

21. B. L. Casper and J. C. Campbell, "Multigigabit-per-second avalanche photodiode lightwave receivers," *J. Lightwave Tech.*, vol. LT-5, pp. 1351–1364, Oct. 1987.

22. F. Osaka, T. Mikawa, and T. Kaneda, "Impact ionization coefficients for electrons and holes in (100)-oriented GaInAsP," *IEEE J. Quantum Electron.*, vol. QE-21, pp. 1326–1338, Sept. 1985.

23. F. Capasso, M. B. Panish, S. Sumski, and P. W. Foy, "Very high quantum efficiency GaSb mesa photodetectors between 1.3 and 1.6 μm," *Appl. Phys. Lett.*, vol. 36, pp. 165–167, Jan. 1980.

24. Y. Nagao, T. Hariu, and Y. Shibata, "GaSb Schottky diodes for infrared detectors," *IEEE Trans. Electron. Devices*, vol. ED-28, pp. 407–411, Apr. 1981.

25. L. R. Tomasetta, H. D. Law, R. C. Eden, I. Deyhimy, and K. Nakano, "High sensitivity optical receivers for 1.0–1.4 μm fiber optic systems," *IEEE J. Quantum Electron*, vol. QE-14, pp. 800–804, Nov. 1978.

26. (*a*) R. Alabedra, B. Orsal, G. Lecoy, G. Pichard, J. Meslage, and P. Fragnon, "An HgCdTe avalanche photodiode for optical-fiber transmission systems at 1.3 μm," *IEEE Trans. Electron. Devices*, vol. 32, pp. 1302–1306, July 1985.

    (*b*) B. Orsal, R. Alabedra, M. Valenza, G. Lecoy, J. Meslage, and C. Y. Boisrobert," "HgCdTe 1.55-μm avalanche photodiode noise analysis in the vicinity of resonant impact ionization connected with the spin-orbit split-off band," *IEEE Trans. Electron. Devices*, vol. 35, pp. 101–107, Jan. 1988.

27. J. G. Bauer and R. Trommer, "Long-term operation of planar InGaAs/InP *p-i-n* photodiodes," *IEEE Trans. Electron Devices*, vol. 35, pp. 2349–2354, Dec. 1988.

28. B. F. Levine, "Optimization of 10–20 GHz avalanche photodiodes," *IEEE Photonics Tech. Lett.*, vol. 8, pp. 1528–1530, Nov. 1996.

29. M. C. Brain and T. P. Lee, "Optical receivers for lightwave communication systems," *J. Lightwave Tech.*, vol. LT-3, pp. 1281–1300, Dec. 1985.

30. N. Susa, Y. Yamauchi, and H. Kanbe, "Vapor phase epitaxially grown InGaAs photodiodes," *IEEE Trans. Electron. Devices*, vol. ED-27, pp. 92–98, Jan. 1980.

31. C. P. Skrimshire, J. R. Farr, D. F. Sloan, M. J. Robertson, P. A. Putland, J. C. D. Stokoe, and R. R. Sutherland, "Reliability of mesa and planar InGaAs *pin* photodiodes," *IEE Proc.*, vol. 137, pp. 74–78, Feb. 1990.

32. W. Wu, A. R. Hawkins, and J. E. Bowers, "Design of silicon hetero-interface photodetectors," *J. Lightwave Tech.*, vol. 15, pp. 1608–1615, Aug. 1997.

33. I. Watanabe, M. Tsuji, M. Hayashi, K. Makita, and K. Taguchi, "Design and performance of InAlGaAs/InAlAs superlattice avalanche photodiodes," *J. Lightwave Tech.*, vol. 15, pp. 1012–1019, June 1997.

34. M. Makiuchi, M. Norimatsu, C. Sakurai, K. Kondo, N. Yamamoto, and M. Yano, "Flip-chip planar GaInAs/InP *pin* photodiodes—fabrication and characteristics," *J. Lightwave Tech.*, vol. 13, pp. 2270–2275, Nov. 1995.

35. B. M. Oliver, "Thermal and quantum noise," *Proc. IEEE*, vol. 53, pp. 436–454, May 1965.

36. W. M. Hubbard, "Utilization of optical-frequency carriers for low and moderate bandwidth channels," *Bell Sys. Tech. J.*, vol. 52, pp. 731–765, May/June 1973.

37. W. W. Gaertnar, "Depletion-layer photoeffects in semiconductors," *Phys. Rev.*, vol. 116, pp. 84–87, Oct. 1959.

38. R. S. Fyath and J. J. O'Reilly, "Performance degradation of APD-optical receivers due to dark current generated within the multiplication region," *J. Lightwave Tech.*, vol. 7, pp. 62–67, Jan. 1989.

39. S. D. Personick, "Statistics of a general class of avalanche detectors with applications to optical communications," *Bell Sys. Tech. J.*, vol. 50, pp. 3075–3096, Dec. 1971.

40. R. J. McIntyre, "The distribution of gains in uniformly multiplying avalanche photodiodes: Theory," *IEEE Trans. Electron. Devices*, vol. ED-19, pp. 703–713, June 1972.

41. J. Conradi, "The distribution of gains in uniformly multiplying avalanche photodiodes: Experimental," *IEEE Trans. Electron. Devices*, vol. ED-19, pp. 713–718, June 1972.

42. J. N. Haralson II, J. W. Parks, K. F. Brennan, W. Clark, and L. E. Tarof, "Numerical simulation of avalanche breakdown within InP–InGaAs SAGCM standoff avalanche photodiodes," *J. Lightwave Tech.*, vol. 15, pp. 2137–2140, Nov. 1997.

43. H. Nie, K. A. Anselm, C. Lenox, P. Yuan, C. Hu, G. Kinsey, B. G. Streetman, and J. C. Campbell, "Resonant-cavity separate absorption, charge and multiplication APDs with high-speed and high gain-bandwidth product," *IEEE Photonics Tech. Lett.*, vol. 10, pp. 409–411, Mar. 1998.

44. T. Mikawa, S. Kagawa, T. Kaneda, Y. Toyama, and O. Mikami, "Crystal orientation dependence of ionization rates in germanium," *Appl. Phys. Lett.*, vol. 37, pp. 387–389, Aug. 1980.

45. C. R. Crowell and S. M. Sze, "Temperature dependence of avalanche multiplication in semiconductors," *Appl. Phys. Lett.*, vol. 9, pp. 242–244, Sept. 1966.

46. J. Conradi, "Temperature effects in silicon avalanche photodiodes," *Solid State Electron.*, vol. 17, pp. 99–106, Jan. 1974.

47. S. L. Miller, "Avalanche breakdown in germanium," *Phys. Rev.*, vol. 99, pp. 1234–1241, Aug. 1955.

48. M. S. Tyagi, "Zener and avalanche breakdown in silicon alloyed *p–n* junction," *Solid State Electron.*, vol. 11, pp. 99–115, Feb. 1968.

49. N. Susa, H. Nakagome, H. Ando, and H. Kanbe, "Characteristics in InGaAs/InP avalanche photodiodes with separated absorption and multiplication regions," *IEEE J. Quantum Electron*, vol. QE-17, pp. 243–250, Feb. 1981.

CHAPTER
7

OPTICAL RECEIVER OPERATION

Having discussed the characteristics and operation of photodetectors in the previous chapter, we now turn our attention to the optical receiver. An optical receiver consists of a photodetector, an amplifier, and signal-processing circuitry. It has the task of first converting the optical energy emerging from the end of a fiber into an electric signal, and then amplifying this signal to a large enough level so that it can be processed by the electronics following the receiver amplifier.

In these processes, various noises and distortions will unavoidably be introduced, which can lead to errors in the interpretation of the received signal. As we saw in the previous chapter, the current generated by the photodetector is generally very weak and is adversely affected by the random noises associated with the photodetection process. When this electric signal output from the photodiode is amplified, additional noises arising from the amplifier electronics will further corrupt the signal. Noise considerations are thus important in the design of optical receivers, since the noise sources operating in the receiver generally set the lowest limit for the signals that can be processed.

In designing a receiver, it is desirable to predict its performance based on mathematical models of the various receiver stages. These models must take into account the noises and distortions added to the signal by the components in each stage, and they must show the designer which components to choose so that the desired performance criteria of the receiver are met.

The most meaningful criterion for measuring the performance of a digital communication system is the average error probability. In an analog system the fidelity criterion is usually specified in terms of a peak signal-to-rms-noise ratio. The calculation of the error probability for a digital optical communication receiver differs from that of conventional electric systems. This is because of the discrete quantum nature of the optical signal and also because of the probabilistic character of the gain process when an avalanche photodiode is used. Various authors[1-8] have used different numerical methods to derive approximate predictions for receiver performance. In carrying out these predictions, a tradeoff results between simplicity of the analysis and accuracy of the approximation. General reviews and concepts of optical receiver designs are given in Refs. 9–18.

In this chapter, we first present an overview of the fundamental operational characteristics of the various stages of an optical receiver. This consists of tracing the path of a digital signal through the receiver and showing what happens at each step along the way. Section 7.2 then outlines the fundamental probability methods for determining the bit-error rate or probability of error of a digital receiver based on signal-to-noise considerations. The mathematical details for this are given in Sec. 7.3. These derivations and Sec. 7.4 on receiver preamplifiers encompass advanced material that can be skipped without loss of continuity; these sections are designated by a star (★). The final discussion in Sec. 7.5 addresses analog receivers, which play an important part in many applications, such as extensions of microwave and satellite links, CATV, and video transmission systems.

7.1 FUNDAMENTAL RECEIVER OPERATION

The design of an optical receiver is much more complicated than that of an optical transmitter because the receiver must first detect weak, distorted signals and then make decisions on what type of data was sent based on an amplified version of this distorted signal. To get an appreciation of the function of the optical receiver, we first examine what happens to a signal as it is sent through the optical data link shown in Fig. 7-1. Since most fiber optic systems use a two-level binary digital signal, we shall analyze receiver performance by using this signal form first. Analog receivers are discussed in Sec. 7.5.

7.1.1 Digital Signal Transmission

A typical digital fiber transmission link is shown in Fig. 7-1. The transmitted signal is a two-level binary data stream consisting of either a 0 or a 1 in a time slot of duration T_b. This time slot is referred to as a *bit period*. Electrically, there are many ways of sending a given digital message.[19-21] One of the simplest (but not necessarily the most efficient) techniques for sending binary data is *amplitude-shift keying*, wherein a voltage level is switched between two values, which are usually *on* or *off*. The resultant signal wave thus consists of a voltage pulse of amplitude V relative to the zero voltage level when a binary 1 occurs and a zero-

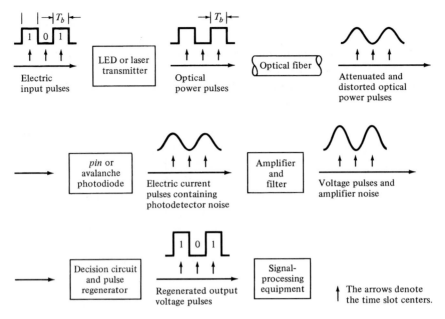

FIGURE 7-1
Signal path through an optical data link. (Adapted with permission from Personick et al.,[4] © 1977, IEEE.)

voltage-level space when a binary 0 occurs. Depending on the coding scheme to be used, a 1 may or may not fill the time slot T_b. For simplicity, here we assume that when a 1 is sent, a voltage pulse of duration T_b occurs, whereas for a 0 the voltage remains at its zero level. A discussion of more efficient transmission codes is given in Chap. 8.

The function of the optical transmitter is to convert the electric signal to an optical signal. As shown in Chap. 4, an electric current $i(t)$ can be used to modulate directly an optical source (either an LED or a laser diode) to produce an optical output power $P(t)$. Thus, in the optical signal emerging from the transmitter, a 1 is represented by a pulse of optical power (light) of duration T_b, whereas a 0 is the absence of any light.

The optical signal that gets coupled from the light source to the fiber becomes attenuated and distorted as it propagates along the fiber waveguide. Upon reaching the receiver, either a *pin* or an avalanche photodiode converts the optical signal back to an electrical format. After the electric signal produced by the photodetector is amplified and filtered, a decision circuit compares the signal in each time slot with a certain reference voltage known as the *threshold level*. If the received signal level is greater than the threshold level, a 1 is said to have been received. If the voltage is below the threshold level, a 0 is assumed to have been received.

In some cases, an optical amplifier is placed ahead of the photodiode to boost the optical signal level before photodetection. This is done so that the signal-to-noise ratio degradation caused by thermal noise in the receiver electronics can be suppressed. Compared with other front-end devices, such as avalanche photodiodes or optical heterodyne detectors, an optical preamplifier provides a larger gain factor and a broader bandwidth. However, this process also introduces additional noise to the optical signal. Chapter 11 addresses optical amplifiers and their effects on system performance.

7.1.2 Error Sources

Errors in the detection mechanism can arise from various noises and disturbances associated with the signal detection system, as shown in Fig. 7-2. The term *noise* is used customarily to describe unwanted components of an electric signal that tend to disturb the transmission and processing of the signal in a physical system, and over which we have incomplete control. The noise sources can be either external to the system (e.g., atmospheric noise, equipment-generated noise) or internal to the system. Here, we shall be concerned mainly with internal noise, which is present in every communication system and represents a basic limitation on the transmission or detection of signals. This noise is caused by the spontaneous fluctuations of current or voltage in electric circuits. The two most common samples of these spontaneous fluctuations are shot noise and thermal noise. Shot noise arises in electronic devices because of the discrete nature of current flow in the device. Thermal noise arises from the random motion of electrons in a conductor. Detailed treatments of electric noise may be found in Ref. 21.

As discussed in Chap. 6, the random arrival rate of signal photons produces a quantum (or shot) noise at the photodetector. Since this noise depends on the signal level, it is of particular importance for *pin* receivers that have large optical input levels and for avalanche photodiode receivers. When using an avalanche photodiode, an additional shot noise arises from the statistical nature of the multiplication process. This noise level increases with increasing avalanche gain M. Additional photodetector noises come from the dark current and leakage current. These are independent of the photodiode illumination and can generally be made very small in relation to other noise currents by a judicious choice of components.

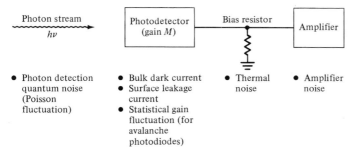

FIGURE 7-2
Noise sources and disturbances in the optical pulse detection mechanism.

Thermal noises arising from the detector load resistor and from the amplifier electronics tend to dominate in applications with low signal-to-noise ratio when a *pin* photodiode is used. When an avalanche photodiode is used in low-optical-signal-level applications, the optimum avalanche gain is determined by a design tradeoff between the thermal noise and the gain-dependent quantum noise.

Since the thermal noises are of a gaussian nature, they can be readily treated by standard techniques. This is shown in Sec. 7.3. The analysis of the noises and the resulting error probabilities associated with the primary photocurrent generation and the avalanche multiplication are complicated, since neither of these processes is gaussian. The primary photocurrent generated by the photodiode is a time-varying Poisson process resulting from the random arrival of photons at the detector. If the detector is illuminated by an optical signal $P(t)$, then the average number of electron–hole pairs $\bar{N}$ generated in a time τ is

$$\bar{N} = \frac{\eta}{h\nu} \int_0^\tau P(t)\ dt = \frac{\eta E}{h\nu} \tag{7-1}$$

where η is the detector quantum efficiency, $h\nu$ is the photon energy, and E is the energy received in a time interval τ. The actual number of electron–hole pairs n that are generated fluctuates from the average according to the Poisson distribution

$$P_r(n) = \bar{N}^n \frac{e^{-\bar{N}}}{n!} \tag{7-2}$$

where $P_r(n)$ is the probability that n electrons are emitted in an interval τ. The fact that it is not possible to predict exactly how many electron–hole pairs are generated by a known optical power incident on the detector is the origin of the type of shot noise called *quantum noise*. The random nature of the avalanche multiplication process gives rise to another type of shot noise. Recall from Chap. 6 that, for a detector with a mean avalanche gain M and an ionization rate ratio k, the excess noise factor $F(M)$ for electron injection is

$$F(M) = kM + \left(2 - \frac{1}{M}\right)(1 - k)$$

This equation is often approximated by the empirical expression

$$F(M) \simeq M^x \tag{7-3}$$

where the factor x ranges between 0 and 1.0 depending on the photodiode material.

A further error source is attributed to *intersymbol interference* (ISI), which results from pulse spreading in the optical fiber. When a pulse is transmitted in a given time slot, most of the pulse energy will arrive in the corresponding time slot at the receiver, as shown in Fig. 7-3. However, because of pulse spreading induced by the fiber, some of the transmitted energy will progressively spread into neighboring time slots as the pulse propagates along the fiber. The presence of this energy in adjacent time slots results in an interfering signal, hence the term *inter-*

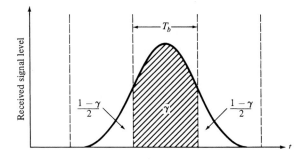

FIGURE 7-3
Pulse spreading in an optical signal that leads to intersymbol interference.

symbol interference. In Fig. 7-3 the fraction of energy remaining in the appropriate time slot is designated by γ, so that $1 - \gamma$ is the fraction of energy that has spread into adjacent time slots. Section 7.3 gives the details of ISI effects on system performance.

7.1.3 Receiver Configuration

A schematic diagram of a typical optical receiver is shown in Fig. 7-4. The three basic stages of the receiver are a photodetector, an amplifier, and an equalizer. The photodetector can be either an avalanche photodiode with a mean gain M or a *pin* photodiode for which $M = 1$. The photodiode has a quantum efficiency η and a capacitance C_d. The detector bias resistor has a resistance R_b which generates a thermal noise current $i_b(t)$.

The amplifier has an input impedance represented by the parallel combination of a resistance R_a and a shunt capacitance C_a. Voltages appearing across this impedance cause current to flow in the amplifier output. This amplifying function is represented by the voltage-controlled current source which is characterized by a *transconductance* g_m (given in amperes/volt, or *siemens*). There are two amplifier noise sources. The input noise current source $i_a(t)$ arises from the thermal noise of the amplifier input resistance R_a, whereas the noise voltage source $e_a(t)$ represents

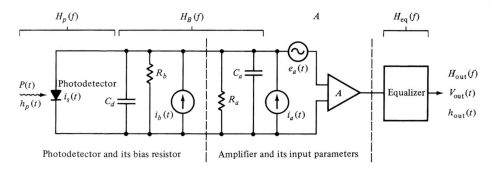

FIGURE 7-4
Schematic diagram of a typical optical receiver.

the thermal noise of the amplifier channel. These noise sources are assumed to be gaussian in statistics, flat in spectrum (which characterizes *white* noise), and uncorrelated (statistically independent). They are thus completely described by their noise spectral densities[19] S_I and S_E (see App. E).

The equalizer that follows the amplifier is normally a linear frequency-shaping filter that is used to mitigate the effects of signal distortion and intersymbol interference. Ideally,[20] it accepts the combined frequency response of the transmitter, the transmission medium, and the receiver, and transforms it into a signal response that is suitable for the following signal-processing electronics. In some cases, the equalizer may be used only to correct for the electric frequency response of the photodetector and the amplifier.

To account for the fact that the rectangular digital pulses that were sent out by the transmitter arrive rounded and distorted at the receiver, the binary digital pulse train incident on the photodetector can be described by

$$P(t) = \sum_{n=-\infty}^{\infty} b_n h_p(t - nT_b) \tag{7-4}$$

Here, $P(t)$ is the received optical power, T_b is the bit period, b_n is an amplitude parameter representing the nth message digit, and $h_p(t)$ is the received pulse shape, which is positive for all t. For binary data the parameter b_n can take on the two values, b_{on} and b_{off}, corresponding to a binary 1 and 0, respectively. If we let the nonnegative photodiode input pulse $h_p(t)$ be normalized to have unit area

$$\int_{-\infty}^{\infty} h_p(t) \, dt = 1 \tag{7-5}$$

then b_n represents the energy in the nth pulse.

The mean output current from the photodiode at time t resulting from the pulse train given in Eq. (7-4) is (neglecting dc components arising from dark noise currents)

$$\langle i(t) \rangle = \frac{\eta q}{h\nu} MP(t) = \mathcal{R}_0 M \sum_{n=-\infty}^{\infty} b_n h_p(t - nT_b) \tag{7-6}$$

where $\mathcal{R}_0 = \eta q / h\nu$ is the photodiode responsivity as given in Eq. (6-6). This current is then amplified and filtered to produce a mean voltage at the output of the equalizer.

7.1.4   Fourier Transform Representation★

To evaluate the specifics of equalizer output voltage, we need to use Fourier transform techniques. We describe these briefly here. The mean voltage at the output of the equalizer is given by the convolution of the current with the amplifier impulse response (see App. E):

$$\langle v_{out}(t)\rangle = A\mathcal{R}_0 MP(t) * h_B(t) * h_{eq}(t)$$

$$= \mathcal{R}_0 GP(t) * h_B(t) * h_{eq}(t) \tag{7-7}$$

Here, A is the amplifier gain, we define $G = AM$ for brevity, $h_B(t)$ is the impulse response of the bias circuit, $h_{eq}(t)$ is the equalizer impulse response, and $*$ denotes convolution.

From Fig. 7-4, $h_B(t)$ is given by the inverse Fourier transform of the bias circuit transfer function $H_B(f)$:

$$h_B(t) = F^{-1}[H_B(f)] = \int_{-\infty}^{\infty} -H_B(f)\, e^{j2\pi ft} df \tag{7-8}$$

where F denotes the Fourier transform operation. The bias current transfer function $H_B(f)$ is simply the impedance of the parallel combination of R_b, R_a, C_d, and C_a:

$$H_B(f) = \frac{1}{1/R + j2\pi fC} \tag{7-9}$$

where

$$\frac{1}{R} = \frac{1}{R_a} + \frac{1}{R_b} \tag{7-10}$$

and

$$C = C_a + C_d \tag{7-11}$$

Analogous to Eq. (7-4), the mean voltage output from the equalizer can be written in the form

$$\langle v_{out}(t)\rangle = \sum_{n=-\infty}^{\infty} b_n h_{out}(t - nT_b) \tag{7-12}$$

where

$$h_{out}(t) = \mathcal{R}_0 G h_p(t) * h_B(t) * h_{eq}(t) \tag{7-13}$$

is the shape of an isolated amplified and filtered pulse. The Fourier transform of Eq. (7-13) can be written as[19] (see App. E)

$$H_{out}(f) = \int_{-\infty}^{\infty} h_{out}(t)\, e^{-j2\pi ft} dt = \mathcal{R}_0 G H_p(f) H_B(f) H_{eq}(f) \tag{7-14}$$

Here, $H_p(f)$ is the Fourier transform of the received pulse shape $h_p(t)$, and $H_{eq}(f)$ is the transfer function of the equalizer.

7.2 DIGITAL RECEIVER PERFORMANCE

In a digital receiver the amplified and filtered signal emerging from the equalizer is compared with a threshold level once per time slot to determine whether or not a pulse is present at the photodetector in that time slot. Ideally, the output signal

$v_{out}(t)$ would always exceed the threshold voltage when a 1 is present and would be less than the threshold when no pulse (a 0) was sent. In actual systems, deviations from the average value of $v_{out}(t)$ are caused by various noises, interference from adjacent pulses, and conditions wherein the light source is not completely extinguished during a zero pulse.

7.2.1   Probability of Error

In practice, there are several standard ways of measuring the rate of error occurrences in a digital data stream.[22] One common approach is to divide the number N_e of errors occuring over a certain time interval t by the number N_t of pulses (ones and zeros) transmitted during this interval. This is called either the *error rate* or the *bit-error rate*, which is commonly abbreviated BER. Thus, we have

$$\text{BER} = \frac{N_e}{N_t} = \frac{N_e}{Bt} \qquad (7\text{-}15)$$

where $B = 1/T_b$ is the bit rate (i.e., the pulse transmission rate). The error rate is expressed by a number, such as 10^{-9}, for example, which states that, on the average, one error occurs for every billion pulses sent. Typical error rates for optical fiber telecommunication systems range from 10^{-9} to 10^{-12}. This error rate depends on the signal-to-noise ratio at the receiver (the ratio of signal power to noise power). The system error rate requirements and the receiver noise levels thus set a lower limit on the optical signal power level that is required at the photodetector.

To compute the bit-error rate at the receiver, we have to know the probability distribution[23] of the signal at the equalizer output. Knowing the signal probability distribution at this point is important because it is here that the decision is made as to whether a 0 or a 1 is sent. The shapes of two signal probability distributions are shown in Fig. 7-5. These are

$$P_1(v) = \int_{-\infty}^{v} p(y|1) \, dy \qquad (7\text{-}16)$$

which is the probability that the equalizer output voltage is less than v when a logical 1 pulse is sent, and

$$P_0(v) = \int_{v}^{\infty} p(y|0) \, dy \qquad (7\text{-}17)$$

which is the probability that the output voltage exceeds v when a logical 0 is transmitted. Note that the different shapes of the two probability distributions in Fig. 7-5 indicate that the noise power for a logical 0 is usually not the same as that for a logical 1. This occurs in optical systems because of signal distortion from transmission impairments (e.g., dispersion, optical amplifier noise, and distortion from nonlinear effects) and from noise and ISI contributions at the receiver. The functions $p(y|1)$ and $p(y|0)$ are the conditional probability distribu-

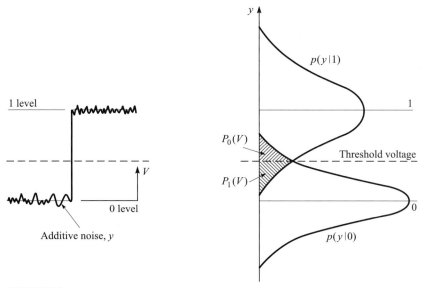

FIGURE 7-5
Probability distributions for received logical 0 and 1 signal pulses. The different widths of the two distributions are caused by various signal distortion effects.

tion functions;[9,23] that is, $p(y|x)$ is the probability that the output voltage is y, given that an x was transmitted.

If the threshold voltage is v_{th} then the error probability P_e is defined as

$$P_e = aP_1(v_{th}) + bP_0(v_{th}) \tag{7-18}$$

The weighting factors a and b are determined by the a priori distribution of the data. That is, a and b are the probabilities that either a 1 or a 0 occurs, respectively. For unbiased data with equal probability of 1 and 0 occurrences, $a = b = 0.5$. The problem to be solved now is to select the decision threshold at that point where P_e is minimum.

To calculate the error probability we require a knowledge of the mean-square noise voltage $\langle v_N^2 \rangle$ which is superimposed on the signal voltage at the decision time. The statistics of the output voltage at the sampling time are very complicated, so that an exact calculation is rather tedious to perform. A number of different approximations[1-18] have therefore been used to calculate the performance of a binary optical fiber receiver. In applying these approximations, we have to make a tradeoff between computational simplicity and accuracy of the results. The simplest method is based on a gaussian approximation. In this method, it is assumed that, when the sequence of optical input pulses is known, the equalizer output voltage $v_{out(t)}$ is a gaussian random variable. Thus, to calculate the error probability, we need only to know the mean and standard deviation of $v_{out}(t)$. Other approximations that have been investigated are more involved[4-6,16-18] and will not be discussed here.

Thus, let us assume that a signal s (which can be either a noise disturbance or a desired information-bearing signal) has a gaussian probability distribution function with a mean value m. If we sample the signal voltage level $s(t)$ at any arbitrary time t_1, the probability that the measured sample $s(t_1)$ falls in the range s to $s + ds$ is given by

$$f(s)\,ds = \frac{1}{\sqrt{2\pi\sigma^2}} e^{-(s-m)^2/2\sigma^2}\,ds \tag{7-19}$$

where $f(s)$ is the *probability density function*, σ^2 is the noise variance, and its square root σ is the *standard deviation*, which is a measure of the width of the probability distribution. By examining Eq. (7-19) we can see that the quantity $2\sqrt{2}\sigma$ measures the full width of the probability distribution at the point where the amplitude is $1/e$ of the maximum.

We can now use the probability density function to determine the probability of error for a data stream in which the 1 pulses are all of amplitude V. As shown in Fig. 7-6, the mean and variance of the gaussian output for a 1 pulse are b_{on} and σ_{on}^2, respectively, whereas for a 0 pulse they are b_{off} and σ_{off}^2, respectively. Let us first consider the case of a 0 pulse being sent, so that no pulse is present at the decoding time. The probability of error in this case is the probability that the noise will exceed the threshold voltage v_{th} and be mistaken for a 1 pulse. This probability of error $P_0(v)$ is the chance that the equalizer output voltage $v(t)$ will fall somewhere between v_{th} and ∞. Using Eqs. (7-17) and (7-19), we have

$$P_0(v_{th}) = \int_{v_{th}}^{\infty} p(y|0)dy = \int_{v_{th}}^{\infty} f_0(y)dy$$
$$= \frac{1}{\sqrt{2\pi}\sigma_{off}} \int_{v_{th}}^{\infty} \exp\left[-\frac{(v - b_{off})^2}{2\sigma_{off}^2}\right]dv \tag{7-20}$$

where the subscript 0 denotes the presence of a 0 bit.

Similarly, we can find the probability of error that a transmitted 1 is misinterpreted as a 0 by the decoder electronics following the equalizer. This prob-

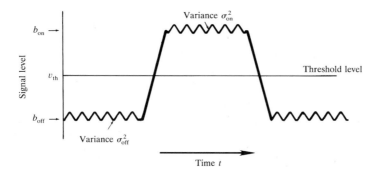

FIGURE 7-6
Gaussian noise statistics of a binary signal showing variances about the on and off signal levels.

ability of error is the likelihood that the sampled signal-plus-noise pulse falls below v_{th}. From Eqs. (7-16) and (7-19), this is simply given by

$$P_1(v_{th}) = \int_{-\infty}^{v_{th}} p(y|1)dy = \int_{-\infty}^{v_{th}} f_1(v)dv$$

$$= \frac{1}{\sqrt{2\pi}\sigma_{on}} \int_{-\infty}^{v_{th}} \exp\left[-\frac{(b_{on} - v)^2}{2\sigma_{on}^2}\right]dv \qquad (7\text{-}21)$$

where the subscript 1 denotes the presence of a 1 bit.

If we assume that the probabilities of 0 and 1 pulses are equally likely, then, using Eqs. (7-20) and (7-21), the bit-error rate (BER) or the error probability P_e given by Eq. (7-18) becomes

$$\text{BER} = P_e(Q) = \frac{1}{\sqrt{\pi}} \int_{Q/\sqrt{2}}^{\infty} e^{-x^2} dx$$

$$= \frac{1}{2}\left[1 - \text{erf}\left(\frac{Q}{\sqrt{2}}\right)\right] \approx \frac{1}{\sqrt{2\pi}} \frac{e^{-Q^2/2}}{Q} \qquad (7\text{-}22)$$

The approximation is obtained from the asymptotic expansion of $\text{erf}(x)$. Here, the parameter Q is defined as

$$Q = \frac{v_{th} - b_{off}}{\sigma_{off}} = \frac{b_{on} - v_{th}}{\sigma_{on}} \qquad (7\text{-}23)$$

and

$$\text{erf}(x) = \frac{2}{\sqrt{\pi}} \int_0^x e^{-y^2} dy \qquad (7\text{-}24)$$

is the *error function*, which is tabulated in various mathematical handbooks.[24] The factor Q is widely used to specify receiver performance, since it is related to the signal-to-noise ratio required to achieve a specific bit-error rate.[25] In particular, it takes into account that in optical fiber systems the variances in the noise powers generally are different for received logical 0 and 1 pulses. Figure 7-7 shows how the BER varies with Q. The approximation for P_e given in Eq. (7-22) and shown by the dashed line in Fig. 7-7 is accurate to 1 percent for $Q \approx 3$ and improves as Q increases. A commonly quoted Q value is 6, since this corresponds to a BER $= 10^{-9}$.

Let us consider the special case when $\sigma_{off} = \sigma_{on} = \sigma$ and $b_{off} = 0$, so that $b_{on} = V$. Then, from Eq. (7-23) we have that the threshold voltage $v_{th} = V/2$, so that $Q = V/2\sigma$. Since σ is usually called the *rms noise*, the ratio V/σ is the *peak signal-to-rms-noise ratio*. In this case, Eq. (7-22) becomes

$$P_e(\sigma_{on} = \sigma_{off}) = \frac{1}{2}\left[1 - \text{erf}\left(\frac{V}{2\sqrt{2}\sigma}\right)\right] \qquad (7\text{-}25)$$

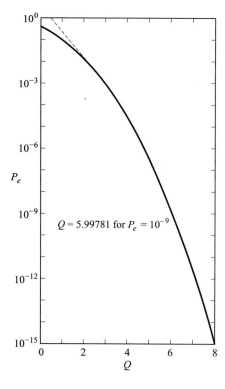

FIGURE 7-7
Plot of the BER (P_e) versus the factor Q. The approximation from Eq. (7-22) is shown by the dashed line.

Example 7-1. Figure 7-8 shows a plot of the BER expression from Eq. (7-25) as a function of the signal-to-noise ratio. Let us look at two cases of transmission rates.

(a) For a signal-to-noise ratio of 8.5 (18.6 dB) we have $P_e = 10^{-5}$. If this is the received signal level for a standard DS1 telephone rate of 1.544 Mb/s, this BER results in a misinterpreted bit every 0.065 s, which is highly unsatisfactory. However, by increasing the signal strength so that $V/\sigma = 12.0$ (21.6 dB), the BER decreases to $P_e = 10^{-9}$. For the DS1 case, this means that a bit is misinterpreted every 650 s (or 11 min), which, in general, is tolerable.

(b) For high-speed SONET links, say the OC-12 rate which operates at 622 Mb/s, BERs of 10^{-11} or 10^{-12} are required. This means that we need to have at least $V/\sigma = 13.0$ (22.3 dB).

Example 7-1 demonstrates the exponential behavior of the probability of error as a function of the signal-to-noise ratio. Here, we saw that by increasing V/σ by $\sqrt{2}$, that is, doubling S/N (a 3-dB power increase), the BER decreased by 10^4. Thus, there exists a narrow range of signal-to-noise ratios above which the error rate is tolerable and below which a highly unacceptable number of errors occur. The signal-to-noise ratio at which this transition occurs is called the *threshold level*. In general, a performance safety margin of 3–6 dB is included in the transmission link design to ensure that this threshold level is not exceeded when

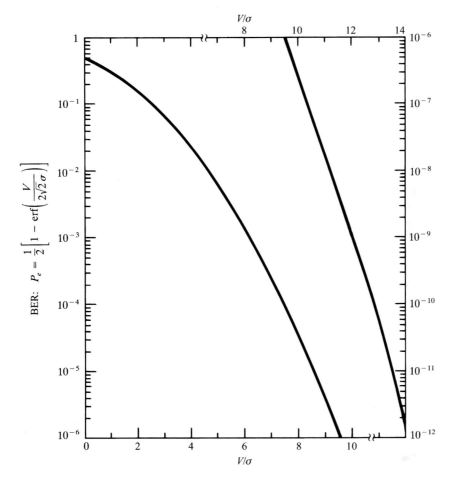

FIGURE 7-8
Bit-error rate as a function of signal-to-noise ratio when the standard deviations are equal ($\sigma_{on} = \sigma_{off}$) and $b_{off} = 0$.

system parameters such as transmitter output, line attenuation, or noise floor vary with time.

For simplicity, in the analyses in Sec. 7.3, we shall first assume that the optical power is completely extinguished in those time slots where a 0 occurs. In practice, there is usually some optical signal in the 0 time slot of a baseband binary signal, since the optical source is often biased slightly on at all times. As we saw in Chap. 4, this is done to increase the response speed of the light source. Biasing the light source slightly on during a 0 time slot results in a nonzero *extinction ratio* ϵ. This is defined as the ratio of the optical power in a 0 pulse to the power in a 1 pulse. Its effect is a power penalty in receiver sensitivity (see Sec. 7.3.5 where Fig. 7-16 shows a 1-dB penalty for $\epsilon = 0$).

7.2.2   The Quantum Limit

In designing an optical system, it is useful to know what the fundamental physical bounds are on the system performance. Let us see what this bound is for the photodetection process. Suppose we have an ideal photodetector which has unity quantum efficiency and which produces no dark current; that is, no electron–hole pairs are generated in the absence of an optical pulse. Given this condition, it is possible to find the minimum received optical power required for a specific bit-error-rate performance in a digital system. This minimum received power level is known as the *quantum limit*, since all system parameters are assumed ideal and the performance is limited only by the photodetection statistics.

Assume that an optical pulse of energy E falls on the photodetector in a time interval τ. This can only be interpreted by the receiver as a 0 pulse if no electron–hole pairs are generated with the pulse present. From Eq. (7-2) the probability that $n = 0$ electrons are emitted in a time interval τ is

$$P_r(0) = e^{-\bar{N}} \tag{7-26}$$

where the average number of electron–hole pairs, $\bar{N}$, is given by Eq. (7-1). Thus, for a given error probability $P_r(0)$, we can find the minimum energy E required at a specific wavelength λ.

Example 7-2. A digital fiber optic link operating at 850 nm requires a maximum BER of 10^{-9}.

(a) Let us first find the quantum limit in terms of the quantum efficiency of the detector and the energy of the incident photon. From Eq. (7-26) the probability of error is

$$P_r(0) = e^{-\bar{N}} = 10^{-9}$$

Solving for $\bar{N}$, we have $\bar{N} = 9 \ln 10 = 20.7 \simeq 21$. Hence, an average of 21 photons per pulse is required for this BER. Using Eq. (7-1) and solving for E, we get

$$E = 20.7 \frac{h\nu}{\eta}$$

(b) Now let us find the minimum incident optical power P_0 that must fall on the photodetector to achieve a 10^{-9} BER at a data rate of 10 Mb/s for a simple binary-level signaling scheme. If the detector quantum efficiency $\eta = 1$, then

$$E = P_0\tau = 20.7h\nu = 20.7 \frac{hc}{\lambda}$$

where $1/\tau$ is one-half the data rate B; that is, $1/\tau = B/2$. (Note: This assumes an equal number of 0 and 1 pulses.) Solving for P_0,

$$P_0 = 20.7 \frac{hcB}{2\lambda}$$

$$= \frac{20.7(6.626 \times 10^{-34} \text{J} \cdot \text{s})(3.0 \times 10^8 \text{ m/s})(10 \times 10^6 \text{ bits/s})}{2(0.85 \times 10^{-6} \text{ m})}$$

$$= 24.2 \text{ pW}$$

or, when the reference power level is 1 mW,

$$P_0 = -76.2 \, \text{dBm}$$

In practice, the sensitivity of most receivers is around 20 dB higher than the quantum limit because of various nonlinear distortions and noise effects in the transmission link. Furthermore, when specifying the quantum limit, one has to be careful to distinguish between average power and peak power. If one uses average power, the quantum limit given in Example 7-2 would be only 10 photons per bit for a 10^{-9} BER. Sometimes, the literature quotes the quantum limit based on these average powers. However, this can be misleading since the limitation on real components is based on peak and not average power.

7.3 DETAILED PERFORMANCE CALCULATIONS★

This section gives the mathematical details of one approach for evaluating digital receiver performance. For completeness of the discussion and a better understanding of the material, a few of the basic equations from Sec. 7.2 are repeated here.

7.3.1 Receiver Noises★

We now turn our attention to calculating the noise voltages or, equivalently, the noise currents. If $v_N(t)$ is the noise voltage causing $v_{\text{out}}(t)$ to deviate from its average value, then the actual equalizer output voltage is of the form

$$v_{\text{out}}(t) = \langle v_{\text{out}}(t) \rangle + v_N(t) \tag{7-27a}$$

The noise voltage at the equalizer output for the receiver shown in Fig. 7-4 can be represented by

$$v_N^2(t) = v_{\text{shot}}^2(t) + v_R^2(t) + v_I^2(t) + v_E^2(t) \tag{7-27b}$$

where

$v_{\text{shot}}(t)$ is the quantum (or shot) noise resulting from the random multiplied Poisson nature of the photocurrent $i_s(t)$ produced by the photodetector

$v_R(t)$ is the thermal (or Johnson) noise associated with the bias resistor R_b

$v_I(t)$ results from the amplifier input noise current source $i_a(t)$

$v_E(t)$ results from the amplifier input voltage noise source $e_a(t)$

The amplifier noise sources will be assumed independent of each other and gaussian in their statistics. The amplifier input current and voltage noise sources are also referred to as a shunt noise current and a series noise voltage, respectively.

Here, we are interested in the mean-square noise voltage $\langle v_N^2 \rangle$, which is given by

$$\langle v_N^2 \rangle = \langle [v_{\text{out}}(t) - \langle v_{\text{out}}(t) \rangle]^2 \rangle$$

$$= \langle v_{\text{out}}^2(t) \rangle - \langle v_{\text{out}}(t) \rangle^2$$

$$= \langle v_{\text{shot}}^2(t) \rangle + \langle v_R^2(t) \rangle + \langle v_I^2(t) \rangle + \langle v_E^2(t) \rangle \tag{7-28}$$

We shall first evaluate the last three thermal noise terms of Eq. (7-28) at the output of the equalizer. The thermal noise of the load resistor R_b is[19]

$$\langle v_R^2(t) \rangle = \frac{4k_B T}{R_b} B_{bae} R^2 A^2 \tag{7-29}$$

Here, $k_B T$ is Boltzmann's constant multiplied by the temperature in K, R is given by Eq. (7-10), A is the amplifier gain, and B_{bae} is the noise equivalent bandwidth of the bias circuit, amplifier, and equalizer defined for positive frequencies only:[19]

$$B_{bae} = \frac{1}{|H_B(0)H_{\text{eq}}(0)|^2} \int_0^\infty |H_B(f)H_{\text{eq}}(f)|^2 df$$

$$= \frac{1}{|H_{\text{out}}(0)/H_p(0)|^2} \int_0^\infty \left| \frac{H_{\text{out}}(f)}{H_p(f)} \right|^2 df \tag{7-30}$$

where we have used Eq. (7-14) for the last equality.

Since the thermal noise contributions from the amplifier input noise current source $i_a(t)$ and from the amplifier input noise voltage source $e_a(t)$ are assumed to be gaussian and independent, they are completely characterized by their noise spectral densities.[19] Thus,

$$\langle v_I^2(t) \rangle = S_I B_{bae} R^2 A^2 \tag{7-31}$$

and

$$\langle v_E^2(t) \rangle = S_E B_e A^2 \tag{7-32}$$

where S_I is the spectral density of the amplifier input noise current source (measured in amperes squared per hertz), S_E is the spectral density of the amplifier noise voltage source (measured in volts squared per hertz), and

$$B_e = \frac{1}{|H_{\text{eq}}(0)|^2} \int_0^\infty |H_{\text{eq}}(f)|^2 df$$

$$= \frac{R^2}{|H_{\text{out}}(0)/H_p(0)|^2} \int_0^\infty \left| \frac{H_{\text{out}}(f)}{H_p(f)} \left(\frac{1}{R} + j2\pi fC \right) \right|^2 df \tag{7-33}$$

is the noise equivalent bandwidth of the equalizer. The last equality comes from Eq. (7-14). The noise spectral densities are described further in Sec. 7.4.

7.3.2 Shot Noise★

The nongaussian nature of the photodetection process and the avalanche multiplication noise makes the evaluation of the shot noise $\langle v_s^2(t) \rangle$ more difficult than that of the thermal noise. Personick[1] carried out a detailed analysis that evaluated the shot noise as a function of time within the bit slot. This results in an accurate estimate of the shot noise contribution to the equalizer output noise voltage, but at the expense of computational difficulty.

Smith and Garrett[8] subsequently proposed a simplification of Personick's expressions by relating the mean-square shot noise voltage $\langle v_{\mathrm{shot}}^2(t) \rangle$ at the decision time to the average unity gain photocurrent $\langle i_0 \rangle$ over the bit time T_b through the shot noise expression[11,21]

$$\langle v_{\mathrm{shot}}^2(t) \rangle = 2q \langle i_0 \rangle \langle m^2 \rangle B_{bae} R^2 A^2 \tag{7-34}$$

Here, $\langle m^2 \rangle$ is the mean-square avalanche gain [Eq. (6-32)], which we shall assume takes the form M^{2+x} with $0 < x \le 1.0$. The other terms are as defined in Eq. (7-29). The factor of 2 arises because there is an equal contribution to the noise from the generation and from the recombination of carriers traversing the photoconductive channel.

We now calculate $\langle i_0 \rangle$ at the decision time within a particular bit slot. For this, we must take into account not only the shot noise contribution from a pulse within this particular time slot but also the shot noises resulting from all other pulses that overlap into this bit period. The shot noise within a time slot will thus depend on the shape of the received pulse (i.e., how much of it has spread into adjacent time slots, as shown in Fig. 7-3) and on the data sequence (the distribution of 1 and 0 pulses in the data stream). The worst case of shot noise in any particular time slot occurs when all neighboring pulses are 1, since this causes the greatest amount of intersymbol interference. For this case, the mean unity gain photocurrent over a bit time T_b for a 1 pulse is

$$\langle i_0 \rangle_1 = \sum_{n=-\infty}^{\infty} \frac{\eta q}{h\nu} b_{\mathrm{on}} \frac{1}{T_b} \int_{-T_b/2}^{T_b/2} h_p(t - nT_b) \, dt$$

$$= \frac{\eta q \, b_{\mathrm{on}}}{h\nu \, T_b} \int_{-\infty}^{\infty} h_p(t) \, dt = \frac{\eta q \, b_{\mathrm{on}}}{h\nu \, T_b} \tag{7-35}$$

where we have made use of Eq. (7-5).

For a 0 pulse (with all adjacent pulses being 1), we assume $b_{\mathrm{off}} = 0$, so that

$$\langle i_0 \rangle_0 = \sum_{n \ne 0} \frac{\eta q}{h\nu} b_{\mathrm{on}} \frac{1}{T_b} \int_{-T_b/2}^{T_b/2} h_p(t - nT_b) \, dt$$

$$= \frac{\eta q \, b_{\mathrm{on}}}{h\nu \, T_b} \left[\sum_{n=-\infty}^{\infty} \int_{-T_b/2}^{T_b/2} h_p(t - nT_b) \, dt - \int_{-T_b/2}^{T_b/2} h_p(t) \, dt \right]$$

$$= \frac{\eta q \, b_{\mathrm{on}}}{h\nu \, T_b} (1 - \gamma) \tag{7-36}$$

The parameter

$$\gamma = \int_{-T_b/2}^{T_b/2} h_p(t)\, dt \tag{7-37}$$

is the fractional energy of a 1 pulse that is contained with its bit period, as shown by the shaded area in Fig. 7-3. The factor $1 - \gamma$ is thus the fractional energy of a pulse that has spread outside of its bit period as it traveled through the optical fiber.

Equations (7-35) and (7-36) can now be substituted back into Eq. (7-34) to find the worst-case shot noise for a 1 and a 0 pulse, respectively.

7.3.3 Receiver Sensitivity Calculation★

To calculate the sensitivity of an optical receiver, we first simplify the noise voltage expressions by using the notation of Personick.[1] We begin by assuming that the equalized pulse stream has no intersymbol interference at the sampling times nT_b, as shown in Fig. 7-9, and that the maximum value of $h_{\text{out}}(t)$ at $t = 0$ is unity. This means that

$$h_{\text{out}}(t = 0) = 1$$
$$h_{\text{out}}(t = nT_b) = 0 \qquad \text{for} \qquad n \neq 0 \tag{7-38}$$

Substituting this into Eq. (7-12), we then have from Eq. (7-27a) that the actual equalizer output voltage at the sampling times $t = nT_b$ is

$$v_{\text{out}} = b_n h_{\text{out}}(0) + v_N(nT_b) \tag{7-39}$$

This shows that the noise $v_N(t)$ depends on all the b_n values and on the time t.

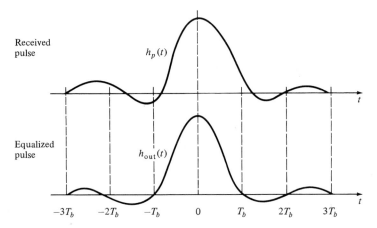

FIGURE 7-9
Equalized output pulse with no intersymbol interference at the decision time.

Furthermore, we introduce the dimensionless time and frequency variables $\tau = t/T_b$ and $\phi = fT_b$ to make the bandwidth integrals of Eqs. (7-30) and (7-33) independent of the bit period T_b. This means that the numerical value will depend only on the shapes of the received and equalized pulses and not on their scale. Using these values, it can be shown by considering the Fourier transforms of $h_p(\tau)$ and $h_{\text{out}}(\tau)$ that the normalized transforms, denoted by $H'_p(\phi)$ and $H'_{\text{out}}(\phi)$, are related to $H_p(f)$ and $H_{\text{out}}(f)$ by

$$H'_p(\phi) = H_p(f)$$

$$H'_{\text{out}}(\phi) = \frac{1}{T_b} H_{\text{out}}(f) \tag{7-40}$$

Thus, it follows that the *normalized bandwidth integrals* are[18]

$$I_2 = \frac{1}{T_b} \int_0^\infty \left| \frac{H_{\text{out}}(f)}{H_p(f)} \right|^2 df = \int_0^\infty \left| \frac{H'_{\text{out}}(\phi)}{H'_p(\phi)} \right|^2 d\phi \tag{7-41}$$

and

$$I_3 = \frac{1}{T_b} \int_0^\infty \left| \frac{H_{\text{out}}(f)}{H_p(f)} \right|^2 f^2 df = \int_0^\infty \left| \frac{H'_{\text{out}}(\phi)}{H'_p(\phi)} \right|^2 \phi^2 d\phi \tag{7-42}$$

Thus, using Eqs. (7-41) and (7-42), the bandwidth integrals in Eqs. (7-30) and (7-33) become

$$B_{bae} = \frac{I_2}{T_b} = I_2 B \tag{7-43}$$

and

$$B_e = \frac{I_2}{T_b} + \frac{(2\pi RC)^2}{T_b^3} I_3 = I_2 B + (2\pi RC)^2 I_3 B^3 \tag{7-44}$$

In evaluating Eqs. (7-30) and (7-33), we used the normalization conditions on $h_p(t)$ and $h_{\text{out}}(t)$ given by Eqs. (7-5) and (7-38), respectively, to derive the relationships

$$H_p(0) = 1 = H'_p(0) \quad \text{and} \quad H_{\text{out}}(0) = T_b$$

so that $H'_{\text{out}}(0) = 1$.

Using these expressions for B_{bae} and B_e, and from Eqs. (7-29), (7-31), (7-32), and (7-34), the total mean-square noise voltage in Eq. (7-28) becomes

$$\langle v_N^2 \rangle = R^2 A^2 \left(2q\langle i_0 \rangle M^{2+x} + \frac{4k_B T}{R_b} + S_I + \frac{S_E}{R^2} \right) I_2 B + (2\pi RC)^2 A^2 S_E I_3 B^3$$

$$= (qRAB)^2 \left(\frac{2\langle i_0 \rangle}{q} M^{2+x} T_b I_2 + W \right) \tag{7-45}$$

where

$$W = \frac{1}{q^2 B}\left(S_I + \frac{4k_B T}{R_b} + \frac{S_E}{R^2}\right)I_2 + \frac{(2\pi C)^2}{q^2}S_E I_3 B \qquad (7\text{-}46)$$

is a dimensionless parameter characterizing the thermal noise of the receiver. We shall call this parameter the *thermal noise characteristic* of the receiver amplifier.

Since the signal and noise voltages in Eq. (7-45) are each proportional to the resistance R, we can rewrite all of the expressions making up Eq. (7-45) in terms of input signal and noise *currents*. Doing so yields the results shown in Table 7-1.

Our task now is to find the minimum energy per pulse that is required to achieve a prescribed maximum bit-error rate. For this, we shall assume that the output voltage is approximately a gaussian variable. This is the *signal-to-noise ratio approximation*. Although the shot noise has a Poisson distribution, the inaccuracy resulting from the gaussian approximation is small.[7] The mean and variance of the gaussian output for a 1 pulse are b_{on} and σ_{on}^2, whereas for a 0 pulse they are b_{off} and σ_{off}^2. This is illustrated in Fig. 7-6. The variances σ_{on}^2 and σ_{off}^2 are defined as the worst-case values of $\langle v_N^2 \rangle$, which are obtained by substituting Eq. (7-35) and (7-36), respectively, for $\langle i_0 \rangle$ into Eq. (7-34):

$$\sigma_{on}^2 = \left(\frac{h\nu}{\eta}\right)^2\left(\frac{\eta M^x}{h\nu}b_{on}I_2 + \frac{W}{M^2}\right) \qquad (7\text{-}47)$$

$$\sigma_{off}^2 = \left(\frac{h\nu}{\eta}\right)^2\left[\frac{\eta M^x}{h\nu}b_{on}I_2(1-\gamma) + \frac{W}{M^2}\right] \qquad (7\text{-}48)$$

If the decision threshold voltage v_{th} is set so that there is an equal error probability for 0 and 1 pulses, and if we assume that there are an equal number of 0 and 1 pulses [i.e., $a = b = \frac{1}{2}$ in Eq. (7-18)], then from Eqs. (7-16) and (7-17),

$$P_0(v_{th}) = P_1(v_{th}) = \tfrac{1}{2}P_e$$

TABLE 7-1
Input signal and noise currents in an optical receiver

Shot noise	$\langle i_s^2 \rangle = 2q\langle i_0 \rangle \langle m^2 \rangle A^2 I_2 B$
Thermal noise	$\langle i_R^2 \rangle = \dfrac{4k_B T}{R_b}A^2 I_2 B$
Shunt noise	$\langle i_I^2 \rangle = S_I A^2 I_2 B$
Series noise	$\langle i_E^2 \rangle = S_E A^2\left[\dfrac{I_2 B}{R^2} + (2\pi C)^2 I_3 B^3\right]$
Total noise	$\langle i_N^2 \rangle = \langle i_s^2 \rangle + \langle i_R^2 \rangle + \langle i_I^2 \rangle + \langle i_E^2 \rangle$
	$\quad = A^2(2q\langle i_0 \rangle \langle m^2 \rangle I_2 B + q^2 W B^2)$

Assuming that the equalizer output is a gaussian variable, the bit-error rate (BER) or the error probability P_e is given by Eqs. (7-20) and (7-21):

$$\text{BER} = P_e = \frac{1}{\sqrt{2\pi}\sigma_{\text{off}}} \int_{v_{\text{th}}}^{\infty} \exp\left[-\frac{(v - b_{\text{off}})^2}{2\sigma_{\text{off}}^2}\right] dv$$

$$= \frac{1}{\sqrt{2\pi}\sigma_{\text{on}}} \int_{-\infty}^{v_{\text{th}}} \exp\left[\frac{-(-v + b_{\text{on}})^2}{2\sigma_{\text{on}}^2}\right] dv \qquad (7\text{-}49)$$

Defining the parameter Q as

$$Q = \frac{v_{\text{th}} - b_{\text{off}}}{\sigma_{\text{off}}} = \frac{b_{\text{on}} - v_{\text{th}}}{\sigma_{\text{on}}} \qquad (7\text{-}50)$$

then Eq. (7-49) becomes

$$\text{BER} = P_e(Q) = \frac{1}{\sqrt{\pi}} \int_{Q/\sqrt{2}}^{\infty} e^{-x^2} dx$$

$$= \frac{1}{2}\left[1 - \text{erf}\left(\frac{Q}{\sqrt{2}}\right)\right] \qquad (7\text{-}51)$$

where $\text{erf}(x)$ is the *error function* which is defined in Eq. (7-24). To an excellent approximation, Eq. (7-51) can be replaced by[26]

$$P_e(Q) = \frac{1}{\sqrt{2\pi}} \frac{e^{-Q^2/2}}{Q} \qquad (7\text{-}52)$$

The parameter Q is related to the signal-to-noise ratio required to achieve the desired bit-error rate.[26] Equation (7-51) states that relative to the noise at b_{off} the threshold voltage v_{th} must be at least Q standard deviations above b_{off}, or, equivalently, relative to the noise at b_{on} the threshold voltage must be no more than Q standard deviations below b_{on} to have the desired error rate.

Example 7-3. When there is little intersymbol interference, γ is small, so that $\sigma_{\text{on}}^2 \simeq \sigma_{\text{off}}^2$. Then, by letting $b_{\text{off}} = 0$, we have from Eq. (7-50) that

$$Q = \frac{b_{\text{on}}}{2\sigma_{\text{on}}} = \frac{1}{2}\frac{S}{N}$$

which is one-half the signal-to-noise ratio. In this case, $v_{\text{th}} = b_{\text{on}}/2$, so that the optimum decision threshold is midway between the 0 and 1 signal levels.

Example 7-4. For an error rate of 10^{-9} we have from Eq. (7-51) that

$$P_e(Q) = 10^{-9} = \frac{1}{2}\left[1 - \text{erf}\left(\frac{Q}{\sqrt{2}}\right)\right]$$

From Fig. 7-6 we have that $Q \simeq 6$ (an exact evaluation yields $Q = 5.99781$), which gives a signal-to-noise ratio of 12, or 10.8 dB [i.e., $10\log(S/N) = 10\log 12 = 10.8$ dB].

Using the expression in Eq. (7-50), the receiver sensitivity is given by

$$b_{on} - b_{off} = Q(\sigma_{on} + \sigma_{off}) \tag{7-53}$$

If b_{off} is zero, the required energy per pulse that is needed to achieve a desired bit-error rate characterized by the parameter Q is

$$b_{on} = \frac{Q}{M} \frac{h\nu}{\eta} \left\{ \left(M^{2+x} \frac{\eta}{h\nu} b_{on} I_2 + W \right)^{1/2} + \left[M^{2+x} \frac{\eta}{h\nu} b_{on} I_2 (1 - \gamma) + W \right]^{1/2} \right\} \tag{7-54}$$

We can now determine the optimum value of the avalanche gain, M_{opt}, by differentiating Eq. (7-54) with respect to M and putting $db_{on}/dM = 0$. After going through some lengthy but straightforward algebra, we obtain[8]

$$M_{opt}^{2+x} b_{on} = \frac{h\nu}{\eta} \frac{W}{2I_2} \left(\frac{2 - \gamma}{1 - \gamma} \right) K \tag{7-55}$$

where

$$K = -1 + \left[1 + 16 \frac{1 + x}{x^2} \frac{1 - \gamma}{(2 - \gamma)^2} \right]^{1/2} \tag{7-56}$$

The minimum energy per pulse necessary to achieve a bit-error rate characterized by Q can then be found by substituting Eq. (7-55) into Eq. (7-54) and solving for b_{on}. Doing so yields[8]

$$b_{on,min} = Q^{(2+x)/(1+x)} \frac{h\nu}{\eta} W^{x/(2+2x)} I_2^{1/(1+x)} L \tag{7-57}$$

where

$$L = \left[\frac{2(1 - \gamma)}{K(2 - \gamma)} \right]^{1/(1+x)} \left\{ \left[\frac{(2 - \gamma)K}{2(1 - \gamma)} + 1 \right]^{1/2} + [\tfrac{1}{2}(2 - \gamma)K + 1]^{1/2} \right\}^{(2+x)/(1+x)} \tag{7-58}$$

The parameter L has a somewhat involved expression, but it has the feature of depending only on the fraction γ of the pulse energy contained within a bit period T_b and on the avalanche photodiode factor x. Values for L are typically between 2 and 3. Recalling from Chap. 6 that x takes on values between 0 and 1.0 (e.g., 0 for *pin* photodiodes, 0.3 for silicon APDs, 0.7 for InGaAs APDs, and 1.0 for Ge APDs), we plot L as a function of γ in Fig. 7-10 for three different values of x. Note that these curves give L for any received pulse shape, since L depends only on x and γ.

The optimum gain at the desired bit-error rate characterized by Q can be found by substituting Eq. (7-57) into Eq. (7-55) to obtain

$$M_{opt}^{1+x} = \frac{W^{1/2}}{QI_2} \left[\frac{(2 - \gamma)K}{2(1 - \gamma)L} \right]^{(1+x)/(2+x)} \tag{7-59}$$

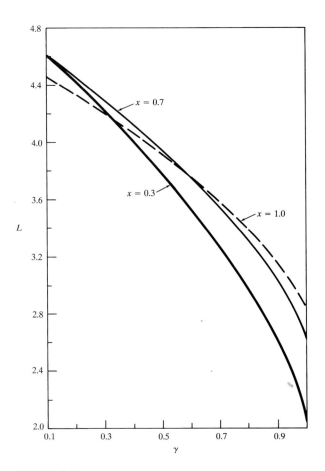

FIGURE 7-10
Relationship between the parameter L and the fraction γ of received optical energy of a pulse in a time slot T_b for $x = 0.3$ (Si), 0.7 (InGaAs), and 1.0 (Ge).

The optimum avalanche gain becomes, when $\gamma = 1$ (no intersymbol interference),

$$M_{\text{opt}}^{1+x} = \frac{2W^{1/2}}{xQI_2} \qquad (7\text{-}60)$$

The proof of this is left as an exercise.

7.3.4 Performance Curves★

Using Eq. (7-57) we can calculate the effect of intersymbol interference on the required energy per pulse at the optimum gain for any received and equalized pulse shape. The minimum optical power required occurs for very narrow optical input pulses.[1] Ideally, this is a unit impulse or delta function. More power is necessary for other received pulse shapes. The additional or *excess power* ΔP

required for pulse shapes other than impulses is normally defined as a *power penalty* measured in decibels. Thus,

$$\Delta P = 10 \log \frac{b_{on,nonimpulse}}{b_{on,impulse}} \qquad (7\text{-}61)$$

As an example, we shall calculate the case for which the amplifier resistance R given by Eq. (7-10) is sufficiently large so that the term

$$\frac{(2\pi C)^2}{T_b q^2} S_E I_3$$

dominates the thermal noise in Eq. (7-46). In this case,

$$\Delta P = 10 \log \frac{I_{3n}^{x/(2+2x)} I_{2n}^{1/(1+x)} L_n}{I_{3i}^{x/(2+2x)} I_{2i}^{1/(1+x)} L_i} \qquad (7\text{-}62)$$

where the subscripts n and i refer to *nonimpulse* and *impulse*, respectively.

For the input pulse shape $h_p(t)$ to the receiver, we shall choose a gaussian pulse,

$$h_p(t) = \frac{1}{\sqrt{2\pi}\alpha T_b} e^{-t^2/2\alpha^2 T_b^2} \qquad (7\text{-}63)$$

the normalized Fourier transform of which is

$$H'_p(\phi) = e^{-(2\pi\alpha\phi)^2/2} \qquad (7\text{-}64)$$

As shown in Fig. 7-11, the parameter αT_b, where T_b is the bit period, defines the variance or spread of the pulse.

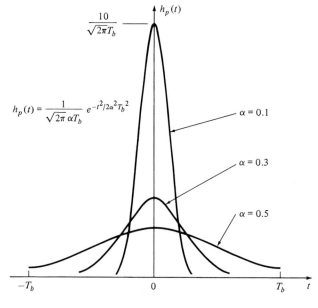

$$\frac{10}{\sqrt{2\pi T_b}}$$

$$h_p(t) = \frac{1}{\sqrt{2\pi}\,\alpha T_b} e^{-t^2/2\alpha^2 T_b{}^2}$$

$\alpha = 0.1$

$\alpha = 0.3$

$\alpha = 0.5$

FIGURE 7-11
Shape of a gaussian pulse as a function of the parameter α.

For the equalizer output waveform $h_{out}(t)$, we choose the commonly used raised-cosine pulse

$$h_{out}(t) = \frac{\sin \pi\tau}{\pi\tau} \frac{\cos \pi\beta\tau}{1 - (2\beta\tau)^2} \tag{7-65}$$

where $\tau = t/T_b$. A plot of $h_{out}(t)$ with $\beta = 0$, 0.5, and 1.0 is shown in Fig. 7-12. The normalized Fourier transform is

$$H'_{out}(\phi) = \begin{cases} 1 & \text{for} & 0 < |\phi| \le \dfrac{1-\beta}{2} \\[2mm] \dfrac{1}{2}\left[1 - \sin\left(\dfrac{\pi\phi}{\beta} - \dfrac{\pi}{2\beta}\right)\right] & \text{for} & \dfrac{1-\beta}{2} < |\phi| \le \dfrac{1+\beta}{2} \\[2mm] 0 & \text{otherwise} \end{cases} \tag{7-66}$$

The parameter β varies between 0 and 1 and determines the bandwidth used by the pulse, as shown in Fig. 7-12. A β value of unity indicates that the bandwidth is $2/T_b$, whereas $\beta = 0$ means that the minimum bandwidth of $1/T_b$ is used.

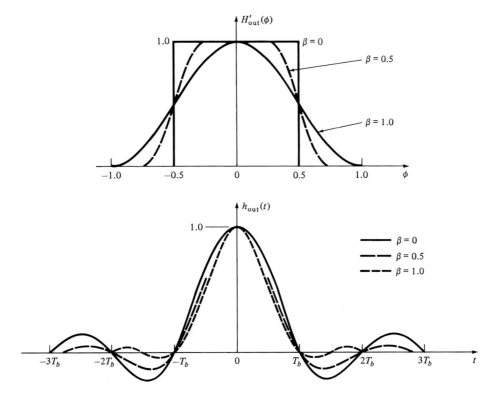

FIGURE 7-12
Shape of a raised-cosine pulse and its Fourier transform for three different values of the parameter β.

Although less bandwidth is used as β decreases, the tails of $h_{\text{out}}(t)$ become larger, and signal timing and equalization become more difficult. For simplicity, we shall choose $\beta = 1$ in the examples given here.

The unit impulse or Dira delta function is characterized by

$$h_{pi}(t) = \delta(t) \tag{7-67}$$

where

$$\delta(t) = 0 \qquad \text{for} \qquad t \neq 0 \tag{7-68}$$

and

$$\int_{-\infty}^{\infty} \delta(t)\, dt = 1$$

Thus, the Fourier transform of the impulse function is

$$H'_{pi}(\phi) = F[\delta(t)] = \int_{-\infty}^{\infty} \delta(t)\, e^{j2\pi ft} dt = 1 \tag{7-69}$$

Using Eqs. (7-41), (7-66), and (7-69), we have for an impulse input $H'_p(\phi)$ and a raised-cosine output,

$$I_{2i} = \int_0^{\infty} |H'_{\text{out}}(\phi)|^2 d\phi = \frac{1}{2}\left(1 - \frac{\beta}{4}\right) \tag{7-70}$$

which, for $\beta = 1$, becomes

$$I_{2i} = \tfrac{3}{8} \tag{7-71}$$

From Eq. (7-42),

$$I_{3i} = \int_0^{\infty} |H'_{\text{out}}(\phi)|^2 \phi^2\, d\phi$$

$$= \frac{\beta^3}{16}\left(\frac{1}{\pi^2} - \frac{1}{6}\right) + \beta^2\left(\frac{1}{8} - \frac{1}{\pi^2}\right) - \frac{\beta}{32} + \frac{1}{24} \tag{7-72}$$

For $\beta = 1$, we have

$$I_{3i} = 0.03001$$

We now evaluate I_{2n} and I_{3n} in Eq. (7-62) for the gaussian input pulse shape given by Eq. (7-64) and for a raised-cosine output. With $\beta = 1$ in Eq. (7-66), we have

$$I_{2n} = \int_0^{\infty} \left|\frac{H'_{\text{out}}(\phi)}{H'_p(\phi)}\right|^2 d\phi$$

$$= \frac{2}{\pi} \int_0^{\pi/2} e^{16\alpha^2 x^2} \cos^4 x\, dx \tag{7-73}$$

The results of a numerical evaluation of Eq. (7-73) as a function of α are given in Fig. 7-13. With the same assumptions,

$$I_{3n} = \left(\frac{2}{\pi}\right)^3 \int_0^{\pi/2} x^2 \, e^{16\alpha^2 x^2} \cos^4 x \, dx \qquad (7\text{-}74)$$

The numerical evaluation of I_{3n} as a function of α is shown in Fig. 7-13.

Two more parameters (L_i and L_n) now remain to be evaluated in Eq. (7-62) in order to determine the receiver power penalty. Since all the pulse energy is contained within the bit period for a unit impulse input, L_i is determined by taking the limit of L in Eq. (7-58) as γ goes to unity. Thus,

$$L_i = \lim_{\gamma \to 1} L = (1 + x)\left(\frac{2}{x}\right)^{x/(1+x)} \qquad (7\text{-}75)$$

The proof of this is left as an exercise.

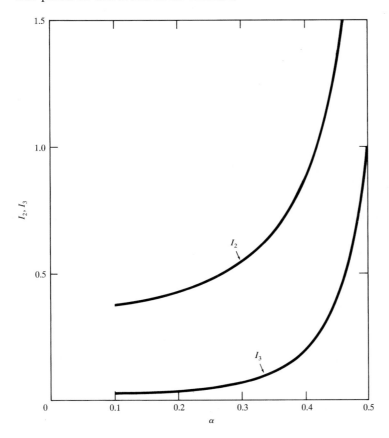

FIGURE 7-13
Plots of the normalized dimensionless bandwidth variables I_2 and I_3 as a function of α for a gaussian input pulse. The output pulse is a raised cosine with $\beta = 1.0$.

The parameter L_n is given by Eq. (7-58). It depends on the pulse energy per bit period (γ) and on the excess noise coefficient x of the avalanche process. For a gaussian input pulse, γ, in turn, depends on the parameter α in Eq. (7-63). The relation between γ and α is found from Eqs. (7-37) and (7-63):

$$\gamma = \int_{-T_b/2}^{T_b/2} h_p(t)dt = \frac{2}{\sqrt{\pi}} \int_0^{1/(2\sqrt{2}\alpha)} e^{-x^2} dx$$

$$= \text{erf}\left(\frac{1}{2\sqrt{2}\alpha}\right) \tag{7-76}$$

where the error function erf(x) is defined in Eq. (7-24). The relationship between γ and α given by Eq. (7-76) is shown in Fig. 7-14.

We are now finally ready to evaluate Eq. (7-62). Choosing $x = 0.3$, which is characteristic of silicon avalanche photodiodes, Eq. (7-75) yields $L_i = 2.38$. For InGaAs we have $x = 0.7$ and $L_i = 2.620$. What we wish to plot is the penalty in minimum received power ΔP (required for a certain bit-error rate) as a function of the fraction of pulse energy $1 - \gamma$ that has spread outside of the bit period T_b. For a given value of γ we read the corresponding value of L_n from Fig. 7-10. To find I_{2n} and I_{3n}, we first find the value of α corresponding to γ from Fig. 7-14, and then we find the values of I_{2n} and I_{3n} corresponding to this value of α from Fig. 7-13. Substituting all these values into Eq. (7-62) for various values of γ and letting $x = 0.3$ for Si and 0.7 for InGaAs, we obtain the results shown in Fig. 7-15.

The effect of intersymbol interference (or bandwidth limitation) on the receiver power penalty is readily deduced from Fig. 7-15. As the fraction of pulse energy outside the bit period increases, there is a steep rise in the power penalty curve. This curve gives a clear implication as to the effects of attempting to operate an optical fiber system at such high data rates that bandwidth limitations arise. Intersymbol interference becomes more pronounced at higher data rates, since the individual data pulses start to overlap significantly as the data rate

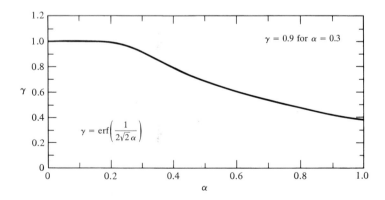

FIGURE 7-14
A plot of the fraction of pulse energy γ as a function of the gaussian-pulse-shape parameter α.

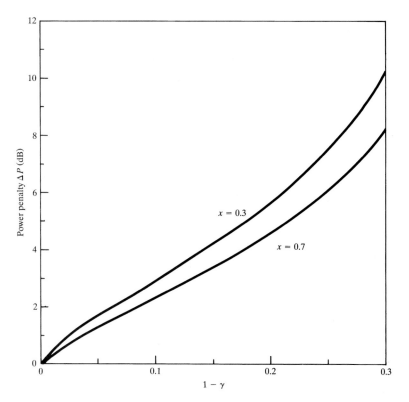

FIGURE 7-15
Plots of Eq. (7-62) that show the penalty in minimum received optical power (receiver sensitivity) arising from pulse spreading outside of the bit period for gaussian received pulses.

approaches the system bandwidth limit. Since the receiver power penalty increases rapidly for larger pulse overlaps, operating a fiber optic system much beyond its bandwidth is generally not worthwhile, even though it may be possible to correct for intersymbol interference through the use of an equalization circuit.

7.3.5   Nonzero Extinction Ratio★

In the previous section we assumed that there is no optical power incident on the photodetector during a 0 pulse, so that $b_{off} = 0$. In actual systems, the light source may be biased slightly on at all times in order to obtain a shorter light source turn-on time (see Sec. 4.3.7). Thus, some optical power is also emitted during a 0 pulse. This is of particular importance for laser diodes, since it is generally desirable to bias them on to just below the lasing threshold. This means that during a zero pulse the laser acts like an LED and can launch a significant amount of optical power into a fiber.

The ratio ϵ of the optical energy emitted in the 0 pulse state to that emitted during a 1 pulse is called the *extinction ratio*:

$$\epsilon = \frac{b_{\text{off}}}{b_{\text{on}}} \tag{7-77}$$

The extinction ratio as defined here thus varies between 0 and 1. Note: Equation (7-77) is not a universal definition for the extinction ratio. In many cases in the literature, the reciprocal of Eq. (7-77) is used; that is, the extinction ratio is also defined as $b_{\text{on}}/b_{\text{off}}$. In that case, the extinction ratio ranges between 1 and ∞. Thus, when looking at the literature, the reader can easily see which definition is being used. A receiver performance calculation[27] identical to that of Sec. 7.3.4 can be carried out for a nonzero extinction ratio simply by using b_{off} from Eq. (7-77) and by replacing γ by $\gamma' = \gamma(1 - \epsilon)$ in Eq. (7-36). With these replacements, Eq. (7-53) yields

$$b_{\text{on}}(1 - \epsilon) = \frac{Q}{M} \frac{h\nu}{\eta} \left\{ \left(M^{2+x} \frac{\eta}{h\nu} b_{\text{on}} I_2 + W \right)^{1/2} \right.$$
$$\left. + \left[M^{2+x} \frac{\eta}{h\nu} b_{\text{on}} I_2 (1 - \gamma') + W \right]^{1/2} \right\} \tag{7-78}$$

Analogous to the derivation of Eq. (7-57), we differentiate Eq. (7-78) for b_{on} with respect to M to find the minimum energy $b_{\text{on,min}}(\epsilon)$ per pulse required at the optimum gain M_{opt}, which results in

$$b_{\text{on,min}}(\epsilon) = Q^{(2+x)/(1+x)} \frac{h\nu}{\eta} W^{x/(2+2x)} I_2^{1/(1+x)} L' \left(\frac{1}{1-\epsilon} \right)^{(2+x)/(1+x)} \tag{7-79}$$

where

$$L'^{(1+x)} = \frac{2(1 - \gamma')}{K'(2 - \gamma')} \left\{ \left(\frac{1}{2} \frac{2 - \gamma'}{1 - \gamma'} K' + 1 \right)^{1/2} + [\tfrac{1}{2}(2 - \gamma')K' + 1]^{1/2} \right\}^{2+x} \tag{7-80}$$

with

$$K' = -1 + \left[1 + 16 \frac{1+x}{x^2} \frac{1 - \gamma'}{(2 - \gamma')^2} \right]^{1/2} \tag{7-81}$$

If a data stream has an equal probability of 1 and 0 pulses, then the minimum received power (or the receiver sensitivity), $P_{r,\text{min}}$, is given by the average energy detected per pulse multiplied by the pulse rate $1/T_b$:

$$P_{r,\text{min}} = \frac{b_{\text{on}} + b_{\text{off}}}{2T_b} = \frac{1 + \epsilon}{2T_b} b_{\text{on}} \tag{7-82}$$

where the last equality was obtained by using Eq. (7-77). The extinction ratio penalty (i.e., the penalty in receiver sensitivity as a function of the extinction ratio) is

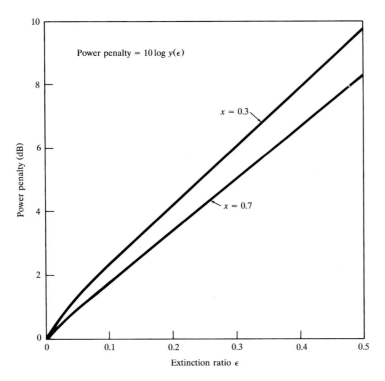

Power penalty = 10 log y(ε)

FIGURE 7-16
Plots of Eq. (7-83) that give the penalty in receiver sensitivity as a function of the extinction ratio ϵ for $\gamma = 1.0$ and $x = 0.3$ (Si) and 0.7 (InGaAs).

$$y(\epsilon) = \frac{P_{r,\min}(\epsilon)}{P_{r,\min}(0)} = (1 + \epsilon)\left(\frac{1}{1 - \epsilon}\right)^{(2+x)/(1+x)} \frac{L'}{L} \qquad (7\text{-}83)$$

Both L and L' are given by Fig. 7-10 on using γ and γ', respectively, for the abscissa. Plots of Eq. (7-83) in the form $10 \log y(\epsilon)$ are given in Fig. 7-16 for $x = 0.3$ (Si) and $x = 0.7$ (InGaAs) in the special case where $\gamma = 1.0$.

7.4 PREAMPLIFIER TYPES★

Having examined the performance characteristics of a general class of receivers, we now turn our attention to some specific types of receiver preamplifiers. Since the sensitivity and bandwidth of a receiver are dominated by the noise sources at the front end (the preamplifier stage), the major emphasis in the literature has been on the design of a low-noise preamplifier. The goals generally are to maximize the receiver sensitivity while maintaining a suitable bandwidth.

Preamplifiers used in optical fiber communication receivers can be classified into three broad categories. These categories are not actually distinct, since a continuum of intermediate designs are possible, but they serve to illustrate the design approaches. The three categories encompass the low-impedance, the high-impedance, and the transimpedance preamplifiers.

The *low-impedance (LZ) preamplifier* is the most straightforward, but not necessarily the optimum preamplifier design. In this design, a photodiode operates into a low-impedance amplifier (e.g., 50 Ω). Here, a bias or load resistor R_b is used to match the amplifier impedance (to suppress standing waves for uniform frequency response). The value of the bias resistor, in conjunction with the amplifier input capacitance, is such that the preamplifier bandwidth is equal to or greater than the signal bandwidth. Although low-impedance preamplifiers can operate over a wide bandwidth, they do not provide a high receiver sensitivity, because only a small signal voltage can be developed across the amplifier input impedance and the resistor R_b. This limits their use to special short-distance applications where high sensitivity is not a major concern.

In the *high-impedance (HZ) preamplifier* design shown in Fig. 7-4, the goal is to reduce all sources of noise to the absolute minimum. This is accomplished by reducing the input capacitance through the selection of low-capacitance high-frequency devices, by selecting a detector with low dark currents, and by minimizing the thermal noise contributed by the biasing resistors. The thermal noise can be reduced by using a high-impedance amplifier [e.g., a bipolar transistor or a field-effect transistor (FET)] together with a large photodetector bias resistor R_b, which is why this design is referred to as a high-impedance preamplifier. Since the high impedance produces a large input RC time constant, the front-end bandwidth is less than the signal bandwidth. Thus, the input signal is integrated, and equalization techniques must be employed to compensate for this.

The *transimpedance preamplifier* design largely overcomes the drawbacks of the high-impedance preamplifier. This is done by utilizing a low-noise high-impedance amplifier with a negative-feedback resistor R_f with an equivalent thermal noise current $i_f(t)$ shunting the input as shown in Fig. 7-17. The amplifier has an input equivalent series voltage noise source $e_a(t)$, an equivalent shunt current noise $i_a(t)$, and an input impedance given by the parallel combination of R_a and C_a.

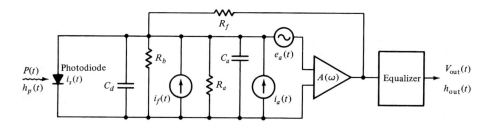

FIGURE 7-17
An equivalent circuit of a transimpedance receiver design. (Reproduced with permission from Personick, Rhodes, Hanson, and Chan, *Proc. IEEE*, vol. 68, p. 1260, Oct. 1980, © 1980, IEEE.)

What we are interested in here is to evaluate the noise current given by Eq. (7-45) for different amplifier designs. To this end, we shall examine only the parameter W of Eq. (7-46). This is a very useful figure of merit for a receiver, since it measures the noisiness of the amplifier. From Eq. (7-46) it can be seen that the noise is minimized if the amplifier and bias resistances, R_a and R_b, respectively, are large; the total capacitance C at the amplifier input is small; and the noise current and voltage spectral heights, S_I and S_E, respectively, are small. In general, these parameters are not independent, so that tradeoffs have to be made among them to minimize the noise. In addition, the freedom of the designer to optimize the device parameters is often restricted by the limited variety of components available. We shall examine several high-impedance and transimpedance amplifier configurations to illustrate some of the design considerations that must be taken into account.

7.4.1 High-Impedance FET Amplifiers★

A number of different FETs (field-effect transistors) can be used for front-end receiver designs.[11,14,16,28-31] For gigabit-per-second data rates, for example, the lowest-noise receivers are made using GaAs MESFET (metal semiconductor field-effect transistor) preamplifiers. At lower frequencies, silicon MOSFETs (metal-oxide semiconductor field-effect transistors) or JFETs (junction field-effect transistors) are generally used. The circuit of a simple FET amplifier is shown in Fig. 7-18. Typical FETs have very large input resistances R_a (usually greater than 10^6 Ω), so, for practical purposes $R_a = \infty$. The total resistance R given by Eq. (7-10) then reduces to the value of the detector bias resistor R_b.

The principal noise sources are thermal noise associated with the FET channel conductance, thermal noise from the load or feedback resistor, and shot noise arising from gate-leakage current. A fourth noise source is FET $1/f$ noise. This was not included in the above analyses because it contributes to the overall noise only at very low bit rates[11] (see Prob. 7-26). Since the amplifier input resistance is very large, the input current noise spectral density S_I is

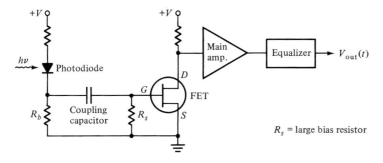

FIGURE 7-18
Simple high-impedance preamplifier design using a FET.

$$S_{I,\text{FET}} = \frac{4k_B T}{R_a} + 2q I_{\text{gate}}$$

$$\simeq 2q I_{\text{gate}} \tag{7-84}$$

where I_{gate} is the gate-leakage current of the FET. In an FET the thermal noise of the conducting-channel resistance is characterized by the transconductance g_m. The voltage noise spectral density is[32]

$$S_E = \frac{4k_B T \Gamma}{g_m} \tag{7-85}$$

where the FET *channel-noise factor* Γ is a numerical constant that accounts for thermal noise and gate-induced noise plus the correlation between these two noises. The thermal noise characteristics W [Eq. (7-46)] at the equalizer output is then

$$W = \frac{1}{q^2 B} \left(2q I_{\text{gate}} + \frac{4k_B T}{R_b} + \frac{4k_B T \Gamma}{g_m R_b^2} \right) I_2 + \left(\frac{2\pi C}{q} \right)^2 \frac{4k_B T \Gamma}{g_m} I_3 B \tag{7-86}$$

Some typical values of the various parameters for GaAs MESFETs, Si MOSFETs, and Si JFETs are given in Table 7-2. Here, C_{gs} and C_{gd} are the FET gate-source and gate-drain capacitances, respectively. For a typical FET and a good photodiode, we can expect values of $C = C_a + C_d + C_{gs} + C_{gd} = 10\,\text{pF}$. The $1/f$-noise corner frequency f_c is defined as the frequency at which $1/f$ noise, which dominates the FET noise at low frequencies and has a $1/f$ power spectrum, becomes equal to the high-frequency channel noise described by Γ.

To minimize the noise in a high-impedance design, the bias resistor should be very large. The effect of this is that the detector output signal is integrated by the amplifier input resistance. We can compensate for this by differentiation in the equalizing filter. This integration–differentiation approach is known as the *high-impedance amplifier design* technique. It yields low noise, but also results in a low *dynamic range* (the range of signal levels that can be processed with high quality). An alternative method to deal with this is described in Sec. 7.4.3.

TABLE 7-2

Typical values of various parameters for GaAs MESFETs, Si MOSFETs, and Si JFETs

Parameter	Si JFET	Si MOSFET	GaAs MESFET
g_m (mS)	5–10	20–40	15–50
C_{gs} (pF)	3–6	0.5–1.0	0.2–0.5
C_{gd} (pF)	0.5–1.0	0.05–0.1	0.01–0.05
Γ	0.7	1.5–3.0	1.1–1.75
I_{gate} (nA)	0.01–0.1	0	1–1000
f_c (MHz)	< 0.1	1–10	10–100

As the signal frequency reaches values of about 25–50 MHz, the gain of a silicon FET approaches unity. Much higher frequencies (4 Gb/s and above) can be achieved with either a GaAs MESFET or a silicon bipolar transistor.[29,33]

7.4.2 High-Impedance Bipolar Transistor Amplifiers★

The circuit of a simple bipolar grounded-emitter transistor amplifier is shown in Fig. 7-19. The input resistance of a bipolar transistor is given by[32,34]

$$R_{\text{in}} = \frac{k_B T}{q I_{BB}} \tag{7-87}$$

where I_{BB} is the base bias current. For a bipolar transistor amplifier the input resistance R_a is given by the parallel combination of the bias resistors R_1 and R_2 and the transistor input resistance R_{in}. For a low-noise design, R_1 and R_2 are chosen to be much greater than R_{in}, so that $R_a \simeq R_{\text{in}}$. Thus, in contrast to the FET amplifier, R_a for a transistor amplifier is adjustable by the designer.

The spectral density (in A^2/Hz) of the input noise current source results from the shot noise of the base current:[32,34]

$$S_I = 2q I_{BB} = \frac{2k_B T}{R_{\text{in}}} \tag{7-88}$$

where the last equality comes from Eq. (7-87). The spectral height (in V^2/Hz) of the noise voltage source is[32,34]

$$S_E = \frac{2k_B T}{g_m} \tag{7-89}$$

Here, the transconductance g_m is related to the shot noise by virtue of the collector current I_c:

$$g_m = \frac{q I_c}{k_B T} = \frac{\beta}{R_{\text{in}}} \tag{7-90}$$

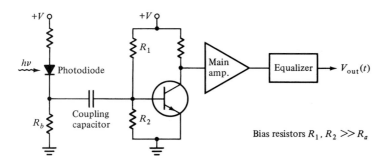

FIGURE 7-19
Simple high-impedance preamplifier design using a bipolar transistor.

where Eq. (7-87) has been used in the last equality to express g_m in terms of the current gain $\beta = I_c/I_{BB}$ and the input resistance R_{in}.

Substituting Eqs. (7-87) through (7-89) into Eq. (7-46), we have

$$W = \frac{T_b}{q^2} 2k_BT\left[\left(\frac{1}{R_{\text{in}}} + \frac{2}{R_b} + \frac{R_{\text{in}}}{\beta R^2}\right)I_2 + \frac{(2\pi C)^2}{T_b^2}\frac{R_{\text{in}}}{\beta}I_3\right] \qquad (7\text{-}91)$$

The contribution C_a to C from the bipolar transistor is a few picofarads. If the photodetector bias resistor R_b is much larger than the amplifier resistance R_a, then, from Eq. (7-10), $R \simeq R_a \simeq R_{\text{in}}$, so that

$$W = \frac{2k_BT}{q^2}\left[\frac{T_b}{R_{\text{in}}}\frac{\beta+1}{\beta}I_2 + \frac{(2\pi C)^2}{\beta T_b}R_{\text{in}}I_3\right] \qquad (7\text{-}92)$$

As in the case with a high-impedance FET preamplifier, the impedance loading the photodetector integrates the detector output signal. Again, to compensate for this, the amplified signal is differentiated in the equalizing filter.

7.4.3 Transimpedance Amplifier★

Although a high-impedance design produces the lowest-noise amplifier, it has two limitations: (*a*) for broadband applications, equalization is required; and (*b*) it has a limited dynamic range. An alternative design is the *transimpedance amplifier*[17,29,35-37] shown in Fig. 7-17. This is basically a high-gain high-impedance amplifier with feedback provided to the amplifier input through the feedback resistor R_f. This design yields both low noise and a large dynamic range.

To compare the nonfeedback and the feedback designs, we make the restriction that both have the same transfer function $H_{\text{out}}(f)/H_p(f)$. For the transimpedance amplifier the thermal noise characteristic W_{TZ} at the equalizer output is therefore simply found by replacing R_b in Eq. (7-46) with R_b', where[35]

$$R_b' = \frac{R_bR_f}{R_b + R_f} \qquad (7\text{-}93)$$

is the parallel combination of R_b and R_f. Thus, from Eq. (7-46),

$$W_{TZ} = \frac{T_b}{q^2}\left(S_I + \frac{4k_BT}{R_b'} + \frac{S_E}{(R')^2}\right)I_2 + \frac{(2\pi C)^2}{q^2 T_b}S_EI_3 \qquad (7\text{-}94)$$

where, from Eq. (7-10),

$$\frac{1}{R'} = \frac{1}{R} + \frac{1}{R_f} = \frac{1}{R_a} + \frac{1}{R_b} + \frac{1}{R_f} \qquad (7\text{-}95)$$

In practice, the feedback resistance R_f is much greater than the amplifier input resistance R_a. Consequently, $R' \simeq R$ in Eq. (7-95), so that

$$W_{TZ} = W_{HZ} + \frac{T_b}{q^2}\frac{4k_BT}{R_f}I_2 \qquad (7\text{-}96)$$

where W_{HZ} is the high-impedance amplifier noise characteristic given by either Eq. (7-86) for FET designs or by Eq. (7-92) for the bipolar transistor case. The thermal noise of the transimpedance amplifier is thus modeled as the sum of the output noise of a nonfeedback amplifier plus the thermal noise associated with the feedback resistance. In practice, the noise considerations tend to be more involved, since R_f has an effect on the frequency response of the amplifier. More details are given by Smith and Personick.[18]

We now compare the bandwidths of the two designs. From Eq. (7-9), the transfer function of the nonfeedback amplifier is (in V/A)

$$H(f) = \frac{AR}{1 + j2\pi RCf} \tag{7-97}$$

where R and C are given by Eqs. (7-10) and (7-11), respectively, and A is the frequency-independent gain of the amplifier. Using Eq. (E-10), this yields a bandwidth of $(4RC)^{-1}$. For the transimpedance amplifier the transfer function $H_{TZ}(f)$ is

$$H_{TZ} = \frac{1}{1 + j2\pi RCf/A} \tag{7-98}$$

which yields a bandwidth of

$$B_{TZ} = \frac{A}{4RC} \tag{7-99}$$

which is A times that of the high-impedance design. This makes the equalization task simpler in the feedback amplifier case.

In summary, the benefits of a transimpedance amplifier are as follows:

1. It has a wide dynamic range compared with the high-impedance amplifier.
2. Usually, little or no equalization is required because the combination of R_{in} and the feedback resistor R_f is very small, which means the time constant of the detector is also small.
3. The output resistance is small, so that the amplifier is less susceptible to pickup noise, crosstalk, electromagnetic interference (EMI), etc.
4. The transfer characteristic of the amplifier is actually its transimpedance, which is the feedback resistor. Therefore, the transimpedance amplifier is very easily controlled and stable.
5. Although the transimpedance amplifier is less sensitive than the high-impedance amplifier (since $W_{TZ} > W_{HZ}$), this difference is usually only about 2–3 dB for most practical wideband designs.

7.4.4 High-Speed Circuits

Improvements in component performance, cost, and reliability have led to major applications of fiber optic technology for long-distance carriers, local telephone services, and local area networks. To utilize the wide bandwidth available, there has been an increased implementation of high-speed systems for both digital and analog links.[38-43] Along with this has come the miniaturization of transmitters and receivers into integrated circuit formats. Many different types of receivers with operating speeds up to multigigahertz rates are thus commercially available from a wide variety of vendors.

7.5 ANALOG RECEIVERS

In addition to the wide usage of fiber optics for the transmission of digital signals, there are many potential applications for analog links. These range from individual 4-kHz voice channels to microwave links operating in the multigigahertz region.[38-40] In the previous sections we discussed digital receiver performance in terms of error probability. For an analog receiver, the performance fidelity is measured in terms of a *signal-to-noise ratio*. This is defined as the ratio of the mean-square signal current to the mean-square noise current.

The simplest analog technique is to use amplitude modulation of the source.[2] In this scheme, the time-varying electric signal $s(t)$ is used to modulate directly an optical source about some bias point defined by the bias current I_B, as shown in Fig. 7-20. The transmitted optical power $P(t)$ is thus of the form

$$P(t) = P_t[1 + ms(t)] \tag{7-100}$$

where P_t is the average transmitted optical power, $s(t)$ is the analog modulation signal, and m is the modulation index defined by (see Sec. 4.4)

$$m = \frac{\Delta I}{I_B} \tag{7-101}$$

Here, ΔI is the variation in current about the bias point. In order not to introduce distortion into the optical signal, the modulation must be confined to the linear

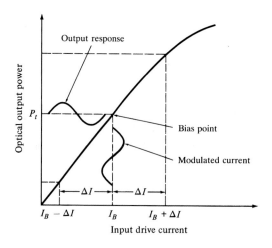

FIGURE 7-20
Direct analog modulation of an LED source.

region of the light source output curve shown in Fig. 7-20. Also, if $\Delta I > I_B$, the lower portion of the signal gets cut off and severe distortion results.

At the receiver end, the photocurrent generated by the analog optical signal is

$$i_s(t) = \mathcal{R}_0 M P_r[1 + ms(t)]$$

$$= I_p M[1 + ms(t)] \tag{7-102}$$

where $\mathcal{R}_0$ is the detector responsivitity, P_r is the average received optical power, $I_p = \mathcal{R}_0 P_r$ is the primary photocurrent, and M is the photodetector gain. If $s(t)$ is a sinusoidally modulated signal, then the mean-square signal current at the photodetector output is (ignoring a dc term)

$$\langle i_s^2 \rangle = \tfrac{1}{2}(\mathcal{R}_0 M m P_r)^2 = \tfrac{1}{2}(M m I_p)^2 \tag{7-103}$$

Recalling from Eq. (6-18) that the mean-square noise current for a photodiode receiver is the sum of the mean-square quantum noise current, the equivalent-resistance thermal noise current, the dark noise current, and the surface-leakage noise current, we have

$$\langle i_N^2 \rangle = 2q(I_p + I_D)M^2 F(M)B + 2qI_L B + \frac{4k_B T B}{R_{\text{eq}}} F_t \tag{7-104}$$

where I_p = primary (unmultiplied) photocurrent = $\mathcal{R}_0 P_r$
       I_D = primary bulk dark current
       I_L = surface-leakage current
       $F(M)$ = excess photodiode noise factor $\simeq M^x (0 < x \le 1)$
       B = effective noise bandwidth
       R_{eq} = equivalent resistance of photodetector load and amplifier
       F_t = noise figure of the baseband amplifier

By a suitable choice of the photodetector, the leakage current can be rendered negligible. With this assumption, the signal-to-noise ratio S/N is

$$\frac{S}{N} = \frac{\langle i_s^2 \rangle}{\langle i_N^2 \rangle} = \frac{\tfrac{1}{2}(\mathcal{R}_0 M m P_r)^2}{2q(\mathcal{R}_0 P_r + I_D)M^2 F(M)B + (4k_B T B / R_{\text{eq}})F_t}$$

$$= \frac{\tfrac{1}{2}(I_p M m)^2}{2q(I_p + I_D)M^2 F(M)B + (4k_B T B / R_{\text{eq}})F_t} \tag{7-105}$$

For a *pin* photodiode we have $M = 1$. When the optical power incident on the photodiode is small, the circuit noise term dominates the noise current, so that

$$\frac{S}{N} \simeq \frac{\tfrac{1}{2}m^2 I_p^2}{(4k_B T B / R_{\text{eq}})F_t} = \frac{\tfrac{1}{2}m^2 \mathcal{R}_0^2 P_r^2}{(4k_B T B / R_{\text{eq}})F_t} \tag{7-106}$$

Here, the signal-to-noise ratio is directly proportional to the square of the photodiode output current and inversely proportional to the thermal noise of the circuit.

For large optical signals incident on a *pin* photodiode, the quantum noise associated with the signal detection process dominates, so that

$$\frac{S}{N} \simeq \frac{m^2 I_p}{4qB} = \frac{m^2 \mathcal{R}_0 P_r}{4qB} \Bigg) \, 7.12 \tag{7-107}$$

Since the signal-to-noise ratio in this case is independent of the circuit noise, it represents the fundamental or quantum limit for analog receiver sensitivity.

When an avalanche photodiode is employed at low signal levels and with low values of gain M, the circuit noise term dominates. At a fixed low signal level, as the gain is increased from a low value, the signal-to-noise ratio increases with gain until the quantum noise term becomes comparable to the circuit noise term. As the gain is increased further beyond this point, the signal-to-noise ratio *decreases* as $F(M)^{-1}$. Thus, for a given set of operating conditions, there exists an optimum value of the avalanche gain for which the signal-to-noise ratio is a maximum. Since an avalanche photodiode increases the signal-to-noise ratio for small optical signal levels, it is the preferred photodetector for this situation.

For very large optical signal levels, the quantum noise term dominates the receiver noise. In this case, an avalanche photodiode serves no advantage, since the detector noise increases more rapidly with increasing gain M than the signal level. This is shown in Fig. 7-21, where we compare the signal-to-noise ratio for a *pin* and an avalanche photodiode receiver as a function of the received optical power. The signal-to-noise ratio for the avalanche photodetector is at the optimum gain (see Probs. 7-28 and 7-29). The parameter values chosen for this

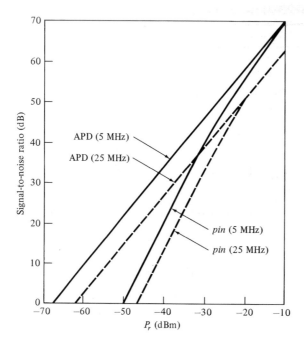

FIGURE 7-21

Comparison of the signal-to-noise ratio for *pin* and avalanche photodiodes as a function of received optical power for bandwidths of 5 and 25 MHz.

example are $B = 5$ MHz and 25 MHz, $x = 0.5$ for the avalanche photodiode and 0 for the *pin* diode, $m = 80$ percent, $\mathcal{R}_0 = 0.5$ A/W, and $R_{eq}/F_t = 10^4 \, \Omega$. We see that for low signal levels an avalanche photodiode yields a higher signal-to-noise ratio, whereas at large received optical power levels a *pin* photodiode gives better performance.

PROBLEMS

7-1. In avalanche photodiodes the ionization ratio k is approximately 0.02 for silicon, 0.35 for indium gallium arsenide, and 1.0 for germanium. Show that, for gains up to 100 in Si and up to 25 in InGaAs and Ge, the excess noise factor $F(M)$ can be approximated to within 10 percent by M^x, where x is 0.3 for Si, 0.7 for InGaAs, and 1.0 for Ge.

7-2. Find the Fourier transform $h_B(t)$ of the bias circuit transfer function $H_B(f)$ given by Eq. (7-9).

7-3. Show that the following pulse shapes satisfy the normalization condition

$$\int_{-\infty}^{\infty} h_p(t)dt = 1$$

(*a*) Rectangular pulse ($\alpha = $ constant)

$$h_p(t) = \begin{cases} \dfrac{1}{\alpha T_b} & \text{for} & \dfrac{-\alpha T_b}{2} < t < \dfrac{\alpha T_b}{2} \\ 0 & \text{otherwise} \end{cases}$$

(*b*) Gaussian pulse

$$h_p(t) = \frac{1}{\sqrt{2\pi}\,\alpha T_b} e^{-t^2/2(\alpha T_b)^2}$$

(*c*) Exponential pulse

$$h_p(t) = \begin{cases} \dfrac{1}{\alpha T_b} e^{-t/\alpha T_b} & \text{for } 0 \leq t \leq \infty \\ 0 & \text{otherwise} \end{cases}$$

7-4. The mathematical operation of convolving two real-valued functions of the same variable is defined as

$$p(t) * q(t) = \int_{-\infty}^{\infty} p(x)q(t - x)\, dx$$

$$= \int_{-\infty}^{\infty} q(x)p(t - x)\, dx$$

where $*$ denotes convolution. If $P(f)$ and $Q(f)$ are the Fourier transforms of $p(t)$ and $q(t)$, respectively, show that

$$F[p(t) * q(t)] = P(f)Q(f) = F[p(t)]F[q(t)]$$

That is, the convolution of two signals in the time domain corresponds to the multiplication of their Fourier transforms in the frequency domain.

7-5. Derive Eq. (7-25) from Eq. (7-18).

7-6. A transmission system sends out information at 200,000 b/s. During the transmission process, fluctuation noise is added to the signal so that at the decoder output the signal pulses are 1 V in amplitude and the rms noise voltage is 0.2 V.

(*a*) Assuming that ones and zeros are equally likely to be transmitted, what is the average time in which an error occurs?

(*b*) How is this time changed if the voltage amplitude is doubled with the rms noise voltage remaining the same?

7-7. Consider the probability distributions shown in Fig. 7-5, where the signal voltage for a binary 1 is V_1 and $v_{th} = V_1/2$.

(*a*) If $\sigma = 0.20V_1$ for $p(y|0)$ and $\sigma = 0.24V_1$ for $p(y|1)$, find the error probabilities $P_0(v_{th})$ and $P_1(v_{th})$.

(*b*) If $a = 0.65$ and $b = 0.35$, find P_e.

(*c*) If $a = b = 0.5$, find P_e.

7-8. An LED operating at 1300 nm injects $25\,\mu W$ of optical power into a fiber. If the attenuation between the LED and the photodetector is 40 dB and the photodetector quantum efficiency is 0.65, what is the probability that fewer than 5 electron–hole pairs will be generated at the detector in a 1-ns interval?

7-9. Derive the expression for the mean-square noise voltage at the equalizer output given by Eq. (7-28).

7-10. (*a*) Shows that Eqs. (7-30) and (7-33) can be rewritten as Eqs. (7-43) and (7-44).

(*b*) Show that Eq. (7-28) can be rewritten as Eq. (7-45).

7-11. Show that, by using Eq. (7-50), the error-probability expressions given by Eq. (7-49) both reduce to Eq. (7-51).

7-12. A useful approximation to $\frac{1}{2}(1 - \mathrm{erf}\,x)$ for values of x greater than 3 is given by

$$\tfrac{1}{2}(1 - \mathrm{erf}\,x) \simeq \frac{\exp(-x^2)}{2\sqrt{\pi}x} \qquad x > 3$$

Using this approximation, consider an on–off binary system that transmits the signal levels 0 and A with equal probability in the presence of gaussian noise. Let the signal amplitude A be K multiplied by the standard deviation of the noise.

(*a*) Calculate the net probability of error if $K = 10$.

(*b*) Find the value of K required to give a net error probability of 10^{-5}.

7-13. Derive Eq. (7-55) by differentiating both sides of Eq. (7-54) with respect to M and setting $db_{on}/dM = 0$.

7-14. Verify the expression given by Eq. (7-57) for the minimum energy per pulse needed to achieve a bit-error rate characterized by Q.

7-15. Show that, when there is no intersymbol interference ($\gamma = 1$), the optimum avalanche gain given by Eq. (7-59) reduces to Eq. (7-60).

7-16. Derive Eq. (7-70) for I_2 and Eq. (7-72) for I_3 (impulse input and raised-cosine output).

7-17. Verify the expressions given by Eqs. (7-73) and (7-74) for I_2 and I_3, respectively, for a gaussian input pulse and a raised-cosine output.

7-18. Plot I_2 versus α for a gaussian input pulse for values of $\beta = 0.1, 0.5$, and 1.0.

7-19. Plot I_3 versus α for a gaussian input pulse for values of $\beta = 0.1, 0.5$, and 1.0.

7-20. Derive Eq. (7-75).

7-21. Compare the values of $y(\epsilon)$ in Eq. (7-83) (in decibels for 10-percent extinction ratio ($\epsilon = 0.10$) when (a) $\gamma = 0.90$, $x = 0.5$; (b) $\gamma = 0.90$, $x = 1.0$.

7-22. (a) Plot the values of the thermal noise characteristic W for a high-impedance FET amplifier for data rates $1/T_b$ ranging from 1 to 50 Mb/s. Let $T = 300$ K, $g_m = 0.005$ S, $R_b = 10^5$ Ω, $C = 10$ pF, and $\gamma = 0.90$. Use Fig. 7-13 to find I_2 and I_3. Recall that I_2 and I_3 depend on α, which, in turn, depends on γ.

(b) Plot the values of W for a high-impedance bipolar transistor preamplifier for data rates ranging from 20 to 100 Mb/s. Let $T = 300$ K, $\beta = 100$, $I_{BB} = 5\,\mu$A, $C = 10$ pF, and $\gamma = 0.90$.

7-23. The receiver sensitivity P_r is given by the average energy b_{on} detected per pulse multiplied by the pulse rate $1/T_b$ (if $b_{\text{off}} = 0$):

$$P_r = \frac{b_{\text{on}}}{T_b}$$

Find the sensitivity in dBm (see App. D) of an avalanche photodiode receiver with an FET preamplifier at a 10-Mb/s data rate. Let the required bit-error rate be 10^{-9} and take $T = 300$ K, $x = 0.5$, $\gamma = 0.9$, $\eta q/h\nu = 0.7$ A/W (the detector responsivity), $g_m = 0.005$ S, $R_b = 10^5$ Ω, and $C = 10$ pF.

7-24. If a transimpedance amplifier has a feedback resistance of 5000 Ω, by how much does W_{TZ} differ from W_{HZ} at a 10-Mb/s data rate? Assume $\gamma = 0.9$ and $T = 300$ K. Compared with a high-impedance amplifier, what is the decrease in receiver sensitivity (in dB) for this transimpedance amplifier at 10 Mb/s if $x = 0.5$ and $W_{HZ} = 1 \times 10^6$?

7-25. (a) To calculate the receiver sensitivity P_r as a function of gain M, we need to solve Eq. (7-54) for b_{on}. Show that, for $\gamma = 1.0$ and $b_{\text{off}} = 0$, this becomes

$$b_{\text{on}} = \frac{h\nu}{\eta}\left(M^x Q^2 I_2 + \frac{2QW^{1/2}}{M}\right)$$

(b) Consider a receiver operating at 50 Mb/s. Let the receiver have an avalanche photodiode with $x = 0.5$ and a bipolar transistor front end (preamplifier). Assume $W = 2 \times 10^6$, $Q = 6$ for a 10^{-9} bit-error rate, $I_2 = 1.08$, and $\eta q/h\nu = 0.7$ A/W. Using the foregoing expression for b_{on}, plot $P_r = b_{\text{on}}/T_b$ in dBm as a function of gain M for values of M ranging from 30 to 120.

7-26. For $1/f$ noise the noise current $\langle i_f^2 \rangle$ is given by

$$\langle i_f^2 \rangle = \frac{8k_B T\Gamma(2\pi C)^2 f_C I_f B^2}{g_m}$$

where

$$I_f = \frac{1}{T_b}\int_0^\infty \left|\frac{H_{\text{out}}(f)}{H_p(f)}\right|^2 f\,df = \int_0^\infty \left|\frac{H'_{\text{out}}(\phi)}{H'_p(\phi)}\right|^2 \phi\,d\phi$$

and f_c is the corner frequency.

(a) Show that

$$I_f = \left(\frac{2}{\pi}\right)^2 \int_0^{\pi/2} x\,e^{16\alpha^2 x^2}\cos^4 x\,dx$$

TABLE P7-26

Parameter	Value
FET gate leakage I_{gate}	50 nA
Transconductance g_m	30 mS
Noise figure Γ	1.1
Capacitance C	1.5 pF
$1/f$ corner frequency f_c	20 MHz
Resistance $R_b = R$	400 Ω
I_f	0.12
I_2	0.50
I_3	0.085

for a gaussian input pulse and a raised-cosine output.

(b) For a GaAs FET at 20°C, compare the values of $\langle i_R^2 \rangle$, $\langle i_I^2 \rangle$, and $\langle i_f^2 \rangle$ for bit rates B ranging from 50 Mb/s to 5 Gb/s. Letting the amplifier gain $A = 1$, use the performance parameter values in Table P7-26.

7-27. Using Eq. (E-10) for the bandwidth definition, show that the bandwidths of the transfer functions given by Eqs. (7-97) and (7-98) are $1/4RC$ and $A/4RC$.

7-28. Show that the signal-to-noise ratio given by Eq. (7-105) is a maximum when the gain is optimized at

$$M_{\text{opt}}^{2+x} = \frac{4k_B TF_t / R_{\text{eq}}}{q(I_p + I_D)x}$$

7-29. (a) Show that, when the gain M is given by the expression in Prob. 7-28, the signal-to-noise ratio given by Eq. (7-105) can be written as

$$\frac{S}{N} = \frac{xm^2}{2B(2+x)} \frac{I_p^2}{[q(I_p + I_D)x]^{2/(2+x)}} \left(\frac{R_{\text{eq}}}{4k_B TF_t} \right)^{x/(2+x)}$$

(b) Show that, when I_p is much larger than I_D, the foregoing expression becomes

$$\frac{S}{N} = \frac{m^2}{2Bx(2+x)} \left[\frac{(xI_p)^{2(1+x)}}{q^2(4k_B TF_t / R_{\text{eq}})^x} \right]^{1/(2+x)}$$

7-30. Consider the signal-to-noise ratio expression given in Prob. 7-29a. Analogous to Figure 7-21, plot S/N in dB [i.e., $10\log(S/N)$] as a function of the received power level P_r in dBm when the dark current $I_D = 10$ nA and $x = 1.0$. Let $B = 5$ MHz, $m = 0.8$, $\mathscr{R}_0 = 0.5$ A/W, $T = 300$ K, and $R_{\text{eq}}/F_t = 10^4$ Ω. Recall that $I_p = \mathscr{R}_0 P_r$.

REFERENCES

1. S. D. Personick, "Receiver design for digital fiber optic communication systems," *Bell Sys. Tech. J.*, vol. 52, pp. 843–886, July/Aug. 1973.
2. W. M. Hubbard, "Utilization of optical-frequency carriers for low and moderate bandwidth channels," *Bell Sys. Tech. J.*, vol. 52, pp. 731–765, May/June 1973.
3. J. E. Mazo and J. Salz, "On optical data communication via direct detection of light pulses," *Bell Sys. Tech. J.*, vol. 55, pp. 347–369, Mar. 1976.

4. S. D. Personick, P. Balaban, J. Bobsin, and P. Kumer, "A detailed comparison of four approaches to the calculation of the sensitivity of optical fiber receivers," *IEEE Trans. Commun.*, vol. COM-25, pp. 541–548, May 1977.

5. (*a*) G. L. Cariolaro, "Error probability in digital optical fiber communication systems," *IEEE Trans. Inform. Theory*, vol. IT-24, pp. 213–221, Mar. 1978.

   (*b*) R. Dogliotti, A. Luvison, and G. Pirani, "Error probability in optical fiber transmission systems," *IEEE Trans. Inform. Theory*, vol. IT-25, pp. 170–178, Mar. 1979.

   (*c*) D. G. Messerschmitt, "Minimum MSE equalization of digital fiber optic systems," *IEEE Trans. Commun.*, vol. COM-26, pp. 1110–1118, July 1978.

   (*d*) W. Hauk, F. Bross, and M. Ottka, "The calculation of error rates for optical fiber systems," *IEEE Trans. Commun.*, vol. COM-26, pp. 1119–1126, July 1978.

6. M. Kavehrad and M. Joseph, "Maximum entropy and the method of moments in performance evaluation of digital communication systems," *IEEE Trans. Commun.*, vol. 34, pp. 1183–1189, Dec. 1986.

7. G. Einarsson and M. Sundelin, "Performance analysis of optical receivers by gaussian approximation," *J. Opt. Commun.*, vol. 16, pp. 227–232, Dec. 1995.

8. D. R. Smith and I. Garrett, "A simplified approach to digital optical receiver design," *Opt. Quantum Electron.*, vol. 10, pp. 211–221, 1978.

9. Y. K. Park and S. W. Granlund, "Optical preamplifier receivers: Application to long-haul digital transmission," *Opt. Fiber Technol.*, vol. 1, pp. 59–71, Oct. 1994.

10. (*a*) T. V. Muoi, "Receiver design for high-speed optical-fiber systems," *J. Lightwave Tech.*, vol. LT-2, pp. 243–267, June 1984.

    (*b*) S. R. Forrest, "The sensitivity of photoconductive receivers for long-wavelength optical communications," *J. Lightwave Tech.*, vol. LT-3, pp. 347–360, April 1985.

11. M. Brain and T.-P. Lee, "Optical receivers for lightwave communication systems," *J. Lightwave Tech.*, vol. LT-3, pp. 1281–1300, Dec. 1985.

12. B. L. Casper and J. C. Campbell, "Multigigabit-per-second avalanche photodiode lightwave receivers," *J. Lightwave Tech.*, vol. LT-5, pp. 1351–1364, Oct. 1987.

13. S. B. Alexander, *Optical Communication Receiver Design*, SPIE Optical Engineering Press, Bellingham, WA, 1997.

14. G. F. Williams, "Lightwave receivers," in T. Li, ed., *Topics in Lightwave Transmission Systems*, Academic, New York, 1991.

15. S. D. Personick, "Receiver design," in *Optical Fiber Telecommunications*, S. E. Miller and A. G. Chynoweth, eds., Academic, New York, 1979.

16. B. L. Kasper, "Receiver design," in *Optical Fiber Telecommunications—II*, S. E. Miller and I. P. Kaminow, eds., Academic, New York, 1988.

17. D. A. Fishman and B. S. Jackson, "Transmitter and receiver design for amplified lightwave systems," in I. P. Kaminow and T. L. Koch, eds., *Optical Fiber Telecommunications—III*, vol. B, Academic, New York, 1997, chap. 3, pp. 69–114.

18. R. G. Smith and S. D. Personick, "Receiver design for optical fiber communication systems," *in Semiconductor Devices for Optical Communications*, H. Kressel, ed., Springer-Verlag, New York, 2nd ed., 1982.

19. See any basic book on communication systems, for example:

    (*a*) A. B. Carlson, *Communication Systems*, McGraw-Hill, New York, 3rd ed., 1986.

    (*b*) J. G. Proakis, *Digital Communications*, McGraw-Hill, New York, 3rd ed., 1995.

    (*c*) L. W. Couch II, *Digital and Analog Communication Systems*, Prentice Hall, Upper Saddle River, NJ, 5th ed., 1997.

20. E. A. Lee and D. G. Messerschmitt, *Digital Communication*, Kluwer Academic, Boston, 2nd ed., 1993.

21. (*a*) H. Taub and D. L. Schilling, *Principles of Communication Systems*, McGraw-Hill, New York, 2nd ed., 1986.

    (*b*) R. E. Ziemer and W. H. Tranter, *Principles of Communications: Systems, Modulation, and Noise*, Wiley, New York, 4th ed., 1995.

    (*c*) A. van der Ziel, *Noise in Solid State Devices and Circuits*, Wiley, New York, 1986.

22. E. A. Newcombe and S. Pasupathy, "Error rate monitoring for digital communications," *Proc. IEEE*, vol. 70, pp. 805–828, Aug. 1982.

23. (*a*) A. Papoulis, *Probability, Random Variables, and Stochastic Processes*, McGraw-Hill, New York, 3rd ed., 1991.

   (*b*) P. Z. Peebles, Jr., *Probability, Random Variables, and Random Signal Principles*, McGraw-Hill, New York, 3rd ed., 1993.

24. (*a*) M. Kurtz, *Handbook of Applied Mathematics for Engineers and Scientists*, McGraw-Hill, New York, 1991.

   (*b*) D. Zwillinger, ed., *Standard Mathematical Tables and Formulae*, CRC Press, Boca Raton, FL, 30th ed., 1995.

25. N. S. Bergano, F. W. Kerfoot, and C. R. Davidson, "Margin measurements in optical amplifier systems," *IEEE Photonics Tech. Lett.*, vol. 5, pp. 304–306, Aug. 1993.

26. J. M. Wozencroft and I. M. Jacobs, *Principles of Communication Engineering*, Wiley, New York, 1965.

27. R. C. Hooper and P. B. White, "Digital optical receiver design for non-zero extinction ratio using a simplified approach," *Opt. Quantum Electron.*, vol. 10, pp. 279–282, 1978.

28. J. E. Goell, "An optical repeater with high impedance input amplifier," *Bell Sys. Tech. J.*, vol. 53, pp. 629–643, Apr. 1974.

29. R. A. Minasian, "Optimum design of a 4-Gb/s GaAs MESFET optical preamplifier," *J. Lightwave Tech.*, vol. LT-5, pp. 373–379, Mar. 1987.

30. A. A. Abidi, "On the choice of optimum FET size in wide-band transimpedance amplifiers," *J. Lightwave Tech.*, vol. 6, pp. 64–66, Jan. 1988.

31. R. M. Jopson, A. H. Gnauck, B. L. Kasper, R. E. Tench, N. A. Olsson, C. A. Burrus, and A. R. Chraplyvy, "8 Gbit/s 1.3 μm receiver using optical preamplifier," *Electron. Lett.*, vol. 25, pp. 233–235, Feb. 2, 1989.

32. (*a*) A. van der Ziel, *Introductory Electronics*, Prentice Hall, Englewood Cliffs, NJ, 1974.

   (*b*) D. Schilling and C. Belove, *Electronic Circuits: Discrete and Integrated*, McGraw-Hill, New York, 3rd ed., 1989.

33. T. T. Ha, *Solid State Microwave Amplifier Design*, Wiley, New York, 1981.

34. (*a*) E. Yang, *Microelectronic Devices*, McGraw-Hill, New York, 1988.

   (*b*) M. Zambuto, *Semiconductor Devices*, McGraw-Hill, New York, 1989.

35. J. L. Hullett and T. V. Muoi, "A feedback receiver amplifier for optical transmission systems," *IEEE Trans. Commun.*, vol. COM-24, pp. 1180–1185, Oct. 1976.

36. B. L. Kasper, A. R. McCormick, C. A. Burrus, and J. R. Talman, "An optical-feedback trans-impedance receiver for high sensitivity and wide dynamic range at low bit rates," *J. Lightwave Tech.*, vol. 6, pp. 329–338, Feb. 1988.

37. G. F. Williams and H. P. LeBlanc, "Active feedback lightwave receivers," *J. Lightwave Tech.*, vol. LT-4, pp. 1502–1508, Oct. 1986.

38. R. G. Swartz, "High performance integrated circuits for lightwave systems," in S. E. Miller and I. P. Kaminow, eds., *Optical Fiber Telecommunications—II*, Academic, New York, 1988.

39. See special issue on "Applications of RF and Microwave Subcarriers to Optical Fiber Transmission in Present and Figure Broadband Networks," *J. Lightwave Tech.*, vol. 8, 1990.

40. See special issue on "Ultrafast Electronics, Photonics, and Optoelectronics," *IEEE J. Sel. Topics Quantum Electron.*, vol. 2, Sept. 1996.

41. See special issue on "Microwave and Millimeter-Wave Photonics," *IEEE Trans. Microwave Theory Tech.*, vol. 45, part 2, Aug. 1997.

42. T. T. Ha, G. E. Keiser, and R. L. Borchardt, "Bit error probabilities of OOK lightwave systems with optical amplifiers," *J. Opt. Commun.*, vol. 18, pp. 151–155, Aug. 1997.

43. C.-S. Li, F. Tong, and G. Berkowitz, "Variable bit-rate receivers for WDMA/WDM systems," *IEEE Photonics Tech. Lett.*, vol. 9, pp. 1158–1160, Aug. 1997.

CHAPTER

8

DIGITAL TRANSMISSION SYSTEMS

The preceding chapters have presented the fundamental characteristics of the individual building blocks of an optical fiber transmission link. These include the optical fiber transmission medium, the optical source, the photodetector and its associated receiver, and the connectors used to join individual fiber cables to each other and to the source and detector. Now we shall examine how these individual parts can be put together to form a complete optical fiber transmission link. In particular, we shall study basic digital links in this chapter, and analog links in Chap. 9. More complex transmission links are examined in Chap. 12.

The first discussion involves the simplest case of a point-to-point link. This will include examining the components that are available for a particular application and seeing how these components relate to the system performance criteria (such as dispersion and bit-error rate). For a given set of components and a given set of system requirements, we then carry out a power budget analysis to determine whether the fiber optic link meets the attenuation requirements or if amplifiers are needed to boost the power level. The final step is to perform a system rise-time-analysis to verify that the overall system performance requirements are met.

We next turn our attention to line-coding schemes that are suitable for digital data transmission over optical fibers. These coding schemes are used to introduce randomness and redundancy into the digital information stream to ensure efficient timing recovery and to facilitate error monitoring at the receiver.

To increase the end-to-end fidelity of an optical transmission line, forward error correction (FEC) can be used if the bit-error rate is limited by optical noise and dispersion. Section 8.3 describes the basics of FEC.

As one moves to higher-speed (> 400 Mb/s) single-mode applications, a variety of system and component noise factors affect the fiber optic transmission quality. These include modal noise, mode-partition noise, laser chirping, and reflection noise, which are the topics of Sec. 8.4.

8.1 POINT-TO-POINT LINKS

The simplest transmission link is a point-to-point line that has a transmitter on one end and a receiver on the other, as is shown in Fig. 8-1. This type of link places the least demand on optical fiber technology and thus sets the basis for examining more complex system architectures.[1-8]

The design of an optical link involves many interrelated variables among the fiber, source, and photodetector operating characteristics, so that the actual link design and analysis may require several iterations before they are completed satisfactorily. Since performance and cost constraints are very important factors in fiber optic communication links, the designer must carefully choose the components to ensure that the desired performance level can be maintained over the expected system lifetime without overspecifying the component characteristics.

The following key system requirements are needed in analyzing a link:

1. The desired (or possible) transmission distance
2. The data rate or channel bandwidth
3. The bit-error rate (BER)

To fulfill these requirements the designer has a choice of the following components and their associated characteristics:

1. Multimode or single-mode optical fiber
   (a) Core size
   (b) Core refractive-index profile
   (c) Bandwidth or dispersion
   (d) Attenuation
   (e) Numerical aperture or mode-field diameter

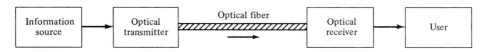

FIGURE 8-1
Simplex point-to-point link.

2. LED or laser diode optical source

    (*a*) Emission wavelength

    (*b*) Spectral line width

    (*c*) Output power

    (*d*) Effective radiating area

    (*e*) Emission pattern

    (*f*) Number of emitting modes

3. *pin* or avalanche photodiode

    (*a*) Responsivity

    (*b*) Operating wavelength

    (*c*) Speed

    (*d*) Sensitivity

Two analyses are usually carried out to ensure that the desired system performance can be met: these are the *link power budget* and the system *rise-time budget* analyses. In the link power budget analysis one first determines the power margin between the optical transmitter output and the minimum receiver sensitivity needed to establish a specified BER. This margin can then be allocated to connector, splice, and fiber losses, plus any additional margins required for possible component degradations, transmission-line impairments, or temperature effects. If the choice of components did not allow the desired transmission distance to be achieved, the components might have to be changed or amplifiers might have to be incorporated into the link.

Once the link power budget has been established, the designer can perform a system rise-time analysis to ensure that the desired overall system performance has been met. We shall now examine these two analyses in more detail.

8.1.1 System Considerations

In carrying out a link power budget, we first decide at which wavelength to transmit and then choose components that operate in this region. If the distance over which the data are to be transmitted is not too far, we may decide to operate in the 800-to-900-nm region. On the other hand, if the transmission distance is relatively long, we may want to take advantage of the lower attenuation and dispersion that occurs at wavelengths around 1300 or 1550 nm.

Having decided on a wavelength, we next interrelate the system performances of the three major optical link building blocks; that is, the receiver, transmitter, and optical fiber. Normally, the designer chooses the characteristics of two of these elements and then computes those of the third to see if the system performance requirements are met. If the components have been over- or under-specified, a design iteration may be needed. The procedure we shall follow here is first to select the photodetector. We then choose an optical source and see how far data can be transmitted over a particular fiber before an amplifier is needed in the line to boost up the power level of the optical signal.

In choosing a particular photodetector, we mainly need to determine the minimum optical power that must fall on the photodetector to satisfy the bit-error rate (BER) requirement at the specified data rate. In making this choice, the designer also needs to take into account any design cost and complexity constraints. As noted in Chaps. 6 and 7, a *pin* photodiode receiver is simpler, more stable with changes in temperature, and less expensive than an avalanche photodiode receiver. In addition, *pin* photodiode bias voltages are normally less than 5 V, whereas those of avalanche photodiodes range from 40 V to several hundred volts. However, the advantages of *pin* photodiodes may be overruled by the increased sensitivity of the avalanche photodiode if very low optical power levels are to be detected.

The system parameters involved in deciding between the use of an LED and a laser diode are signal dispersion, data rate, transmission distance, and cost. As shown in Chap. 4, the spectral width of the laser output is much narrower than that of an LED. This is of importance in the 800-to-900-nm region, where the spectral width of an LED and the dispersion characteristics of silica fibers limit the data-rate–distance product to around $150 \, (\text{Mb/s}) \cdot \text{km}$. For higher values [up to $2500 \, (\text{Mb/s}) \cdot \text{km}$], a laser must be used at these wavelengths. At wavelengths around $1.3 \, \mu\text{m}$, where signal dispersion is very low, bit-rate–distance products of at least $1500 \, (\text{Mb/s}) \cdot \text{km}$ are achievable with LEDs. For InGaAsP lasers, this figure is in excess of $25 \, (\text{Gb/s}) \cdot \text{km}$ at $1.3 \, \mu\text{m}$. A single-mode fiber can provide the ultimate bit-rate–distance product, with values of over $500 \, (\text{Gb/s}) \cdot \text{km}$ having been demonstrated at 1550 nm.

Since laser diodes typically couple from 10 to 15 dB more optical power into a fiber than an LED, greater repeaterless transmission distances are possible with a laser. This advantage and the lower dispersion capability of laser diodes may be offset by cost constraints. Not only is a laser diode itself more expensive than an LED, but also the laser transmitter circuitry is much more complex, since the lasing threshold has to be dynamically controlled as a function of temperature and device aging.

For the optical fiber, we have a choice between single-mode and multimode fiber, either of which could have a step- or a graded-index core. This choice depends on the type of light source used and on the amount of dispersion that can be tolerated. Light-emitting diodes (LEDs) tend to be used with multimode fibers, although, as we saw in Chap. 5, edge-emitting LEDs can launch sufficient optical power into a single-mode fiber for transmission at data rates greater than 500 Mb/s over several kilometers. The optical power that can be coupled into a fiber from an LED depends on the core–cladding index difference Δ, which, in turn, is related to the numerical aperture of the fiber (for $\Delta = 0.01$, the numerical aperture NA $\simeq 0.21$). As Δ increases, the fiber-coupled power increases correspondingly. However, since dispersion also becomes greater with increasing Δ, a tradeoff must be made between the optical power that can be launched into the fiber and the maximum tolerable dispersion.

When choosing the attenuation characteristics of a cabled fiber, the excess loss that results from the cabling process must be considered in addition to the attenuation of the fiber itself. This must also include connector and splice losses as well as environmental-induced losses that could arise from temperature variations, radiation effects, and dust and moisture on the connectors.

8.1.2   Link Power Budget

An optical power loss model for a point-to-point link is shown in Fig. 8-2. The optical power received at the photodetector depends on the amount of light coupled into the fiber and the losses occurring in the fiber and at the connectors and splices. The link loss budget is derived from the sequential loss contributions of each element in the link. Each of these loss elements is expressed in decibels (dB) as

$$\text{loss} = 10 \log \frac{P_{\text{out}}}{P_{\text{in}}} \tag{8-1}$$

where P_{in} and P_{out} are the optical powers emanating into and out of the loss element, respectively.

In addition to the link loss contributors shown in Fig. 8-2, a link power margin is normally provided in the analysis to allow for component aging, temperature fluctuations, and losses arising from components that might be added at future dates. A link margin of 6–8 dB is generally used for systems that are not expected to have additional components incorporated into the link in the future.

The link loss budget simply considers the total optical power loss P_T that is allowed between the light source and the photodetector, and allocates this loss to cable attenuation, connector loss, splice loss, and system margin. Thus, if P_S is the optical power emerging from the end of a fiber flylead attached to the light source, and if P_R is the receiver sensitivity, then

$$P_T = P_S - P_R$$
$$= 2l_c + \alpha_f L + \text{system margin} \tag{8-2}$$

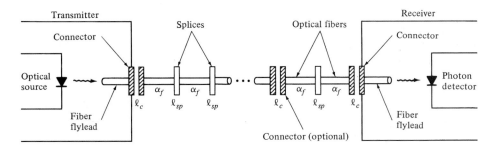

FIGURE 8-2
Optical power loss model for a point-to-point link. The losses occur at connectors (l_c), at splices (l_{sp}), and in the fiber (α_f).

Here, l_c is the connector loss, α_f is the fiber attenuation (dB/km), L is the transmission distance, and the system margin is nominally taken as 6 dB. Here, we assume that the cable of length L has connectors only on the ends and none in between. The splice loss is incorporated into the cable loss for simplicity.

Example 8-1. To illustrate how a link loss budget is set up, let us carry out a specific design example. We shall begin by specifying a data rate of 20 Mb/s and a bit-error rate of 10^{-9} (i.e., at most one error can occur for every 10^9 bits sent). For the receiver, we shall choose a silicon *pin* photodiode operating at 850 nm. Figure 8-3 shows that the required receiver input signal is −42 dBm (42 dB below 1 mW). We next select a GaAlAs LED that can couple a 50-μW (−13-dBm) average optical power level into a fiber flylead with a 50-μm core diameter. We thus have a 29-dB allowable power loss. Assume further that a 1-dB loss occurs when the fiber flylead is connected to the cable and another 1-dB connector loss occurs at the cable–photodetector interface. Including a 6-dB system margin, the possible transmission distance for a cable with an attenuation of α_f dB/km can be found from Eq. (8-2):

$$P_T = P_S - P_R = 29\,\text{dB}$$
$$= 2(1\,\text{dB}) + \alpha_f L + 6\,\text{dB}$$

If $\alpha_f = 3.5$ dB/km, then a 6.0-km transmission path is possible.

The link power budget can be represented graphically as is shown in Fig. 8-4. The vertical axis represents the optical power loss allowed between the transmitter and the receiver. The horizontal axis give the transmission distance. Here, we show a

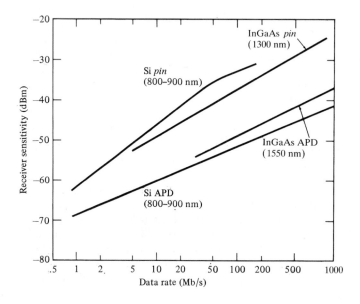

FIGURE 8-3
Receiver sensitivities as a function of bit rate. The Si *pin*, Si APD, and InGaAs *pin* curves are for a 10^{-9} BER. The InGaAs APD curve is for a 10^{-11} BER.

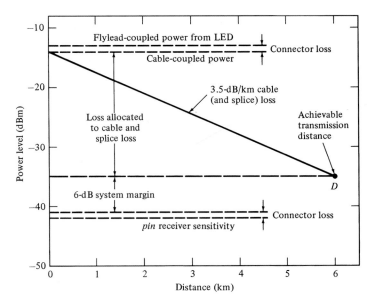

FIGURE 8-4

Graphical representation of a link-loss budget for an 800-nm LED/*pin* system operating at 20 Mb/s.

silicon *pin* receiver with a sensitivity of −42 dBm (at 20 Mb/s) and an LED with an output power of −13 dBm coupled into a fiber flylead. We subtract a 1-dB connector loss at each end, which leaves a total margin of 27 dB. Subtracting a 6-dB system safety margin leaves us with a tolerable loss of 21 dB that can be allocated to cable and splice loss. The slope of the line shown in Fig. 8-4 is the 3.5-dB/km cable (and splice, in this case) loss. This line starts at the −14-dBm point (which is the optical power coupled into the cabled fiber) and ends at the −35-dBm level (the receiver sensitivity minus a 1-dB connector loss and a 6-dB system margin). The intersection point *D* then defines the maximum possible transmission path length.

A convenient procedure for calculating the power budget is to use a tabular or spreadsheet form. We will illustrate this by way of an example for a SONET OC-48 (2.5 Gb/s) link.

Example 8-2. Consider a 1550-nm laser diode that launched a +3-dBm (2-mW) optical power level into a fiber flylead, an InGaAs APD with a −32-dBm sensitivity at 2.5 Gb/s, and a 60-km long optical cable with a 0.3-dB/km attenuation. Assume that here, because of the way the equipment is arranged, a short optical jumper cable is needed at each end between the end of the transmission cable and the SONET equipment rack. Assume that each jumper cable introduces a loss of 3 dB. In addition, assume a 1-dB connector loss occurs at each fiber joint (two at each end because of the jumper cables).

Table 8-1 lists the components in column 1 and the associated optical output, sensitivity, or loss in column 2. Column 3 gives the power margin available after subtracting the component loss from the total optical power loss that is allowed

TABLE 8-1
Example of a spreadsheet for calculating an optical-link power budget

Component/loss parameter	Output/sensitivity/loss	Power margin (dB)
Laser output	3 dBm	
APD sensitivity at 2.5 Gb/s	−32 dBm	
Allowed loss [3 − (−32)]		35
Source connector loss	1 dB	34
Jumper + connector loss	3 + 1 dB	30
Cable attenuation (60 km)	18 dB	12
Jumper + connector loss	3 + 1 dB	8
Receiver connector loss	1 dB	7 (final margin)

between the light source and the photodetector, which, in this case, is 35 dB. Adding all the losses results in a final power margin of 7 dB.

8.1.3 Rise-Time Budget

A rise-time budget analysis is a convenient method for determining the dispersion limitation of an optical fiber link. This is particularly useful for digital systems. In this approach, the total rise time t_{sys} of the link is the root sum square of the rise times from each contributor t_i to the pulse rise-time degradation:

$$t_{sys} = \left(\sum_{i=1}^{N} t_i^2 \right)^{1/2} \tag{8-3}$$

The four basic elements that may significantly limit system speed are the transmitter rise time t_{tx}, the group-velocity dispersion (GVD) rise time t_{GVD} of the fiber, the modal dispersion rise time t_{mod} of the fiber, and the receiver rise time t_{rx}. Single-mode fibers do not experience modal dispersion, so in these fibers the rise time is related only to GVD. Generally, the total transition-time degradation of a digital link should not exceed 70 percent of an NRZ (non-return-to-zero) bit period or 35 percent of a bit period for RZ (return-to-zero) data, where one bit period is defined as the reciprocal of the data rate (NRZ and RZ data formats are discussed in more detail in Sec. 8.2).

The rise times of transmitters and receivers are generally known to the designer. The transmitter rise time is attributable primarily to the light source and its drive circuitry. The receiver rise time results from the photodetector response and the 3-dB electrical bandwidth of the receiver front end. The response of the receiver front end can be modeled by a first-order lowpass filter having a step response[9]

$$g(t) = [1 - \exp(-2\pi B_{rx}t)]u(t)$$

where B_{rx} is the 3-dB electrical bandwidth of the receiver and $u(t)$ is the unit step function which is 1 for $t \geq 0$ and 0 for $t < 0$. The rise time t_{rx} of the receiver is usually defined as the time interval between $g(t) = 0.1$ and $g(t) = 0.9$. This is known as the *10- to 90-percent rise time*. Thus, if B_{rx} is given in megahertz, then the receiver front-end rise time in nanoseconds is (see Prob. 8-3)

$$t_{rx} = \frac{350}{B_{rx}} \tag{8-4}$$

In practice, an optical fiber link seldom consists of a uniform, continuous, jointless fiber. Instead, a transmission link nominally is formed from several concatenated (tandemly joined) fibers that may have different dispersion characteristics. This is especially true for dispersion-compensated links operating at 10 Gb/s and higher (see Chap. 12). In addition, multimode fibers experience modal distributions at fiber-to-fiber joints owing to misaligned joints, different core index profiles in each fiber, and/or different degrees of mode mixing in individual fibers. Determining the fiber rise times resulting from GVD and modal dispersion then becomes more complex than for the case of a single uniform fiber.

The fiber rise time t_{GVD} resulting from GVD over a length L can be approximated by Eq. (3-54) as

$$t_{GVD} \approx |D| L \sigma_\lambda \tag{8-5}$$

where σ_λ is the half-power spectral width of the source, and the dispersion D is given by Eq. (3-57) for a non-dispersion-shifted fiber and by Eq. (3-59) for a dispersion-shifted fiber. Since the dispersion value generally changes from fiber section to section in a long link, an average value should be used for D in Eq. (8-5).

The difficulty in predicting the bandwidth (and hence the modal rise time) of a series of concatenated multimode fibers arises from the observation that the total route bandwidth can be a function of the order in which fibers are joined. For example, instead of randomly joining together arbitrary (but very similar) fibers, an improved total link bandwidth can be obtained by selecting adjoining fibers with alternating over- and undercompensated refractive-index profiles to provide some modal delay equalization. Although the ultimate concatenated fiber bandwidth can be obtained by judiciously selecting adjoining fibers for optimum modal delay equalization, in practice this is unwieldy and time-consuming, particularly since the initial fiber in the link appears to control the final link characteristics.

A variety of empirical expressions for modal dispersion have thus been developed.[10-15] From practical field experience, it has been found that the bandwidth B_M in a link of length L can be expressed to a reasonable approximation by the empirical relation

$$B_M(L) = \frac{B_0}{L^q} \tag{8-6}$$

where the parameter q ranges between 0.5 and 1, and B_0 is the bandwidth of a 1-km length of cable. A value of $q = 0.5$ indicates that a steady-state modal equilibrium has been reached, whereas $q = 1$ indicates little mode mixing. Based on field experience, a reasonable estimate is $q = 0.7$.

Another expression that has been proposed for B_M, based on curve fitting of experimental data, is

$$\frac{1}{B_M} = \left[\sum_{n=1}^{N} \left(\frac{1}{B_n} \right)^{1/q} \right]^q \qquad (8\text{-}7)$$

where the parameter q ranges between 0.5 (quadrature addition) and 1.0 (linear addition), and B_n is the bandwidth of the nth fiber section. Alternatively, Eq. (8-7) can be written as

$$t_M(N) = \left[\sum_{n=1}^{N} (t_n)^{1/q} \right]^1 \qquad (8\text{-}8)$$

where $t_M(N)$ is the pulse broadening occurring over N cable sections in which the individual pulse broadenings are given by t_n.

We now need to find the relation between the fiber rise time and the 3-dB bandwidth. For this, we use a variation of the expression derived by Midwinter.[16] We assume that the optical power emerging from the fiber has a gaussian temporal response described by

$$g(t) = \frac{1}{\sqrt{2\pi}\sigma} e^{-t^2/2\sigma^2} \qquad (8\text{-}9)$$

where σ is the rms pulse width.

The Fourier transform of this function is

$$G(\omega) = \frac{1}{\sqrt{2\pi}} e^{-\omega^2\sigma^2/2} \qquad (8\text{-}10)$$

From Eq. (8-9) the time $t_{1/2}$ required for the pulse to reach its half-maximum value; that is, the time required to have

$$g(t_{1/2}) = 0.5g(0) \qquad (8\text{-}11)$$

is given by

$$t_{1/2} = (2\ln 2)^{1/2}\sigma \qquad (8\text{-}12)$$

If we define the time t_{FWHM} as the full width of the pulse at its half-maximum value, then

$$t_{\text{FWHM}} = 2t_{1/2} = 2\sigma(2\ln 2)^{1/2} \qquad (8\text{-}13)$$

The 3-dB optical bandwidth $B_{3\text{dB}}$ is defined as the modulation frequency $f_{3\text{dB}}$ at which the received optical power has fallen to 0.5 of the zero frequency value. Thus, from Eqs. (8-10) and (8-13), we find that the relation between the full-width half-maximum rise time t_{FWHM} and the 3-dB optical bandwidth is

$$f_{3dB} = B_{3dB} = \frac{0.44}{t_{FWHM}} \tag{8-14}$$

Using Eq. (8-6) for the 3-dB optical bandwidth of the fiber link and letting t_{FWHM} be the rise time resulting from modal dispersion, then, from Eq. (8-14),

$$t_{mod} = \frac{0.44}{B_M} = \frac{0.44L^q}{B_0} \tag{8-15}$$

If t_{mod} is expressed in nanoseconds and B_M is given in megahertz, then

$$t_{mod} = \frac{440}{B_M} = \frac{440L^q}{B_0} \tag{8-16}$$

Substituting Eqs. (3-20), (8-4), and (8-16) into Eq. (8-3) gives a total system rise time of

$$t_{sys} = \left[t_{tx}^2 + t_{mod}^2 + t_{GVD}^2 + t_{rx}^2 \right]^{1/2}$$

$$= \left[t_{tx}^2 + \left(\frac{440L^q}{B_0} \right)^2 + D^2 \sigma_\lambda^2 L^2 + \left(\frac{350}{B_{rx}} \right)^2 \right]^{1/2} \tag{8-17}$$

where all the times are given in nanoseconds, σ_λ is the half-power spectral width of the source, and the dispersion D [expressed in ns/(nm · km)] is given by Eq. (3-57) for a non-dispersion-shifted fiber and by Eq. (3-59) for a dispersion-shifted fiber. In the 800-to-900-nm region, D is about 0.07 ns/(nm · km), which is principally due to material dispersion, so that $t_{GVD}^2 \approx t_{mat}^2 = D_{mat}^2 \sigma_\lambda^2 L^2$. Much smaller values of D are seen in the 1300- and 1550-nm window (see Fig. 3-26).

Example 8-3. As an example of a rise-time budget for a multimode link, let us continue the analysis of the link we started to examine in Sec. 8.1.2. We shall assume that the LED together with its drive circuit has a rise time of 15 ns. Taking a typical LED spectral width of 40 nm, we have a material-dispersion-related rise-time degradation of 21 ns over the 6-km link. Assuming the receiver has a 25-MHz bandwidth, then from Eq. (8-4) the contribution to the rise-time degradation from the receiver is 14 ns. If the fiber we select has a 400-MHz · km bandwidth–distance product and with $q = 0.7$ in Eq. (8-6), then from Eq. (8-15) the modal-dispersion-induced fiber rise time is 3.9 ns. Substituting all these values back into Eq. (8-17) results in a link rise time of

$$t_{sys} = (t_{tx}^2 + t_{mat}^2 + t_{mod}^2 + t_{rx}^2)^{1/2}$$

$$= \left[(15\,\text{ns})^2 + (21\,\text{ns})^2 + (3.9\,\text{ns})^2 + (14\,\text{ns})^2 \right]^{1/2}$$

$$= 30\,\text{ns}$$

This value falls below the maximum allowable 35-ns rise-time degradation for our 20-Mb/s NRZ data stream. The choice of components was thus adequate to meet our system design criteria.

Analogous to power-budget calculations, a convenient procedure for keeping track of the various rise-time values in the rise-time budget is to use a tabular or spreadsheet form. We will illustrate this by way of an example for the SONET OC-48 (2.5 Gb/s) link we looked at in Example 8-2.

Example 8-4. Assume that the laser diode together with its drive circuit has a rise time of 0.025 ns (25 ps). Taking a 1550-nm laser diode spectral width of 0.1 nm and an average dispersion of 2 ps/(nm · km) for the fiber, we have a GVD-related rise-time degradation of 12 ps (0.012 ns) over a 60-km long optical cable. Assuming the InGaAs-APD-based receiver has a 2.5-GHz bandwidth, then from Eq. (8-4) the receiver rise time is 0.14 ns. Using Eq. (8-17) to add up the various contributions, we have a total rise time of 0.14 ns.

Table 8-2 lists the components in column 1 and the associated rise times in column 2. Column 3 gives the allowed system rise-time budget of 0.28 ns for a 2.5-Gb/s NRZ data stream at the top. This is found from the expression $0.7/B_{\text{NRZ}}$ where B_{NRZ} is the bit rate for the NRZ signal. The calculated system rise time of 0.14 ns is shown at the bottom. The system rise time, in this case, is dominated by the receiver and is well within the required limits.

8.1.4 First-Window Transmission Distance

Figure 8-5 shows the attenuation and dispersion limitation on the repeaterless transmission distance as a function of data rate for the short-wavelength (800–900-nm) LED/*pin* combination. The BER was taken as 10^{-9} for all data rates. The fiber-coupled LED output power was assumed to be a constant −13 dBm for all data rates up to 200 Mb/s. The attenuation limit curve was then derived by using a fiber loss of 3.5 dB/km and the receiver sensitivities shown in Fig. 8-3. Since the minimum optical power required at the receiver for a given BER becomes higher for increasing data rates, the attenuation limit curve slopes downward to the right. We have also included a 1-dB connector-coupling loss at each end and a 6-dB system operating margin.

The dispersion limit depends on material and modal dispersion. Material dispersion at 800 nm is taken as 0.07 ns/(nm · km) or 3.5 ns/km for an LED with a

TABLE 8-2
Example of a tabular form for keeping track of component contributions to an optical-link rise-time budget

Component	Rise time	Rise-time budget
Allowed rise-time budget		$t_{\text{sys}} = 0.7/B_{\text{NRZ}} = 0.28$ ns
Laser transmitter	25 ps	
GVD in fiber	12 ps	
Receiver rise time	0.14 ns	
System rise time [Eq. (8-17)]		0.14 ns

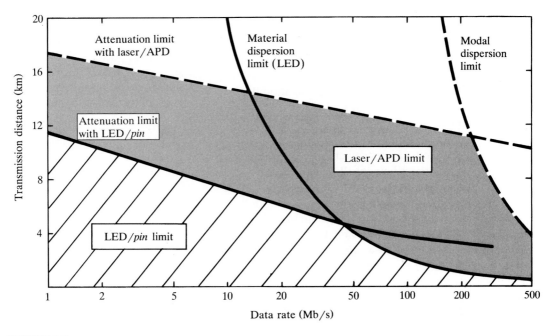

FIGURE 8-5

Transmission distance limits as a function of data rate for an 800-MHz · km fiber, an 800-nm LED source with a Si *pin* photodiode combination, and an 850-nm laser diode with a Si APD.

50-nm spectral width. The curve shown is the material dispersion limit in the absence of modal dispersion. This limit was taken to be the distance at which t_{mat} is 70 percent of a bit period. The modal dispersion was derived from Eq. (8-15) for a fiber with an 800-MHz · km bandwidth–distance product and with $q = 0.7$. The modal dispersion limit was then taken to be the distance at which t_{mod} is 70 percent of a bit period. The achievable transmission distances are those that fall below the attenuation limit curve and to the left of the dispersion line, as indicated by the hatched area. The transmission distance is attenuation-limited up to about 40 Mb/s, after which it becomes material-dispersion-limited.

Greater transmission distances are possible when a laser diode is used in conjunction with an avalanche photodiode. Let us consider an AlGaAs laser emitting a 850 nm with a spectral width of 1 nm that couples 0 dBm (1 mW) into a fiber flylead. The receiver uses an APD with a sensitivity depicted in Fig. 8-3. The fiber is the same as described in Sec. 8.1.4. In this case, the material-dispersion-limit curve lies off the graph to the right of the modal-dispersion-limit curve, and the attenuation limit (with an 8-dB system margin) is as shown in Fig. 8-5. The achievable transmission distances now include those indicated by the shaded area.

8.1.5 Transmission Distance for Single-Mode Links

At the other extreme from that shown in Fig. 8-5, let us examine a single-mode link operating at 1550 nm. In this case, the dispersion in the fiber is due only to GVD effects, since there is no modal dispersion. We take the dispersion to be $D = 2.5\,\text{ps}/(\text{nm} \cdot \text{km})$ and the attenuation to be 0.30 dB/km at 1550 nm. For the source we first choose a laser which couples 0 dBm of optical power into the fiber and which has a large spectral width $\sigma_\lambda = 3.5$ nm. Then we select a laser with $\sigma_\lambda = 1$ nm. The receiver can use either an InGaAs avalanche photodiode (APD) with a sensitivity of $P_r = 11.5 \log B - 71.0$ dBm or an InGaAs *pin* photodiode with a sensitivity of $P_r = 11.5 \log B - 60.5$ dBm, where B is the data rate in Mb/s. The attenuation-limited transmission distances for these two photodiodes are shown in Fig. 8-6 with the inclusion of an 8-dB system margin.

For the dispersion limit we examine two cases. First, for the $\sigma_\lambda = 3.5$ nm case, Fig. 8-6 shows the limit for NRZ data where the product $D\sigma_\lambda L$ is equal to 70 percent of the bit period and where for RZ data the product $D\sigma_\lambda L$ is equal to 35 percent of the bit period. Next we show the limit of NRZ data for the $\sigma_\lambda = 1$ nm case. Note the dramatic change for the narrower linewidth. These curves are for

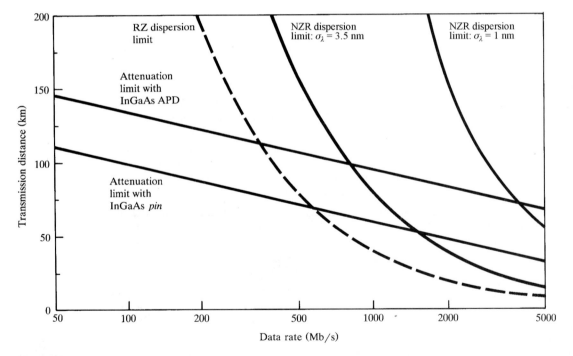

FIGURE 8-6

Transmission-distance limits as a function of data rate for 1550-nm laser diodes having spectral widths $\sigma_\lambda = 3.5$ and 1.0 nm, an InGaAs APD, and a single-mode fiber with $D = 2.5\,\text{ps}/(\text{nm} \cdot \text{km})$ and a 0.3-dB/km attenuation.

the ideal case. In reality, various noise effects due to laser instabilities coupled with chromatic dispersion in the fiber can greatly decrease the dispersion-limited distance. Section 8.4 discusses these factors and their effect on system performance.

8.2 LINE CODING

In designing an optical fiber link, an important consideration is the format of the transmitted optical signal. This is of importance because, in any practical digital optical fiber data link, the decision circuitry in the receiver must be able to extract precise *timing information* from the incoming optical signal. The three main purposes of timing are to allow the signal to be sampled by the receiver at the time the signal-to-noise ratio is a maximum, to maintain the proper pulse spacing, and to indicate the start and end of each timing interval. In addition, since errors resulting from channel noise and distortion mechanisms can occur in the signal-detection process, it may be desirable for the optical signal to have an inherent error-detecting capability. These features can be incorporated into the data stream by restructuring (or encoding) the signal. This is generally done by introducing extra bits into the raw data stream at the transmitter on a regular and logical basis and extracting them again at the receiver.

Signal encoding uses a set of rules for arranging the signal symbols in a particular pattern. This process is called *channel* or *line coding*. The purpose of this section is to examine the various types of line codes that are well suited for digital transmission on an optical fiber link. The discussion here is limited to binary codes, because they are the most widely used electrical codes and also because they are the most advantageous codes for optical systems.

One of the principal functions of a line code is to introduce redundancy into the data stream for the purpose of minimizing errors that result from channel interference effects. Depending on the amount of redundancy introduced, any degree of error-free transmission of digital data can be achieved, provided that the data rate that includes this redundancy is less than the channel capacity. This is a result of the well-known Shannon channel-coding theory.[17,18]

Although large system bandwidths are attainable with optical fibers, the signal-to-noise considerations of the receiver discussed in Chap. 7 show that larger bandwidths result in larger noise contributions. Thus, from noise considerations, minimum bandwidths are desirable. However, a larger bandwidth may be needed to have timing data available from the bit stream. In selecting a particular line code, a tradeoff must therefore be made between timing and noise bandwidth.[19] Normally, these are largely determined by the expected characteristics of the raw data stream.

The three basic types of two-level binary line codes that can be used for optical fiber transmission links are the non-return-to-zero (NRZ) format, the return-to-zero (RZ) format, and the phase-encoded (PE) format. In NRZ codes a transmitted data bit occupies a full bit period. For RZ formats the pulse width is less than a full bit period. In the PE format both full-width and half-width data bits are present. Multilevel binary (MLB) signalling[20] is also possible, but it is used much less frequently than the popular NRZ and RZ codes. A brief description of some NRZ and RZ codes will be given here. Additional details can be found in various communications books and articles[9,21-25]

8.2.1 NRZ Codes

A number of different NRZ codes are widely used, and their bandwidths serve as references for all other code groups. The simplest NRZ code is NRZ-level (or NRZ-L), shown in Fig. 8-7. For a serial data stream, an on–off (or unipolar) signal represents a 1 by a pulse of current or light filling an entire bit period, whereas for a 0 no pulse is transmitted. These codes are simple to generate and decode, but they possess no inherent error-monitoring or correcting capabilities and they have no self-clocking (timing) features.

The minimum bandwidth is needed with NRZ coding, but the average power input to the receiver is dependent on the data pattern. For example, the high level of received power occurring in a long string of consecutive 1 bits can result in a *baseline wander* effect, as shown in Fig. 8-8. This effect results from the accumulation of pulse tails that arise from the low-frequency characteristics of the ac-coupling filter in the receiver. If the receiver recovery to the original threshold is slow after the long string of 1 bits has ended, an error may occur if the next 1 bit has a low amplitude.

In addition, a long string of NRZ ones or zeros contains no timing information, since there are no level transitions. Thus, unless the timing clocks in the system are extremely stable, a long string of N identical bits could be misinterpreted as either $N - 1$ or $N + 1$ bits. However, the use of highly stable clocks increases system costs and requires a long system startup time to achieve synchronization. Two common techniques for restricting the longest time interval in which no level transitions occur are the use of block codes (see Sec. 8.2.3) and scrambling.[26-28] *Scrambling* produces a random data pattern by the modulo-2 addition of a known bit sequence with the data stream. At the receiver, the same known bit sequence is again modulo-2 added to the received data, and the original bit sequence is recovered. Although the randomness of scrambled NRZ data ensures an adequate amount of timing information, the penalty for its use is an increase in the complexity of the NRZ encoding and decoding circuitry.

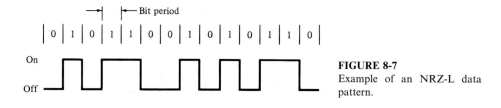

| 0 | 1 | 0 | 1 | 1 | 0 | 0 | 1 | 0 | 1 | 0 | 1 | 1 | 0 |

FIGURE 8-7
Example of an NRZ-L data pattern.

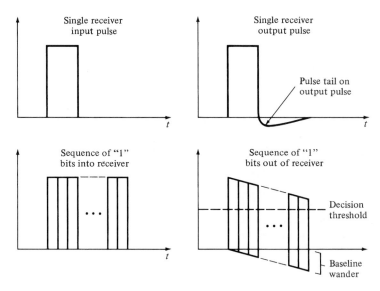

FIGURE 8-8
Baseline wander at the receiver resulting from the transmission of long strings of NRZ 1 bits.

8.2.2 RZ Codes

If an adequate bandwidth margin exists, each data bit can be encoded as two optical line code bits. This is the basis of RZ codes. In these codes, a signal level transition occurs during either some or all of the bit periods to provide timing information. A variety of RZ code types exist, some of which are shown in Fig. 8-9. The baseband (NRZ-L) data are shown in Fig. 8-9a. In the unipolar RZ data, a 1 bit is represented by a half-period optical pulse that can occur in either the first or second half of the bit period. A 0 is represented by no signal during the bit period.

A disadvantage of the unipolar RZ format is that long strings of 0 bits can cause loss of timing synchronization. A common data format not having this limitation is the *biphase* or *optical Manchester* code shown in Fig. 8-9d. Note that this is a unipolar code, which is in contrast to the conventional bipolar Manchester code used in wire lines. The optical Manchester signal is obtained by direct modulo-2 addition of the baseband (NRZ-L) signal and a clock signal (Fig. 8-9b). In this code, there is a transition at the center of each bit interval. A negative-going transition indicates a 1-bit, whereas a positive-going transition means a 0 bit was sent. The Manchester code is simple to generate and decode. Since it is an RZ-type code, it requires twice the bandwidth of an NRZ code. In addition, it has no inherent error-detecting or correcting capability.

Coaxial or wire-pair cable systems commonly use the bipolar RZ or alternate mark inversion (AMI) coding scheme. These wire line codes have also been adapted to unipolar optical systems.[29] The two-level AMI optical pulse formats

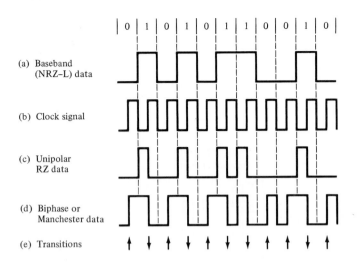

FIGURE 8-9

Examples of RZ data formats. (*a*) NRZ-L baseband data, (*b*) clock signal, (*c*) unipolar RZ data, (*d*) biphase or optical Manchester, (*e*) transitions occurring within a bit period for Manchester data.

require twice the transmission bandwidth of NRZ codes, but they provide timing information in the data stream, and the redundancy of the encoded information (which is inherent in these codes) allows for direct in-service error monitoring. Other more complex schemes have been tried for high-speed links.[30-34]

8.2.3 Block Codes

An efficient category of redundant binary codes is the *mBnB* block code class.[35-38] In this class of codes, blocks of *m* binary bits are converted to longer blocks of $n > m$ binary bits. These new blocks are then transmitted in NRZ or RZ format. As a result of the additional redundant bits, the increase in bandwidth using this scheme is given by the ratio n/m. At the expense of this increased bandwidth, the *mBnB* block codes provide adequate timing and error-monitoring information, and they do not have baseline wander problems, since long strings of ones and zeros are eliminated.

A convenient concept used for block codes is the *accumulated* or *running disparity*, which is the cumulative difference between the numbers of 1 and 0 bits. A simple means of measuring this is with an up–down counter. The key factors in selecting a particular block code are low disparity and a limit in the disparity variation (the difference between the maximum and minimum values of the accumulated disparity). A low disparity allows the dc component of the signal to be cancelled. A bound on the accumulated disparity avoids the low-frequency spectral content of the signal and facilitates error monitoring by detecting the disparity overflow. Generally, one chooses codes that have an even *n* value, since for odd values of *n* there are no coded words with zero disparity.

TABLE 8-3
A comparison of several *m*B*n*B codes

Code	n/m	N_{max}	D	W (%)
3B4B	1.33	4	±3	25
6B8B	1.33	6	±3	75
5B6B	1.20	6	±4	28
7B8B	1.14	9	±7	27
9B10B	1.11	11	±8	24

A comparison of several *m*B*n*B codes is given in Table 8-3. The following parameters are shown in this table:

1. The ratio n/m, which gives the bandwidth increase.
2. The longest number N_{max} of consecutive identical symbols (small values of N_{max} allow for easier clock recovery).
3. The bounds on the accumulated disparity D.
4. The percentage W of n-bit words that are not used (the detection of invalid words at the receiver permits character reframing).

Suitable codes for high data rates are the 3B4B, 4B5B, 5B6B, and 8B10B codes. If simplicity of the encoder and decoder circuits is the main criterion, the 3B4B is the most convenient code. The 5B6B code is the most advantageous if bandwidth reduction is the major concern. In local-area network (LAN) applications, the widely used Fiber Distributed Data Interface (FDDI) uses 4B5B coding and Fibre Channel employs an 8B10B code.[38]

8.3 ERROR CORRECTION

For high-speed broadband networks, the data-transmission reliability provided by the network may be lower than the reliability requested by an application. In this case, the transport protocol of the network must compensate for the difference in the bit-loss rate. The two basic schemes for improving the reliability are *automatic repeat request* (ARQ) and *forward error correction* (FEC).[39-44] The ARQ schemes have been used for many years and are widely implemented. As shown in Fig. 8-10, the technique uses a feedback channel between the receiver and the transmitter to request message retransmission in case errors are detected at the receiver. Since each such retransmission adds at least one roundtrip time of latency, ARQ may not be feasible for applications that require low latency. Among these applications are voice and video services that involve human interaction, process control, and remote sensing in which data must arrive within a certain time in order to be useful.

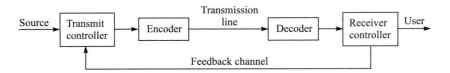

FIGURE 8-10
Basic setup for an automatic-repeat-request (ARQ) error-correction scheme.

Forward error correction avoids the shortcomings of ARQ for high bandwidth optical networks that require low delays. In FEC techniques, redundant information is transmitted along with the original information. If some of the original data is lost or received in error, the redundant information is used to reconstruct it. Typically, the amount of redundant information is small, so the FEC scheme does not use up much additional bandwidth and thus remains efficient. Depending on the application, some considerations of FEC code properties include the ability to accommodate self-synchronous scramblers (with characteristics polynomial $1 + x^{43}$) used in SONET (see Chap. 12), the 4B5B line code used in FDDI, or the 8B10B line code used in Fibre Channel.

The most popular error-correction codes are *cyclic codes*. These are designated by the notation (n, m), where n equals the number of original bits m plus the number of redundant bits. Some examples that have been used include a (224, 216) shortened Hamming code,[40] a (192, 190) Reed-Solomon code,[42] a (255, 239) Reed-Solomon code,[43] and (18880, 18865) and (2370, 2358) shortened Hamming codes.[44]

The results of the (224, 216) code are shown in Figs. 8-11 and 8-12. Figure 8-11 is a plot of the FEC-decoded BER versus the primary BER. The data is from an experiment using a 565-Mb/s multimode laser system operating at 1300 nm. The laser had a FWHM of 4 nm at the one-tenth-maximum point. Various levels of AWGN (additive white gaussian noise) were injected at the receiver decision point in order to vary the signal-to-noise ratio. The relative performance improvement using FEC increases as the error probability decreases. For example, at a 10^{-4} primary BER the performance improves by a factor of about 25, whereas at a 10^{-6} primary error level the BER resulting from FEC decreases by a factor of about 3000.

Figure 8-12 shows the measured BER performance as a function of the received power level with and without FEC. Analogous to Fig. 8-11, the performance improvement with FEC is significant.

8.4   NOISE EFFECTS ON SYSTEM PERFORMANCE

In the analysis in Sec. 8.1 we assumed that the optical power falling on the photodetector is a clearly defined function of time within the statistical nature of the quantum detection process. In reality, as noted in Sec. 4.5, various inter-

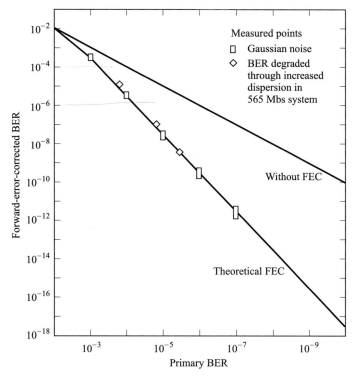

FIGURE 8-11
Calculated and measured forward-error-correction (FEC) characteristics using a (224, 216) code. (Reproduced with permission from Grover,[40] © 1988, OSA.)

actions between spectral imperfections in the propagating optical power and the dispersive waveguide give rise to variations in the optical power level falling on the photodetector. These variations create receiver output noises and hence give rise to optical power penalties, which are particularly serious for high-speed links. The main penalties are due to modal noise, wavelength chirp, spectral broadening induced by optical reflections back into the laser, and mode-partition noise. Modal noise is not present in single-mode links; however, mode-partition noise, chirping, and reflection noise are critical in these systems.

8.4.1 Modal Noise

Modal noise arises when the light from a coherent laser is coupled into a multi-mode fiber.[12,44-50] This is generally not a problem for links operating below 100 Mb/s, but becomes disastrous at speeds around 400 Mb/s and higher. The following factors can produce modal noise in an optical fiber link:

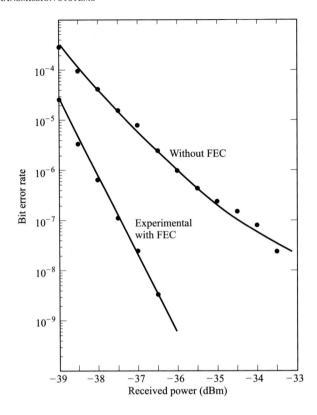

FIGURE 8-12

Measured performance of a 565-Mb/s transmission system with and without FEC. (Reproduced with permission from Grover,[40] © 1988, OSA.)

1. Mechanical disturbances along the link, such as vibrations, connectors, splices, microbends, and source or detector coupling, can result in differential mode delay or modal and spatial filtering of the optical power. This produces temporal fluctuations in the speckle pattern at the receiving end, thus creating modal noise in the receiver.

2. Fluctuations in the frequency of an optical source can also give rise to inter-modal delays. A coherent source forms speckle patterns when its coherence time is greater than the intermodal dispersion time δT within the fiber. If the source has a frequency width δv, then its coherence time is $1/\delta v$. Modal noise occurs when the speckle pattern fluctuates; that is, when the source coherence time becomes much less than the intermodal dispersion time. The modal distortion resulting from interference between a single pair of modes will appear as a sinusoidal ripple of frequency

$$v = \delta T \frac{dv_{\text{source}}}{dt} \tag{8-18}$$

where dv_{source}/dt is the rate of change of optical frequency.

Several researchers have examined how modal noise degrades the bit-error rate (BER) performance of a digital link.[48-50] As an example, Fig. 8-13 illustrates the error rates with the addition of modal noise to an avalanche-photodiode receiver system.[49] The analysis is for 280 Mb/s at 1200 nm with a gaussian-shaped received pulse. The factor M' in this figure is related to the number of speckles falling on the photodetector. For a very large number of speckles ($M' = 2910$), the error-rate curve is very close to the case when there is no modal noise. As the number of speckles decreases, the performance degrades. When $M' = 50$, one needs an additional 1.0 dB of received optical power to maintain an error rate of 10^{-6}. When $M' = 20$, one must have 2.0 dB more power to achieve a 10^{-6} BER than in the case of no modal noise. This number becomes 4.9 dB when $M' = 4$.

The performance of a high-speed, laser-based multimode fiber link is difficult to predict, since the degree of modal noise which can appear depends greatly on the particular installation. Thus, the best policy is to take steps to avoid it. This can be done by the following measures:

1. Use LEDs (which are incoherent sources). This totally avoids modal noise.
2. Use a laser which has a large number of longitudinal modes (10 or more). This increases the graininess of the speckle pattern, thus reducing intensity fluctuations at mechanical disruptions in the link.
3. Use a fiber with a large numerical aperture, since it supports a large number of modes and hence gives a greater number of speckles.

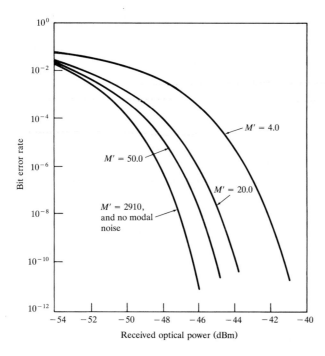

FIGURE 8-13
Error-rate curves for 280-Mb/s avalanche-photodiode-based system with the addition of modal noise. The factor M' corresponds to the number of speckles. (Reproduced with permission from Chan and Tjhung,[49] © 1989, IEEE.)

4. Use a single-mode fiber, since it supports only one mode and thus has no modal interference.

The last point needs some further explanation. If a connector or splice point couples some of the optical power from the fundamental mode into the first higher-order mode (the LP_{11} mode), then a significant amount of power could exist in the LP_{11} mode in a short section of fiber between two connectors or at a repair splice.[46,50] Figure 8-14 illustrates this effect. In a single-mode system, modal noise could occur in short connectorized patch-cords, in laser diode flyleads, or when two high-loss splices are a very short distance apart. To circumvent this problem, one should specify that the effective cutoff wavelength of short patch-cord and flylead fiber lengths is well below the system operating wavelength. Thus, mode coupling is not a problem in links that have long fiber lengths between connectors and splices, since the LP_{11} mode is usually sufficiently attenuated over the link length.

8.4.2   Mode-Partition Noise

As noted in Sec. 4.5, *mode-partition noise* is associated with intensity fluctuations in the longitudinal modes of a laser diode;[51-59] that is, the side modes are not

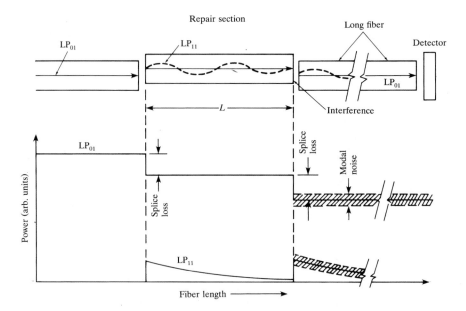

FIGURE 8-14
Repair sections can produce modal noise in a single-mode fiber link. This arises through the interchange of optical power between the LP_{01} and the LP_{11} modes at splice or connector joints. (Reproduced with permission from Sears, White, Kummer, and Stone,[46] © 1986, IEEE.)

sufficiently suppressed. This is the dominant noise in single-mode fibers. Intensity fluctuations can occur among the various modes in a multimode laser even when the total optical output is constant, as exhibited in Fig. 8-15. This power distribution can vary significantly both within a pulse and from pulse to pulse.

Because the output pattern of a laser diode is highly directional, the light from these fluctuating modes can be coupled into a single-mode fiber with high efficiency. Each of the longitudinal modes that is coupled into the fiber has a different attenuation and time delay, because each is associated with a slightly different wavelength (see Sec. 3.3). Since the power fluctuations among the dominant modes can be quite large, significant variations in signal levels can occur at the receiver in systems with high fiber dispersion.

The signal-to-noise ratio due to mode-partition noise is independent of signal power, so that the overall system error rate cannot be improved beyond the limit set by this noise. This is an important difference from the degradation of receiver sensitivity normally associated with chromatic dispersion, which one can compensate for by increasing the signal power.

The power penalty in decibels caused by laser mode-partition noise can be approximated by[56]

$$ P_{mpn} = -5\frac{x+2}{x+1}\log\left[1 - \frac{k^2 Q^2}{2}(\pi BLD\sigma_\lambda)^4\right] \qquad (8\text{-}19) $$

where x is the excess noise factor of an APD, Q is the signal-to-noise factor (see Fig. 7-7), B is the bit rate in Gb/s, L is the fiber length in km, D is the fiber chromatic dispersion in ps/(nm · km), σ_λ is the rms spectral width of the source in

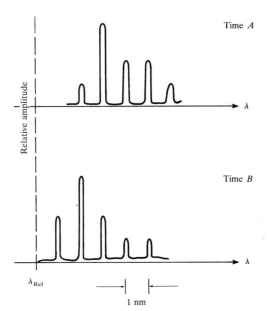

FIGURE 8-15
Time-resolved dynamic spectra of a laser diode. Different modes or groups of modes dominate the optical output at different times. The modes are approximately 1 nm apart.

nm, and k is the mode-partition-noise factor. The parameter k is difficult to quantify since it can vary from 0 to 1 depending on the laser. However, experimental values of k range from 0.6 to 0.8. To keep the power penalty less than 0.5 dB, a well-designed system should have the quantity $BLD\sigma_\lambda < 0.1$.

Mode-partition noise becomes more pronounced for higher bit rates. The errors due to mode-partition noise can be reduced and sometimes eliminated by setting the bias point of the laser above threshold. However, raising the bias power level reduces the available signal-pulse power, thereby reducing the achievable signal-to-thermal-noise ratio.

In attempts to describe the effects of mode-partition noise, researchers have tried to identify a figure of merit for the laser diode spectrum that could be measured experimentally, yet give an accurate theoretical prediction of system performance. One approach applies to lasers that have many lasing modes,[56] whereas another addresses two-mode lasers where the side mode is below the lasing threshold.[53,54] The second case is of interest in practice, since the distribution of mode-partition fluctuations is exponential rather than gaussian. This means that the fluctuations can cause very high error rates in all lasers except those wherein the nonlasing modes are greatly suppressed.

Figure 8-16 shows the result of a tradeoff analysis[54] between the mode-partition-noise BER and the system BER in the absence of mode-partition

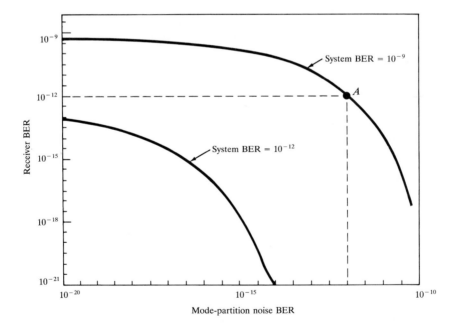

FIGURE 8-16

Example of a tradeoff analysis between the mode-partition-noise BER and the system BER in the absence of mode-partition noise. (Reproduced with permission from Basch, Kearns, and Brown,[54] © 1986, IEEE.)

noise. The curves represent the total system performance for error probabilities of 10^{-9} and 10^{-12}. As an example, to maintain a total system BER of 10^{-9} and also have a receiver error probability of 10^{-12}, the required error rate for mode partitioning is less than 10^{-12}, as shown by point A. This is equivalent to a mode-intensity ratio of $I_0/J_0 \simeq 50$, where I_0 is the average intensity of the main lasing mode and J_0 is the average intensity of the strongest nonlasing mode.

To prevent the occurrence of a high system bit-error rate due to power partitioning among insufficiently suppressed side modes, one must select lasers carefully. To evaluate the dynamics of the side modes, one can measure either the time-resolved photon statistics of the laser output, or the bit-error rate characteristics under realistic biasing conditions.

8.4.3 Chirping

A laser which oscillates in a single longitudinal mode under CW operation may experience dynamic line broadening when the injection current is directly modulated.[59-66] This line broadening is a frequency "chirp" associated with modulation-induced changes in the carrier density. Laser chirping can lead to significant dispersion effects for intensity-modulated pulses when the laser emission wavelength is displaced from the zero-dispersion wavelength of the fiber. This is particularly true in systems operating at 1550 nm, where dispersion in standard non-dispersion-shifted fibers is much greater than at 1300 nm.

To a good approximation, the time-dependent frequency change $\Delta\nu(t)$ of the laser can be given in terms of the output optical power $P(t)$ as[63]

$$\Delta\nu(t) = \frac{-\alpha}{4\pi}\left[\frac{d}{dt}\ln P(t) + \kappa P(t)\right] \tag{8-20}$$

where α is the *linewidth enhancement factor*[66] and κ is a frequency-independent factor that depends on the laser structure.[63] The factor α ranges from -3.5 to -5.5 for AlGaAs lasers[67] and from -6 to -8 for InGaAsP lasers.[68]

One approach to minimize chirp is to increase the bias level of the laser so that the modulation current does not drive it below threshold where $\ln P$ and P change rapidly. However, this results in a lower extinction ratio (in this case, the ratio of on-state power to off-state power), which leads to an extinction-ratio power penalty at the receiver because of a reduced signal-to-background noise ratio. This penalty could be several decibels. Figure 8-17 gives examples of this for two types of laser structures. For higher extinction ratios (bias points progressively lower than threshold), the extinction-ratio power penalty decreases. However, the chirping-induced power penalty increases with lower bias levels.

When the effect of laser chirp is small, the eye closure Δ can be approximated by[62]

$$\Delta = \left(\tfrac{4}{3}\pi^2 - 8\right)t_{\text{chirp}}DLB^2\,\delta\lambda[1 + \tfrac{2}{3}(DL\,\delta\lambda - t_{\text{chirp}})] \tag{8-21}$$

where t_{chirp} is the chirp duration, B is the bit rate, D is the fiber chromatic dispersion, L is the fiber length, and $\delta\lambda$ is the chirp-induced wavelength excursion.

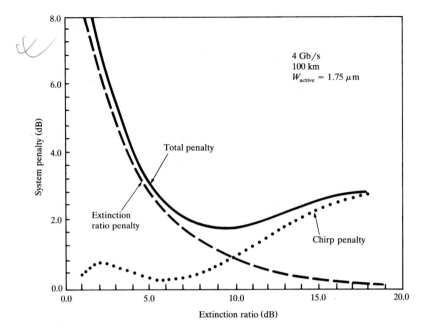

FIGURE 8-17

Extinction-ratio, chirping, and total-system power penalties at 1550 nm for a 4-Gb/s 100-km-long single-mode link having a fiber with a dispersion $D = 17$ ps/(nm · km) and a DFB laser with an active layer width of 1.75 μm. (Reproduced with permission from Corvini and Koch,[63] © 1987, IEEE.)

The power penalty for an APD system can be estimated from the signal-to-noise ratio degradation (in dB) due to the signal amplitude decrease as

$$p_{chirp} = -10\frac{x+2}{x+1}\log(1 - \Delta) \qquad (8\text{-}22)$$

where x is the excess noise factor of an APD.

The best approach to minimizing chirp effects is to choose the laser emission wavelength close to the zero-dispersion wavelength of the fiber. Experiments of this type[28] have shown no degradation in receiver sensitivity due to chromatic dispersion.

Figure 8-18 illustrates the effects of chirping at a 5-Gb/s transmission rate in different single-mode fiber links.[59] Here, the laser side-mode suppression is greater than 30 dB, the back-reflected optical power is more than 30 dB below the transmitted signal, and the extinction ratio is about 8 dB. At 1536 nm the standard non-dispersion-shifted fiber has a dispersion $D = 17.3$ ps/(nm · km) and the dispersion-shifted fiber has $D = -1.0$ ps/(nm · km). The combined-fiber link consists of concatenated standard positive-dispersion and negative-dispersion fibers. This leads to a spectral compression of the signal so that dispersion compensation occurs. Thus, Fig. 8-18 shows the dramatic reduction in chirping penalty when using a dispersion-shifted fiber, or when combining fibers with positive and negative dispersion.

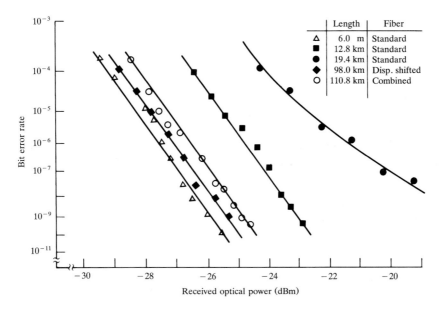

	Length	Fiber
△	6.0 m	Standard
■	12.8 km	Standard
●	19.4 km	Standard
◆	98.0 km	Disp. shifted
○	110.8 km	Combined

FIGURE 8-18

The effects of chirping at 5-Gb/s in different single-mode fiber links. The laser side-mode suppression is > 30 dB, the reflected power is more than 30 dB below the transmitted signal, and the extinction ratio is ≈ 8 dB. At 1536 nm the standard fiber has $D = 17.3 \, \text{ps/(nm} \cdot \text{km)}$ and the dispersion-shifted fiber has $D = -1.0 \, \text{ps/(nm} \cdot \text{km)}$. (Reproduced with permission from Heidemann,[59] © 1988, IEEE.)

8.4.4   Reflection Noise

When light travels through a fiber link, some optical power gets reflected at refractive-index discontinuities such as in splices, couplers, and filters, or at air–glass interfaces in connectors. The reflected signals can degrade both transmitter and receiver performance.[41,59,69-71] In high-speed systems, this reflected power causes optical feedback which can induce laser instabilities. These instabilities show up as either intensity noise (output power fluctuations), jitter (pulse distortion), or phase noise in the laser, and they can change its wavelength, linewidth, and threshold current. Since they reduce the signal-to-noise ratio, these effects cause two types of power penalties in receiver sensitivities. First, as shown in Fig. 8-19*a*, multiple reflection points set up an interferometric cavity that feeds power back into the laser cavity, thereby converting phase noise into intensity noise. A second effect created by multiple optical paths is the appearance of spurious signals arriving at the receiver with variable delays, thereby causing intersymbol interference. Figure 8-19*b* illustrates this.

Unfortunately, these effects are signal-dependent, so that increasing the transmitted or received optical power does not improve the bit-error-rate performance. Thus, one has to find ways to eliminate reflections. Let us first look at their magnitudes. As we saw from Eq. (5-10), a cleaved silica-fiber end face in air typically will reflect about

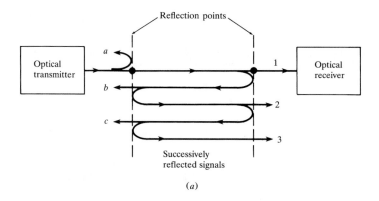

(a)

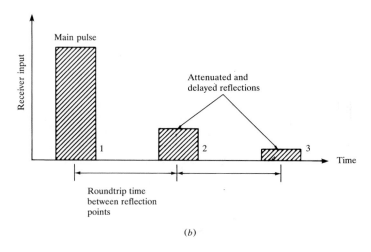

(b)

FIGURE 8-19
(a) Two refractive-index discontinuities can set up multiple reflections in a fiber link. (b) Each roundtrip between reflection points of the back-transmitted portion of a light pulse creates another attenuated and delayed pulse, which can cause intersymbol interference.

$$R = \left(\frac{1.47 - 1.0}{1.47 + 1.0}\right)^2 = 3.6 \text{ percent}$$

This corresponds to an optical return loss of 14.4 dB down from the incident signal. Polishing the fiber ends can create a thin surface layer with an increased refractive index of about 1.6. This increases the reflectance to 5.3 percent (a 12.7-dB optical return loss). A further increase in the optical feedback level occurs when the distance between multiple reflection points equals an integral number of half-wavelengths of the transmitted wavelength. In this case, all roundtrip distances equal an integral number of in-phase wavelengths, so that constructive interference arises. This quadruples the reflection to 14 percent or 8.5 dB for unpolished end faces and to over 22 percent (a 6.6-dB optical return loss) for polished end faces.

The power penalties can be reduced to a few tenths of a decibel by keeping the return losses below values ranging from -15 to -32 dB for data rates varying from 500 Mb/s to 4 Gb/s, respectively.[69] Techniques for reducing optical feedback include the following:

1. Prepare fiber end faces with a curved surface or an angle relative to the laser emitting facet. This directs reflected light away from the fiber axis, so it does not re-enter the waveguide. Return losses of 45 dB or higher can be achieved with end-face angles of 5–15°. However, this increases both the insertion loss and the complexity of the connector.

2. Use index-matching oil or gel at air–glass interfaces. The return loss with this technique is usually greater than 30 dB. However, this may not be practical or recommended if connectors need to be remated often, since contaminants can collect on the interface.

3. Use connectors in which the end faces make physical contact (the so-called *PC connectors*). Return losses of 25–40 dB have been measured with these connectors.

4. Use optical isolators within the laser transmitter module. These devices easily achieve 25-dB return losses, but they also can introduce up to 1 dB of forward loss in the link.

PROBLEMS

8-1. Make a graphical comparison, as in Fig. 8-4, and a spreadsheet calculation of the maximum attenuation-limited transmission distance of the following two systems operating at 100 Mb/s:

System 1 operating at 850 nm

(a) GaAlAs laser diode: 0-dBm (1-mW) fiber-coupled power.

(b) Silicon avalanche photodiode: -50-dBm sensitivity.

(c) Graded-index fiber: 3.5-dB/km attenuation at 850 nm.

(d) Connector loss: 1 dB/connector.

System 2 operating at 1300 nm

(a) InGaAsP LED: -13-dBm fiber-coupled power.

(b) InGaAs *pin* photodiode: -38-dBm sensitivity.

(c) Graded-index fiber: 1.5-dB/km attenuation at 1300 nm.

(d) Connector loss: 1 dB/connector.

Allow a 6-dB system operating margin in each case.

8-2. An engineer has the following components available:

(a) GaAlAs laser diode operating at 850 nm and capable of coupling 1 mW (0 dBm) into a fiber.

(b) Ten sections of cable each of which is 500 m long, as a 4-dB/km attenuation, and has connectors on both ends.

(c) Connector loss of 2 dB/connector.

(d) A *pin* photodiode receiver.

(e) An avalanche photodiode receiver.

Using these components, the engineer wishes to construct a 5-km link operating at 20 Mb/s. If the sensitivities of the *pin* and APD receivers are −45 and −56 dBm, respectively, which receiver should be used if a 6-dB system operating margin is required?

8-3. Using the step response $g(t)$, show that the 10- to 90-percent receiver rise time is given by Eq. (8-4).

8-4. (a) Verify Eq. (8-12).

(b) Show that Eq. (8-14) follows from Eqs. (8-10) and (8-13).

8-5. Show that, if t_e is the full width of the gaussian pulse in Eq. (8-9) at the $1/e$ points, then the relationship between the 3-dB optical bandwidth and t_e is given by

$$f_{3dB} = \frac{0.53}{t_e}$$

8-6. A 90-Mb/s NRZ data transmission system that sends two DS3 channels uses a GaAlAs laser diode that has a 1-nm spectral width. The rise time of the laser transmitter output is 2 ns. The transmission distance is 7 km over a graded-index fiber that has an 800-MHz · km bandwidth–distance product.

(a) If the receiver bandwidth is 90 MHz and the mode-mixing factor $q = 0.7$, what is the system rise time? Does this rise time meet the NRZ data requirement of being less than 70 percent of a pulse width?

(b) What is the system rise time if there is no mode mixing in the 7-km link; that is, $q = 1.0$?

8-7. Verify the plot in Fig. 8-5 of the transmission distance versus data rate of the following system. The transmitter is a GaAlAs laser diode operating at 850 nm. The laser power coupled into a fiber flylead is 0 dBm (1 mW), and the source spectral width is 1 nm. The fiber has a 3.5-dB/km attenuation at 850 nm and a bandwidth of 800 MHz · km. The receiver uses a silicon avalanche photodiode which has the sensitivity versus data rate shown in Fig. 8-3. For simplicity, the receiver sensitivity (in dBm) can be approximated from curve fitting by

$$P_R = 9 \log B - 68.5$$

where B is the data rate in Mb/s. For the data rate range of 1–1000 Mb/s, plot the attenuation-limited transmission distance (including a 1-dB connector loss at each end and a 6-dB system margin), the modal dispersion limit for full mode mixing ($q = 0.5$), the modal dispersion limit for no mode mixing ($q = 1.0$), and the material dispersion limit.

8-8. Make a plot analogous to Fig. 8-5 of the transmission distance versus data rate of the following system. The transmitter is an InGaAsP LED operating at 1300 nm. The fiber-coupled power from this source is −13 dBm (50 μW), and the source spectral width is 40 nm. The fiber has a 1.5-dB/km attenuation at 1300 nm and a bandwidth of 800 MHz · km. The receiver uses an InGaAs *pin* photodiode which has the sensitivity versus data rate shown in Fig. 8-3. For simplicity, the receiver sensitivity (in dBm) can be approximated from curve fitting by

$$P_R = 11.5 \log B - 60.5$$

where B is the data rate in Mb/s. For the data rate range of 10–1000 Mb/s, plot the attenuation-limited transmission distance (including a 1-dB connector loss at each end and a 6-dB system margin), the modal dispersion limit for no mode mixing ($q = 1.0$), and the modal dispersion limit for full mode mixing ($q = 0.5$). Note that the material dispersion is negligible in this case, as can be seen from Fig. 3-13.

8-9. A 1550-nm single-mode digital fiber optic link needs to operate at 622 Mb/s over 80 km without amplifiers. A single-mode InGaAsP laser launches an average optical power of 13 dBm into the fiber. The fiber has a loss of 0.35 dB/km, and there is a splice with a loss of 0.1 dB every kilometer. The coupling loss at the receiver is 0.5 dB, and the receiver uses an InGaAs APD with a sensitivity of −39 dBm. Excess-noise penalties are predicted to be 1.5 dB. Set up an optical power budget for this link and find the system margin. What is the system margin at 2.5 Gb/s with an APD sensitivity of −31 dBm?

8-10. A popular RZ code used in fiber optic systems is the optical Manchester code. This is formed by direct modulo-2 addition of the baseband (NRZ-L) signal and a double-frequency clock signal as is shown in Fig. 8-9. Using this scheme, draw the pulse train for the data string 001101111001.

8-11. Design the encoder logic for an NRZ-to-optical Manchester converter.

8-12. Consider the encoder shown in Fig. P8-12 that changes NRZ data into a PSK (phase-shift-keyed) waveform. Using this encoder, draw the NRZ and PSK waveforms for the data sequence 0001011101001101.

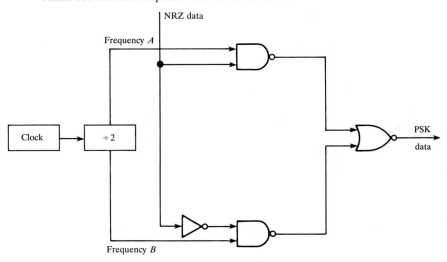

FIGURE P8-12

8-13. A 3B4B code converts blocks of three binary symbols into blocks of four binary digits according to the translation rules shown in Table P8-13. When there are two or more consecutive blocks of three zeros, the coded binary blocks 0010 and 1101 are used alternately. Likewise, the coded blocks 1011 and 0100 are used alternately for consecutive blocks of three ones.

(*a*) Using these translation rules, find the coded bit stream for the data input

010001111111101000000001111110

TABLE P8-13

| Original code | 3B4B Code | |
	Mode 1	Mode 2
000	0010	1101
001	0011	
010	0101	
011	0110	
100	1001	
101	1010	
110	1100	
111	1011	0100

    (*b*) What is the maximum number of consecutive identical bits in the coded pattern?

8-14. Consider Eq. (8-19) for the power penalty caused by laser mode-partition noise.

    (*a*) Plot the power penalty (in dB) as a function of the factor $BLD\sigma_\lambda$ (ranging from 0 to 0.2) at a 10^{-10} BER for mode-partition-noise factors $k = 0.4, 0.6, 0.8$, and 1, when using an InGaAs APD with $x = 0.7$.

    (*b*) Given that a multimode laser has a spectral width of 2.0 nm, what are the minimum dispersions required for 100-km spans operating at 155 Mb/s and 622 Mb/s with a 0.5-dB power penalty?

8-15. (*a*) Using Eq. (8-21), plot the chirp-induced power penalty in decibels as a function of the factor $DL\delta\lambda$ (product of the total dispersion and wavelength excursion) for the following parameter values (let $DL\delta\lambda$ range from 0 to 1.5 ns):

    (1) $t_{chirp} = 0.1$ ns and $B = 2.5$ Gb/s

    (2) $t_{chirp} = 0.1$ ns and $B = 622$ Mb/s

    (3) $t_{chirp} = 0.05$ ns and $B = 2.5$ Gb/s

    (4) $t_{chirp} = 0.05$ ns and $B = 622$ Mb/s

    (*b*) Find the distance limitation at 2.5 Gb/s if a 0.5-dB power penalty is allowed with $D = 1.0$ ps/(nm · km) and $\delta\lambda = 0.5$ nm.

REFERENCES

1. K. Ogawa, L. D. Tzeng, Y. K. Park, and E. Sano, "Advances in high bit-rate transmission systems," in I. P. Kaminow and T. L. Koch, eds., *Optical Fiber Telecommunications — IIIA*, Academic, New York, 1997, pp. 336–372.

2. P. S. Henry, R. A. Linke, and A. H. Gnauck, "Introduction to lightwave systems," in S. E. Miller and I. P. Kaminow, eds., *Optical Fiber Telecommunications—II*, Academic, New York, 1988.

3. I. Jacobs, "Design considerations for long-haul lightwave systems," *IEEE J. Sel. Areas Commun.*, vol. 4, pp. 1389–1395, Dec. 1986.

4. D. H. Rice and G. E. Keiser, "Applications of fiber optics to tactical communication systems," *IEEE Commun. Mag.*, vol. 23, pp. 46–57, May 1985.

5. T. Kimura, "Factors affecting fiber-optic transmission quality," *J. Lightwave Tech.*, vol. 6, pp. 611–619, May 1988.

6. D. J. H. Maclean, *Optical Line Systems*, Wiley, New York, 1996.

7. P. Bell, K. Cobb, and J. Peacock, "Optical power budget calculating tool for fibre in the access network," *BT Technol. J.*, vol. 14, pp. 116–120, Apr. 1996.

8. J. Powers, *An Introduction to Fiber Optic Systems*, Irwin, Chicago, 2nd ed., 1997.

9. A. B. Carlson, *Communication Systems*, McGraw-Hill, New York, 3rd ed., 1986.

10. M. Eve, "Multipath time dispersion theory of an optical network," *Opt. Quantum Electron.*, vol. 10, pp. 45–51, Jan. 1978.

11. J. V. Wright and B. P. Nelson, "Bandwidth studies of concatenated multimode fiber links," *Tech. Dig—Symp. on Optical Fiber Measurements*, NBS Special Publ. 641, pp. 9–12, Oct. 1982.

12. T. Kanada, "Evaluation of modal noise in multimode fiber-optic systems," *J. Lightwave Tech.*, vol. 2, pp. 11–18, Feb. 1984.

13. D. A. Nolan, R. M. Hawk, and D. B. Keck, "Multimode concatenation modal group analysis," *J. Lightwave Tech.*, vol. 5, pp. 1727–1732, Dec. 1987.

14. R. D. de la Iglesia and E. T. Azpitarte, "Dispersion statistics in concatenated single-mode fibers," *J. Lightwave Tech.*, vol. 5, pp. 1768–1772, Dec. 1987.

15. P. M. Rodhe, "The bandwidth of a multimode fiber chain," *J. Lightwave Tech.*, vol. 3, pp. 145–154, Feb. 1985.

16. J. Midwinter, *Optical Fibers for Transmission*, Wiley, New York, 1979.

17. C. E. Shannon, "A mathematical theory of communication," *Bell Sys. Tech. J.*, part 1 in vol. 27, pp. 379–423, July 1948 and part 2 in vol. 27, pp. 623–656, Oct. 1948; "Communication in the presence of noise," *Proc. IRE*, vol. 37, pp. 10–21, Jan. 1949.

18. R. C. Houts and T. A. Green, "Comparing bandwidth requirements for binary baseband signals," *IEEE Trans. Commun.*, vol. COM-21, pp. 776–781, June 1973.

19. A. J. Jerri, "The Shannon sampling theory—its various extensions and applications: A tutorial review," *Proc. IEEE.*, vol. 65, pp. 1565–1596, Nov. 1977.

20. T. V. Muoi and J. L. Hullett, "Receiver design for multilevel digital fiber optic systems," *IEEE Trans. Commun.*, vol. COM-23, pp. 987–994, Sept. 1975.

21. L. W. Couch II, *Digital and Analog Communication Systems*, Prentice Hall, Upper Saddle River, NJ, 5th ed., 1997.

22. E. A. Lee and d. G. Messerschmitt, *Digital Communications*, Kluwer Academic, Boston, 2nd ed., 1993.

23. H. Taub and D. L. Schilling, *Principles of Communication Systems*, McGraw-Hill, New York, 2nd ed., 1986.

24. S. Haykin, *Communication Systems*, Wiley, New York, 3rd ed., 1994.

25. T. Matsuda, A. Naka, and S. Saito, "Comparison between NRZ and RZ signal formats for in-line amplifier transmission in the zero-dispersion region," *J. Lightwave Tech.*, vol. 16, pp. 340–348, Mar. 1998.

26. J. E. Savage, "Some simple self-synchronizing digital data assemblers," *Bell Sys. Tech. J.*, vol. 46, pp. 449–487, Feb. 1967.

27. R. D. Gitlin and J. F. Hayes, "Timing recovery and scramblers in data transmission," *Bell Sys. Tech. J.*, vol. 54, pp. 569–593, Mar. 1975.

28. B. Sklar, *Digital Communications*, Prentice Hall, Upper Saddle River, NJ, 1988.

29. Y. Takasaki, M. Tanaka, N. Maeda, K. Yamashita, and K. Nagano, "Optical pulse formats for fiber optic digital communications," *IEEE Trans. Commun.*, vol. COM-24, pp. 404–413, Apr. 1976.

30. E. Meissner, H. Rodler, and M. Lades, "Pattern independent 2.5-Gb/s AMI-CPFSK transmission with −44 dBm receiver sensitivity," *Electron. Lett.*, vol. 30, pp. 345–346, Feb. 1994.

31. S. P. Majumder, R. Gangopadhyay, and G. Prati, "Effect of line coding on heterodyne FSK optical systems with nonuniform laser FM response," *IEE Proc.—Optoelectron.*, vol. 141, pp. 200–208, June 1994.

32. N. L. Swenson and J. M. Cioffi, "Sliding-block line codes to increase dispersion-limited distance of optical fiber channels," *IEEE J. Sel. Areas Commun.*, vol. 13, pp. 485–498, Apr. 1995.

33. A. J. Phillips, R. A. Cryan, and J. M. Senior, "Optically preamplified pulse-position modulation for fibre-optic communication systems," *IEE Proc.—Optoelectron.*, vol. 143, pp. 153–159, Apr. 1996.

34. W. A. Kryzmien, "Transmission performance analysis of a new class of line codes for optical fiber systems," *IEEE Trans. Commun.*, vol. 37, pp. 402–404, Apr. 1989.

35. M. Rousseau, "Block codes for optical fibre communications," *Electron. Lett.*, vol. 12, pp. 478–479, Sept. 1976.

36. R. Petrovic, "5B6B optical-fiber line code bearing auxiliary signals," *Electron. Lett.*, vol. 24, pp. 274–275, Mar. 1988.

37. R. Petrovic, "Low redundancy optical-fibre line code," *J. Opt. Commun.*, vol. 9, pp. 108–111, Sept. 1988.

38. S. Saunders, *The McGraw-Hill High-Speed LANs Handbook*, McGraw-Hill, New York, 1996.

39. A. M. Michelson and A. H. Levesque, *Error-Control Techniques for Digital Communications*, Wiley, New York, 1985.

40. W. D. Grover, "Forward error correction in dispersion-limited lightwave systems," *J. Lightwave Tech.*, vol. 6, pp. 643–654, May 1988.

41. E. W. Biersack, "Performance of forward error correction in an ATM environment," *IEEE J. Sel. Areas Commun.*, vol. 11, pp. 631–640, May 1993.

42. S.-M. Lei, "Forward error correction codes for MPEG2 over ATM," *IEEE Trans. Circuits Sys. for Video Tech.*, vol. 4, pp. 200–203, Apr. 1994.

43. K.-P. Ho and C. Lin, "Performance analysis of optical transmission system with polarization-mode dispersion and FEC," *IEEE Photonics Tech. Lett.*, vol. 9, pp. 1288–1290, Sept. 1997.

44. (*a*) M. Tomizawa, Y. Yamabayashi, K. Murata, T. Ono, Y. Kobayashi, and K. Hagimoto, "FEC code for arbitrary multiplexing levels in SDH fibre-optic transmission systems," *Electron. Lett.*, vol. 31, pp. 662–663, Apr. 1995; "FEC codes in synchronous fiber optic transmission systems," *J. Lightwave Tech.*, vol. 15, pp. 43–52, Jan. 1997.

    (*b*) M. Tomizawa, K. Murata, Y. Miyamoto, Y. Yamabayashi, Y. Kobayashi, and K. Hagimoto, "STM-64 linearly repeatered optical transmission experiment using forward error correcting codes," *Electron. Lett.*, vol. 31, pp. 1001–1003, June 1995.

45. P. E. Couch and R. E. Epworth, "Reproducible modal-noise measurements in system design and analysis," *J. Lightwave Tech.*, vol. LT-1, pp. 591–595, Dec. 1983.

46. F. M. Sears, I. A. White, R. B. Kummer, and F. T. Stone, "Probability of modal noise in single-mode lightguide systems," *J. Lightwave Tech.*, vol. LT-4, pp. 652–655, June 1986.

47. K. Petermann and G. Arnold, "Noise and distortion characteristics of semiconductor lasers in optical fiber communication systems," *IEEE J. Quantum Electron.*, vol. QE-18, pp. 543–554, Apr. 1982.

48. A. M. J. Koonen, "Bit-error-rate degradation in a multimode fiber optic transmission link due to modal noise," *IEEE J. Sel. Areas. Commun.*, vol. SAC-4, pp. 1515–1522, Dec. 1986.

49. P. Chan and T. T. Tjhung, "Bit-error-rate performance for optical fiber systems with modal noise," *J. Lightwave Tech.*, vol. 7, pp. 1285–1289, Sept. 1989.

50. P. M. Shankar, "Bit-error-rate degradation due to modal noise in single-mode fiber optic communication systems," *J. Opt. Commun.*, vol. 10, pp. 19–23, Mar. 1989.

51. N. H. Jensen, H. Olesen, and K. E. Stubkjaer, "Partition noise in semiconductor lasers under CW and pulsed operation," *IEEE J. Quantum Electron.*, vol. QE-18, pp. 71–80, Jan. 1987.

52. M. Ohtsu and Y. Teramachi, "Analyses of mode partition and mode hopping in semiconductor lasers," *IEEE J. Quantum Electron.*, vol. 25, pp. 31–38, Jan. 1989.

53. C. H. Henry, P. S. Henry, and M. Lax, "Partition fluctuations in nearly single longitudinal mode lasers," *J. Lightwave Tech.*, vol. LT-2, pp. 209–216, June 1984.

54. E. E. Basch, R. F. Kearns, and T. G. Brown, "The influence of mode partition fluctuations in nearly single-longitudinal-mode lasers on receiver sensitivity, *J. Lightwave Tech.*, vol. LT-4, pp. 516–519, May 1986.

55. J. C. Cartledge, "Performance implications of mode partition fluctuations in nearly single longitudinal mode lasers," *J. Lightwave Tech.*, vol. 6, pp. 626–635, May 1988.

56. K. Ogawa, "Analysis of mode partition noise in laser transmission systems," *IEEE J. Quantum Electron.*, vol. QE-18, pp. 849–855, May 1982.

57. N. A. Olsson, W. T. Tsang, H. Temkin, N. K. Dutta, and R. A. Logan, "Bit-error-rate saturation due to mode-partition noise induced by optical feedback in 1.5 μm single longitudinal-mode C^3 and DFB semiconductor lasers," *J. Lightwave Tech.*, vol. LT-3, pp. 215–218, Apr. 1985.

58. S. E. Miller, "On the injection laser contribution to mode partition noise in fiber telecommunication systems," *IEEE J. Quantum Electron.*, vol. 25, pp. 1771–1781, Aug. 1989.

59. R. Heidemann, "Investigations on the dominant dispersion penalties occurring in multigigabit direct detection systems," *J. Lightwave Tech.*, vol. 6, pp. 1693–1697, Nov. 1988.

60. R. A. Linke, "Modulation induced transient chirping in single frequency lasers," *IEEE J. Quantum Electron.*, vol. QE-21, pp. 593–597, June 1985.

61. Y. Yoshikuni and G. Motosugi, "Multielectrode distributed feedback laser for pure frequency modulation and chirping suppressed amplitude modulation," *J. Lightwave Tech.*, vol. LT-5, pp. 516–522, Apr. 1987.

62. S. Yamamoto, M. Kuwazuru, H. Wakabayashi, and Y. Iwamoto, "Analysis of chirp power penalty in 1.55-μm DFB-LD high-speed optical fiber transmission systems," *J. Lightwave Tech.*, vol. LT-5, pp. 1518–1524, Oct. 1987.

63. P. J. Corvini and T. L. Koch, "Computer simulation of high-bit-rate optical fiber transmission using single-frequency lasers," *J. Lightwave Tech.*, vol. LT-5, pp. 1591–1595, Nov. 1987.

64. J. C. Cartledge and G. S. Burley, "The effect of laser chirping on lightwave system performance," *J. Lightwave Tech.*, vol. 7, pp. 568–573, Mar. 1989.

65. G. Yabre, "Effect of relatively strong light injection on the chirp-to-power ratio and the 3-dB bandwidth of directly modulated semiconductor lasers," *J. Lightwave Tech.*, vol. 14, pp. 2367–2373, Oct. 1996.

66. C. H. Henry, "Theory of the linewidth of semiconductor lasers," *IEEE J. Quantum Electron.*, vol. QE-18, pp. 259–264, Feb. 1982.

67. C. H. Harder, K. Vahala, and A. Yariv, "Measurement of the linewidth enhancement factor of semiconductor lasers," *Appl. Phys. Lett.*, vol. 42, pp. 328–330, Apr. 1983.

68. R. Schimpe, J. E. Bowers, and T. L. Koch, "Characterization of frequency response of 1.5-μm InGaAsP DFB laser diode and InGaAs *PIN* photodiode by heterodyne measurement technique," *Electron. Lett.*, vol. 22, pp. 453–454, Apr. 24, 1986.

69. M. Shikada, S. Takano, S. Fujita, I. Mito, and K. Minemura, "Evaluation of power penalties caused by feedback noise of distributed feedback laser diodes," *J. Lightwave Tech.*, vol. 6, pp. 655–659, May 1988.

70. M. Nakazawa, "Rayleigh backscattering theory for single-mode fibers," *J. Opt. Soc. Amer.*, vol. 73, pp. 1175–1180, Sept. 1983.

71. P. Wan and J. Conradi, "Impact of double Rayleigh backscatter noise on digital and analog fiber systems," *J. Lightwave Tech.*, vol. 14, pp. 288–297, Mar. 1996.

CHAPTER
9

ANALOG
SYSTEMS

In telecommunication networks the trend has been to link telephone exchanges with digital circuits. A major reason for this was the introduction of digital integrated-circuit technology which offered a reliable and economic method of transmitting both voice and data signals. Since the initial applications of fiber optics were to telecommunication networks, its first widespread usage has involved digital links. However, in many instances, it is more advantageous to transmit information in analog form instead of first converting it to a digital format. Some examples of this are microwave-multiplexed signals,[1] subscriber services using hybrid fiber/coax (HFC),[2] video distribution,[3,4] antenna remoting,[5,6] and radar signal processing.[7-9] For most analog applications, one uses laser diode transmitters, so we shall concentrate on this optical source here.

When implementing an analog fiber optic system, the main parameters one needs to consider are the carrier-to-noise ratio, bandwidth, and signal distortion resulting from nonlinearities in the transmission system. Section 9.1 describes the general operational aspects and components of an analog fiber optic link. Traditionally, in an analog system, a carrier-to-noise ratio analysis is used instead of a signal-to-noise ratio analysis, since the information signal is normally superimposed on a radio-frequency (RF) carrier. Thus, in Sec. 9.2 we examine carrier-to-noise ratio requirements. This is first done for a single channel under the assumption that the information signal is directly modulated onto an optical carrier.

For transmitting multiple signals over the same channel, one can use a subcarrier modulation technique. In this method, which is described in Sec. 9.3, the information signals are first superimposed on ancillary RF subcarriers. These carriers are then combined and the resulting electrical signal is used to modulate the optical carrier. A limiting factor in these systems is the signal impairment arising from harmonic and intermodulation distortions.

9.1 OVERVIEW OF ANALOG LINKS

Figure 9-1 shows the basic elements of an analog link. The transmitter contains either an LED or a laser diode optical source. As noted in Sec. 4.4 and shown in Fig. 4-35, in analog applications, one first sets a bias point on the source approximately at the midpoint of the linear output region. The analog signal can then be sent using one of several modulation techniques. The simplest form for optical fiber links is direct intensity modulation, wherein the optical output from the source is modulated simply by varying the current around the bias point in proportion to the message signal level. Thus, the information signal is transmitted directly in the baseband.

A somewhat more complex but often more efficient method is to translate the baseband signal onto an electrical subcarrier prior to intensity modulation of the source. This is done using standard amplitude-modulation (AM), frequency-modulation (FM), or phase-modulation (PM) techniques.[10,11] No matter which method is implemented, one must pay careful attention to signal impairments in the optical source. These include harmonic distortions, intermodulation products, relative intensity noise (RIN) in the laser, and laser clipping.[12]

In relation to the fiber-optic element shown in Fig. 9-1, one must take into account the frequency dependence of the amplitude, phase, and group delay in the fiber. Thus, the fiber should have a flat amplitude and group-delay response within the passband required to send the signal free of linear distortion. In addition, since modal-distortion-limited bandwidth is difficult to equalize, it is best to choose a single-mode fiber. The fiber attenuation is also important, since the carrier-to-noise performance of the system will change as a function of the received optical power.

The use of an optical amplifier in the link leads to additional noise, known as amplified spontaneous emission (ASE), as is described in Chap. 11. In the

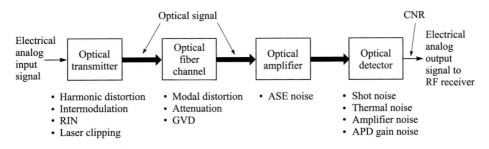

FIGURE 9-1
Basic elements of an analog link and the major noise contributors.

optical receiver, the principal impairments are quantum or shot noise, APD gain noise, and thermal noise.

9.2 CARRIER-TO-NOISE RATIO

In analyzing the performance of analog systems, one usually calculates the ratio of rms carrier power to rms noise power at the input of the RF receiver following the photodetection process. This is known as the *carrier-to-noise ratio* (CNR). Let us look at some typical CNR values for digital and analog data. For digital data, consider the use of frequency-shift keying (FSK). In this modulation scheme, the amplitude of a sinusoidal carrier remains constant, but the phase shifts from one frequency to another to represent binary signals. For FSK, BERs of 10^{-9} and 10^{-15} translate into CNR values of 36 (15.6 dB) and 64 (18.0 dB), respectively. The analysis for analog signals is more complex, since it sometimes depends on user perception of the signal quality, such as in viewing a television picture. A widely used analog signal is a 525-line studio-quality television signal. Using amplitude modulation (AM) for such a signal requires a CNR of 56 dB, since the need for bandwidth efficiency leads to a high signal-to-noise ratio. Frequency modulation (FM), on the other hand, only needs CNR values of 15–18 dB.

If CNR_i represents the carrier-to-noise ratio related to a particular signal contaminant (e.g., shot noise), then for N signal-impairment factors the total CNR is given by

$$\frac{1}{CNR} = \sum_{i=1}^{N} \frac{1}{CNR_i} \tag{9-1}$$

For links in which only a single information channel is transmitted, the important signal impairments include laser intensity noise fluctuations, laser clipping, photodetector noise, and optical-amplifier noise. When multiple message channels operating at different carrier frequencies are sent simultaneously over the same fiber, then harmonic and intermodulation distortions arise. Furthermore, the inclusion of an optical amplifier gives rise to ASE noise. In principle, the three dominant factors that cause signal impairments in a fiber link are shot noise, optical-amplifier noise, and laser clipping.[12] Most other degradation effects can be sufficiently reduced or eliminated.

In this section, we shall first examine a simple single-channel amplitude-modulated signal sent at baseband frequencies. Section 9.3 addresses multichannel systems in which intermodulation noise becomes important. Problem 9-10 gives expressions for the effects of laser clipping and ASE noise.

9.2.1 Carrier Power

To find the carrier power, let us first look at the signal generated at the transmitter. As shown in Fig. 9-2, the drive current through the optical source is the sum of the fixed bias current and a time-varying sinusoid. The source acts as a

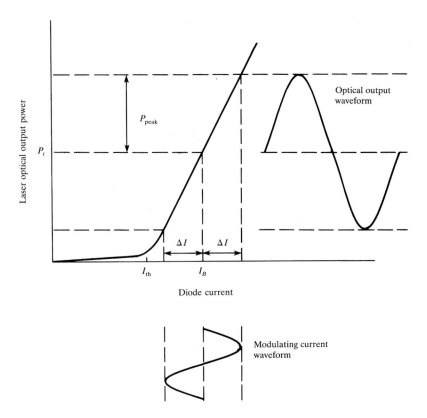

FIGURE 9-2
Biasing conditions of a laser diode and its response to analog signal modulation.

square-law device, so that the envelope of the output optical power $P(t)$ has the same form as the input drive curent. If the time-varying analog drive signal is $s(t)$, then

$$P(t) = P_t[1 + ms(t)] \qquad (9\text{-}2)$$

where P_t is the optical output power at the bias current level and the modulation index m is defined by Eq. (4-54). In terms of optical power, the modulation index is given by

$$m = \frac{P_{\text{peak}}}{P_t} \qquad (9\text{-}3)$$

where P_{peak} and P_t are defined in Fig. 9-2. Typical values of m for analog applications range from 0.25 to 0.50.

For a sinusoidal received signal, the carrier power C at the output of the receiver (in units of A^2) is

$$C = \tfrac{1}{2}(m\mathscr{R}_0 M \bar{P})^2 \qquad (9\text{-}4)$$

where $\mathscr{R}_0$ is the unity gain responsivity of the photodetector, M is the photo-detector gain ($M = 1$ for *pin* photodiodes), and $\bar{P}$ is the average received optical power.

9.2.2   Photodetector and Preamplifier Noises

The expressions for the photodiode and preamplifier noises are given by Eqs. (6-16) and (6-17), respectively. That is, for the photodiode noise we have

$$\langle i_N^2 \rangle = \sigma_N^2 \approx 2q(I_p + I_D)M^2 F(M)B \qquad (9\text{-}5)$$

Here, as defined in Chap. 6, $I_p = \mathscr{R}_0 \bar{P}$ is the primary photocurrent, I_D is the detector bulk dark current, M is the photodiode gain with $F(M)$ being its associated noise figure, and B is the receiver bandwidth. Then, the CNR for the photodetector only is $\text{CNR}_{\text{det}} = C/\sigma_N^2$.

Generalizing Eq. (6-17) for the preamplifier noise, we have

$$\langle i_T^2 \rangle = \sigma_T^2 = \frac{4k_B T}{R_{\text{eq}}} B F_t \qquad (9\text{-}6)$$

Here, R_{eq} is the equivalent resistance of the photodetector load and the pre-amplifier, and F_t is the noise factor of the preamplifier. Then, the CNR for the preamplifier only is $\text{CNR}_{\text{preamp}} = C/\sigma_T^2$.

9.2.3   Relative Intensity Noise (RIN)

Within a semiconductor laser, fluctuations in the amplitude or intensity of the output produce optical intensity noise. These fluctuations could arise from temperature variations or from spontaneous emission contained in the laser output. The noise resulting from the random intensity fluctuations is called *relative intensity noise* (RIN), which may be defined in terms of the mean-square intensity variations. The resultant mean-square noise current is given by

$$\langle i_{\text{RIN}}^2 \rangle = \sigma_{\text{RIN}}^2 = \text{RIN}(\mathscr{R}_0 \bar{P})B \qquad (9\text{-}7)$$

Then, the CNR due to laser amplitude fluctuations only is $\text{CNR}_{\text{RIN}} = C/\sigma_{\text{RIN}}^2$. Here, the RIN, which is measured in dB/Hz, is defined by the noise-to-signal power ratio

$$\text{RIN} = \frac{\langle (\Delta P_L)^2 \rangle}{\bar{P}_L^2} \qquad (9\text{-}8)$$

where $\langle (\Delta P_L)^2 \rangle$ is the mean-square intensity fluctuation of the laser output and $\bar{P}_L$ is the average laser light intensity. This noise decreases as the injection-current level increases according to the relationship

$$\text{RIN} \propto \left(\frac{I_B}{I_{\text{th}}} - 1 \right)^{-3} \qquad (9\text{-}9)$$

Example 9-1. Figure 9-3 shows an example of Eq. (9-9) for two buried-heterostructure lasers.[13] The noise level was measured at 100 MHz. For injection currents sufficiently above threshold (i.e., for $I_B/I_{th} > 1.2$), the RIN of these index-guided lasers lies between −140 and −150 dB/Hz.

Example 9-2. Figure 9-4 shows the RIN of an InGaAsP buried-heterostructure laser as a function of modulation frequency at several different bias levels.[1] The relative intensity noise is essentially independent of frequency below several hundred megahertz, and it peaks at the resonant frequency. In this case, at a bias level of 60 mA, which gives a 5-mW output, the RIN is typically less than −135 dB/Hz for modulation frequencies up to 8 GHz. For received optical signal levels of −13 dBm (50 μW) or less, the RIN of buried-heterostructure InGaAsP lasers lies sufficiently below the noise level of a 50-Ω amplifier with a 3-dB noise figure.

Vendor data sheets for 1550-nm DFB lasers typically quote RIN values of −152 to −158 dB/Hz.

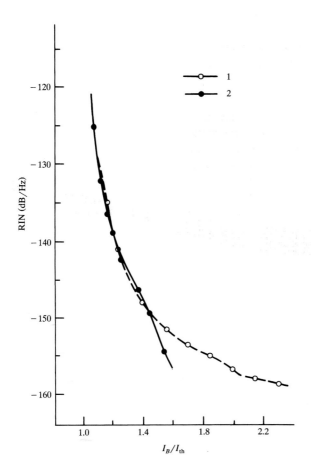

FIGURE 9-3
Example of the relative intensity noise (RIN) for two buried-heterostructure laser diodes. The noise level was measured at 100 MHz. (Reproduced with permission from Sato,[13] © 1983, IEEE.)

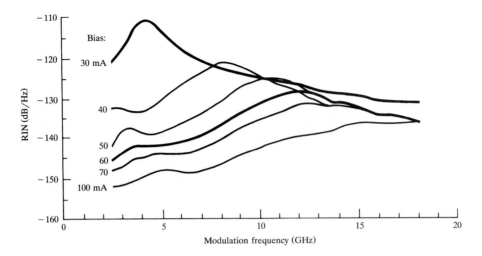

FIGURE 9-4

The RIN of an InGaAsP buried-heterostructure laser as a function of modulation frequency at several different bias levels. (Reproduced with permission from Olshansky, Lanzisera, and Hill,[1] © 1989, IEEE.)

Substituting the CNRs resulting from Eqs. (9-4) through (9-7) into Eq. (9-1) yields the following carrier-to-noise ratio for a single-channel AM system:

$$\frac{C}{N} = \frac{\frac{1}{2}(m\mathscr{R}_0 M\bar{P})^2}{\text{RIN}(\mathscr{R}_0\bar{P})^2 B + 2q(I_p + I_D)M^2 F(M)B + (4k_B T/R_{eq})BF_t} \tag{9-10}$$

9.2.4 Reflection Effects on RIN

In implementing a high-speed analog link, one must take special precautions to minimize optical reflections back into the laser.[2] Back-reflected signals can increase the RIN by 10–20 dB as shown in Fig. 9-5. These curves show the increase in relative intensity noise for bias points ranging from 1.24 to 1.62 times the threshold-curent level. The feedback power ratio in Fig. 9-5 is the amount of optical power reflected back into the laser relative to the light output from the source. As an example, the dashed line shows that at $1.33I_{th}$ the feedback ratio must be less than -60 dB in order to maintain an RIN of less than -140 dB/Hz.

9.2.5 Limiting Conditions

Let us now look at some limiting conditions. When the optical power level at the receiver is low, the preamplifier circuit noise dominates the system noise. For this, we have

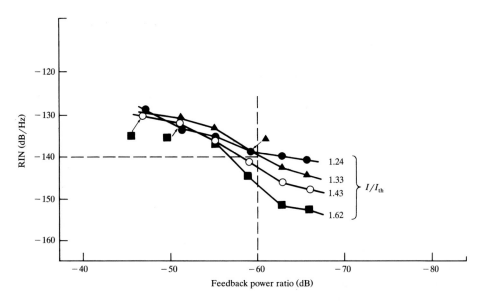

FIGURE 9-5
The increase in RIN due to back-reflected optical signals. (Reproduced with permission from Sato,[13] © 1983, IEEE.)

$$\left(\frac{C}{N}\right)_{\text{limit 1}} = \frac{\frac{1}{2}(m\mathscr{R}_0 M \bar{P})^2}{(4k_B T / R_{\text{eq}})BF_t} \tag{9-11}$$

In this case, the carrier-to-noise ratio is directly proportional to the square of the received optical power, so that for each 1-dB variation in received optical power, C/N will change by 2 dB.

For well-designed photodiodes, the bulk and surface dark currents are small compared with the shot (quantum) noise for intermediate optical signal levels at the receiver. Thus, at intermediate power levels the quantum-noise term of the photodiode will dominate the system noise. In this case, we have

$$\left(\frac{C}{N}\right)_{\text{limit 2}} = \frac{\frac{1}{2}m^2\mathscr{R}_0 \bar{P}}{2qF(M)B} \tag{9-12}$$

so that the carrier-to-noise ratio will vary by 1 dB for every 1-dB change in the received optical power.

If the laser has a high RIN value so that the reflection noise dominates over other noise terms, then the carrier-to-noise ratio becomes

$$\left(\frac{C}{N}\right)_{\text{limit 3}} = \frac{\frac{1}{2}(mM)^2}{\text{RIN }B} \tag{9-13}$$

which is a constant. In this case, the performance cannot be improved unless the modulation index is increased.

Example 9-3. As an example of the limiting conditions, consider a link with a laser transmitter and a *pin* photodiode receiver having the following characteristics:

Transmitter	Receiver
$m = 0.25$	$\mathcal{R}_0 = 0.6$ A/W
RIN $= -143$ dB/Hz	$B = 10$ MHz
$P_c = 0$ dBm	$I_D = 10$ nA
	$R_{\text{eq}} = 750\ \Omega$
	$F_t = 3$ dB

where P_c is the optical power coupled into the fiber. To see the effects of the different noise terms on the carrier-to-noise ratio, Fig. 9-6 shows a plot of C/N as a function of the optical power level at the receiver. In this case, we see that at high received powers the source noise dominates to give a constant C/N. At intermediate levels, the quantum noise is the main contributor, with a 1-dB drop in C/N for every 1-dB decrease in received optical power. For low light levels, the thermal noise of the receiver is the limiting noise term, yielding a 2-dB rolloff in C/N for each 1-dB drop in received optical power. It is important to note that the limiting factors can vary significantly depending on the transmitter and receiver characteristics. For example, for low-impedance amplifiers the thermal noise of the receiver can be the dominating performance limiter for all practical link lengths (see Prob. 9-1).

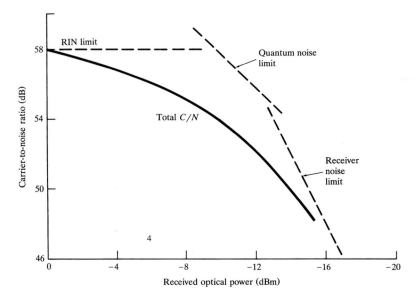

FIGURE 9-6
Carrier-to-noise ratio as a function of optical power level at the receiver. In this case, RIN dominates at high powers, quantum noise gives a 1-dB drop in C/N for each 1-dB power decrease at intermediate levels, and receiver thermal noise yields a 2-dB C/N rolloff per 1-dB drop in received power at low light levels.

9.3 MULTICHANNEL TRANSMISSION TECHNIQUES

So far, we have examined only the case of a single signal being transmitted over a channel. In broadband analog applications, such as cable television (CATV) supertrunks, one needs to send multiple analog signals over the same fiber. To do this, one can employ a multiplexing technique where a number of baseband signals are superimposed on a set of N subcarriers that have different frequencies $f_1, f_2, \ldots, f_N$. These modulated subcarriers are then combined electrically through frequency-divison multiplexing (FDM) to form a composite signal that directly modulates a single optical source. Methods for achieving this include vestigial-sideband amplitude modulation (VSB-AM), frequency modulation (FM), and subcarrier multiplexing (SCM).

Of these, AM is simple and cost-effective in that it is compatible with the equipment interfaces of a large number of CATV customers, but its signal is very sensitive to noise and nonlinear distortion. Although FM requires a larger bandwidth than AM, it provides a higher signal-to-noise ratio and is less sensitive to source nonlinearities. Microwave SCM operates at higher frequencies than AM or FM and is an interesting approach for broadband distribution of both analog and digital signals. To simplify the interface with existing coaxial cable systems, current fiber links in CATV networks primarily use the AM-VSB scheme described in Sec. 9.3.1.

9.3.1 Multichannel Amplitude Modulation

The initial widespread application of analog fiber optic links, which started in the late 1980s, was to CATV networks.[14-17] These coax-based television networks operate in a frequency range from 50 to 88 MHz and from 120 to 550 MHz. The band from 88 to 120 MHz is not used, since it is reserved for FM radio broadcast. The CATV networks can deliver over 80 amplitude-modulated vestigial-sideband (AM-VSB) video channels, each having a noise bandwidth of 4 MHz within a channel bandwidth of 6 MHz, with signal-to-noise ratios exceeding 47 dB. To remain compatible with existing coax-based networks, a multichannel AM-VSB format was also chosen for the fiber optic system.

Figure 9-7 depicts the technique for combining N independent messages. An information-bearing signal on channel i amplitude-modulates a carrier wave that has a frequency f_i, where $i = 1, 2 \ldots, N$. An RF power combiner then sums these N amplitude-modulated carriers to yield a composite frequency-division-multiplexed (FDM) signal which intensity-modulates a laser diode. Following the optical receiver, a bank of parallel bandpass filters separates the combined carriers back into individual channels. The individual message signals are recovered from the carriers by standard RF techniques.

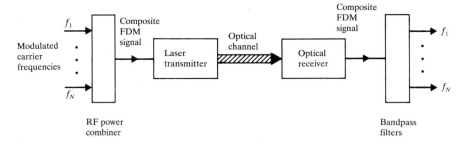

FIGURE 9-7
Standard technique for frequency-division multiplexing of N independent information-bearing signals.

For a large number of FDM carriers with random phases, the carriers add on a power basis. Thus, for N channels the optical modulation index m is related to the per-channel modulation index m_i by

$$m = \left(\sum_{i=1}^{N} m_i^2 \right)^{1/2} \tag{9-14a}$$

If each channel modulation index m_i has the same value m_c, then

$$m = m_c N^{0.5} \tag{9-14b}$$

As a result, when N signals are frequency-multiplexed and used to modulate a single optical source, the carrier-to-noise ratio of a single channel is degraded by $10 \log N$. If only a few channels are combined, the signals will add in voltage rather than power, so that the degradation will have a $20 \log N$ characteristic.

When multiple carrier frequencies pass through a nonlinear device such as a laser diode, signal products other than the original frequencies can be produced. As noted in Sec. 4.4, these undesirable signals are called *intermodulation products* and they can cause serious interference in both in-band and out-of-band channels. The result is a degradation of the transmitted signal. Among the intermodulation products, generally only the second-order and third-order terms are considered, since higher-order products tend to be significantly smaller.

Third-order intermodulaton (IM) distortion products at frequencies $f_i + f_j - f_k$ (which are known as *triple-beat IM products*) and $2f_i - f_j$ (which are known as *two-tone third-order IM products*) are the most dominant, since many of these fall within the bandwidth of a multichannel system. For example, a 50-channel CATV network operating over a standard frequency range of 55.25–373.25 MHz has 39 second-order IM products at 54.0 MHz and 786 third-order IM tones at 229.25 MHz. The amplitudes of the triple-beat products are 3 dB higher than the two-tone third-order IM products. In addition, since there are $N(N-1)(N-2)/2$ triple-beat terms compared with $N(N-1)$ two-tone third-order terms, the triple-beat products tend to be the major source of IM noise.

If a signal passband contains a large number of equally spaced carriers, several IM terms will exist at or near the same frequency. This so-called *beat*

TABLE 9-1
Distribution of the number of third-order triple-beat intermodulation products for the number of channels N ranging from 1 to 8

					r			
N	1	2	3	4	5	6	7	8
1	0							
2	0	0						
3	0	1	0					
4	1	2	2	1				
5	2	4	4	4	2			
6	4	6	7	7	6	4		
7	6	9	10	11	10	9	6	
8	9	12	14	15	15	14	12	9

stacking is additive on a power basis. For example, for N equally spaced equal-amplitude carriers, the number of third-order IM products that fall right on the rth carrier is given by[18,19]

$$D_{1,2} = \tfrac{1}{2}\{N - 2 - \tfrac{1}{2}[1 - (-1)^N](-1)^r\} \tag{9-15}$$

for two-tone terms of the type $2f_i - f_j$, and by

$$D_{1,1,1} = \frac{r}{2}(N - r + 1) + \tfrac{1}{4}\{(N - 3)^2 - 5 - \tfrac{1}{2}[1 - (-1)^N](-1)^{N+r}\} \tag{9-16}$$

for triple-beat terms of the type $f_i + f_j - f_k$.

Whereas the two-tone third-order terms are fairly evenly spread through the operating passband, the triple-beat products tend to be concentrated in the middle of the channel passband, so that the center carriers receive the most intermodulation interference. Tables 9-1 and 9-2 show the distributions of the third-order triple-beat and two-tone IM products for the number of channels N ranging from 1 to 8.

TABLE 9-2
Distribution of the number of third-order two-tone intermodulation products for the number of channels N ranging from 1 to 8

					r			
N	1	2	3	4	5	6	7	8
1	0							
2	0	0						
3	1	0	1					
4	1	1	1	1				
5	2	1	2	1	2			
6	2	2	2	2	2	2		
7	3	2	3	2	3	2	3	
8	3	3	3	3	3	3	3	3

The results of beat stacking are commonly referred to as *composite second order (CSO)* and *composite triple beat (CTB)*, and are used to describe the performance of multichannel AM links. These are defined as[20]

$$CSO = \frac{\text{peak carrier power}}{\text{peak power in composite 2nd-order IM tone}} \qquad (9\text{-}17)$$

and

$$CTB = \frac{\text{peak carrier power}}{\text{peak power in composite 3rd-order IM tone}} \qquad (9\text{-}18)$$

Example 9-4. Figures 9-8 and 9-9 show the predicted relative second-order and third-order intermodulation performance, respectively, for 60 CATV channels in the frequency range 50–450 MHz. The effect of CSO is most significant at the passband edges, whereas CTB contributions are most critical at the center of the band.

9.3.2 Multichannel Frequency Modulation

The use of AM-VSB signals for transmitting multiple analog channels is, in principle, straightforward and simple. However, it has a C/N requirement (or,

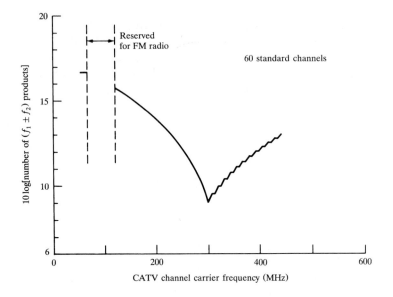

FIGURE 9-8
Predicted relative CSO performance for 60 amplitude-modulated CATV channels. The 88-to-120-MHz band is reserved for FM radio broadcast. (Reproduced with permission from Darcie, Lipson, Roxlo, and McGrath,[4] © 1990, IEEE.)

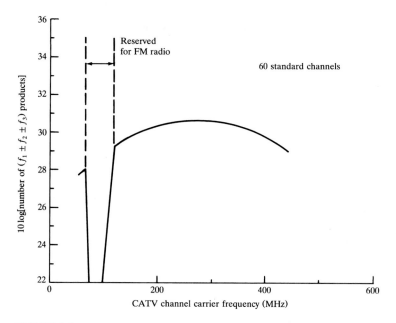

FIGURE 9-9
Predicted relative CTB performance for 60 amplitude-modulated CATV channels. The 88-to-120-MHz band is reserved for FM radio broadcast. (Reproduced with permission from Darcie, Lipson, Roxlo, and McGrath,[4] © 1990, IEEE.)

equivalently, for AM, an S/N requirement) of at least 40 dB for each AM channel, which places very stringent requirements on laser and receiver linearity. An alternative technique is frequency modulation (FM), wherein each subcarrier is frequency-modulated by a message signal.[2,21,22] This requires a wider bandwidth (30 MHz versus 4 MHz for AM), but yields a signal-to-noise ratio improvement over the carrier-to-noise ratio.

The S/N at the output of an FM detector is much larger than the C/N at the input of the detector. The improvement is given by[2]

$$\left(\frac{S}{N}\right)_{out} = \left(\frac{C}{N}\right)_{in} + 10\log\left[\frac{3}{2}\frac{B}{f_v}\left(\frac{\Delta f_{pp}}{f_v}\right)^2\right] + w \tag{9-19}$$

where B is the required bandwidth, Δf_{pp} is the peak-to-peak frequency deviation of the modulator, f_v is the highest video frequency, and w is a weighting factor used to account for the nonuniform response of the eye pattern to white noise in the video bandwidth. The total S/N improvement depends on the system design, but is generally in the range 36–44 dB.[23,24] The reduced C/N requirements thus make an FM system much less susceptible to laser and receiver noises than an AM system.

Example 9-5. Figure 9-10 shows a plot of RIN versus optical modulation index per channel, comparing AM and FM broadcast TV systems.[2] The following assumptions were made in this calculation:

RIN noise dominates

$S/N = C/N + 40$ dB for the FM system

AM bandwidth per channel $= 4$ MHz

FM bandwidth per channel $= 30$ MHz

If the per-channel optical modulation index is 5 percent, then a RIN of less than -120 dB/Hz is needed for each FM TV program to have studio-quality reception, requiring $S/N \geq 56$ dB. This is easily met with a typical packaged laser diode which has a nominal RIN value of -130 dB/Hz. On the other hand, for an AM system a laser with an RIN value of -140 dB/Hz can barely meet the CATV reception requirement of $S/N \geq 40$ dB.

Example 9-6. Another performance factor of AM transmission compared with FM is the limited power margin of AM. Figure 9-11 depicts the calculated power budget versus the optical modulation index (OMI) per channel for distribution of multi-channel AM and FM video signals. The curves are given for different signal-to-noise ratios. The following assumptions were made in this calculation:

Laser power coupled into single-mode fiber $= 0$ dBm

RIN $= -140$ dB/Hz

pin Photodiode receiver with a 50-Ω front end

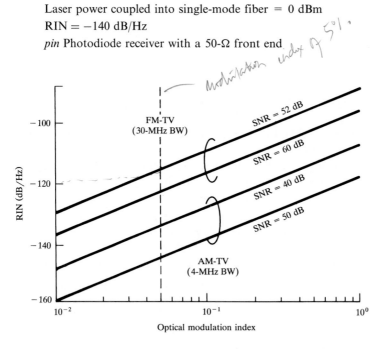

FIGURE 9-10
RIN versus the optical modulation index per channel for AM and FM video signals for several different signal-to-noise ratios (SNR). (Reproduced with permission from Way,[2] © 1989, IEEE.)

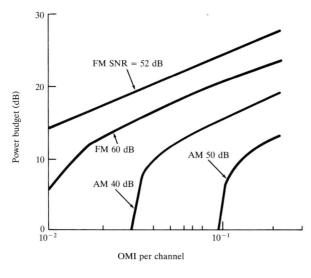

FIGURE 9-11
Power budget versus the optical modulation index (OMI) per channel for distribution of multichannel AM and FM video signals. (Reproduced with permission from Way,[2] © 1989, IEEE.)

Preamplifer noise figure = 2 dB
AM bandwidth per channel = 4 MHz
FM bandwidth per channel = 30 MHz

Again, assuming a per-channel optical modulation index of 5 percent, the AM system has a power margin of about 10 dB for a 40-dB signal-to-noise ratio, whereas the FM system has a power margin of 20 dB for $S/N = 52$ dB.

9.3.3 Subcarrier Multiplexing

There is also great interest in using RF or *microwave subcarrier multiplexing* for high-capacity lightwave systems.[1,2,25-28] The term *subcarrier multiplexing* (SCM) is used to describe the capability of multiplexing both multichannel analog and digital signals within the same system.

Figure 9-12 shows the basic concept of an SCM system. The input to the transmitter consists of a mixture of N independent analog and digital baseband signals. These signals can carry either voice, data, video, digital audio, high-definition video, or any other analog or digital information. Each incoming signal $s_i(t)$ is mixed with a local oscillator (LO) having a frequency f_i. The local oscillator frequencies employed are in the 2-to-8-GHz range and are known as the *subcarriers*. Combining the modulated subcarriers gives a composite frequency-division-multiplexed signal which is used to drive a laser diode.

At the receiving end, the optical signal is directly detected with a high-speed wideband InGaAs *pin* photodiode and reconverted to a microwave signal. For long-distance links, one can also employ a wideband InGaAs avalanche photodiode with a 50- to 80-GHz gain–bandwidth product or use an optical preamplifier. For amplifying the received microwave signal, one can use a commercially available wideband low-noise amplifier or a *pin*-FET receiver.

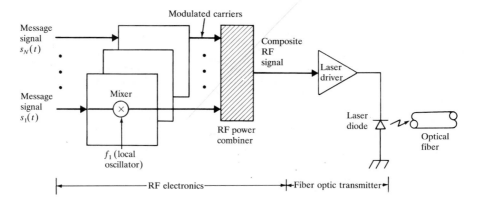

FIGURE 9-12
Basic concept of subcarrier multiplexing. One can simultaneously send analog and digital signals by frequency-division multiplexing them on different subcarrier frequencies.

PROBLEMS

9-1. Commercially available wideband receivers have equivalent resistance $R_{eq} = 75\,\Omega$. With this value of R_{eq} and letting the remaining transmitter and receiver parameters be the same as in Example 9-3, plot the total carrier-to-noise ratio and its limiting expressions, as given by Eqs. (9-10) through (9-13), for received power levels ranging from 0 to -16 dBm. Show that the thermal noise of the receiver dominates over the quantum noise at all power levels when $R_{eq} = 75\,\Omega$.

9-2. Consider a five-channel frequency-division-multiplexed (FDM) system having carriers at $f_1, f_2 = f_1 + \Delta, f_3 = f_1 + 2\Delta, f_4 = f_1 + 3\Delta$, and $f_5 = f_1 + 4\Delta$, where Δ is the spacing between carriers. On a frequency plot, show the number and location of the triple-beat and two-tone third-order intermodulation products.

9-3. Suppose we want to frequency-division multiplex 60 FM signals. If 30 of these signals have a per-channel modulation index $m_i = 3$ percent and the other 30 signals have $m_i = 4$ percent, find the optical modulation index of the laser.

9-4. Consider an SCM system having 120 channels, each modulated at 2.3 percent. The link consists of 12 km of single-mode fiber having a loss of 1 dB/km, plus a connector having a 0.5-dB loss on each end. The laser source couples 2 mW of optical power into the fiber and has RIN $= -135$ dB/Hz. The *pin* photodiode receiver has a responsivity of 0.6 A/W, $B = 5$ GHz, $I_D = 10$ nA, $R_{eq} = 50\,\Omega$, and $F_t = 3$ dB. Find the carrier-to-noise ratio for this system.

9-5. What is the carrier-to-noise ratio for the system described in Prob. 9-4 if the *pin* photodiode is replaced with an InGaAs avalanche photodiode having $M = 10$ and $F(M) = M^{0.7}$?

9.6. Consider a 32-channel FDM system with a 4.4-percent modulation index per channel. Let RIN $= -135$ dB/Hz, and assume the *pin* photodiode receiver has a responsivity of 0.6 A/W, $B = 5$ GHz, $I_D = 10$ nA, $R_{eq} = 50\,\Omega$, and $F_t = 3$ dB.
(*a*) Find the carrier-to-noise ratio for this link if the received optical power is -10 dBm.

(b) Find the carrier-to-noise ratio if the modulation index is increased to 7 percent per channel and the received optical power is decreased to -13 dBm.

9-7. For a fiber optic link using a single-longitudinal-mode laser with a 3-dB linewidth of Δv and having two fiber connectors with reflectivities R_1 and R_2, the worst-case RIN occurs when the direct and doubly reflected optical fields interfere in quadrature.[29] If τ is the light-travel time in the fiber, this is described by

$$\text{RIN}(f) = \frac{4R_1 R_2}{\pi} \frac{\Delta v}{f^2 + \Delta v^2} [1 + e^{-4\pi\Delta v\tau} - 2e^{-2\pi\Delta v\tau} \cos(2\pi f\tau)]$$

where f is the carrier frequency. Show that this expression reduces to

$$\text{RIN}(f) = \frac{16R_1 R_2}{\pi} \Delta v\, \tau^2 \qquad \text{for} \qquad \Delta v \cdot \tau \ll 1$$

and

$$\text{RIN}(f) = \frac{4R_1 R_2}{\pi} \frac{\Delta v}{f^2 + \Delta v^2} \qquad \text{for} \qquad \Delta v \cdot \tau \gg 1$$

9-8. A typical DFB laser has a linewidth $1\,\text{MHz} < \Delta v < 40\,\text{MHz}$, and with a 1- to 10-m optical jumper cable we have $0.005 < \Delta v \cdot \tau < 2$. With a 1-m jumper fiber and letting $\Delta v = f = 10\,\text{MHz}$, use the expression in Prob. 9-7 to show that to achieve an RIN of less than -140 dB/Hz, the average reflectivity per connector should be less than -30 dB.

9-9. It is possible to achieve CATV supertrunk applications with lengths greater than 40 km by cascading standard fiber-optic CATV transmitters, which act as amplifiers. These systems are typically limited by the required CTB performance.[30] When two amplifiers are cascaded, their individual CTB products add as

$$\text{CTB}_{\text{cascade}} = x \log(10^{\text{CTB}_1/x} + 10^{\text{CTB}_2/x})$$

where CTB_i is the composite triple beat of amplifier i.

(a) For identical amplifiers, $x = 20$. What is the CTB power penalty in this case?

(b) When the amplifiers are different, the value of x can vary from 0 (cancellation of beats between the two amplifiers) to 20 (all beats of the two amplifiers are in phase). Find the values of x for the following experimental measurements on dissimilar amplifiers where dBc is the power relative to the carrier:

CTB$_1$ (dBc)	CTB$_2$ (dBc)	CTB$_{\text{cascade}}$ (dBc)
-75.2	-69.9	-70.5
-74.7	-71.4	-71.0
-72.1	-71.3	-66.7

9-10. Consider a subcarrier-multiplexed CATV distribution system that has N channels. The CNRs of the three fundamental irreducible sources of signal degradation in a fiber link containing an optical amplifier are as follows:

(1) Nonlinear distortion resulting from clipping of the laser output,

$$\text{CNR}_{clip} = \sqrt{2\pi} \frac{(1 + 6\mu^2)}{\mu^3} e^{1/2\mu^2}$$

Here, the rms modulation index $\mu = m\sqrt{N/2}$ where m is the modulation depth per channel. This distortion arises when the modulation depth m is increased to the point where the signal starts getting clipped at the lasing threshold level.

(2) Shot noise in the photodetector,

$$CNR_{shot} = \frac{\mathscr{R}m^2 G P_{in} L}{4qB_e}$$

where $\mathscr{R}$ is the photodetector responsivity, G is the optical amplifier gain, P_{in} is the input signal power, L is the postamplifer loss, q is the electron charge, and B_e is the receiver noise bandwidth.

(3) Signal-spontaneous emission beat noise,

$$CNR_{sig\text{-}sp} = \frac{m^2 G P_{in}}{8hvn_{sp}(G-1)B_e}$$

where n_{sp} is the population-inversion factor in the optical amplifier and $v = c/\lambda$ is the signal frequency. As discussed in Chap. 11, this noise arises when the amplified-spontaneous-emission (ASE) noise generated by the optical amplifier beats with the optical signal in the photodetector.

The total CNR at the receiver is then given by Eq. (9-1). For simplicity of calculation, this can be written in terms of the noise-to-carrier ratio $NCR = 1/CNR$, so that Eq. (9-1) becomes

$$NCR_{total} = NCR_{clip} + NCR_{shot} + NCR_{sig\text{-}sp}$$

Using the parameter values given in Table P9-10, plot expressions (*a*) to (*c*) on the same graph as a function of the rms modulation index over the range $0.04 \leq \mu \leq 0.36$:

(*a*) NCR_{clip}
(*b*) $NCR_{shot} + NCR_{sig\text{-}sp}$
(*c*) NCR_{total}
(*d*) Explain where the minimum NCR (maximum CNR) occurs on the plot.

TABLE P9-10

Parameter	Value
P_{in}	1 mW
m	$0.68\sqrt{N}$
G	100 (20 dB)
n_{sp}	2
λ	1551 nm
B_e	1×10^9 Hz
$\mathscr{R}$	0.6 A/W
L	40/3
N	60

REFERENCES

1. R. Olshansky, V. A. Lanzisera, and P. M. Hill, "Subcarrier multiplexed lightwave systems for broadband distribution," *J. Lightwave Tech.*, vol. 7, pp. 1329–1342, Sept. 1989.

2. (*a*) W. I. Way, "Subcarrier multiplexed lightwave system design considerations for subscripter loop applications," *J. Lightwave Tech.*, vol. 7, pp. 1806–1818, Nov. 1989.

   (*b*) W. I. Way, *Broadband Hybrid Fiber/Coax Access System Technologies*, Academic, New York, 1998.

3. A. S. Andrawis and I. Jacobs, "A new compound modulation technique for multichannel analog video transmission on fiber," *J. Lightwave Tech.*, vol. 11, pp. 49–54, Jan. 1993.

4. T. E. Darcie, J. Lipson, C. B. Roxlo, and C. J. McGrath, "Fiber optic device technology for broadband analog video systems," *IEEE Mag. Lightwave Commun.*, vol. 1, pp. 46–52, Feb. 1990.

5. K. J. Williams, L. T. Nichols, and R. D. Esman, "Photodetector nonlinearity limitations on a high-dynamic range 3-GHz fiber optic link," *J. Lightwave Tech.*, vol. 16, pp. 192–199, Feb. 1998.

6. E. I. Ackerman and A. S. Daryoush, "Broadband external modulation fiber-optic links for antenna-remoting applications," *IEEE Trans. Microwave Theory Tech.*, vol. 45, pp. 1436–1442, Aug. 1997.

7. Special Issue on "Applications of Lightwave Technology to Microwave Devices, Circuits, and Systems," *IEEE Trans. Microwave Theory Tech.*, vol. 38, May 1990.

8. J. J. Lee, R. Y. Loo, S. Livingston, V. I. Jones, J. B. Lewis, H.-W. Yen, G. L. Tangonan, and M. Wechsberg, "Photonic wideband array antennas," *IEEE Trans. Antennas Propagat.*, vol. 43, pp. 966–982, Sept. 1995.

9. C. Cox III, E. I. Ackerman, R. Helkey, and G. E. Betts, "Techniques and performance of intensity-modulation direct-detection analog optical links," *IEEE Trans. Microwave Theory Tech.*, vol. 45, pp. 1375–1383, Aug. 1997.

10. E. A. Lee and D. G. Messerschmitt, *Digital Communications*, Kluwer Academic, Boston, 2nd ed., 1993.

11. S. Haykin, *Communication Systems*, Wiley, New York, 3rd ed., 1994.

12. (*a*) C. J. Chung and I. Jacobs, "Practical TV channel capacity of lightwave multichannel AM SCM systems limited by the threshold nonlinearity of laser diodes," *IEEE Photonics Tech. Lett.*, vol. 4, pp. 289–291, Mar. 1992.

    (*b*) A. J. Rainal, "Laser clipping distortion in analog and digital channels," *J. Lightwave Tech.*, vol. 15, pp. 1805–1807, Oct. 1997.

    (*c*) B. H. Wang, P.-Y. Chiang, M.-S. Kao, and W. I. Way, "Large-signal spurious-free dynamic range due to static and dynamic clipping in direct and external modulation systems," *J. Lightwave Tech.*, vol. 16, pp. 1773–1785, Oct. 1998.

13. K. Sato, "Intensity noise of semiconductor laser diodes in fiber optic analog video transmission," *IEEE J. Quantum Electron.*, vol. 19, pp. 1380–1391, Sept. 1983.

14. W. S. Ciciora, "An introduction to cable television in the United States," *IEEE Mag. Lightwave Commun.*, vol. 1, pp. 19–25, Feb. 1990.

15. T. E. Darcie and G. E. Bodeep, "Lightwave carrier CATV transmission systems," *IEEE Trans. Microwave Theory Tech.*, vol. 31, pp. 524–533, 1990.

16. E. Yoneda, K. Suto, K. Kikushima, and H. Yoshinaga, "All-fiber video distribution systems using SCM and EDFA techniques," *J. Lightwave Tech.*, vol. 11, pp. 128–137, Jan. 1993.

17. I. M. I. Habbab and A. A. M. Saleh, "Fundamental limitations in EDFA-based subcarrier-multiplexed AM-VSB CATV systems," *J. Lightwave Tech.*, vol. 11, pp. 42–48, Jan. 1993.

18. T. T. Ha, *Digital Satellite Communications*, McGraw-Hill, New York, 1990.

19. J. H. Schaffner and W. B. Bridges, "Intermodulation distortion in high dynamic range microwave fiber-optic links with linearized modulators," *J. Lightwave Tech.*, vol. 11, pp. 3–6, Jan. 1993.

20. (*a*) *NCTA Recommended Practices for Measurements on Cable Television Systems*, National Cable Television Association, 2nd ed., 1993.

    (*b*) J. A. Chiddix, H. Laor, D. M. Pangrac, L. D. Williamson, and R. W. Wolfe, "AM video on fiber in CATV systems: Need and implementation," *IEEE J. Sel. Areas Commun.*, vol. 8, pp. 1229–1239, Sept. 1990.

(*c*) C.-K. Chan and L.-K. Chen, "A correction scheme for measurement accuracy improvement in multichannel CATV systems," *IEEE Trans. Broadcasting*, vol. 42, pp. 122–129, June 1996.

21. W. I. Way, R. S. Wolff, and M. Krain, "A 35-km fiber-optic microwave multicarrier transmission system for satellite earth stations," *J. Lightwave Tech.*, vol. 5, pp. 1325–1332, Sept. 1987.

22. R. Olshansky and V. A. Lanzisera, "60-channel FM video subcarrier-multiplexed optical communication system," *Electron. Lett.*, vol. 23, pp. 1196–1198, 1987.

23. F. V. C. Mendis and P. A. Rosher, "CNR requirements for subcarrier-multiplexed multichannel video FM transmission in optical fibre," *Electron. Lett.*, vol. 25, pp. 72–74, Jan. 1989.

24. L. W. Couch II, *Digital and Analog Communication Systems*, Prentice Hall, Upper Saddle River, NJ, 5th ed., 1997.

25. O. K. Tonguz and H. Jung, "Personal communications access networks using subcarrier-multiplexed optical links," *J. Lightwave Tech.*, vol. 14, pp. 1400–1409, June 1996.

26. T. Iwai, K. Sato, and K. Suto, "Reduction of dispersion-induced distortion in SCM transmission systems," *J. Lightwave Tech.*, vol. 15, pp. 169–178, Feb. 1997.

27. C.-K. Ko and S.-Y. Kuo, "Multiaccess processor interconnection using subcarrier and wavelength division multiplexing," *J. Lightwave Tech.*, vol. 15, pp. 228–241, Feb. 1997.

28. Y. Aburakawa and H. Ohtsuka, "Signal extraction with frequency arrangement and SSM schemes in fiber-oriented wireless access systems," *J. Lightwave Tech.*, vol. 15, pp. 2223–2231, Dec. 1997.

29. R. W. Tkach and A. R. Chraplyvy, "Phase noise and linewidth in an InGaAsP DFB laser," *J. Lightwave Tech.*, vol. 4, pp. 1711–1716, Nov. 1986.

30. K. D. LaViolette, "CTB performance of cascaded externally modulated and directly modulated CATV transmitters," *IEEE Photonics Tech. Lett.*, vol. 8, pp. 281–283, Feb. 1996.

CHAPTER
10

WDM CONCEPTS
AND COMPONENTS

A powerful aspect of an optical communication link is that many different wavelengths can be sent along a single fiber simultaneously in the 1300-to-1600-nm spectral band. The technology of combining a number of wavelengths onto the same fiber is known as *wavelength-division multiplexing* or WDM.[1-7] Conceptually, the WDM scheme is the same as frequency-division multiplexing (FDM) used in microwave radio and satellite systems. Just as in FDM, the wavelengths (or optical frequencies) in WDM must be properly spaced to avoid interchannel interference. The key system features of WDM are as follows:

- *Capacity upgrade*. The classical application of WDM has been to upgrade the capacity of existing point-to-point fiber optic transmission links. If each wavelength supports an independent network signal of perhaps a few gigabits per second, then WDM can increase the capacity of a fiber network dramatically.

- *Transparency*. An important aspect of WDM is that each optical channel can carry any transmission format. Thus, using different wavelengths, fast or slow asynchronous and synchronous digital data and analog information can be sent simultaneously, and independently, over the same fiber, without the need for a common signal structure.

- *Wavelength routing*. In addition to using multiple wavelengths to increase link capacity and flexibility, the use of wavelength-sensitive optical routing devices makes it possible to use wavelength as another dimension, in addition to time and space, in designing communication networks and switches. Wavelength-routed networks use the actual wavelength of a signal as the intermediate or final address.

● *Wavelength switching.* Whereas wavelength-routed networks are based on a rigid fiber infrastructure, wavelength-switched architectures allow reconfigurations of the optical layer. Key components for implementing these networks include optical add/drop multiplexers, optical cross connects, and wavelength converters. Chapter 12 gives a detailed treatment of wavelength routing and switching.

This chapter addresses the operating principles of WDM and describes the components needed for its realization. These components range in complexity from simple passive optical splitters or combiners to sophisticated tunable optical sources and wavelength filters. Before we look at the optical networks that can be created with these devices, Chap. 11 discusses the optical amplifiers that are needed for boosting the power levels of several signals simultaneously, where each signal uses a different wavelength. Following this, Chap. 12 shows how to put everything together to form multiwavelength networks that use a minimum of electronic processing to route high-capacity optical signals.

10.1   OPERATIONAL PRINCIPLES OF WDM

In standard point-to-point links a single fiber line has one optical source at its transmitting end and one photodetector at the receiving end. Signals from different light sources use separate and uniquely assigned optical fibers. Since an optical source has a narrow linewidth, this type of transmission makes use of only a very narrow portion of the transmission bandwidth capability of a fiber.

To see the potential of WDM, let us first examine the characteristics of a high-quality optical source. As an example, the modulated output of a DFB laser has a frequency spectrum of 10–50 MHz, which is equivalent to a nominal spectral linewidth of 10^{-3} nm. When using such a source, a guard band of 0.4–1.6 nm is typically employed. This is done to take into account possible drifts of the peak wavelength due to aging or temperature effects, and to give both the manufacturer and the user some leeway in specifying and choosing the precise peak emission wavelength. With such spectral bandwidths, simplex systems make use of only a small portion of the transmission bandwidth capability of a standard fiber. This can be seen from Fig. 10-1, which depicts the attenuation of light in a silica fiber as a function of wavelength. The curve shows that the two low-loss regions of a single-mode fiber extend over the wavelengths ranging from about 1270 to 1350 nm (the 1310-nm window) and from 1480 to 1600 nm (the 1550-nm window). However, note the wider window of AllWave® fibers in Fig. 3-1.

We can view these regions either in terms of spectral width (the wavelength band occupied by the light signal and its guard band) or by means of optical bandwidth (the frequency band occupied by the light signal). To find the optical bandwidth corresponding to a particular spectral width in these regions, we use

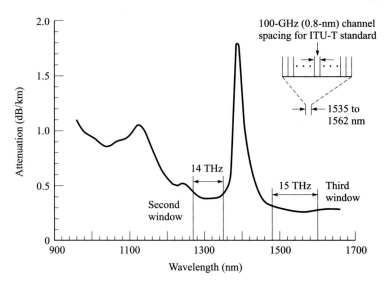

FIGURE 10-1

The transmission bandwidths in the 1310-nm and 1550-nm windows allow the use of many simultaneous channels for sources with narrow spectral widths. The ITU-T standard for WDM specifies channels with 100-GHz spacings. Also, see Fig. 3-1 for water-free AllWave® fiber.

the relationship $c = \lambda \nu$, which relates the wavelength λ to the carrier frequency ν, where c is the speed of light. Differentiating this we have for $\Delta\lambda << \lambda^2$

$$|\Delta\nu| = \left(\frac{c}{\lambda^2}\right)|\Delta\lambda| \tag{10-1}$$

where the deviation in frequency $\Delta\nu$ corresponds to the wavelength deviation $\Delta\lambda$ around λ. From Eq. (10-1), the optical bandwidth is $\Delta\nu = 14$ THz for a usable spectral band $\Delta\lambda = 80$ nm in the 1310-nm window. Similarly, $\Delta\nu = 15$ THz for a usable spectral band $\Delta\lambda = 120$ nm in the 1550-nm window. This yields a total available fiber bandwidth of about 30 THz in the two low-loss windows.

Since the spectral width of a high-quality source occupies only a narrow optical bandwidth, the two low-loss windows provide many additional operating regions. By using a number of light sources, each emitting at a different peak wavelength that is sufficiently spaced from its neighbor so as not to create interference, the integrities of the independent messages from each source are maintained for subsequent conversion to electrical signals at the receiving end.

Example 10-1. If one takes a spectral band of 0.8 nm (or, equivalently, a mean frequency spacing of 100 GHz) within which narrow-linewidth lasers are transmitting, then one can send 50 independent signals in the 1525-to-1565-nm band on a single fiber.

Since WDM is essentially frequency-division multiplexing at optical carrier frequencies, the WDM standards developed by the International Telecommunication Union (ITU) specify channel spacings in terms of frequency.[8] A key reason for selecting a fixed frequency spacing, rather than a constant wavelength spacing, is that when locking a laser to a particular operating mode it is the frequency of the laser that is fixed.[9] The ITU-T Recommendation G.692 specifies selecting the channels from a grid of frequencies referenced to 193.100 THz (1552.524 nm) and spacing them 100 GHz (0.8 nm at 1552 nm) apart. Suggested alternative spacings include 50 GHz (0.4 nm) and 200 GHz (1.6 nm).

The literature often uses the term *dense WDM* (DWDM), in contrast to conventional or regular WDM. This term does not denote a precise operating region or implementation condition, but, instead, is a historically derived designation. In general, it refers to the spacings denoted by ITU-T G.692. The original use of WDM was to upgrade the capacity of installed point-to-point transmission links. Typically, this was achieved by adding wavelengths that were separated by several tens, or even hundreds, of nanometers, in order not to impose strict requirements on the different laser sources and the receiving optical wavelength splitters. In the late 1980s, with the advent of tunable lasers that have extremely narrow linewidths, one then could have very closely spaced signal bands. This is the basis of dense WDM.

A key feature of WDM is that the discrete wavelengths form an orthogonal set of carriers that can be separated, routed, and switched without interfering with each other. This holds as long as the optical intensity is kept sufficiently low to prevent nonlinear effects, such as stimulated Brillouin scattering and four-wave mixing processes, from degrading the link performance.[10-12]

The implementation of WDM networks requires a variety of passive and/or active devices to combine, distribute, isolate, and amplify optical power at different wavelengths. Passive devices require no external control for their operation, so they are somewhat limited in their application in WDM networks. These components are mainly used to split and combine or tap off optical signals. The performance of active devices can be controlled electronically, thereby providing a large degree of network flexibility. Active WDM components include tunable optical filters, tunable sources, and optical amplifiers.

Figure 10.2 shows the use of such components in a typical WDM link containing various types of optical amplifiers (these are discussed in Chap. 11). At the transmitting end, there are several independently modulated light sources, each emitting signals at a unique wavelength. Here, a *multiplexer* is needed to combine these optical outputs into a serial spectrum of closely spaced wavelength signals and couple them onto a single fiber. At the receiving end, a *demultiplexer* is required to separate the optical signals into appropriate detection channels for signal processing.[13] At the transmitting end, the basic design challenge is to have the multiplexer provide a low-loss path from each optical source to the multiplexer output. Since the optical signals that are combined generally do not emit any significant amount of optical power outside of the designated channel spectral

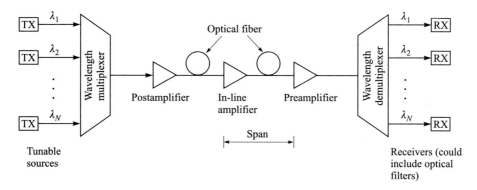

FIGURE 10-2
Implementation of a typical WDM network containing various types of optical amplifiers.

width, interchannel crosstalk factors are relatively unimportant at the transmitting end.

A different requirement exists for the demultiplexer, since photodetectors are usually sensitive over a broad range of wavelengths, which could include all the WDM channels. To prevent spurious signals from entering a receiving channel (i.e., to give good channel isolation of the different wavelengths being used), the demultiplexer must exhibit narrow spectral operation, or very stable optical filters with sharp wavelength cutoffs must be used. The tolerable interchannel crosstalk levels can vary widely depending on the application. In general, a −10-dB level is not satisfactory, whereas a level of −30 dB is acceptable. In principle, any optical demultiplexer can also be used as a multiplexer. For simplicity, the word "multiplexer" is used as a general term to refer to both combining and separating functions, except when it is necessary to distinguish the two devices or functions.

10.2 PASSIVE COMPONENTS

Passive devices operate completely in the optical domain to split and combine light streams. They include $N \times N$ couplers (with $N \geq 2$), power splitters, power taps, and star couplers. These components can be fabricated either from optical fibers or by means of planar optical waveguides using material such as lithium niobate ($LiNbO_3$) or InP.

Basically, most passive WDM devices are variations of a star-coupler concept. Figure 10-3 shows a generic star coupler, which can perform both power combining and splitting. In the broadest application, star couplers combine the light streams from two or more input fibers and divide them among several output fibers. In the general case, the splitting is done uniformly for all wavelengths, so that each of the N outputs receives $1/N$ of the power entering the device. A common fabrication method for an $N \times N$ splitter is to fuse together the cores of N single-mode fibers over a length of a few millimeters. The optical power inserted through one of the N fiber entrance ports gets divided uniformly into the

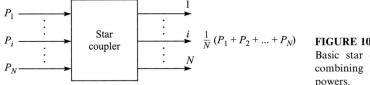

FIGURE 10-3
Basic star coupler concept for combining or splitting optical powers.

cores of the N output fibers through evanescent power coupling in the fused region.

Any size star coupler can be made, in principle, provided that all fibers can be heated uniformly during the coupler-fabrication process. Couplers with 64 inputs and outputs are possible, although, more commonly, the size tends to be less than 10. One simple device is a power tap. Taps are nonuniform 2×2 couplers which are used to extract a small portion of optical power from a fiber line for monitoring signal quality.

The three fundamental technologies for making passive components are based on optical fibers, integrated optical waveguides, and bulk micro-optics.[14] Researchers have examined many different component designs using these techniques.[15] The next sections describe the physical principles of several simple examples of fiber-based and integrated-optic devices to illustrate the fundamental operating principles. Couplers using micro-optic designs are not widely used because the strict tolerances required in the fabrication and alignment processes affect their cost, performance, and robustness. See Prob. 10-14 for an example of a micro-optic-based multiplexer using a plane reflection grating.

10.2.1 The 2×2 Fiber Coupler

When discussing couplers and splitters, it is customary to refer to them in terms of the number of input and output ports on the device. For example, a device with two inputs and two outputs would be called a "2×2 coupler." In general, an $N \times M$ coupler has N inputs and M outputs.

The 2×2 coupler[16-19] is a simple fundamental device that we will use here to demonstrate the operational principles. A common construction is the fused-fiber coupler. This is fabricated by twisting together, melting, and pulling two single-mode fibers so they get fused together over a uniform section of length W, as shown in Fig. 10-4. Each input and output fiber has a long tapered section of length L, since the transverse dimensions are gradually reduced down to that of the coupling region when the fibers are pulled during the fusion process. The total draw length is $\mathscr{L} = L + W$. This device is known as a *fused biconical tapered coupler*. Here, P_0 is the input power, P_1 is the throughout power, and P_2 is the power coupled into the second fiber. The parameters P_3 and P_4 are extremely low signal levels (-50 to -70 dB below the input level) resulting from backward reflections and scattering due to bending in and packaging of the device.[20]

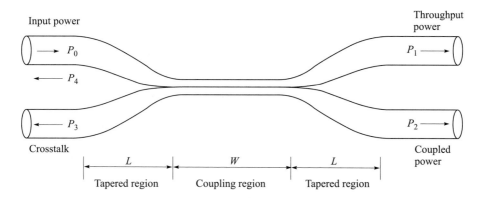

FIGURE 10-4
Cross-sectional view of a fused-fiber coupler having a coupling region W and two tapered regions of length L. The total span $\mathscr{L} = 2L + W$ is the coupler draw length.

As the input light P_0 propagates along the taper in fiber 1 and into the coupling region W, there is a significant decrease in the V number owing to the reduction in the ratio r/λ [see Eq. (2-58)], where r is the reduced fiber radius. Consequently, as the signal enters the coupling region, an increasingly larger portion of the input field now propagates outside the core of the fiber. Depending on the dimensioning of the coupling region, any desired fraction of this decoupled field can be recoupled into the other fiber. By making the tapers very gradual, only a negligible fraction of the incoming optical power is reflected back into either of the input ports. Thus, these devices are also known as *directional couplers.*

The optical power coupled from one fiber to another can be varied through three parameters: the axial length of the coupling region over which the fields from the two fibers interact; the size of the reduced radius r in the coupling region; and Δr, the difference in the radii of the two fibers in the coupling region. In making a fused fiber coupler, the coupling length W is normally fixed by the width of the heating flame, so that only L and r change as the coupler is elongated. Typical values for W and L are a few millimeters, the exact values depending on the coupling ratios desired for a specific wavelength, and $\Delta r/r$ is around 0.015. Assuming that the coupler is lossless, the expression for the power P_2 coupled from one fiber to another over an axial distance z is [14-17]

$$P_2 = P_0 \sin^2(\kappa z) \tag{10-2}$$

where κ is the *coupling coefficient* describing the interaction between the fields in the two fibers. By conservation of power, for identical-core fibers we have

$$P_1 = P_0 - P_2 = P_0[1 - \sin^2(\kappa z)] = P_0 \cos^2(\kappa z) \tag{10-3}$$

This shows that the phase of the driven fiber always lags 90° behind the phase of the driving fiber, as Fig. 10-5a illustrates. Thus, when power is launched

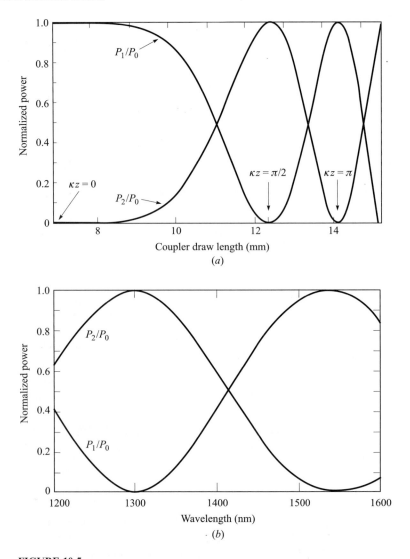

FIGURE 10-5

(*a*) Normalized coupled powers P_2/P_0 and P_1/P_0 as a function of the coupler draw length for a 1300-nm power level P_0 launched into fiber 1. (*b*) Dependence on wavelength of the coupled powers in the completed 15-mm-long coupler. (Adapted with permission from Eisenmann and Weidel,[19] © 1988, IEEE.)

into fiber 1, at $z = 0$ the phase in fiber 2 lags 90° behind that in fiber 1. This lagging phase relationship continues for increasing z, until at a distance that satisfies $\kappa z = \pi/2$, all of the power has been transferred from fiber 1 to fiber 2. Now fiber 2 becomes the driving fiber, so that for $\pi/2 \leq \kappa z \leq \pi$ the phase in fiber 1 lags behind that in fiber 2, and so on. As a result of this phase relationship, the

2×2 coupler is a *directional coupler*. That is, no energy can be coupled into a wave traveling backward in the negative-z direction in the driven waveguide.

Figure 10-5b shows how κ varies with wavelength for the final 15-mm-long coupler. Thus, different-performance couplers can be made by varying the parameters W, L, r, and Δr for a specific wavelength λ.

In specifying the performance of an optical coupler, one usually indicates the percentage division of optical power between the output ports by means of the *splitting ratio* or *coupling ratio*. Referring to Fig. 10-4, with P_0 being the input power and P_1 and P_2 the output powers, then

$$\text{Splitting ratio} = \left(\frac{P_2}{P_1 + P_2}\right) \times 100\% \qquad (10\text{-}4)$$

By adjusting the parameters so that power is divided evenly, with half of the input power going to each output, one creates a *3-dB coupler*. A coupler could also be made in which almost all the optical power at 1500 nm goes to one port and almost all the energy around 1300 nm goes to the other port (see Prob. 10-3).

In the above analysis, we have assumed, for simplicity, that the device is lossless. However, in any practical coupler there is always some light that is lost when a signal goes through it. The two basic losses are excess loss and insertion loss. The *excess loss* is defined as the ratio of the input power to the total output power. Thus, in decibels, the excess loss for a 2×2 coupler is

$$\text{Excess loss} = 10 \log\left(\frac{P_0}{P_1 + P_2}\right) \qquad (10\text{-}5)$$

The *insertion loss* refers to the loss for a particular port-to-port path. For example, for the path from input port i to output port j, we have, in decibels,

$$\text{Insertion loss} = 10 \log\left(\frac{P_i}{P_j}\right) \qquad (10\text{-}6)$$

Another performance parameter is *crosstalk*, which measures the degree of isolation between the input at one port and the optical power scattered or reflected back into the other input port. That is, it is a measure of the optical power level P_3 shown in Fig. 10-4:

$$\text{Crosstalk} = 10 \log\left(\frac{P_3}{P_0}\right) \qquad (10\text{-}7)$$

Example 10-2. A 2×2 biconical tapered fiber coupler has an input optical power level of $P_0 = 200\,\mu\text{W}$. The output powers at the other three ports are $P_1 = 90\,\mu\text{W}$, $P_2 = 85\,\mu\text{W}$, and $P_3 = 6.3\,\text{nW}$. From Eq. 10-4, the coupling ratio is

$$\text{Coupling ratio} = \left(\frac{85}{90 + 85}\right) \times 100\% = 48.6\%$$

From Eq. 10-5, the excess loss is

$$\text{Excess loss} = 10 \log\left(\frac{200}{90 + 85}\right) = 0.58 \, \text{dB}$$

Using Eq. 10-6, the insertion losses are

$$\text{Insertion loss (port 0 to port 1)} = 10 \log\left(\frac{200}{90}\right) = 3.47 \, \text{dB}$$

$$\text{Insertion loss (port 0 to port 2)} = 10 \log\left(\frac{200}{85}\right) = 3.72 \, \text{dB}$$

The crosstalk is given by Eq. 10-7 as

$$\text{Crosstalk} = 10 \log\left(\frac{6.3 \times 10^{-3}}{200}\right) = -45 \, \text{dB}$$

10.2.2 Scattering Matrix Representation

One can also analyze a 2×2 guided-wave coupler as a four-terminal device that has two inputs and two outputs, as shown in Fig. 10-6. Either all-fiber or integrated-optics devices can be analyzed in terms of the *scattering matrix* (also called the *propagation matrix*) **S**, which defines the relationship between the two input field strengths a_1 and a_2, and the two output field strengths b_1 and b_2. By definition,[21-24]

$$\mathbf{b} = \mathbf{Sa}, \quad \text{where } \mathbf{b} = \begin{bmatrix} b_1 \\ b_2 \end{bmatrix}, \quad \mathbf{a} = \begin{bmatrix} a_1 \\ a_2 \end{bmatrix}, \quad \text{and } \mathbf{S} = \begin{bmatrix} s_{11} & s_{12} \\ s_{21} & s_{22} \end{bmatrix} \quad (10\text{-}8)$$

Here, $s_{ij} = |s_{ij}| \exp(j\phi_{ij})$ represents the *coupling coefficient* of optical power transfer from input port i to output port j, with $|s_{ij}|$ being the magnitude of s_{ij} and ϕ_{ij} being its phase at port j relative to port i.

For an actual physical device, two restrictions apply to the scattering matrix **S**. One is a result of the reciprocity condition arising from the fact that Maxwell's equations are invariant for time inversion; that is, they have two solutions in

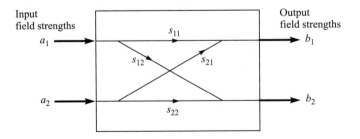

FIGURE 10-6

Generic 2×2 guided-wave coupler. Here, a_i and b_j represent the field strengths of input port i and output port j, respectively, and the s_{ij} are the scattering matrix parameters.

opposite propagating directions through the device, assuming single-mode operation. The other restriction arises from energy-conservation principles under the assumption that the device is lossless. From the first condition, it follows that

$$s_{12} = s_{21} \qquad (10\text{-}9)$$

From the second restriction, if the device is lossless, the sum of the output intensities I_o must equal the sum of the input intensities I_i:

$$I_o = b_1^* b_1 + b_2^* b_2 = I_i = a_1^* a_1 + a_2^* a_2 \qquad \text{or} \qquad b^+ b = a^+ a \qquad (10\text{-}10)$$

where the superscript $*$ means the complex conjugate and the superscript $+$ indicates the transpose conjugate. Substituting Eqs. (10-8) and (10-9) into Eq. (10-10) yields the following set of three equations:

$$s_{11}^* s_{11} + s_{12}^* s_{12} = 1 \qquad (10\text{-}11)$$

$$s_{11}^* s_{12} + s_{12}^* s_{22} = 0 \qquad (10\text{-}12)$$

$$s_{22}^* s_{22} + s_{12}^* s_{12} = 1 \qquad (10\text{-}13)$$

If we now assume that the coupler has been constructed so that the fraction $(1 - \epsilon)$ of the optical power from input 1 appears at output port 1, with the remainder ϵ going to port 2, then we have $s_{11} = \sqrt{1 - \epsilon}$, which is a real number between 0 and 1. Here, we have assumed, without loss of generality, that the electric field at output 1 has a zero phase shift relative to the input at port 1; that is $\phi_{11} = 0$. Since we are interested in the phase change that occurs when the coupled optical power from input 1 emerges from port 2, we make the simplifying assumption that the coupler is symmetric. Then, analogous to the effect at port 1, we have $s_{22} = \sqrt{1 - \epsilon}$ with $\phi_{22} = 0$. Using these expressions, we can determine the phases ϕ_{12} of the coupled outputs relative to the input signals and find the constraints on the composite outputs when both input ports are receiving signals.

Inserting the expressions for s_{11} and s_{22} into Eq. (10-12) and letting $s_{12} = |s_{12}| \exp(j\phi_{12})$, where $|s_{12}|$ is the magnitude of s_{12} and ϕ_{12} is its phase, we have

$$\exp(j2\phi_{12}) = -1 \qquad (10\text{-}14)$$

which holds when

$$\phi_{12} = (2n + 1)\frac{\pi}{2} \qquad \text{where } n = 0, 1, 2, \ldots \qquad (10\text{-}15)$$

so that the scattering matrix from Eq. (10-8) becomes

$$\mathbf{S} = \begin{bmatrix} \sqrt{1 - \epsilon} & j\sqrt{\epsilon} \\ j\sqrt{\epsilon} & \sqrt{1 - \epsilon} \end{bmatrix} \qquad (10\text{-}16)$$

Example 10-3. Assume we have a 3-dB coupler, so that half of the input power gets coupled to the second fiber. Then, $\epsilon = 0.5$ and the output field intensities $E_{\text{out},1}$ and $E_{\text{out},2}$ can be found from the input intensities $E_{\text{in},1}$ and $E_{\text{in},2}$ and the scattering matrix in Eq. (10-6):

$$\begin{bmatrix} E_{\text{out},1} \\ E_{\text{out},2} \end{bmatrix} = \frac{1}{\sqrt{2}} \begin{bmatrix} 1 & j \\ j & 1 \end{bmatrix} \begin{bmatrix} E_{\text{in},1} \\ E_{\text{in},2} \end{bmatrix}$$

Letting $E_{\text{in},2} = 0$, we have $E_{\text{out},1} = (1/\sqrt{2})E_{\text{in},1}$ and $E_{\text{out},2} = (j/\sqrt{2})E_{\text{in},1}$. The output powers are then given by

$$P_{\text{out},1} = E_{\text{out},1}E_{\text{out},1}^* = \tfrac{1}{2}E_{\text{in},1}^2 = \tfrac{1}{2}P_0$$

Similarly,

$$P_{\text{out},2} = E_{\text{out},2}E_{\text{out},2}^* = \tfrac{1}{2}E_{\text{in},1}^2 = \tfrac{1}{2}P_0$$

so that half the input power appears at each output of the coupler.

It is also important to note that when we want a large portion of the input power from, say, port 1 to emerge from output 1, we need ϵ be small. However, this, in turn, means that the amount of power at the same wavelength coupled to output 1 from input 2 is small. Consequently, if one is using the same wavelength, it is not possible, in a passive 2×2 coupler, to have all the power from both inputs coupled simultaneously to the same output. The best that can be done is to have half of the power from each input appear at the same output. However, if the wavelengths are different at each input, it is possible to couple a large portion of both power levels onto the same fiber.[18,19]

10.2.3 The 2×2 Waveguide Coupler

More versatile 2×2 couplers are possible with waveguide-type devices.[25-27] Figure 10-7 shows two types of 2×2 waveguide couplers. The uniformly symmetric device has two identical parallel guides in the coupling region, whereas the uniformly asymmetric coupler has one guide wider than the other. Analogous to fused-fiber couplers, waveguide devices have an intrinsic wavelength dependence in the coupling region, and the degree of interaction between the guides can be varied through the guide width w, the gap s between the guides, and the refractive index n_1 between the guides. In Fig. 10-7, the z direction lies along the coupler length and the y axis lies in the coupler plane transverse to the two waveguides.

Let us first consider the symmetric coupler. In real waveguides, with absorption and scattering losses, the propagation constant β_z is a complex number given by

$$\beta_z = \beta_r + j\frac{\alpha}{2} \tag{10-17}$$

where β_r is the real part of the propagation constant and α is the optical loss coefficient in the guide. Hence, the total power contained in both guides decreases by a factor $\exp(-\alpha z)$ along their length. For example, losses in semiconductor waveguide devices fall in the $0.05 < \alpha < 0.3\,\text{cm}^{-1}$ range (or, equivalently, about $0.2 < \alpha < 1\,\text{dB/cm}$), which is substantially higher than the nominal 0.1-dB/km losses in fused-fiber couplers. Recall from Eq. (3-1) the relationship $\alpha(\text{dB/cm}) = 4.343\alpha(\text{cm}^{-1})$.

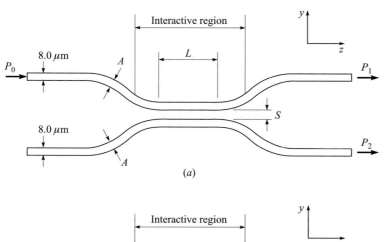

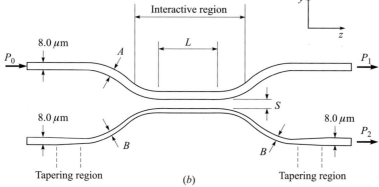

FIGURE 10-7
Cross-sectional top views of (a) a uniformly symmetric directional waveguide coupler with both guides having a width $A = 8\ \mu m$, (b) a uniformly asymmetric directional coupler in which one guide has a narrower width B in the coupling region. (Adapted with permission from Takagi, Jinguji, and Kawachi,[25] © 1992, IEEE.)

The transmission characteristics of the symmetric coupler can be expressed through the coupled-mode theory approach to yield[26,27]

$$P_2 = P_0 \sin^2(\kappa z)\, e^{-\alpha z} \tag{10-18}$$

where the coupling coefficient is

$$\kappa = \frac{2\beta_y^2 q\, e^{-qs}}{\beta_z w(q^2 + \beta_y^2)} \tag{10-19}$$

This is a function of the waveguide propagation constants β_y and β_z (in the y and z directions, respectively), the gap width and separation, and the extinction coefficient q in the y direction (i.e., the exponential falloff in the y direction) outside the waveguide, which is

$$q^2 = \beta_y^2 - k_1^2 \tag{10-20}$$

The theoretical power distribution as a function of the guide length is as shown in Fig. 10-8, where we have used $\kappa = 0.6 \, \text{mm}^{-1}$ and $\alpha = 0.02 \, \text{mm}^{-1}$. Analogous to the fused-fiber coupler, complete power transfer to the second guide occurs when the guide length L is

$$L = \frac{\pi}{2\kappa}(m+1) \qquad \text{with } m = 0, 1, 2, \ldots \qquad (10\text{-}21)$$

Since κ is found to be almost monotonically proportional to wavelength, the coupling ratio P_2/P_0 rises and falls sinusoidally from 0 to 100 percent as a function of wavelength, as Fig. 10-9 illustrates generically (assuming here, for simplicity, that the guide loss is negligible).

Example 10-4. A symmetric waveguide coupler has a coupling coefficient $\kappa = 0.6 \, \text{mm}^{-1}$. Using Eq. (10-21), we find the coupling length for $m = 1$ to be $L = 5.24$ mm.

When the two guides do not have the same widths, as shown in Fig. 10-7b, the amplitude of the coupled power is dependent on wavelength, and the coupling ratio becomes

$$P_2/P_0 = \frac{\kappa^2}{g^2} \sin^2(gz) \, e^{-\alpha z} \qquad (10\text{-}22)$$

where

$$g^2 = \kappa^2 + \left(\frac{\Delta\beta}{2}\right)^2 \qquad (10\text{-}23)$$

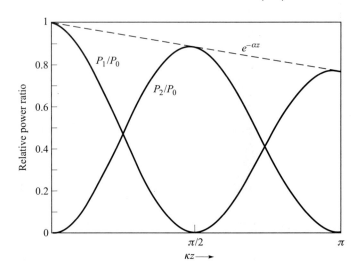

FIGURE 10-8
Theoretical through-path and coupled power distributions as a function of the guide length in a symmetric 2 × 2 guided-wave coupler with $\kappa = 0.6 \, \text{mm}^{-1}$ and $\alpha = 0.02 \, \text{mm}^{-1}$.

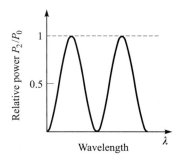

FIGURE 10-9
Wavelength response of the coupled power P_2/P_0 in the symmetric 2×2 guided-wave coupler shown in Fig. 10-7a.

with $\Delta\beta$ being the phase difference between the two guides in the z direction. With this type of configuration, one can fabricate devices that have a flattened response in which the coupling ratio is less than 100 percent in a specific desired wavelength range, as shown in Fig. 10-10. The main cause of the wave-flattened response at the lower wavelength results from suppression by the amplitude term κ^2/g^2. This asymmetric characteristic can be used in a device where only a fraction of power from a specific wavelength should be tapped off. Note also that when $\Delta\beta = 0$, Eq. (10-22) reduces to the symmetric case given by Eq. (10-18).

More complex structures are readily fabricated in which the widths of the guides are tapered.[23] These nonsymmetric structures can be used to flatten the wavelength response over a particular spectral range. It is also important to note that the above analysis based on the coupled-mode theory holds when the indices of the two waveguides are identical, but a more complex analytical treatment is needed for different indices.[28]

10.2.4 Star Couplers

The principal role of all star couplers is to combine the powers from N inputs and divide them equally among M output ports. Techniques for creating star couplers include fused fibers, gratings, micro-optic technologies, and integrated-optics schemes. The fiber-fusion technique has been a popular construction method for $N \times N$ star couplers. For example, 7×7 devices and 1×19 splitters or combiners with excess losses at 1300 nm of 0.4 dB and 0.85 dB, respectively, have been demonstrated.[29,30] However, large-scale fabrication of these devices for $N > 2$ is

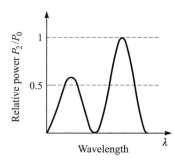

FIGURE 10-10
Wavelength response of the coupled power P_2/P_0 in the asymmetric 2×2 guided-wave coupler shown in Fig. 10-7b.

limited because of the difficulty in controlling the coupling response between the numerous fibers during the heating and pulling process. Figure 10-11 shows a generic 4 × 4 fused-fiber star coupler.

In an ideal star coupler, the optical power from any input is evenly divided among the output ports. The total loss of the device consists of its splitting loss plus the excess loss in each path through the star. The *splitting loss* is given in decibels by

$$\text{Splitting loss} = -10 \log\left(\frac{1}{N}\right) = 10 \log N \qquad (10\text{-}24)$$

Similar to Eq. (10-5), for a single input power P_{in} and N output powers, the excess loss in decibels is given by

$$\text{Fiber star excess loss} = 10 \log\left(\frac{P_{\text{in}}}{\sum_{i=1}^{N} P_{\text{out},i}}\right) \qquad (10\text{-}25)$$

The insertion loss and crosstalk can be found from Eqs. (10-6) and (10-7), respectively.

An alternative is to construct star couplers by cascading 3-dB couplers.[31-33] Figure 10-12 shows an example for an 8 × 8 device formed by using twelve 2 × 2 couplers. This device could be made from either fused-fiber or integrated-optic components. As can be seen from this figure, a fraction $1/N$ of the launched power from each input port appears at all output ports. A limitation to the flexibility or modularity of this technique is that N is a multiple of 2; that is, $N = 2^n$ with the integer $n \geq 1$. The consequence is that if an extra node needs to be added to a fully connected $N \times N$ network, the $N \times N$ star needs to be replaced by a $2N \times 2N$ star, thereby leaving $2(N - 1)$ new ports being unused. Alternatively, one extra 2 × 2 coupler can be used at one port to get $N + 1$ outputs. However, these two new ports have an additional 3-dB loss.

As can be deduced from Fig. 10-12, the number of 3-dB couplers needed to construct an $N \times N$ star is

$$N_c = \frac{N}{2} \log_2 N = \frac{N}{2} \frac{\log N}{\log 2} \qquad (10\text{-}26)$$

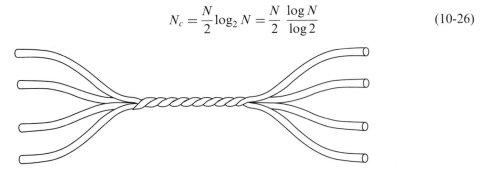

FIGURE 10-11
Generic 4 × 4 fused-fiber star coupler fabricated by twisting, heating, and pulling on four fibers to fuse them together.

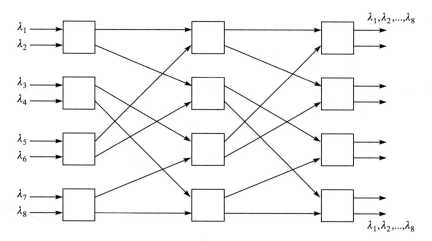

FIGURE 10-12
Example of an 8 × 8 star coupler formed by interconnecting twelve 2 × 2 couplers.

since there are $N/2$ elements in the vertical direction and $\log_2 N = \log N/\log 2$ elements horizontally. (Reminder: We use "$\log x$" to designate the base-10 logarithm of x.)

If the fraction of power traversing each 3-dB coupler element is F_T, with $0 \leq F_T \leq 1$ (i.e., a fraction $1 - F_T$ of power is lost in each 2 × 2 element), then the *excess loss* in decibels is

$$\text{Excess loss} = -10\log(F_T^{\log_2 N}) \tag{10-27}$$

The splitting loss for this star is, again, given by Eq. (10-24). Thus, the total loss experienced by a signal as it passes through the $\log_2 N$ stages of the $N \times N$ star and gets divided into N outputs is, in decibels,

$$\text{Total loss} = \text{splitting loss} + \text{exces loss} = -10\log\left(\frac{F_T^{\log_2 N}}{N}\right)$$

$$= -10\left(\frac{\log N \log F_T}{\log 2} - \log N\right) = 10(1 - 3.322\log F_T)\log N \tag{10-28}$$

This shows that the loss increases logarithmically with N.

▓▓ **Example 10-5.** Consider a commercially available 32 × 32 single-mode coupler made from a cascade of 3-dB fused-fiber 2 × 2 couplers, where 5 percent of the power is lost in each element. From Eq. (10-27), the excess loss is

$$\text{Excess loss} = -10\log(0.95^{\log 32/\log 2}) = 1.1\,\text{dB}$$

and, from Eq. (10-24), the splitting loss is

$$\text{Splitting loss} = 10\log 32 = 15\,\text{dB}$$

Hence, the total loss is 16.1 dB.

10.2.5   Mach-Zehnder Interferometer Multiplexers

Wavelength-dependent multiplexers can also be made using Mach-Zehnder interferometry techniques.[34-37] These devices can be either active or passive. Here, we look first at passive multiplexers. Figure 10-13 illustrates the constituents of an individual Mach-Zehnder interferometer (MZI). This 2×2 MZI consists of three stages: an initial 3-dB directional coupler which splits the input signals, a central section where one of the waveguides is longer by ΔL to give a wavelength-dependent phase shift between the two arms, and another 3-dB coupler which recombines the signals at the output. As we will see in the following derivation, the function of this arrangement is that, by splitting the input beam and introducing a phase shift in one of the paths, the recombined signals will interfere constructively at one output and destructively at the other. The signals then finally emerge from only one output port. For simplicity, the following analysis does not take into account waveguide material losses or bend losses.

The propagation matrix $\mathbf{M}_{\text{coupler}}$ for a coupler of length d is

$$\mathbf{M}_{\text{coupler}} \begin{bmatrix} \cos \kappa d & j \sin \kappa d \\ j \sin \kappa d & \cos \kappa d \end{bmatrix} \tag{10-29}$$

where κ is the coupling coefficient. Since we are considering 3-dB couplers which divide the power equally, then $2\kappa d = \pi/2$, so that

$$\mathbf{M}_{\text{coupler}} = \frac{1}{\sqrt{2}} \begin{bmatrix} 1 & j \\ j & 1 \end{bmatrix} \tag{10-30}$$

In the central region, when the signals in the two arms come from the same light source, the outputs from these two guides have a phase difference $\Delta\phi$ given by

$$\Delta\phi = \frac{2\pi n_1}{\lambda} L - \frac{2\pi n_2}{\lambda}(L + \Delta L) \tag{10-31}$$

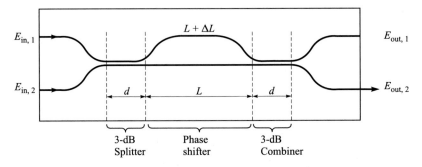

FIGURE 10-13
Layout of a basic 2×2 Mach-Zehnder interferometer.

Note that this phase difference can arise either from a different path length (given by ΔL) or through a refractive index difference if $n_1 \neq n_2$. Here, we take both arms to have the same index and let $n_1 = n_2 = n_{eff}$ (the effective refractive index in the waveguide). Then we can rewrite Eq. (10-31) as

$$\Delta\phi = k\Delta L \tag{10-32}$$

where $k = 2\pi n_{eff}/\lambda$.

For a given phase difference $\Delta\phi$, the propagation matrix $\mathbf{M}_{\Delta\phi}$ for the phase shifter is

$$M_{\Delta\phi} = \begin{bmatrix} \exp(jk\Delta L/2) & 0 \\ 0 & \exp(-jk\Delta L/2) \end{bmatrix} \tag{10-33}$$

The optical output fields $E_{out,1}$ and $E_{out,2}$ from the two central arms can be related to the input fields $E_{in,1}$ and $E_{in,2}$ by

$$\begin{bmatrix} E_{out,1} \\ E_{out,2} \end{bmatrix} = M \begin{bmatrix} E_{in,1} \\ E_{in,2} \end{bmatrix} \tag{10-34}$$

where

$$M = M_{coupler} \cdot M_{\Delta\phi} \cdot M_{coupler} = \begin{bmatrix} M_{11} & M_{12} \\ M_{21} & M_{22} \end{bmatrix} = j \begin{bmatrix} \sin(k\Delta L/2) & \cos(k\Delta L/2) \\ \cos(k\Delta L/2) & -\sin(k\Delta L/2) \end{bmatrix} \tag{10-35}$$

Since we want to build a multiplexer, we need to have the inputs to the MZI at different wavelengths; that is, $E_{in,1}$ is at λ_1 and $E_{in,2}$ is at λ_2. Then, from Eq. (10-34), the output fields $E_{out,1}$ and $E_{out,2}$ are each the sum of the individual contributions from the two input fields:

$$E_{out,1} = j[E_{in,1}(\lambda_1)\sin(k_1\,\Delta L/2) + E_{in,2}(\lambda_2)\cos(k_2\,\Delta L/2)] \tag{10-36}$$

$$E_{out,2} = j[E_{in,1}(\lambda_1)\cos(k_1\,\Delta L/2) - E_{in,2}(\lambda_2)\sin(k_2\,\Delta L/2)] \tag{10-37}$$

where $k_j = 2\pi n_{eff}/\lambda_j$. The output powers are then found from the light intensity, which is the square of the field strengths. Thus,

$$P_{out,1} = E_{out,1}E^*_{out,1} = \sin^2(k_1\,\Delta L/2)P_{in,1} + \cos^2(k_2\,\Delta L/2)P_{in,2} \tag{10-38}$$

$$P_{out,2} = E_{out,2}E^*_{out,2} = \cos^2(k_1\,\Delta L/2)P_{in,1} + \sin^2(k_2\,\Delta L/2)P_{in,2} \tag{10-39}$$

where $P_{in,j} = |E_{in,j}|^2 = E_{in,j} \cdot E^*_{in,j}$. In deriving Eqs. (10-38) and (10-39), the cross terms are dropped because their frequency, which is twice the optical carrier frequency, is beyond the response capability of the photodetector.

From Eqs. (10-38) and (10-39), we see that if we want all the power from both inputs to leave the same output port (e.g., port 2), we need to have $k_1\,\Delta L/2 = \pi$ and $k_2\,\Delta L/2 = \pi/2$, or

$$(k_1 - k_2)\,\Delta L = 2\pi n_{eff}\left(\frac{1}{\lambda_1} - \frac{1}{\lambda_2}\right)\Delta L = \pi \tag{10-40}$$

Hence, the length difference in the interferometer arms should be

$$\Delta L = \left[2n_{\text{eff}}\left(\frac{1}{\lambda_1} - \frac{1}{\lambda_2}\right)\right]^{-1} = \frac{c}{2n_{\text{eff}}\,\Delta\nu} \tag{10-41}$$

where $\Delta\nu$ is the frequency separation of the two wavelengths.

Example 10-6. (*a*) Assume that the input wavelengths of a 2×2 silicon MZI are separated by 10 GHz (i.e., $\Delta\lambda = 0.08$ nm at 1550 nm). With $n_{\text{eff}} = 1.5$ in a silicon waveguide, we have from (Eq. 10-41) that the waveguide length difference must be

$$\Delta L = \frac{3 \times 10^8 \text{ m/s}}{2(1.5)10^{10}/\text{s}} = 10 \text{ mm}$$

(*b*) If the frequency separation is 130 GHz (i.e., $\Delta\lambda = 1$ nm), then $\Delta L = 0.77$ mm.

Using basic 2×2 MZIs, any size $N \times N$ multiplexer (with $N = 2^n$) can be constructed. Figure 10-14 gives an example for a 4×4 multiplexer.[34] Here, the inputs to MZI$_1$ are ν and $\nu + 2\Delta\nu$ (which we will call λ_1 and λ_3, respectively), and the inputs to MZI$_2$ are $\nu + \Delta\nu$ and $\nu + 3\Delta\nu$ (λ_2 and λ_4, respectively). Since the signals in both interferometers of the first stage are separated by $2\Delta\nu$, the path differences satisfy the condition

$$\Delta L_1 = \Delta L_2 = \frac{c}{2n_{\text{eff}}(2\Delta\nu)} \tag{10-42}$$

In the next stage, the inputs are separated by $\Delta\nu$. Consequently, we need to have

$$\Delta L_3 = \frac{c}{2n_{\text{eff}}\,\Delta\nu} = 2\Delta L_1 \tag{10-43}$$

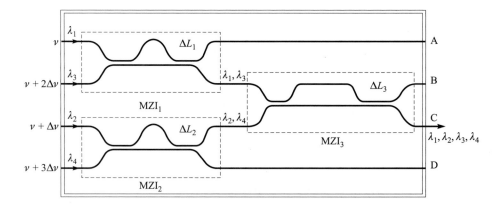

FIGURE 10-14
Example of a four-channel wavelength multiplexer using three 2×2 MZI elements. (Adapted with permission from Verbeek et al.,[34] © 1988, IEEE.)

When these conditions are satisfied, all four input powers will emerge from port C.

From this design example, we can deduce that for an N-to-1 MZI multiplexer, where $N = 2^n$ with the integer $n \geq 1$, the number of multiplexer stages is n and the number of MZIs in stage j is 2^{n-j}. The path difference in an interferometer element of stage j is thus

$$\Delta L_{\text{stage } j} = \frac{c}{2^{n-j} n_{\text{eff}} \Delta \nu} \tag{10-44}$$

The N-to-1 MZI multiplexer can also be used as a 1-to-N demultiplexer by reversing the light-propagation direction. For a real MZI, the ideal case given in these examples needs to be modified to have a slight difference in ΔL_1 and ΔL_2 (see Ref. 32 for details).

10.2.6 Fiber Grating Filters

A grating is an important element in WDM systems for combining and separating individual wavelengths. Basically, a grating is a periodic structure or perturbation in a material. This variation in the material has the property of reflecting or transmitting light in a certain direction depending on the wavelength. Thus, gratings can be categorized as either transmitting or reflecting gratings.

Figure 10-15 defines various parameters for a reflection grating. Here, θ_i is the incident angle of the light, θ_d is the diffracted angle, and Λ is the *period of the grating* (the periodicity of the structural variation in the material). In a transmission grating consisting of a series of equally spaced slits, the spacing between two adjacent slits is called the *pitch* of the grating. Constructive interference at a

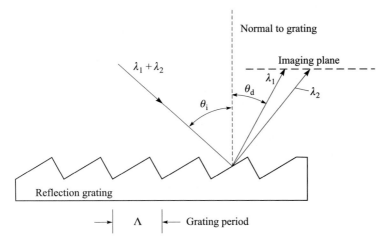

FIGURE 10-15
Basic parameters in reflection grating.

wavelength λ occurs in the imaging plane when the rays diffracted at the angle θ_d satisfy the *grating equation* given by

$$\Lambda(\sin\theta_i - \sin\theta_d) = m\lambda \tag{10-45}$$

Here, m is called the *order* of the grating. In general, only the first-order diffraction condition $m = 1$ is considered. (Note that in some texts the incidence and refraction angles are defined as being measured from the same side of the normal to the grating. In this case, the sign in front of the term $\sin\theta_d$ changes.) A grating can separate individual wavelengths since the grating equation is satisfied at different points in the imaging plane for different wavelengths.

A Bragg grating constructed within an optical fiber constitutes a high-performance device for accessing individual wavelengths in the closely spaced spectrum of dense WDM systems.[38-42] Since this is an all-fiber device, its main advantages are low cost, low loss (around 0.1 dB), ease of coupling with other fibers, polarization insensitivity, low temperature coefficient ($< 0.7\,\text{pm/°C}$), and simple packaging. A fiber grating is a narrowband reflection filter that is fabricated through a photoimprinting process. The technique is based on the observation that germanium-doped silica fiber exhibits high photosensitivity.[38] This means that one can induce a change in the refractive index of the core by exposing it to 244-nm ultraviolet radiation. Optical bandwidths of 100 GHz and less have been demonstrated in such photoinduced gratings.[43,44]

Several methods can be used to create a fiber phase-grating. Figure 10-16 demonstrates the so-called *external-writing technique*. The grating fabrication is

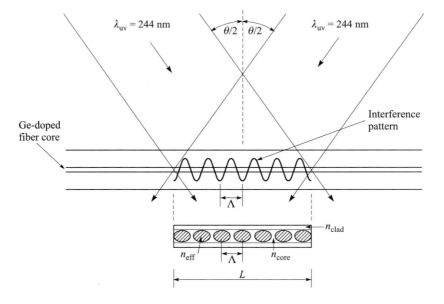

FIGURE 10-16
Formation of a Bragg grating in a fiber core by means of two intersecting ultraviolet light beams.

accomplished by means of two ultraviolet beams transversely irradiating the fiber to produce an interference pattern in the core. Here, the regions of high intensity (denoted by the shaded ovals) cause an increase in the local refractive index of the photosensitive core, whereas it remains unaffected in the zero-intensity regions. A permanent reflective Bragg grating is thus written into the core. When a multi-wavelength signal encounters the grating, those wavelengths that are phase-matched to the Bragg reflection condition are not transmitted.

Using the standard grating equation given by Eq. (10-45), with λ being the wavelength of the ultraviolet light λ_{uv}, the period Λ of the interference pattern (and hence the period of the grating) can be calculated from the angle θ between the two interfering beams of free-space wavelength λ_{uv}. Note from Fig. 10-16 that θ is measured outside of the fiber.

The imprinted grating can be represented as a uniform sinusoidal modulation of the refractive index along the core:

$$n(z) = n_{core} + \delta n \left[1 + \cos\left(\frac{2\pi z}{\Lambda}\right) \right] \tag{10-46}$$

where n_{core} is the unexposed core refractive index and δn is the photoinduced change in the index.

The maximum reflectivity R of the grating occurs when the Bragg condition holds; that is, at a reflection wavelength λ_{Bragg} where

$$\lambda_{Bragg} = 2\Lambda n_{eff} \tag{10-47}$$

and n_{eff} is the mode effective index of the core. At this wavelength, the peak reflectivity R_{max} for the grating of length L and coupling coefficient κ is given by (see Prob. 10-15)

$$R_{max} = \tanh^2(\kappa L) \tag{10-48}$$

The full bandwidth $\Delta\lambda$ over which the maximum reflectivity holds is[42]

$$\Delta\lambda = \frac{\lambda_{Bragg}^2}{\pi n_{eff} L} \left[(\kappa L)^2 + \pi^2 \right]^{1/2} \tag{10-49}$$

An approximation for the full-width half-maximum (FWHM) bandwidth is

$$\Delta\lambda_{FWHM} \approx \lambda_{Bragg} s \left[\left(\frac{\delta n}{2n_{core}}\right)^2 + \left(\frac{\Lambda}{L}\right)^2 \right]^{1/2} \tag{10-50}$$

where $s \approx 1$ for strong gratings with near 100 percent reflectivity, and $s \approx 0.5$ for weak gratings.

For a uniform sinusoidal modulation of the index throughout the core, the coupling coefficient κ is given by

$$\kappa = \frac{\pi \delta n \eta}{\lambda_{Bragg}} \tag{10-51}$$

with η being the fraction of optical power contained in the fiber core. Under the assumption that the grating is uniform in the core, η can be approximated by

$$\eta \approx 1 - V^{-2} \tag{10-52}$$

where V is the V number of the fiber. A more precise evaluation is needed for nonuniform or nonsinusoidal index variations.[45]

Example 10-7. (*a*) The table below shows the values of R_{max} as given by Eq. (10-48) for different values of κL:

κL	R_{max} (%)
1	58
2	93
3	98

(*b*) Consider a fiber grating with the following parameters: $L = 0.5$ cm, $\lambda_{Bragg} = 1530$ nm, $n_{eff} = 1.48$, $\delta n = 2.5 \times 10^{-4}$, and $\eta = 82$ percent. From Eq. (10-51) we have $\kappa = 4.2$ cm^{-1}. Substituting this into Eq. (10-49) then yields $\Delta\lambda = 0.38$ nm.

Figure 10-17 shows a simple concept of a demultiplexing function using a fiber Bragg grating.[46,47] To extract the desired wavelength, a *circulator* is used in conjunction with the grating. In a three-port circulator, an input signal on one port exits at the next port. For example, an input signal at port 1 is sent out at port 2. Here, the circulator takes the four wavelengths entering port 1 and sends them out at port 2. All wavelengths except λ_2 pass through the grating. Since λ_2 satisfies the Bragg condition of the grating, it gets reflected, enters port 2 of the circulator, and exits at port 3. More complex multiplexing and demultiplexing structures with several gratings and several circulators can be realized with this scheme.

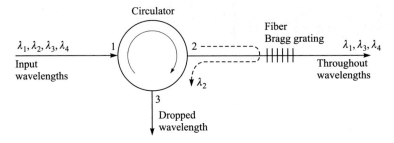

FIGURE 10-17
Simple concept of a demultiplexing function using a fiber grating and an optical circulator.

10.2.7 Phase-Array-Based WDM Devices

A highly versatile WDM device is based on using an arrayed waveguide grating. This device can function as a multiplexer, a demultiplexer, a drop-and-insert element, or a wavelength router. Here, we will look at one of a variety of design concepts that have been examined.[48-53] The arrayed waveguide grating is a generalization of the 2×2 Mach-Zehnder interferometer multiplexer. One popular design consists of M_{in} input and M_{out} output slab waveguides and two identical focusing planar star couplers connected by N uncoupled waveguides with a propagation constant β. The lengths of adjacent waveguides in the central region differ by a constant value ΔL, so that they form a Mach-Zehnder-type grating, as Fig. 10-18 shows. For a pure multiplexer, we can take $M_{in} = N$ and $M_{out} = 1$. The reverse holds for a demultiplexer; that is $M_{in} = 1$ and $M_{out} = N$. In the case of a network routing application, we can have $M_{in} = M_{out} = N$.

Figure 10-19 depicts the geometry of the star coupler. The coupler acts as a lens of focal length L_f so that the object and image planes are located at a distance L_f from the transmitter and receiver slab waveguides, respectively. Both the input and output waveguides are positioned on the focal lines, which are circles of radius $L_f/2$. In Fig. 10-19, x is the center-to-center spacing between the input waveguides and the output waveguides, d is the spacing between the grating array waveguides, and θ is the diffraction angle in the input or output slab waveguide. The refractive indices of the star coupler and the grating array waveguides are n_s and n_c, respectively.

Since any two adjacent grating waveguides have the same length difference ΔL, a phase difference $2\pi n_c \Delta L/\lambda$ results. Then, from the phase-matching condition, the light emitted from the output channel waveguides must satisfy the grating equation

$$n_s d \sin \theta + n_c \Delta L = m\lambda \tag{10-53}$$

where the integer m is the diffraction order of the grating.

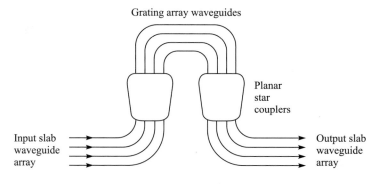

Grating array waveguides

Planar
star
couplers

Input slab
waveguide
array

Output slab
waveguide
array

FIGURE 10-18
Top view of a typical arrayed waveguide grating used as a highly versatile passive WDM device.

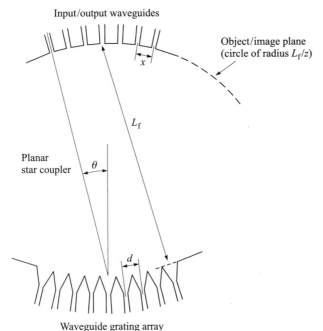

Input/output waveguides

Object/image plane
(circle of radius L_f/z)

x

L_f

Planar
star coupler

θ

d

Waveguide grating array

FIGURE 10-19
Geometry of the star coupler
used in the arrayed waveguide
grating WDM device.

Focusing is achieved by making the path-length difference ΔL between
adjacent array waveguides, measured inside the array, to be an integer multiple
of the central design wavelength of the demultiplexer

$$\Delta L = m \frac{\lambda_c}{n_c} \qquad (10\text{-}54)$$

where λ_c is the central wavelength in vacuum; that is, it is defined as the pass
wavelength for the path from the center input waveguide to the center output
waveguide.

To determine the channel spacing, we need to find the angular dispersion.
This is defined as the incremental lateral displacement of the focal spot along the
image plane per unit frequency change, and is found by differentiating Eq. (10-53)
with respect to frequency. Doing so, and considering the result in the vicinity of
$\theta = 0$, yields

$$\frac{d\theta}{dv} = -\frac{m\lambda^2}{n_s cd} \frac{n_g}{n_c} \qquad (10\text{-}55)$$

where the group index of the grating array waveguide is defined as

$$n_g = n_c - \lambda \frac{dn_c}{d\lambda} \qquad (10\text{-}56)$$

In terms of frequency, the channel spacing Δv is

$$\Delta v = \frac{x}{L_f} \left(\frac{d\theta}{dv}\right)^{-1} = \frac{x}{L_f} \frac{n_s c d}{m\lambda^2} \frac{n_c}{n_g} \qquad (10\text{-}57)$$

or, in terms of wavelength,

$$\Delta\lambda = \frac{x}{L_f} \frac{n_s d}{m} \frac{n_c}{n_g} = \frac{x}{L_f} \frac{\lambda_0 d}{\Delta L} \frac{n_s}{n_g} \qquad (10\text{-}58)$$

Equations (10-57) and (10-58) thus define the pass frequencies or wavelengths for which the multiplexer operates, given that it is designed for a central wavelength λ_c. We also note that by making ΔL large, the device can multiplex and demultiplex optical signals with very small wavelength spacings.

▨ **Example 10-8.** Consider an $N \times N$ waveguide grating multiplexer having $L_f = 10\,\text{mm}$, $x = d = 5\,\mu\text{m}$, $n_c = 1.45$, and a central design wavelength $\lambda_c = 1550\,\text{nm}$. For $m = 1$, the waveguide length difference from Eq. (10-54) is

$$\Delta L = m\frac{\lambda_c}{n_c} = \frac{1550\,\text{nm}}{1.45} = 1.069\,\mu\text{m}$$

If $n_s = 1.45$ and $n_g = 1.47$, then from Eq. (10-58) we have

$$\Delta\lambda = \frac{x}{L_f} \frac{n_s d}{m} \frac{n_c}{n_g} = \frac{5}{10^4} \frac{(1.45)(5)}{1} \frac{1.45}{1.47} \mu\text{m} = 3.58\,\text{nm}$$

Equation (10-53) shows that the phased array is periodic for each path through the device, so that after every change of 2π in θ between adjacent waveguides the field will again be imaged at the same spot. The period between two successive field maxima in the frequency domain is called the *free spectral range* (FSR) and can be represented by the relationship[51]

$$\Delta v_{\text{FSR}} = \frac{c}{n_g(\Delta L + d \sin\theta_i + d \sin\theta_o)}. \qquad (10\text{-}59)$$

where θ_i and θ_o are the diffraction angles in the input and output waveguides, respectively. These angles are generally measured from the center of the array, so that we have $\theta_i = jx/L_f$ and $\theta_o = kx/L_f$ for the jth input port and the kth output port, respectively, on either side of the central port. This shows that the FSR depends on which input and output ports the optical signal utilizes. When the ports are across from each other, so that $\theta_i = \theta_o = 0$, then

$$\Delta v_{\text{FSR}} = \frac{c}{n_g \Delta L} \qquad (10\text{-}60)$$

10.3 TUNABLE SOURCES

Many different laser designs have been proposed to generate the spectrum of wavelengths needed for WDM. One can choose from three basic options: (1)

a series of discrete DFB or DBR lasers, (2) wavelength-tunable (or frequency-tunable) lasers, or (3) a multiwavelength laser array.

The use of discrete single-wavelength lasers is the simplest method. Here, one hand-selects individual sources, each of which operates at a different wavelength. Although it is straightforward, this method can be expensive because of the high cost of individual lasers. In addition, the sources must be carefully controlled and monitored to ensure that their wavelengths do not drift with time and temperature into the spectral region of adjacent sources.

With a frequency-tunable laser, one needs only this one source.[54-59] These devices are based on DFB or DBR structures, which have a waveguide-type grating filter in the lasing cavity (as is described in Chap. 4). Frequency tuning is achieved either by changing the temperature of the device (since the wavelength changes approximately 0.1 nm/°C), or by altering the injection current into the active (gain) section or the passive section (yielding a wavelength change of 0.8×10^{-2} to 4.0×10^{-2} nm/mA, or, equivalently, 1 to 5 GHz/mA). The latter method is generally used. This results in a change in the effective refractive index, which causes a shift in the peak output wavelength. The maximum tuning range depends on the optical output power, with a larger output level resulting in a narrower tuning range. Figure 10-20 illustrates the tuning range of an injection-tunable three-section DBR laser.

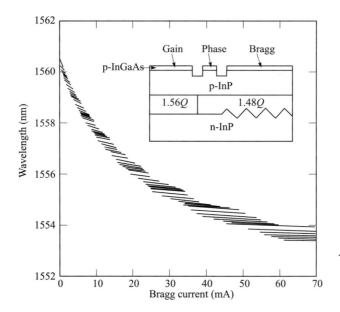

FIGURE 10-20
Tuning range of an injection-tunable three-section DBR laser. (Reproduced with permission from Staring et al.,[59] © 1994, IEEE.)

The tuning range $\Delta\lambda_{\text{tune}}$ can be estimated by

$$\frac{\Delta\lambda_{\text{tune}}}{\lambda} = \frac{\Delta n_{\text{eff}}}{n_{\text{eff}}} \tag{10-61}$$

where Δn_{eff} is the change in the effective refractive index. Practically, the maximum index change is around 1 percent, resulting in a tuning range of 10–15 nm. Figure 10-21 depicts the relationships between tuning range, channel spacing, and source spectral width. To avoid crosstalk between adjacent channels, a channel spacing of 10 times the source spectral width $\Delta\lambda_{\text{signal}}$ is often specified. That is,

$$\Delta\lambda_{\text{channel}} \approx 10\Delta\lambda_{\text{signal}} \tag{10-62}$$

Thus, the maximum number of channels N that can be placed in the tuning range $\Delta\lambda_{\text{tune}}$ is

$$N \approx \frac{\Delta\lambda_{\text{tune}}}{\Delta\lambda_{\text{channel}}} \tag{10-63}$$

Example 10-9. Suppose that the maximum index change of a particular DBR laser operating at 1550 nm is 0.65 percent. Then, the tuning range is

$$\Delta\lambda_{\text{tune}} = \lambda\frac{\Delta n_{\text{eff}}}{n_{\text{eff}}} = (1550\,\text{nm})(0.0065) = 10\,\text{nm}$$

If the source spectral width $\Delta\lambda_{\text{signal}}$ is 0.02 nm for a 2.5-Gb/s signal, then the number of channels that can operate in this tuning range is

$$N \approx \frac{\Delta\lambda_{\text{tune}}}{\Delta\lambda_{\text{channel}}} = \frac{10\,\text{nm}}{10(0.02\,\text{nm})} = 50$$

An array of tunable lasers provides a more versatile implementation in large WDM networks.[60] In addition to greater flexibility of use, these arrays can be integrated on the same substrate as multiplexing components. Most laser arrays are fabricated from a combination of DFB and MQW architectures. The letters MQW stand for *multiple quantum well*, which is a multilayer structure used to enhance the carrier and optical confinement in thin active areas (see Sec. 4.3.5).

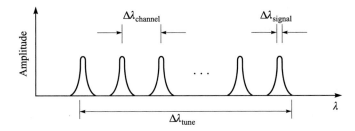

FIGURE 10-21
Relationship between tuning range, channel spacing, and source spectral width.

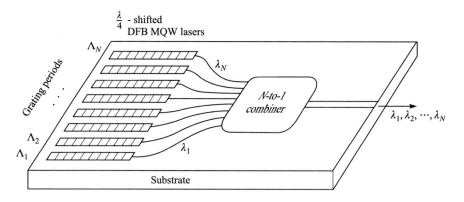

FIGURE 10-22
One possible configuration of an array of tunable MQW-DFB lasers and a light-power combiner all fabricated on the same wafer.

Basically, MQW allows a low threshold current in the active area and DFB is used for frequency tuning. Figure 10-22 shows one possible configuration with several MQW–DFB lasers fabricated on the same wafer. As described in Sec. 4.3.6, by introducing a quarter-wavelength shift in the corrugation at the center of the optical cavity, the lasing wavelength of each DFB laser in the array will coincide with the Bragg wavelength of the grating, which is (for a first-order grating)

$$\lambda_{\text{Bragg}} = 2n_{\text{eff}} \Lambda \qquad (10\text{-}64)$$

where Λ is the grating spatial period. Thus, there are two options for obtaining multiple wavelengths in such an array. One can either vary the grating spatial period of each laser during the manufacturing process; or one can fabricate each laser to have a different waveguide thickness, which creates a different energy band-gap, and hence a different lasing wavelength, in the active region.

Other designs utilize an integrated combination of an optical source (either a broadband laser diode or an LED), a waveguide grating multiplexer, and an optical amplifier.[61-63] In this method, which is known as *spectral slicing*, a broad spectral output (e.g., from an amplified LED) is spectrally sliced by the waveguide grating to produce a comb of precisely spaced optical frequencies, which become an array of constant-output sources. These spectral slices are then fed into a sequence of individually addressable wavelength channels that can be modulated externally.

10.4 TUNABLE FILTERS

Optical filters that are dynamically tunable over a certain optical frequency band can be used to increase the flexibility of a WDM network. Most tunable optical filters operate on the same principles as passive devices. The main difference is that, in active devices, at least one branch of the coupler can have its length or refractive index slightly altered by means of a control mechanism such as a voltage or temperature change. This allows the network operator to select specific optical frequencies to pass through the filter.[64,65]

10.4.1 System Considerations

Figure 10-23 shows the basic concept of a tunable optical filter. Here, the filter operates over a frequency range Δv, and is electrically tuned to allow only one optical frequency band to pass through it. The relevant system parameters include the following:

1. The *tuning range* Δv over which the filter can be tuned. If the filter needs to be tuned over one of the long-wavelength transmission windows at 1300 or 1500 nm, then 25 THz (or $\Delta\lambda = 200$ nm) is a reasonable tuning range. In networks using fiber-based optical amplifiers, which are addressed in Chap. 11, a maximum range of $\Delta\lambda = 35$ nm centered at 1550 nm ($\Delta v = 4.4$ THz centered at 193.1 THz) may be adequate. Gain-flattened amplifiers need wider ranges.

2. The *channel spacing* δv, which is the minimum frequency separation between channels that is required to guarantee a minimum crosstalk degradation.[66,67] The crosstalk signal level from an adjacent channel should generally be about 30 dB below the desired signal in order to have adequate system performance.

3. The *maximum number of channels N*, which is the maximum number of equally spaced channels that can be packed into the tuning range while maintaining an adequately low level of crosstalk between adjacent channels. This is

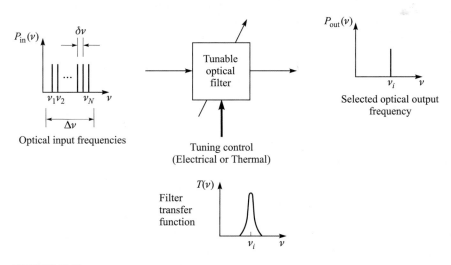

FIGURE 10-23
Basic concept of a tunable optical filter.

defined as the ratio of the total tuning range Δv to the channel spacing δv; that is,

$$N = \frac{\Delta v}{\delta v} \tag{10-65}$$

4. The *tuning speed*, which designates how quickly the filter can be reset from one frequency to another. For applications where a channel is left set up for a relatively long time (minutes to hours), a millisecond tuning speed is sufficient. However, if one wants to switch information packets rapidly, then submicrosecond tuning times are required. For example, at a 2.5-Gb/s channel rate, the time to transmit a 500-bit packet is only $0.2\,\mu$s.

10.4.2 Tunable Filter Types

Many technologies have been examined for creating tunable optical filters. During the evolution of WDM methodologies, interest has moved toward systems that have fixed frequency spacings with channel separations that are multiples of 100 GHz (or 0.8 nm in the 1550-nm transmission window). Thus, here we will concentrate on those tunable optical filters for which $\delta v \leq 100\,$GHz ($\delta \lambda \leq 1$ nm). Such devices include the following:

• *Tunable 2×2 directional couplers* have multiple control electrodes placed on the coupling waveguides.[68] Figure 10-24 illustrates a multielectrode asymmetric directional coupler fabricated on a $LiNbO_3$ crystal, where one arm is thinner than the other. For a wavelength-dropping application in this device, M wavelengths enter input port 1. Applying a specific voltage to the electrodes changes the refractive index of the waveguides, thereby selecting one of the wavelengths, say λ_i, to be coupled to the second waveguide, so that it exists from port 4. The remaining $M - 1$ wavelengths pass through the device and leave from port 3.

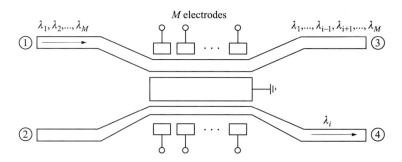

FIGURE 10-24
Concept of a tunable multielectrode asymmetric directional coupler. (Adapted with permission from Brooks and Ruschin,[68] © 1995, IEEE.)

To insert a wavelength and combine it with an input stream entering port 1, one inserts λ_i into port 2, so that it couples across to the top waveguide. Thus, it exits port 3 along with the other wavelengths $\lambda_1, \ldots, \lambda_{i-1}, \lambda_{i+1}, \lambda_M$ that entered port 1.

Tuning ranges for these type of filters are on the order of 60 nm with channel bandwidths of around 1 nm (125 GHz at 1550 nm).

- *Tunable Mach-Zehnder interferometers* use either thermo-optic or electro-optic control mechanisms to change the length of the interferometer arms.[65,69] Thermo-optic tuning in about 1 ms has been demonstrated over 12 channels spaced 5 GHz apart. Electro-optic tuning has been accomplished over eight channels spaced 50 GHz apart in optical frequency (0.4 nm in wavelength at 1550 nm) in less than 50 ns. This was done with a cascade of three tunable MZIs, as shown in Fig. 10-25. Here, the path-length differences in the three MZI stages were 1.4, 0.7, and 0.35 mm (i.e., a ratio of $4 : 2 : 1$), to achieve frequency spacings of 50, 100, and 200 GHz, respectively. Any one of the eight channels can be selected by varying the path-length difference in each MZI stage. This is done by applying the appropriate voltages to electrodes attached to the interferometer arms in each MZI.

- *Fiber Fabry-Perot filters* work on the principle of partial interference of the incident beam with itself in a mirrored resonant cavity to produce transmission peaks and nulls in the frequency domain.[70,71] Figure 10-26 shows an example of a tunable fiber Fabry-Perot filter. The ends of two single-mode fibers are polished and coated with a reflecting dielectric material. The fibers are then mounted between two piezoelectric crystals and are spaced a distance d apart to form a Fabry-Perot cavity. Applying a voltage to the piezoelectric crystals causes them to expand slightly, which changes the spacing between the dielectric mirrors. Thus, the resonant frequency of the device can be changed by adjusting the spacing in the cavity. These devices have narrowband tuning capabilities over a wide spectral

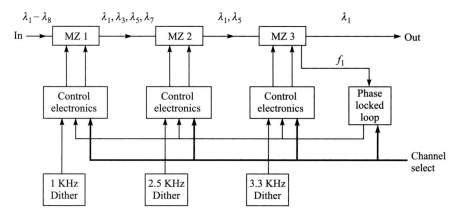

FIGURE 10-25
Three-stage tunable MZI filter. (Reproduced with permission from Wooten et al.,[65] © 1996, IEEE.)

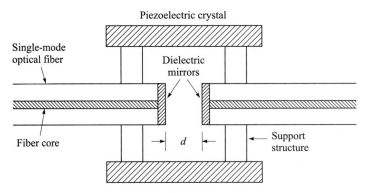

FIGURE 10-26
Operational setup of an optical-fiber Fabry-Perot tunable filter.

range (e.g., 100 channels with 5-GHz bandpasses), but are relatively slow, with switching speeds from one wavelength to another being on the order of $100\,\mu$s.

- *Tunable waveguide arrays* are based on a combination of waveguide grating routers and either optical amplifiers[72] or optical switches[73] at each array output. By appropriately biasing the optical amplifiers to increase or attenuate spectral components, or through specific on/off settings of optical switches, each WDM channel can be selected to either pass through the filter, be equalized in amplitude with the other wavelengths, or be dropped (channelled off) to a receiver. Channel bandwidths of 60 GHz (about 0.5 nm) with channel spacings of 195 GHz have been demonstrated.

- *Liquid-crystal Fabry-Perot filters* are based on the use of high-speed electro-clinic liquid crystals inside a Fabry-Perot cavity.[74,75] In this case, the liquid crystal is positioned between the two fiber end faces, and thus becomes part of the Fabry-Perot cavity. These filters can be widely tuned by applying a voltage across the crystal, which changes the refractive index, and hence the optical path length, in the cavity material. Switching times of less than $10\,\mu$s, a tuning range of 13 nm, and a channel spacing of 0.7 nm (88 GHz) have been achieved when the operating wavelength is centered at 1550 nm.

- *Tunable multigrating filters* can be used to add and drop any number of N different wavelengths.[47,76] Figure 10-27 illustrates the concept, which uses two three-port circulators with a series of N electrically tunable fiber-based reflection gratings placed between them. One grating is used for each wavelength in the system. The demultiplexer separates the dropped wavelengths into individual channels and the multiplexer combines wavelengths for transmission over the fiber trunk line.

    The device operates as follows: a series of up to N wavelengths enter port 1 of the left-hand circulator and exit at port 2. In the untuned state, each fiber grating is transparent to all wavelengths. However, once a grating is tuned to a

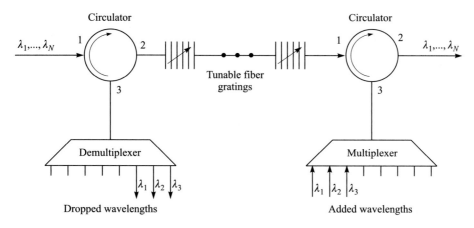

FIGURE 10-27
Multiple tunable fiber gratings used in conjunction with two optical circulators to add and drop any number of N different wavelengths.

specific wavelength, this light will be reflected back, re-enter the left-hand circulator through port 2, and exit from port 3 to the demultiplexer. This can be done for any desired number of channels. All remaining wavelengths that are not reflected pass through to the right-hand circulator. Here, they enter port 1 and exit from port 2. To add or reinsert wavelengths that were dropped, one injects these into port 3 of the right-hand circulator. They first come out of port 1 and travel toward the series of tuned fiber gratings. The tuned gratings reflect each wavelength so that they head back toward the right-hand circulator and pass through it to combine with the other wavelengths.

● *Acousto-optic tunable filters (AOTFs)* operate through the interaction of photons and acoustic waves in a solid such as lithium niobate. Figure 10-28 shows the basic operation.[77-80] Here, an acoustic transducer, which is modulated by a nominal 175-MHz radio-frequency (RF) signal, produces a surface acoustic wave in the $LiNbO_3$ crystal. This wave sets up an artificial grating in the solid, the grating period being determined by the frequency of the driving RF signal. More than one grating can be produced simultaneously by using a number of different driving frequencies. Input wavelengths that match the Bragg condition of the gratings are coupled to the second branch of the AOTF, while the other wavelengths continue on through. Analogous to multigrating filters, conventional AOTFs can pass one or more frequencies spaced more than 500 GHz apart, or they can pass one or several 4-nm spectral bands. Switching speeds are on the order of $10\,\mu s$ with tuning ranges of nominally 145 nm. A new type of tunable AOTF structure featuring triple-stage film-loaded waveguides has been demonstrated to operate with a 3-dB optical bandwidth of 0.37 nm, thus showing the possibility of applying AOTFs to ITU-T-recommended 0.8-nm-spaced systems.[79]

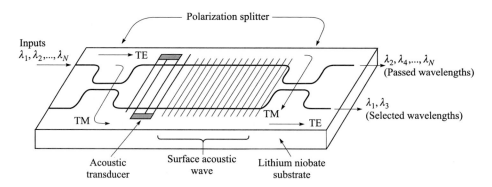

FIGURE 10-28

Basic construction and operation of an acousto-optic tunable filter. Here, λ_1 and λ_3 are selected to be dropped at a node and the other wavelengths pass through. (Adapted with permission from Smith et al.,[77] © 1990, IEEE.)

PROBLEMS

10-1. An optical transmission system is constrained to have 500-GHz channel spacings. How many wavelength channels can be utilized in the 1536-to-1556-nm spectral band?

10-2. A product sheet for a 2×2 single-mode biconical tapered coupler with a 40/60 splitting ratio states that the insertion losses are 2.7 dB for the 60-percent channel and 4.7 dB for the 40-percent channel.

(a) If the input power $P_0 = 200 \, \mu\text{W}$, find the output levels P_1 and P_2.

(b) Find the excess loss of the coupler.

(c) From the calculated values of P_1 and P_2, verify that the splitting ratio is 40/60.

10-3. Consider the coupling ratios as a function of pull lengths as shown in Fig. P10-3 for a fused biconical tapered coupler. The performances are given for 1310-nm and 1540-nm operation. Discuss the behavior of the coupler for each wavelength if its pull length is stopped at the following points: A, B, C, D, E, and each F.

10-4. Consider the 2×2 coupler shown in Fig. 10-6, where **A** and **B** are the matrices representing the field strengths of the input and output propagating waves, respectively. For a given input a_1, we impose the condition that there is no power emerging from the second input port; that is, $a_2 = 0$. Find expressions for the transmissivity T and the reflectivity R in terms of the elements s_{ij} in the scattering matrix **S** given in Eq. (10-8).

10-5. A 2×2 waveguide coupler has $\kappa = 0.4 \text{ mm}^{-1}$, $\alpha = 0.06 \text{ mm}^{-1}$, and $\Delta\beta = 0$. How long should it be to make a 3-dB power divider? If that length is doubled, what fraction of the input power emerges from the second channel?

10-6. Suppose we have two 2×2 waveguide couplers (couplers A and B) that have identical channel geometries and spacings, and are formed on the same substrate material. If the index of refraction of coupler A is larger than that of coupler B, which device has a larger coupling coefficient κ? What does this imply about the device lengths needed in each case to form a 3-dB coupler?

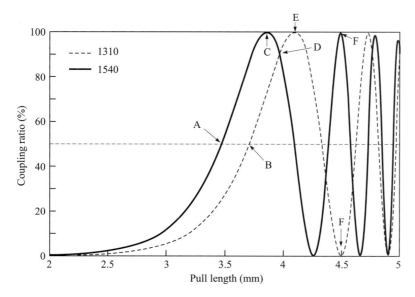

FIGURE P10-3

10-7. Measurements on a 7 × 7 star coupler yield the insertion losses from input port 1 to each output port shown in Table P10-7. Find the total excess loss through the coupler for inputs to port 1.

TABLE P10-7

Exit port no.	1	2	3	4	5	6	7
Insertion loss (dB)	9.33	7.93	7.53	9.03	9.63	8.64	9.04

10-8. Consider an optical fiber transmission star coupler that has seven inputs and seven outputs. Suppose the coupler is constructed by arranging the seven fibers in a circular pattern (a ring of six with one in the center) and butting them against the end of a glass rod that serves as the mixing element.
   (a) If the fibers have 50-μm core diameters and 125-μm outer cladding diameters, what is the coupling loss resulting from light escaping between the fiber cores? Let the rod diameter be 300 μm. Assume the fiber cladding is not removed.
   (b) What is the coupling loss if the fiber ends are arranged in a row and a 50-μm × 800-μm glass plate is used as the star coupler?

10-9. Repeat Prob. 10-8 for seven fibers that have 200-μm core diameters and 400-μm outer cladding diameters. What should the sizes of the glass rod and the glass plate be in this case?

10-10. Suppose an $N \times N$ star coupler is constructed of n 3-dB 2 × 2 couplers, each of which has a 0.1-dB excess loss. Find the maximum value of n and the maximum size N if the power budget for the star coupler is 30 dB.

10-11. Using Eq. (10-29) for the 2×2 coupler propagation matrix, derive the expressions for M_{11}, M_{12}, M_{21}, and M_{22} in Eq. (10-35). From this, find the more general expressions for the output powers given by Eqs. (10-38) and (10-39).

10-12. Consider the 4×4 multiplexer shown in Fig. 10-14.
(a) If $\lambda_1 = 1548$ nm and $\Delta\nu = 125$ GHz, what are the four input wavelengths?
(b) If $n_{\mathrm{eff}} = 1.5$, what are the values of ΔL_1 and ΔL_3?

10-13. Following the same line of analysis as in Example 10-6, use 2×2 Mach-Zehnder interferometers to design an 8-to-1 multiplexer that can handle a channel separation of 25 GHz. Let the shortest wavelength be 1550 nm. Specify the value of ΔL for the 2×2 MZIs in each stage.

10-14. A plane reflection grating can be used as a wavelength-division multiplexer when mounted as shown in Fig. P10-14. The angular properties of this grating are given by the grating equation

$$\sin\phi - \sin\theta = \frac{k\lambda}{n\Lambda}$$

where Λ is the grating period, k is the interference order, n is the refractive index of the medium between the lens and the grating, and ϕ and θ are the angles of the incident and reflected beams, respectively, measured normal to the grating.
(a) Using the grating equation, show that the angular dispersion is given by

$$\frac{d\theta}{d\lambda} = \frac{k}{n\Lambda\cos\theta} = \frac{2\tan\theta}{\lambda}$$

(b) If the fractional beam spread S is given by

$$S = 2(1 + m)\frac{\Delta\lambda}{\lambda}\tan^2\theta$$

where m is the number of wavelength channels, find the upper limit on θ for beam spreading of less than 1 percent given that $\Delta\lambda = 26$ nm, $\lambda = 1350$ nm, and $m = 3$.

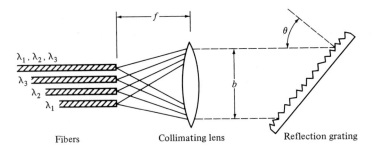

Fibers   Collimating lens   Reflection grating

FIGURE P10-14

10-15. On the same graph, make plots of the reflectivity R_{max} given by Eq. (10-48) and the transmissivity $T = 1 - R_{\mathrm{max}}$ for a fiber Bragg grating as a function of κL for $0 \leq \kappa L \leq 4$. If $\kappa = 0.75$ mm^{-1}, at what grating length does one get 93 percent reflectivity?

10-16. (This problem is best solved using a numerical program on a computer.) Based on coupled-mode theory, the reflectivity of a fiber grating is given by[42]

$$R = \frac{(\kappa L)^2 \sinh^2(SL)}{(\delta\beta L)^2 \sinh^2(SL) + (SL)^2 \cosh^2(SL)} \quad \text{for} \quad (\kappa L)^2 > (\delta\beta L)^2$$

and

$$R = \frac{(\kappa L)^2 \sin^2(QL)}{(\delta\beta L)^2 - (\kappa L)^2 \cos^2(QL)} \quad \text{for} \quad (\kappa L)^2 < (\delta\beta L)^2$$

where

$$SL = (\delta\beta L)\left[\left(\frac{\kappa L}{\delta\beta L}\right)^2 - 1\right]^{1/2} \quad \text{and} \quad QL = (\delta\beta L)\left[1 - \left(\frac{\kappa L}{\delta\beta L}\right)^2\right]^{1/2}$$

Here, $\delta\beta = \beta - p\pi/\Lambda = 2\pi n_{\text{eff}}/\lambda - p\pi/\Lambda$, with Λ being the grating period and p an integer. For values of $\kappa L = 1, 2, 3,$ and 4, plot $R(\kappa L)$ as a function of $\delta\beta L$ for the range $-10 \leq \delta\beta L \leq 10$. Note that R_{max} is found by setting $\delta\beta = 0$.

10-17. Using the expression for $R(\kappa L)$ given in Prob. 10-16, derive Eq. (10-49), which gives the full bandwidth $\Delta\lambda$ measured between the zeros on either side of R_{max}.

10-18. A 0.5-cm-long fiber Bragg grating is constructed by irradiating a single-mode fiber with a pair of 244-nm ultraviolet light beams. The fiber has $V = 2.405$ and $n_{\text{eff}} = 1.48$. The half-angle between the two beams is $\theta/2 = 13.5°$. If the photo-induced index change is 2.5×10^{-4}, find the following:

(*a*) the grating period,
(*b*) the Bragg wavelength,
(*c*) the coupling coefficient,
(*d*) the full bandwidth $\Delta\lambda$ measured between the zeros on either side of R_{max}
(*e*) the maximum reflectivity.

10-19. Show that Eq. (10-55) follows from the differentiation of Eq. (10-53) with respect to frequency.

10-20. Consider a waveguide grating multiplexer that has the values for the operational variables listed in Table P10-20.

(*a*) Find the waveguide length difference.
(*b*) Calculate the channel spacing $\Delta\nu$ and the corresponding pass wavelength differential $\Delta\lambda$.

TABLE P10-20

Symbol	Parameter	Value
L_f	Focal length	9.38 mm
λ_0	Center wavelength	1554 nm
n_c	Array channel index	1.451
n_g	Group index for n_c	1.475
n_s	Slab waveguide index	1.453
x	Input/output waveguide spacing	25 μm
d	Grating waveguide spacing	25 μm
m	Diffraction order	118

(c) What is the free spectral range for diagonally opposite ports in this device?

(d) Letting $\theta_i = jx/L_f$ and $\theta_o = kx/L_f$, what is the FSR for $j = 2$ and $k = 8$?

10-21. Consider a tunable DBR laser operating at 1550 nm that has a linewidth (frequency spread) of 1.25 GHz. If the maximum index change is 0.55 percent, how many wavelength channels can this laser provide if the channel spacing is 10 times the source spectral width?

10-22. Consider a GaInAsP MQW–DBR laser diode that has an effective index of 3.2.

(a) Find the grating period Λ that generates a 1550-nm output.

(b) If a 2-nm channel separation is desired, find the incremental change in Λ that is required between channels.

10-23. Consider a tunable 2×2 MZI that is constructed on an electro-optical crystal having an effective index of 1.5.

(a) If the MZI is used to combine two wavelength channels separated by 0.2 nm, find the required ΔL if the wavelength is centered at 1550 nm.

(b) Assume the parameter ΔL is varied by electrically modulating the refractive index of the electro-optical crystal. What is the index change needed if the waveguide length is 100 mm? Note that the effective optical path length L_{eff} in a waveguide is given by $L_{eff} = n_{eff} L$.

10-24. The transmitted optical intensity I_t in a tunable fiber Fabry-Perot filter is given by

$$I_t = \frac{I_i}{1 + \dfrac{4R \sin^2 kd}{(1 - R)^2}}$$

where I_i is the incident intensity, R is the reflectivity of the dielectric mirrors, k is the propagation constant of the lightwave, and d is the spacing of the mirrors. Then, $2kd$ is the phase of the wave for one roundtrip in the Fabry-Perot cavity. Plot the intensity I_t as a function of kd for values of $R = 0.04$, 0.3, and 0.9 over the range $0 \leq kd \leq 2\pi$.

10-25. The optical frequency ν_o and the acoustic frequency f_a (i.e., the RF driving frequency) in an AOTF are related by

$$\nu_o = \frac{c}{v_a} \left(\frac{1}{\Delta n} \right) f_a$$

where v_a is the speed of sound in the crystal and Δn is the birefringence of the crystal. Consider an AOTF constructed from TeO_2 in which $v_a = 670\,\text{m/s}$ and $\Delta n = 0.11$.

(a) What are the driving frequencies needed to deflect the following wavelengths: 1300 nm, 1546 nm, 1550 nm, and 1554 nm?

(b) What is the wavelength tuning sensitivity with respect to the driving frequency in nm/kHz?

REFERENCES

1. N. K. Cheung, K. Nosu, and G. Winzer (guest eds.), Special Issue on "Dense Wavelength Division Multiplexing," *IEEE J. Sel. Areas Commun.*, vol. 8, Aug. 1990.

2. G. E. Keiser, "A review of WDM technology and applications," *Opt. Fiber Technol.*, vol. 5, pp. 3–39, Jan. 1999.

3. K. Nosu, *Optical FDM Network Technologies*, Artech House, Boston, 1997.

4. J. P. Ryan, "WDM: North American deployment trends," *IEEE Commun. Mag.*, vol. 36, pp. 40–44, Feb. 1998.

5. E. Lowe, "Current European WDM deployment trends," *IEEE Commun. Mag.*, vol. 36, pp. 46–50, Feb. 1998.

6. (a) N. Fujiwara, M. S. Goodman, M. J. O'Mahoney, O. K. Tonguz, and A. E. Willner (guest eds.), Special Issue on "Multiwavelength Optical Technology and Networks," *J. Lightwave Tech.*, vol. 14, June 1996.
   (b) H. Yoshimura, K.-I. Sato, and N. Takachio, "Future photonic transport networks based on WDM technologies," *IEEE Commun. Mag.*, vol. 37, pp. 74–81, Feb. 1999.
   (c) J. M. Senior, M. R. Handley, and M. S. Leeson, "Developments in WDM access networking," *IEEE Commun. Mag.*, vol. 36, pp. 28–36, Dec. 1998.

7. C. A. Brackett, "Dense wavelength division multiplexing networks: Principles and applications," *IEEE J. Sel. Areas Commun.*, vol. 8, pp. 948–964, Aug. 1990.

8. ITU-T Recommendation G.692, *Optical Interfaces for Multichannel Systems with Optical Amplifiers*, Mar. 1997.

9. T. Koch, "Laser sources for amplified and WDM light systems," in I. P. Kaminow and T. L. Koch, eds., *Optical Fiber Telecommunication—III*, vol. B, Academic, New York, 1997, chap. 4, pp. 115–162.

10. E. Iannone, F. Matera, A. Mecozzi, and M. Settembre, *Nonlinear Optical Communication Networks*, Wiley, New York, 1998.

11. N. Shibata, K. Nosu, K. Ieashita, and Y. Azuma, "Transmission limitations due to fiber non-linearities in optical FDM systems," *IEEE J. Sel. Areas Commun.*, vol. 8, pp. 1068–1077, Aug. 1990.

12. A. R. Chraplyvy, "Limitations on lightwave communications imposed by optical-fiber nonlinearities," *J. Lightwave Tech.*, vol. 8, pp. 1548–1567, Oct. 1990.

13. F. Tong, "Multiwavelength receivers for WDM systems," *IEEE Commun. Mag.*, vol. 36, pp. 42–49, Dec. 1998.

14. E. Pennings, G.-D. Khoe, M. K. Smit, and T. Staring, "Integrated-optic versus micro optic devices for fiber-optic telecommunication systems: A comparison," *IEEE J. Sel. Topics Quantum Electron.*, vol. 2, pp. 151–164, June 1996.

15. (a) W. J. Tomlinson, "Passive and low-speed active optical components for fiber systems," in S. E. Miller and I. P. Kaminow, eds., *Optical Fiber Telecommunications—II*, Academic, New York, 1988, chap. 10.
    (b) J. Senior and S. D. Cusworth, "Devices for wavelength multiplexing and demultiplexing," *IEE Proc.*, vol. 136, pt. J, pp. 193–202, June 1989.

16. A. W. Snyder and J. D. Love, *Optical Wavelength Theory*, Chapman & Hall, New York, 1983.

17. A. Ankiewicz, A. W. Snyder, and X.-H. Zheng, "Coupling between parallel optical fiber cores— Critical examination," *J. Lightwave Tech.*, vol. 4, pp. 1317–1323, Sept. 1986.

18. V. J. Tekippe, "Passive fiber optic components made by the fused biconical taper process," *Fiber & Integrated Optics*, vol. 9, no. 2, pp. 97–123, 1990.

19. M. Eisenmann and E. Weidel, "Single-mode fused biconical couplers for WDM with channel spacing between 100 and 300 nm," *J. Lightwave Tech.*, vol. 6, pp. 113–119, Jan. 1988.

20. R. W. C. Vance and J. D. Love, "Back reflection from fused biconic couplers," *J. Lightwave Tech.*, vol. 13, pp. 2282–2289, Nov. 1995.

21. G. L. Abbas, V. W. S. Chan, and T. K. Yee, "A dual detector optical heterodyne receiver for local oscillator noise suppression," *J. Lightwave Tech.*, vol. 3, pp. 1110–1122, Oct. 1985.

22. H. A. Haus, *Waves and Fields in Optoelectronics*, Prentice Hall, New York, 1984.

23. J. Pietzsch, "Scattering matrix analysis of 3 × 3 fiber couplers," *J. Lightwave Tech.*, vol. 7, pp. 303–307, Feb. 1989.

24. P. E. Green, Jr., *Fiber Optic Networks*, Prentice-Hall, New York, 1993.

25. A. Takagi, K. Jinguji, and M. Kawachi, "Wavelength characteristics of 2 × 2 optical channel-type directional couplers with symmetric or nonsymmetric coupling structures," *J. Lightwave Tech.*, vol. 10, pp. 735–746, June 1992.

26. R. G. Hunsperger, *Integrated Optics: Theory and Technology*, Springer Verlag, Heidelberg, 4th ed., 1995.

27. A. Yariv, *Optical Electronics*, Saunders College Publ., Orlando, FL, 4th ed., 1991.

28. S. Srivastava, N. Gupta, M. Saini, and E. K. Sharma, "Power exchange in coupled optical waveguides," *J. Opt. Commun.*, vol. 18, no. 1, pp. 5–9, 1997.

29. J. W. Arkwright and D. B. Mortimore, "7 × 7 monolithic single-mode star coupler," *Electron. Lett.*, vol. 26, pp. 1534–1535, Aug. 30, 1990.

30. J. W. Arkwright, D. B. Mortimore, and R. M. Adams, "Monolithic 1 × 19 single-mode fused fiber couplers," *Electron. Lett.*, vol. 27, pp. 737–738, Apr. 1991.

31. M. E. Marhic, "Hierarchic and combinatorial star couplers," *Opt. Lett.*, vol. 9, pp. 368–370, Aug. 1984.

32. D. B. Mortimore, "Wavelength-flattened 8 × 8 single-mode star coupler," *Electron. Lett.*, vol. 22, pp. 1205–1206, Oct. 1986.

33. K. W. Fussgaenger and R. H. Rossberg, "Uni- and bidirectional 4λ × 560 Mb/s transmission systems using WDM devices based on wavelength-selective fused single-mode fiber couplers," *IEEE J. Sel. Areas Commun.*, vol. 8, pp. 1032–1042, Aug. 1990.

34. B. H. Verbeek, C. H. Henry, N. A. Olsson, K. J. Orlowsky, R. F. Kazarinov, and B. H. Johnson, "Integrated four-channel Mach-Zehnder multi/demultiplexer fabricated with phosphorous doped SiO$_2$ waveguides on Si," *J. Lightwave Tech.*, vol. 6, pp. 1011–1015, June 1988.

35. N. Takato et al., "Silica-based integrated optic Mach-Zehnder multi/demultiplexer family with channel spacing of 0.01–250 nm," *IEEE J. Sel. Areas Commun.*, vol. 8, pp. 1120–1127, Aug. 1990.

36. M. M.-K. Liu, *Principles and Applications of Optical Communications*, Irwin, Chicago, 1996.

37. R. Syms and J. Cozens, *Optical Guided Waves and Devices*, McGraw-Hill, New York, 1992.

38. K. O. Hill, Y. Fujii, D. C. Johnson, and B. S. Kawasaki, "Photosensitivity in optical fiber waveguides: Application to reflection filter fabrication," *Appl. Phys. Lett.*, vol. 32, pp. 647–649, 1978.

39. K. O. Hill, B. Malo, F. Bilodeau, and D. C. Johnson, "Photosensitivity in optical fibers," *Annu. Rev. Mater. Sci.*, vol. 23, pp. 125–157, 1993.

40. C. R. Giles, "Lightwave applications of fiber Bragg gratings," *J. Lightwave Tech.*, vol. 15, pp. 1391–1404, Aug. 1997.

41. (*a*) R. Kashyap, "Photosensitive optical fibers: Devices and applications," *Opt. Fiber Technol.*, vol. 1, pp. 17–34, Oct. 1994.

    (*b*) R. Kashyap, *Fiber Bragg Gratings*, Academic, New York, 1999.

42. I. Bennion, J. A. R. Williams, L. Zhang, K. Sugden, and N. J. Doran, "UV-written in-fibre Bragg gratings: A tutorial review." *Opt. Quantum Electron.*, vol. 28, pp. 93–135, Feb. 1996.

43. P.-Y. Fonjallaz, H. G. Limberger, and R. P. Salathé, "Bragg gratings with efficient and wavelength-selective fiber out-coupling," *J. Lightwave Tech.*, vol. 15, pp. 371–376, Feb. 1997.

44. T. Erdogan, "Fiber grating spectra," *J. Lightwave Tech.*, vol. 15, pp. 1277–1294, Aug. 1997.

45. G. P. Agrawal, *Nonlinear Fiber Optics*, Academic, New York, 2nd ed., 1995.

46. R. Ramaswami and K. N. Sivarajan, *Optical Networks*, Morgan Kaufmann, San Francisco, 1998.

47. Y. Fujii, "High-isolation polarization-independent optical circulator coupled with single-mode fibers," *J. Lightwave Tech.*, vol. 9, pp. 456–460, Apr. 1991.

48. M. K. Smit, "New focusing and dispersive planar components based on an optical phased array," *Electron. Lett.,* vol. 24, pp. 385–386, 1988.

49. M. K. Smit and C. van Dam, "PHASAR-based WDM devices: Principles, design and applications," *IEEE J. Sel. Topics Quantum Electron.*, vol. 2, pp. 236–250, June 1996.

50. C. Dragone, "An *N* × *N* star optical multiplexer using a planar arrangement of two star couplers," *IEEE Photonics Tech. Lett.*, vol. 3, pp. 812–815, Sept. 1991.

51. H. Takahashi, K. Oda, H. Toba, and Y. Inoue, "Transmission characteristics of arrayed wave-guide $N \times N$ wavelength multiplexers," *J. Lightwave Tech.*, vol. 13, pp. 447–455, Mar. 1995.

52. L. H. Spiekman, M. R. Amersfoort, A. H. de Vreede, F. P. G. M. van Ham, A. Kuntze, J. W. Pedersen, P. Demeester, and M. K. Smit, "Design and realization of polarization independent phased array wavelength demultiplexers," *J. Lightwave Tech.*, vol. 14, pp. 991–995, June 1996.

53. W. Lin, H. Li, Y. J. Chen, M. Dagenais, and D. Stone, "Dual-channel-spacing phased-array waveguide grating multi/demultiplexers," *IEEE Photonics Tech. Lett.*, vol. 8, pp. 1501–1503, Nov. 1996.

54. M. Zirngibl, "Multifrequency lasers and applications in WDM networks," *IEEE Commun. Mag.*, vol. 36, pp. 39–41, Dec. 1998.

55. S. Murata and I. Mito, "Tutorial Review: Frequency-tunable semiconductor lasers," *Opt. Quantum Electron.*, vol. 22, pp. 1–15, Jan. 1990.

56. M.-C. Amann and W. Thulke, "Continuously tunable laser diodes: Longitudinal versus transverse tuning scheme," *IEEE J. Sel. Areas Commun.*, vol. 8, pp. 1169–1177, Aug. 1990.

57. T.-P. Lee, "Recent advances in long-wavelength semiconductor lasers for optical communications," *Proc. IEEE*, vol. 79, pp. 253–276, Mar. 1991.

58. B. Mason, S. L. Lee, M. E. Heimbuch, and L. A. Coldren, "Directly modulated sampled grating DBR lasers for long-haul WDM communication systems," *IEEE Photonics Tech. Lett.*, vol. 9, pp. 377–379, Mar. 1997.

59. A. A. M. Staring, J. J. M. Binsma, P. I. Kuindersma, E. J. Jansen, P. J. A. Thijs, T. van Dongen, and G. F. G. Depovere, "Wavelength-independent output power from an injection-tunable DBR laser," *IEEE Photonics Tech. Lett.*, vol. 6, pp. 147–149, Feb. 1994.

60. T.-P. Lee, C.-E. Zah, R. Bhat, W. C. Young, B. Pathak, F. Favire, P. Lin, N. C. Andreadakis, C. Caneau, A. W. Rahjel, M. Koza, J. K. Gamelin, L. Curtis, D. D. Mahoney, and A. Lepore, "Multiwavelength DFB laser array transmitters for ONTC reconfigurable optical network testbed," *J. Lightwave Tech.*, vol. 14, pp. 967–976, June 1996.

61. Y. Tachikawa and K. Okamoto, "Arrayed waveguide grating lasers and their applications to tuning-free wavelength routing," *IEE Proc.—Optoelectron.*, vol. 143, pp. 322–328, Oct. 1996.

62. D. K. Jung, S. K. Shin, C.-H. Lee, and Y. C. Chung, "Wavelength-division-multiplexed passive optical network based on spectrum-slicing techniques," *IEEE Photonics Tech. Lett.*, vol. 10, pp. 1334–1336, Sept. 1998.

63. V. Arya and I. Jacobs, "Optical preamplifier receiver for spectrum-sliced WDM," *J. Lightwave Tech.*, vol. 15, pp. 576–583 Apr. 1997.

64. H. Kobrinski and K.-W. Cheung, "Wavelength-tunable optical filters: Applications and technology," *IEEE Commun. Mag.,* vol. 27, pp. 53–63, Oct. 1989.

65. E. L. Wooten, R. L. Stone, E. W. Miles, and E. M. Bradley, "Rapidly tunable narrowband wavelength filter using LiNbO$_3$ unbalanced Mach-Zehnder interferometers," *J. Lightwave Tech.*, vol. 14, pp. 2530–2536, Nov. 1996.

66. P. A. Humblet and W. M. Hamdy, "Crosstalk analysis and filter optimization of single- and double-cavity Fabry-Perot filters," *IEEE J. Sel. Areas Commun.*, vol. 8, pp. 1095–1107, Aug. 1990.

67. K.-P. Ho and J. M. Kahn, "Methods for crosstalk measurement and reduction in dense WDM systems," *J. Lightwave Tech.*, vol. 14, pp. 1127–1135, June 1996.

68. D. Brooks and S. Ruschin, "Integrated electro-optic multielectrode tunable filter," *J. Lightwave Tech.*, vol. 13, pp. 1508–1513, July 1995.

69. K. Oda, N. Yakato, T. Kominato, and H. Toba, "A 16-channel frequency selection switch for optical FDM distribution systems," *IEEE J. Sel. Areas Commun.*, vol. 8, pp. 1132–1140, Aug. 1990.

70. J. Stone and L. W. Stulz, "High-performance fiber Fabry-Perot filters," *Electron. Lett.*, vol. 27, pp. 2239–2240, Nov. 1991.

71. M. Born and E. Wolf, *Principles of Optics*, Pergamon, New York, 6th ed., 1980.

72. M. Zirngibl, C. H. Joyner, and B. Glance, "Digitally tunable channel-dropping filter/equalizer based on waveguide grating router and optical amplifier integration," *IEEE Photonics Tech. Lett.*, vol. 6, pp. 513–515, Apr. 1994.

73. O. Ishida, H. Takahashi, and Y. Inoue, "Digitally tunable optical filters using AWG multiplexers and optical switches," *J. Lightwave Tech.*, vol. 15, pp. 321–327, Feb. 1997.

74. A. Sneh and K. M. Johnson, "High-speed tunable liquid crystal filter for WDM networks," *J. Lightwave Tech.*, vol. 14, pp. 1067–1080, June 1996.

75. P.-L. Chen, K.-C. Lin, W.-C. Chuang, Y.-C. Tzeng, K.-Y. Lee, and W.-Y. Lee, "Analysis of a liquid crystal Fabry-Perot etalon filter: A novel model," *IEEE Photonics Tech. Lett.*, vol. 9, pp. 467–469, Apr. 1997.

76. S. Akiba and S. Yamamoto, "WDM undersea cable network technology for 100 Gb/s and beyond," *Opt. Fiber Technol.*, vol. 4, pp. 19–33, Jan. 1998.

77. D. A. Smith, J. E. Baran, J. J. Johnson, and K.-W. Cheung, "Integrated-optic acoustically tunable filters for WDM networks," *IEEE J. Sel. Areas Commun.*, vol. 8, pp. 1151–1159, Aug. 1990.

78. D. A. Smith et al., "Evolution of the acousto-optic wavelength routing switch," *J. Lightwave Tech.*, vol. 14, pp. 1005–1019, June 1996.

79. T. Nakazawa, M. Doi, S. Taniguchi, Y. Takasu, and M. Seino, "TiLi:NbO$_3$ AOTF for 0.8 nm channel-spaced WDM systems,"*OFC '98 Postdeadline Paper Proc.*, paper PD1, Feb. 1998.

80. K. Saitoh, M. Koshiba, and Y. Tsuji, "Numerical analysis of integrated acoustooptic tunable filters with weighted coupling," *J. Lightwave Tech.*, vol. 17, pp. 249–254, Feb. 1999.

CHAPTER
11

OPTICAL
AMPLIFIERS

Traditionally, when setting up an optical link, one formulates a power budget and adds repeaters when the path loss exceeds the available power margin. To amplify an optical signal with a conventional repeater, one performs photon-to-electron conversion, electrical amplification, retiming, pulse shaping, and then electron-to-photon conversion. Although this process works well for moderate-speed single-wavelength operation, it can be fairly complex and expensive for high-speed multiwavelength systems. Thus, a great deal of effort has been expended to develop all-optical amplifiers. These devices operate completely in the optical domain to boost the power levels of lightwave signals for the two long-wavelength transmission windows of optical fibers.[1-4]

This chapter first looks at the basic usage of optical amplifiers and classifies the two fundamental amplifier types: semiconductor optical amplifiers (SOAs) and doped-fiber amplifiers (DFAs). Section 11.2 discusses SOAs, which are based on the same operating principles as laser diodes. This discussion includes external pumping principles and gain mechanisms. Next, Sec. 11.3 gives details on erbium-doped fiber amplifiers (EDFAs), which are widely used in the 1550-nm window for optical communication networks. Noise effects are discussed in Sec. 11.4. The topic of Sec. 11.5 is applications of EDFAs in three fundamental configurations. Finally, Sec. 11.6 illustrates how SOAs are used as wavelength-converting devices for deployment in optical networks.

11.1 BASIC APPLICATIONS AND TYPES OF OPTICAL AMPLIFIERS

Optical amplifiers have found widespread use not only in long-distance point-to-point optical fiber links, but also in multi-access networks to compensate for signal-splitting losses. The features of optical amplifiers have led to many diverse applications, each having different design challenges. The basic types of optical amplifiers and their applications are described in this section.

11.1.1 General Applications

Figure 11-1 shows general applications of the following three classes of optical amplifiers:

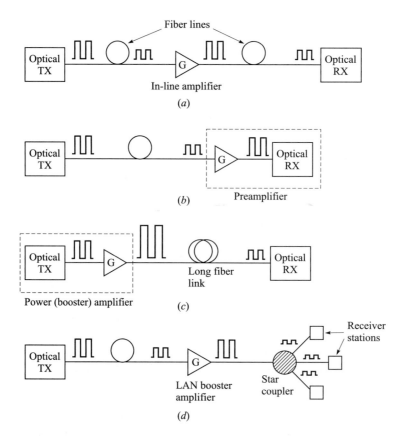

FIGURE 11-1
Four possible applications of optical amplifiers: (*a*) in-line amplifier to increase transmission distance, (*b*) preamplifier to improve receiver sensitivity, (*c*) booster of transmitted power, (*d*) booster of signal level in a local area network.

1. *In-line optical amplifiers.* In a single-mode link, the effects of fiber dispersion may be small so that the main limitation to repeater spacing is fiber attenuation. Since such a link does not necessarily require a complete regeneration of the signal, simple amplification of the optical signal is sufficient. Thus, an optical amplifier can be used to compensate for transmission loss and increase the distance between regenerative repeaters, as Fig. 11-1a illustrates.

2. *Preamplifier.* Figure 11-1b shows an optical amplifier being used as a front-end preamplifier for an optical receiver. Thereby, a weak optical signal is amplified before photodetection so that the signal-to-noise ratio degradation caused by thermal noise in the receiver electronics can be suppressed. Compared with other front-end devices such as avalanche photodiodes or optical heterodyne detectors, an optical preamplifier provides a larger gain factor and a broader bandwidth.

3. *Power amplifier.* Power or booster amplifier applications include placing the device immediately after an optical transmitter to boost the transmitted power, as Fig. 11-1c shows. This serves to increase the transmission distance by 10–100 km depending on the amplifier gain and fiber loss. As an example, using this boosting technique together with an optical preamplifier at the receiving end can enable repeaterless undersea transmission distances of 200–250 km. One can also employ an optical amplifier in a local area network as a booster amplifier to compensate for coupler-insertion loss and power-splitting loss. Figure 11-1d shows an example for boosting the optical signal in front of a star coupler.

11.1.2 Amplifier Types

The two main optical amplifier types can be classified as semiconductor optical amplifiers (SOAs) and active-fiber or doped-fiber amplifiers (DFAs). All optical amplifiers increase the power level of incident light through a stimulated emission process. The mechanism to create the population inversion that is needed for stimulated emission to occur is the same as is used in laser diodes. Although the structure of an optical amplifier is similar to that of a laser, it does not have the optical feedback mechanism that is necessary for lasing to take place. Thus, an optical amplifier can boost incoming signal levels, but it cannot generate a coherent optical output by itself. The basic operation is shown in Fig. 11-2. Here, the device absorbs energy supplied from an external source called the *pump*. The pump supplies energy to electrons in an active medium, which raises them to higher energy levels to produce a population inversion. An incoming signal photon will trigger these excited electrons to drop to lower levels through a stimulated emission process, thereby producing an amplified signal.

Alloys of semiconductor elements from groups III and V (e.g., phosphorus, gallium, indium, and arsenic) make up the active medium in SOAs. The attractiveness of SOAs is that they work in both the 1300-nm and the 1550-nm low-attenuation windows, they can easily be integrated on the same substrate as other

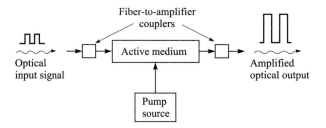

FIGURE 11-2
Basic operation of a generic optical amplifier.

optical devices and circuits (e.g., couplers, optical isolators, and receiver circuits), and compared with DFAs they consume less power, have fewer components, and are more compact. The SOAs have a more rapid gain response, which is on the order of 1 ps to 0.1 ns. This results in both advantages and limitations. The advantage is that SOAs can be implemented when both switching and signal processing are called for in optical networks. The limitation is that the rapid carrier response causes the gain at a particular wavelength to fluctuate with the signal rate for bit rates up to several Gb/s. Since this affects the overall gain, the signal gain at other wavelengths also fluctuates, which gives rise to crosstalk effects when a broad spectrum of wavelengths must be amplified.

In DFAs, the active medium for operation in the 1550-nm window is created by lightly doping a silica fiber core with rare-earth elements such as erbium (Er) or ytterbium (Yb). The DFAs for the 1300-nm window are achieved through doping fluoride-based fibers (rather than silica fibers) with elements such as neodymium (Nd) and praseodymium (Pr). The important features of DFAs include the ability to pump the devices at several different wavelengths, low coupling loss to the compatible fiber transmission medium, and very low dependence of gain on light polarization. In addition, DFAs are highly transparent to signal format and bit rate, since they exhibit slow gain dynamics, with carrier lifetimes on the order of 0.1–10 ms. The result is that, in contrast to SOAs, the gain responses of DFAs are basically constant for signal modulations greater than a few kilohertz. Consequently, they are immune from interference effects (such as crosstalk and intermodulation distortion) between different optical channels within a broad spectrum of wavelengths (e.g., in a 30-nm spectral band ranging from 1530 to 1560 nm) that are injected simultaneously into the amplifier.

11.2 SEMICONDUCTOR OPTICAL AMPLIFIERS

The two major types of SOAs are the resonant, Fabry-Perot amplifier (FPA) and the nonresonant, traveling-wave amplifier (TWA).[5-8] In an FPA, the two cleaved facets of a semiconductor crystal act as partially reflective end mirrors that form a Fabry-Perot cavity.[9,10] The natural reflectivity of the facets is approximately 32 percent. This is sometimes enhanced by means of a reflective dielectric coating deposited on the ends. When an optical signal enters the FPA, it gets amplified as it reflects back and forth between the mirrors until it is emitted at a higher intensity. Although FPAs are easy to fabricate, the optical signal gain is very sensitive to variations in amplifier temperature and input optical frequency. Thus, they require very careful stabilization of temperature and injection current.

The structure of a traveling-wave amplifier is the same as that of an FPA except that the end facets are either antireflection-coated or cleaved at an angle, so that internal reflection does not take place. Thus, the input light gets amplified only once during a single pass through the TWA. These devices have been used more widely than FPAs because they have a large optical bandwidth, high saturation power, and low polarization sensitivity. Since the 3-dB bandwidth of TWAs is about three orders of magnitude greater than that of FPAs, TWAs have become the SOA of choice for networking applications. In particular, TWAs are used as amplifiers in the 1300-nm window and as wavelength converters in the 1550-nm region. For most cases, recent literature on optical fiber systems uses the term "SOA," without qualification, for traveling-wave semiconductor optical amplifiers. In this section, we will concentrate on TWAs only.

11.2.1 External Pumping

External current injection is the pumping method used to create the population inversion needed for having a gain mechanism in SOAs. This is similar to the operation of laser diodes. Thus, from Eq. (4-31), the sum of the injection, stimulated-emission, and spontaneous-recombination rates gives the rate equation that governs the carrier density $n(t)$ in the excited state [5]

$$\frac{\partial n(t)}{\partial t} = R_p(t) - R_{st}(t) - \frac{n(t)}{\tau_r} \tag{11-1}$$

where

$$R_p(t) = \frac{J(t)}{qd} \tag{11-2}$$

is the external pumping rate from the injection current density $J(t)$ into an active layer of thickness d, τ_r is the combined time constant coming from spontaneous emission and carrier-recombination mechanisms, and

$$R_{st}(t) = \Gamma a v_g (n - n_{th}) N_{ph} \equiv g v_g N_{ph} \tag{11-3}$$

is the net stimulated emission rate. Here, v_g is the group velocity of the incident light, Γ is the optical confinement factor, a is a gain constant (which depends on the optical frequency v), n_{th} is the threshold carrier density, N_{ph} is the photon density, and g is the overall gain per unit length. Given that the active area of the optical amplifier has a width w and a thickness d, then for an optical signal of power P_s with photons of energy hv and group velocity v_g, the photon density is

$$N_{ph} = \frac{P_s}{v_g(hv)(wd)} \tag{11-4}$$

Example 11-1. Consider an InGaAsP SOA with $w = 5\,\mu$m and $d = 0.5\,\mu$m. Given that $v_g = 2 \times 10^8$ m/s , if a 1.0-μW optical signal at 1550 nm enters the device, then from Eq. (11-4) the photon density is

$$N_{\text{ph}} = \frac{1 \times 10^{-6}\,\text{W}}{(2 \times 10^8\,\text{m/s})\dfrac{(6.626 \times 10^{-34}\,\text{J}\cdot\text{s})(3 \times 10^8\,\text{m/s})}{1.55 \times 10^{-6}\,\text{m}}(5\,\mu\text{m})(0.5\,\mu\text{m})}$$

$$= 1.56 \times 10^{16}\ \text{photons/m}^3$$

In the steady state, $\partial n(t)/\partial t = 0$, so that Eq. (11-1) becomes

$$R_p = R_{\text{st}} + \frac{n}{\tau_r} \tag{11-5}$$

We now substitute Eq. (11-2) for R_p, the second equality in Eq. (11-3) for R_{st}, and the first equality in Eq. (11-3) solved for n into Eq. (11-5). Solving for g then yields the *steady-state gain per unit length*

$$g = \frac{\dfrac{J}{qd} - \dfrac{n_{\text{th}}}{\tau_r}}{v_g N_{\text{ph}} + 1/(\Gamma a \tau_r)} = \frac{g_0}{1 + N_{\text{ph}}/N_{\text{ph;sat}}} \tag{11-6}$$

where

$$N_{\text{ph;sat}} = \frac{1}{\Gamma a v_g \tau_r} \tag{11-7}$$

is defined as the *saturation photon density,* and

$$g_0 = \Gamma a \tau_r \left(\frac{J}{qd} - \frac{n_{\text{th}}}{\tau_r}\right) \tag{11-8}$$

is the medium gain per unit length in the absence of signal input (when the photon density is zero), which is called the *zero-signal* or *small-signal gain per unit length.*

Example 11-2. Consider the following parameters for a 1300-nm InGaAsP SOA:

Symbol	Parameter	Value
w	Active area width	$3\,\mu$m
d	Active area thickness	$0.3\,\mu$m
L	Amplifier length	$500\,\mu$m
Γ	Confinement factor	0.3
τ_r	Time constant	1 ns
a	Gain coefficient	$2 \times 10^{-20}\,\text{m}^2$
n_{th}	Threshold density	$1.0 \times 10^{24}\,\text{m}^{-3}$

(*a*) If a 100-mA bias current is applied to the device, then, from Eq. (11-2), the pumping rate is

$$R_p = \frac{J}{qd} = \frac{I}{qdwL} = \frac{0.1\,\text{A}}{(1.6 \times 10^{-19}\,\text{C})(0.3\,\mu\text{m})(3\,\mu\text{m})(500\,\mu\text{m})}$$

$$= 1.39 \times 10^{33}\,(\text{electrons/m}^3)/\text{s}$$

(b) From Eq. (11-8), the zero-signal gain is

$$g_0 = 0.3(2.0 \times 10^{-20}\,\text{m}^2)(1\,\text{ns})\left(1.39 \times 10^{33}\,\text{m}^{-3}\,\text{s}^{-1} - \frac{1.0 \times 10^{24}\,\text{m}^{-3}}{1.0\,\text{ns}}\right)$$

$$= 2340\,\text{m}^{-1} = 23.4\,\text{cm}^{-1}$$

11.2.2 Amplifier Gain

One of the most important parameters of an optical amplifier is the *signal gain* or *amplifier gain G*, which is defined as

$$G = \frac{P_{s,\text{out}}}{P_{s,\text{in}}} \tag{11-9}$$

where $P_{s,\text{in}}$ and $P_{s,\text{out}}$ are the input and output powers, respectively, of the optical signal being amplified. As noted in Chap. 4, the radiation intensity at a photon energy $h\nu$ varies exponentially with the distance traversed in a lasing cavity. Hence, using Eq. (4-23), the single-pass gain in the active medium of the SOA is

$$G = \exp[\Gamma(g_m - \bar{\alpha})L] \equiv \exp[g(z)L] \tag{11-10}$$

where Γ is the optical confinement factor in the cavity, g_m is the material gain coefficient, $\bar{\alpha}$ is the effective absorption coefficient of the material in the optical path, L is the amplifier length, and $g(z)$ is the overall gain per unit length.

Equation (11-10) shows that the gain increases with device length. However, the internal gain is limited by gain saturation.[11] This occurs because the carrier density in the gain region of the amplifier depends on the optical input intensity. As the input signal level is increased, excited carriers (electron–hole pairs) are depleted from the active region. When there is a sufficiently large optical input power, further increases in the input signal level no longer yield an appreciable change in the output level, since there are not enough excited carriers to provide an appropriate level of stimulated emission. We note here that the carrier density at any point z in the amplifying cavity depends on the signal level $P_s(z)$ at that point. In particular, near the input where z is small, incremental portions of the device may not have reached saturation at the same time as the sections further down the device, where incremental portions may be saturated because of higher values of $P_s(z)$.

An expression for the gain G as a function of the input power can be derived by examining the gain parameter $g(z)$ in Eq. (11-10). This parameter depends on the carrier density and the signal wavelength. Using Eqs. (11-4) and (11-6), we have that at a distance z from the input end, $g(z)$ is given by

$$g(z) = \frac{g_0}{1 + \dfrac{P_s(z)}{P_{\text{amp,sat}}}} \tag{11-11}$$

where g_0 is the unsaturated medium gain per unit length in the absence of signal input, $P_s(z)$ is the internal signal power at point z, and $P_{\text{amp,sat}}$ is the *amplifier saturation power*, which is defined as the internal power level at which the gain per unit length has been halved. Thus, the gain given by Eq. (11-10) decreases with increasing signal power. In particular, the gain coefficient in Eq. (11-11) is reduced by a factor of 2 when the internal signal power is equal to the amplifier saturation power.

Given that $g(z)$ is the gain per unit length, in an incremental length dz the light power increases by

$$dP = g(z)P_s(z)dz \tag{11-12}$$

Substituting Eq. (11-11) into Eq. (11-12) and rearranging terms gives

$$g_0(z)dz = \left(\frac{1}{P_s(z)} + \frac{1}{P_{\text{amp,sat}}} \right)dP \tag{11-13}$$

Integrating this equation from $z = 0$ to $z = L$ yields

$$\int_0^L g_0 \, dz = \int_{P_{s,\text{in}}}^{P_{s,\text{out}}} \left(\frac{1}{P_s(z)} + \frac{1}{P_{\text{amp,sat}}} \right)dP \tag{11-14}$$

Defining the single-pass gain in the absence of light to be $G_0 = \exp(g_0 L)$, and using Eq. (11-9), we then have

$$G = 1 + \frac{P_{\text{amp,sat}}}{P_{s,\text{in}}} \ln\left(\frac{G_0}{G} \right) \tag{11-15}$$

Figure 11-3 illustrates the dependence of the gain on the input power. Here, the zero-signal gain (or small-signal gain) is $G_0 = 30$ dB, which is a gain factor of 1000. The curve shows that as the input signal power is increased, the gain first stays near the small-signal level and then starts to decrease. After decreasing linearly in the gain saturation region, it finally approaches an asymptotic value of 0 dB (a unity gain) for high input powers. Also shown is the *output saturation power*, which is the point at which the gain is reduced by 3 dB (see Prob. 11-4).

11.3   ERBIUM-DOPED FIBER AMPLIFIERS

The active medium in an optical fiber amplifier consists of a nominally 10- to 30-m length of optical fiber that has been lightly doped (e.g., 1000 parts per million weight) with a rare-earth element, such as erbium (Er), ytterbium (Yb), neodymium (Nd), or praseodymium (Pr). The host fiber material can be either standard silica, a fluoride-based glass, or a multicomponent glass.

The operating regions of these devices depend on the host material and the doping elements. Fluorozirconate glasses doped with Pr or Nd are used for opera-

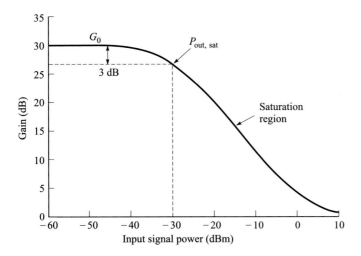

FIGURE 11-3

Typical dependence of the single-pass gain on optical input power for a small-signal gain of $G_0 = 30$ dB (a gain of 1000).

tion in the 1300-nm window, since neither of these ions can amplify 1300-nm signals when embedded in silica glass.[12-14] The most popular material for long-haul telecommunication applications is a silica fiber doped with erbium, which is known as an *erbium-doped fiber amplifier* or EDFA.[15-20] In some cases, Yb is added to increase the pumping efficiency and the amplifier gain.[21] The operation of an EDFA by itself normally is limited to the 1530-to-1560-nm region. However, when combined with a Raman fiber amplifier that boosts the gain at higher wavelengths, a 3-dB gain bandwidth of 75 nm has been achieved over the 1531-to-1616-nm region.[22] For simplicity of discussion in this section, we will use the designation "1550-nm signals" to refer to any particular optical channel in this spectral band.

11.3.1 Amplification Mechanism

Whereas semiconductor optical amplifiers use external current injection to excite electrons to higher energy levels, optical fiber amplifiers use *optical pumping*. In this process, one uses photons to directly raise electrons into excited states. The optical pumping process requires the use of three energy levels. The top energy level to which the electron is elevated must lie energetically above the desired lasing level. After reaching its excited state, the electron must release some of its energy and drop to the desired lasing level. From this level, a signal photon can then trigger it into stimulated emission, whereby it releases its remaining energy in the form of a new photon with a wavelength identical to that of the signal photon. Since the pump photon must have a higher energy than the signal photon, the pump wavelength is shorter than the signal wavelength.

To get a phenomenological understanding of how an EDFA works, we need to look at the energy-level structure of erbium.[15,16] The erbium atoms in silica are actual Er^{3+} ions, which are erbium atoms that have lost three of their outer electrons. In describing the transitions of the outer electrons in these ions to higher energy states, it is common to refer to the process as "raising the ions to higher energy levels." Figure 11–4 shows a simplified energy-level diagram and various energy-level transition processes of these Er^{3+} ions in silica glass. The two principal levels for telecommunication applications are a *metastable level* (the so-called $^4I_{13/2}$ level) and the $^4I_{11/2}$ *pump level*. The term "metastable" means that the lifetimes for transitions from this state to the ground state are very long compared with the lifetimes of the states that led to this level. [Note that, by convention, the possible states of a multielectron atom are referred to by the symbol $^{2S+1}L_J$, where $2S + 1$ is the spin multiplicity, L is the orbital angular momentum, and J is the total angular momentum.] The metastable, the pump, and the ground-state levels are actually bands of closely spaced energy levels that form a manifold due to the effect known as *Stark splitting*. Furthermore, each Stark level is broadened by thermal effects into an almost continuous band.

The metastable band is separated from the bottom of the $^4I_{15/2}$ ground-state level by an energy gap ranging from about 0.814 eV at the bottom of the meta-stable band (corresponding to a 1527-nm photon) to 0.841 eV at the top (corresponding to a 1477-nm photon). The energy band for the pump level exists at a 1.27-eV separation (corresponding to a 980-nm wavelength) from the ground state. The pump band is fairly narrow, so that the pump wavelength must be exact to within a few nanometers. The gap between the top of the $^4I_{15/2}$ level and the bottom of the metastable band is around 0.775 eV (1600 nm).

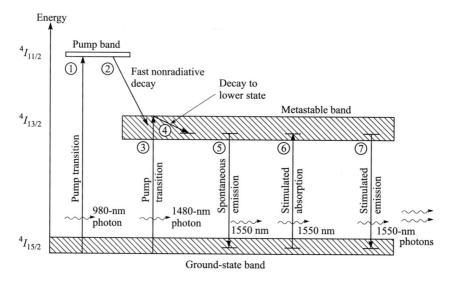

FIGURE 11-4
Simplified energy-level diagrams and various transition processes of Er^{3+} ions in silica.

In normal operation, a pump laser emitting 980-nm photons is used to excite ions from the ground state to the pump level, as shown by transition process 1 in Fig. 11-4. These excited ions decay (relax) very quickly (in about 1 μs) from the pump band to the metastable band, shown as transition process 2. During this decay, the excess energy is released as phonons or, equivalently, mechanical vibrations in the fiber. Within the metastable band, the electrons of the excited ions tend to populate the lower end of the band. Here, they are characterized by a very long fluorescence time of about 10 ms.

Another possible pump wavelength is 1480 nm. The energy of these pump photons is very similar to the signal-photon energy, but slightly higher. The absorption of a 1480-nm pump photon excites an electron from the ground state directly to the lightly populated top of the metastable level, as indicated by transition process 3 in Fig. 11-4. These electrons then tend to move down to the more populated lower end of the metastable level (transition 4). Some of the ions sitting at the metastable level can decay back to the ground state in the absence of an externally stimulating photon flux, as shown by transition process 5. This decay phenomenon is known as *spontaneous emission* and adds to the amplifier noise.

Two more types of transitions occur when a flux of signal photons that have energies corresponding to the band-gap energy between the ground state and the metastable level passes through the device. First, a small portion of the external photons will be absorbed by ions in the ground state, which raises these ions to the metastable level, as shown by transition process 6. Second, in the stimulated emission process (transition process 7) a signal photon triggers an excited ion to drop to the ground state, thereby emitting a new photon of the same energy, wavevector, and polarization as the incoming signal photon. The widths of the metastable and ground-state levels allow high levels of stimulated emissions to occur in the 1530-to-1560-nm range. Beyond 1560 nm, the gain decreases steadily until it reaches 0 dB (unity gain) at around 1616 nm.

11.3.2 EDFA Architecture

An optical fiber amplifier consists of a doped fiber, one or more pump lasers, a passive wavelength coupler, optical isolators, and tap couplers, as shown in Fig. 11-5. The dichroic (two-wavelength) coupler handles either 980/1550-nm or 1480/1550-nm wavelength combinations to couple both the pump and signal optical powers efficiently into the fiber amplifier. The tap couplers are wavelength-insensitive with typical splitting ratios ranging from 99:1 to 95:5. They are generally used on both sides of the amplifier to compare the incoming signal with the amplified output. The optical isolators prevent the amplified signal from reflecting back into the device, where it could increase the amplifier noise and decrease its efficiency.

The pump light is usually injected from the same direction as the signal flow. This is known as *codirectional pumping*. It is also possible to inject the pump power in the opposite direction to the signal flow, which is known as *counter-*

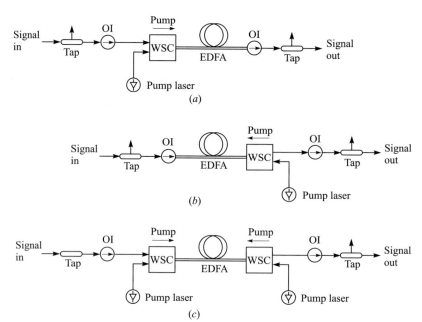

OI: Optical isolator
WSC: Wavelength-selective coupler

FIGURE 11-5
Three possible configurations of an EDFA: (*a*) codirectional pumping, (*b*) counterdirectional pumping, (*c*) dual pumping.

directional pumping. As shown in Fig. 11-5, one can employ either a single pump source or use *dual-pump schemes*, with the resultant gains typically being +17 dB and +35 dB, respectively. Counterdirectional pumping allows higher gains, but codirectional pumping gives better noise performance. In addition, pumping at 980 nm is preferred, since it produces less noise and achieves larger population inversions than pumping at 1480 nm.

11.3.3 EDFA Power-Conversion Efficiency and Gain

As is the case with any amplifier, as the magnitude of the output signal from an EDFA increases, the amplifier gain eventually starts to saturate. The reduction of gain in an EDFA occurs when the population inversion is reduced significantly by a large signal, thereby yielding the typical gain-versus-power performance curve shown in Fig. 11-3.

The input and output powers of an EDFA can be expressed in terms of the principle of energy conservation:[16]

$$P_{s,\text{out}} \leq P_{s,\text{in}} + \frac{\lambda_p}{\lambda_s} P_{p,\text{in}} \tag{11-16}$$

where $P_{p,\text{in}}$ is the input pump power, and λ_p and λ_s are the pump and signal wavelengths, respectively. The fundamental physical principle here is that the amount of signal energy that can be extracted from an EDFA cannot exceed the pump energy that is stored in the device. The inequality in Eq. (11-16) reflects the possibility of effects such as pump photons being lost due to various causes (such as interactions with impurities) or pump energy lost due to spontaneous emission.

From Eq. (11-16), we see that the maximum output signal power depends on the ratio λ_p/λ_s. For the pumping scheme to work, we need to have $\lambda_p < \lambda_s$, and, to have an appropriate gain, it is necessary that $P_{s,\text{in}} \ll P_{p,\text{in}}$. Thus, the *power conversion efficiency* (PCE), defined as

$$\text{PCE} = \frac{P_{s,\text{out}} - P_{s,\text{in}}}{P_{p,\text{in}}} \approx \frac{P_{s,\text{out}}}{P_{p,\text{in}}} \leq \frac{\lambda_p}{\lambda_s} \leq 1 \tag{11-17}$$

is less than unity. The maximum theoretical value of the PCE is λ_p/λ_s. For absolute reference purposes, it is helpful to use the *quantum conversion efficiency* (QCE), which is wavelength-independent and is defined by[16]

$$\text{QCE} = \frac{\lambda_s}{\lambda_p} \text{PCE} \tag{11-18}$$

The maximum value of QCE is unity, in which case all the pump photons are converted to signal photons.

We can also rewrite Eq. (11-16) in terms of the amplifier gain G. Assuming there is no spontaneous emission, then

$$G = \frac{P_{s,\text{out}}}{P_{s,\text{in}}} \leq 1 + \frac{\lambda_p}{\lambda_s} \frac{P_{p,\text{in}}}{P_{s,\text{in}}} \tag{11-19}$$

This shows an important relationship between signal input power and gain. When the input signal power is very large so that $P_{s,\text{in}} \gg (\lambda_p/\lambda_s)P_{p,\text{in}}$, then the maximum amplifier gain is unity. This means that the device is transparent to the signal. From Eq. (11-19), we also see that in order to achieve a specific maximum gain G, the input signal power cannot exceed a value given by

$$P_{s,\text{in}} \leq \frac{(\lambda_p/\lambda_s)P_{p,\text{in}}}{G - 1} \tag{11-20}$$

Example 11-3. Consider an EDFA being pumped at 980 nm with a 30-mW pump power. If the gain at 1550 nm is 20 dB, then, from Eq. (11-20), the maximum input power is

$$P_{s,\text{in}} \leq \frac{(980/1550)(30\,\text{mW})}{100 - 1} = 190\,\mu\text{W}$$

From Eq. (11-16), the maximum output power is

$$P_{s,out}(\max) = P_{s,in}(\max) + \frac{\lambda_p}{\lambda_s} P_{p,in} = 190\,\mu W + 0.63(30\,mW)$$

$$= 19.1\,mW = 12.8\,dBm$$

In addition to pump power, the gain also depends on the fiber length. The maximum gain in a three-level laser medium of length L, such as an EDFA, is given by

$$G_{\max} = \exp(\rho\sigma_e L) \qquad (11\text{-}21)$$

where σ_e is the signal-emission cross section and ρ is the rare-earth element concentration. When determining the maximum gain, Eqs. (11-19) and (11-21) must be considered together. Consequently, the maximum possible EDFA gain is given by the lowest of the two gain expressions:

$$G \leq \min\left\{\exp(\rho\sigma_e L), 1 + \frac{\lambda_p}{\lambda_s}\frac{P_{p,in}}{P_{s,in}}\right\} \qquad (11\text{-}22)$$

Since $G = P_{s,out}/P_{s,in} = \exp(\rho\sigma_e L)$, it follows similarly that the maximum possible EDFA output power is given by the minimum of the two expressions:

$$P_{s,out} \leq \min\left\{P_{s,in}\,\exp(\rho\sigma_e L), P_{s,in} + \frac{\lambda_p}{\lambda_s}P_{p,in}\right\} \qquad (11\text{-}23)$$

Figure 11-6 illustrates the onset of gain saturation for various doped-fiber lengths as the pumping power increases.[23] As the fiber length increases for low

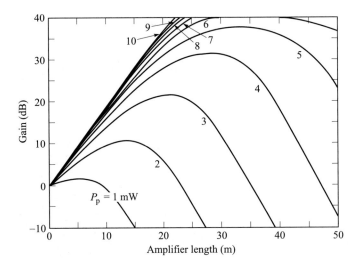

FIGURE 11-6
Calculation of the dependence of EDFA gain on fiber length and pump power for a 1480-nm pump and a 1550-nm signal. (Reproduced with permission from Giles and Desurvire,[23] © 1991, IEEE.)

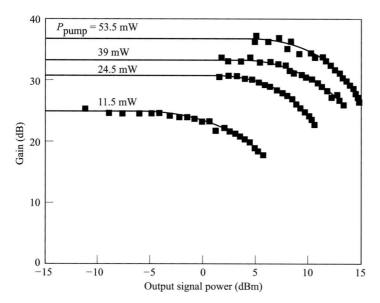

FIGURE 11-7

Gain behavior of an EDFA as a function of output signal power for various pump levels. (Reproduced with permission from Li,[24] © 1993, IEEE.)

pumping powers, the gain starts to decrease after a certain length because the pump does not have enough energy to create a complete population inversion in the downward portion of the amplifier. In this case, the unpumped region of the fiber absorbs the signal, thus resulting in signal loss rather than gain in that section.

Since the metastable level in an EDFA has a relatively long lifetime, it is possible to obtain very high saturated output powers. The *saturated output power* (the power at which gain saturation occurs) is defined as the 3-dB compression point of the small-signal gain.[24] For large signal operation, the saturated gain increases linearly with pump power, as can be inferred from Fig. 11-7. This figure shows that as the input power increases for a given pump level, the amplifier gain remains constant until saturation occurs.

■ **Example 11-4.** From Fig. 11-6 we see that for 1480-nm pumping, a 35-dB gain can be achieved with a pump power of 5 mW for an amplifier length of 30 m.

11.4 AMPLIFIER NOISE

The dominant noise generated in an optical amplifier is *amplified spontaneous emission* (ASE). The origin of this is the spontaneous recombination of electrons and holes in the amplifier medium (transition 5 in Fig. 11-4). This recombination gives rise to a broad spectral background of photons that get amplified along with the optical signal. This effect is shown in Fig. 11-8 for an EDFA amplifying a

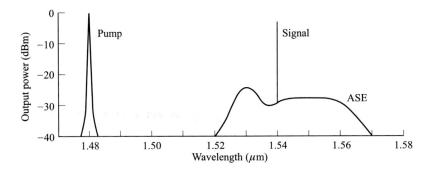

FIGURE 11-8

Representative 1480-nm pump spectrum and a typical output signal at 1540 nm with the associated amplified-spontaneous-emission (ASE) noise.

signal at 1540 nm. The spontaneous noise can be modeled as a stream of random infinitely short pulses that are distributed all along the amplifying medium. Such a random process is characterized by a noise power spectrum that is flat with frequency. The power spectral density of the ASE noise is[7]

$$S_{ASE}(f) = h\nu n_{sp}[G(f) - 1] = P_{ASE}/\Delta\nu_{opt} \qquad (11\text{-}24)$$

where P_{ASE} is the ASE noise power in an optical bandwidth $\Delta\nu_{opt}$ and n_{sp} is the *spontaneous-emission* or *population-inversion factor* is defined as

$$n_{sp} = \frac{n_2}{n_2 - n_1} \qquad (11\text{-}25)$$

where n_1 and n_2 are the fractional densities or populations of electrons in states 1 and 2, respectively. Thus, n_{sp} denotes how complete the population inversion is between two energy levels. From Eq. (11-25) $n_{sp} \geq 1$, with equality holding for an ideal amplifier when the population inversion is complete. Typical values range from 1.4 to 4, depending on the wavelength and the pumping rate.

The ASE noise level depends on whether codirectional or counterdirectional pumping is used. Figure 11-9 shows experimental and calculated data on ASE noise versus pump power for different EDFA lengths.[18]

Since ASE originates ahead of the photodiode, it gives rise to three different noise components in an optical receiver in addition to the normal thermal noise of the photodetector. This occurs because the photocurrent consists of a number of beat signals between the signal and the optical noise fields, in addition to the squares of the signal field and the spontaneous-emission field. If the total optical field is the sum of the signal field E_s and the spontaneous-emission field E_n, then the total photodetector current i_{tot} is proportional to the square of the electric field of the optical signal: $i_{tot} \propto (E_s + E_n)^2 = E_s^2 + E_n^2 + 2E_s \cdot E_n$. Here the first two terms arise purely from the signal and noise, respectively. The third term is a mixing component (a *beat signal*) between the signal and noise, which can fall within the bandwidth of the receiver and degrade the signal-to-noise ratio. First,

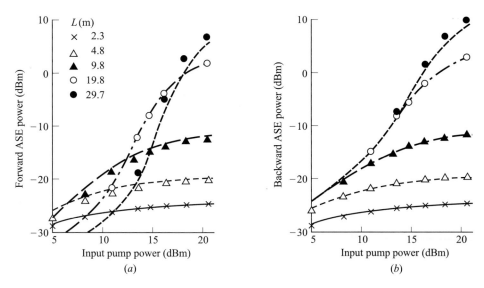

FIGURE 11-9
Experimental and theoretical ASE noise powers versus input pump power for various EDFA lengths arising from (a) codirectional (forward) pumping and (b) counterdirectional (backward) pumping. (Reproduced with permission from Pedersen et al.,[18] © 1990, IEEE.)

taking the ASE photons into account, the optical power incident on the photodetector becomes $P_0 = GP_{s,\text{in}} + S_{\text{ASE}}\Delta\nu_{\text{opt}}$. Note that $\Delta\nu_{\text{opt}}$ can be reduced significantly if an optical filter precedes the photodetector. Substituting this expression for P_0 into Eq. (6-6) then yields the total mean-square shot-noise current

$$\langle i_{\text{shot}}^2 \rangle = \sigma_{\text{shot}}^2 = \sigma_{\text{shot-S}}^2 + \sigma_{\text{shot-ASE}}^2 = 2q\mathscr{R}GP_{s,\text{in}}B + 2q\mathscr{R}S_{\text{ASE}}\,\Delta\nu_{\text{opt}}B \quad (11\text{-}26)$$

where B is the front-end receiver electrical bandwidth.

The other two noises arise from the mixing of the different optical frequencies contained in the light signal and the ASE, which generates two sets of beat frequencies. Since the signal and the ASE have different optical frequencies, the beat noise of the signal with the ASE is

$$\sigma_{s\text{-ASE}}^2 = 4(\mathscr{R}GP_{s,\text{in}})(\mathscr{R}S_{\text{ASE}}B) \quad (11\text{-}27)$$

In addition, since the ASE spans a wide optical frequency range, it can beat against itself giving rise to the noise current (see Sec. 9.3.1)

$$\sigma_{\text{ASE-ASE}}^2 = \mathscr{R}^2 S_{\text{ASE}}^2 (2\Delta\nu_{\text{opt}} - B)B \quad (11\text{-}28)$$

The total mean-square receiver noise current then becomes

$$\langle i_{\text{total}}^2 \rangle = \sigma_{\text{total}}^2 = \sigma_T^2 + \sigma_{\text{shot-}s}^2 + \sigma_{\text{shot-ASE}}^2 + \sigma_{s\text{-ASE}}^2 + \sigma_{\text{ASE-ASE}}^2 \quad (11\text{-}29)$$

where the thermal noise variance σ_T^2 is given by Eq. (6-17).

The last four terms in Eq. (11-29) tend to be of similar magnitudes when the optical bandwidth $\Delta\nu_{\text{opt}}$ is taken to be the optical bandwidth of the spontaneous emission noise, which covers a 30-nm spectrum (see Prob. 11-7). However, one generally uses a narrow optical filter at the receiver, so that $\Delta\nu_{\text{opt}}$ is on the order of 125 GHz (a 1-nm spectral width at 1550 nm) or less. In that case, we can simplify Eq. (11-29) by examining the magnitudes of the various noise components. First, the thermal noise can generally be neglected when the amplifier gain is large enough. Furthermore, since the amplified signal power $GP_{s,\text{in}}$ is much larger than the ASE noise power $S_{\text{ASE}}\Delta\nu_{\text{opt}}$, the ASE–ASE beat noise given by Eq. (11-28) is significantly smaller than the signal–ASE beat noise. This observation reduces Eq. (11-26) to

$$\sigma_{\text{shot}}^2 \approx 2q\mathcal{R}GP_{s,\text{in}}B \tag{11-30}$$

Using these results together with the expression for S_{ASE} from Eq. (11-24) yields the following approximate signal-to-noise ratio (S/N) at the photodetector output:

$$\left(\frac{S}{N}\right)_{\text{out}} = \frac{\sigma_{\text{ph}}^2}{\sigma_{\text{total}}^2} = \frac{\mathcal{R}^2 G^2 P_{s,\text{in}}^2}{\sigma_{\text{total}}^2} \approx \frac{\mathcal{R}P_{s,\text{in}}}{2qB}\frac{G}{1+2\eta n_{\text{sp}}(G-1)} \tag{11-31}$$

where η is the quantum efficiency of the photodetector and, from Eq. (6-11), the mean-square input photocurrent is

$$\langle i_{\text{ph}}^2 \rangle = \sigma_{\text{ph}}^2 = \mathcal{R}^2 G^2 P_{s,\text{in}}^2 \tag{11-32}$$

Note that the term

$$\left(\frac{S}{N}\right)_{\text{in}} = \frac{\mathcal{R}P_{s,\text{in}}}{2qB} \tag{11-33}$$

in Eq. (11-31) is the signal-to-noise ratio of an ideal photodetector at the input to the optical amplifier. From Eq. (11-31) we can then find the noise figure of the optical amplifier, which is a measure of the S/N degradation experienced by a signal after passing through the amplifier. Using the standard definition of *noise figure* as the ratio between the S/N at the input and the S/N at the amplifier output, we have

$$\text{Noise figure} = F = \frac{(S/N)_{\text{in}}}{(S/N)_{\text{out}}} = \frac{1+2\eta n_{\text{sp}}(G-1)}{\eta G} \tag{11-34}$$

When G is large, this becomes $2\eta n_{\text{sp}}$. A perfect amplifier would have $n_{\text{sp}} = 1$, yielding a noise figure of 2 (or 3 dB), assuming $\eta = 1$. That is, using an ideal receiver with a perfect amplifier would degrade the S/N by a factor of 2. In a real EDFA, for example, n_{sp} is around 2, so the input S/N gets reduced by a factor of about 4.

Example 11-5. Figure 11-10 shows measured values of the noise figure for an EDFA under gain saturation for both codirectional and counterdirectional pumping.[25] The pump wavelength was 1480 nm and the signal wavelength was 1558 nm with an input

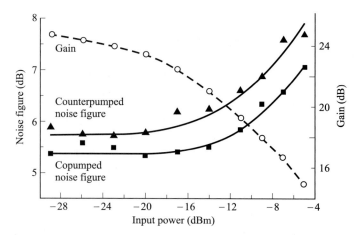

FIGURE 11-10
Measured EDFA noise figure under gain saturation for codirectional pumping and counterdirectional pumping at 1480 nm. The gain was similar for both pump directions. (Reproduced with permission from Walker et al.,[25] © 1991, IEEE.)

power to the amplifier of −60 dBm. Under small-signal conditions, the codirectional pumping noise was about 5.5 dB, which included a 1.5-dB input coupling loss. The noise figure of the optical amplifier itself was thus 4 dB, compared with the theoretical minimum of 3 dB with complete population inversion. The noise figure in the counterdirectional pumping case was about 1 dB higher.

11.5 SYSTEM APPLICATIONS

In designing an optical fiber link that requires optical amplifiers, there are three possible locations where the amplifiers can be placed, as shown in Fig. 11-1. Although the physical amplification process is the same in all three configurations, the various uses require operation of the device over different input power ranges. This, in turn, implies use of different amplifier gains. The complete analysis of the signal-to-noise ratios, taking into account factors such as detailed photon statistics, and discrete amplifier configurations, are fairly involved. Desurvire[16] gives an extensive treatment for readers who need more detail. Here, we will look at simple conceptual analysis and present generic operational values for the three possible locations of EDFAs in an optical link.

11.5.1 Power Amplifiers

For the power amplifier, the input power is high, since the device immediately follows an optical transmitter. High pump powers are normally required for this application.[26] The amplifier inputs are generally −8 dBm or greater, and the power amplifier gain must be greater than 5 dB in order for it to be more advantageous than using a preamplifier at the receiver.

Example 11-6. Consider an EDFA which is used as a power amplifier with a 10-dB gain. Assume the amplifier input is a 0-dBm level from a laser diode transmitter. If the pump wavelength is 980 nm, then from Eq. (11-16), for a 10-dBm output at 1540 nm, the pump power must be at least

$$P_{p,\text{in}} \geq \frac{\lambda_s}{\lambda_p}(P_{s,\text{out}} - P_{s,\text{in}}) = \frac{1540}{980}(10\,\text{mW} - 1\,\text{mW}) = 14\,\text{mW}$$

11.5.2 In-Line Amplifiers

In a long transmission system, optical amplifiers are needed to periodically restore the power level after it has decreased due to attenuation in the fiber. Normally, the gain of each EDFA in this amplifier chain is chosen to compensate exactly for the signal loss incurred in the preceding fiber section of length L, that is, $G = \exp(+\alpha L)$. The accumulated ASE noise is the dominant degradation factor in such a cascaded chain of amplifiers.

Example 11-7. Consider Fig. 11-11, which shows the values of the per-channel signal power, the per-channel ASE noise, and the SNR along a chain of seven optical amplifiers in a WDM link. The input signal level starts out at 6 dBm and decays due to fiber attenuation as it travels along the link. When its power level has dropped to −24 dBm, it gets boosted back to 6 dBm by an optical amplifier. For a given channel transmitted over the link, the SNR starts out at a high level and then decreases at each amplifier as the ASE noise accumulates through the length of the link. For example, following amplifier number 1, the SNR is 28 dB for a 6-dBm

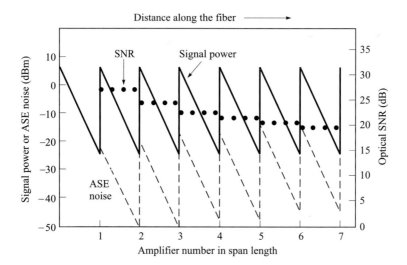

FIGURE 11-11
SNR degradation as a function of link distance over which the ASE noise increases with the number of amplifiers. The curves show the signal level (solid lines), the ASE noise level (dashed lines), and the SNR (dotted lines) for a single channel in a WDM link.

amplified signal level and a −22-dBm ASE noise level. After amplifier number 4, the SNR is 22 dB for a 6-dBm amplified signal level and a −16-dBm ASE noise level. The higher the gain in the amplifier, the faster the ASE noise builds up. However, although the SNR decreases quickly in the first few amplifications, the incremental effect of adding another EDFA diminishes rapidly with an increasing number of amplifiers. As a consequence, although the SNR drops by 3 dB when the number of EDFAs increases from one to two, it also drops by 3 dB when the number of amplifiers is increased from two to four, and by another 3 dB when the number of amplifiers is further increased to eight.

To compensate for the accumulated ASE noise, the signal power must increase at least linearly with the length of the link in order to keep a constant signal-to-noise ratio. If the total system length is $L_{tot} = NL$ and the system contains N optical amplifiers each having a gain $G = \exp(+\alpha L)$, then, using Eq. (11-24), the path-averaged ASE power along a chain of optical amplifiers is[16,27]

$$\langle P_{ASE} \rangle_{path} = \frac{N P_{ASE}}{L} \int_0^L \exp(+\alpha z) \, dz = \alpha L_{tot} \, h\nu \, n_{sp} F_{path}(G) \Delta\nu_{opt} \qquad (11\text{-}35)$$

where α is the fiber attenuation and $F_{path}(G)$ is a penalty factor defined as

$$F_{path}(G) = \frac{1}{G} \left(\frac{G-1}{\ln G} \right)^2 \qquad (11\text{-}36)$$

Basically, $F_{path}(G)$ gives the factor by which the path-average signal energy must be increased (as G increases) in a chain of N cascaded optical amplifiers to maintain a fixed S/N. These optical amplifiers should be placed uniformly along the transmission path to yield the best combination of overall gain and final S/N. The input power levels for these in-line amplifiers nominally ranges from −26 dBm (2.5 μW) to −9 dBm (125 μW), with gains generally greater than 15 dB.

Example 11-8. Consider an optical transmission path containing N cascaded optical amplifiers, each having a 30-dB gain. If the fiber has a loss of 0.2 dB/km, then the span between optical amplifiers is 150 km if there are no other system impairments. As an example, for a 900-km link we would need five amplifiers. From Eq. (11-36), the noise penalty factor over the total path is (in decibels)

$$10 \log F_{path}(G) = 10 \log \left[\frac{1}{1000} \left(\frac{1000-1}{\ln 1000} \right)^2 \right] = 10 \log 20.9 = 13.2 \, \text{dB}$$

If we reduce the gain to 20 dB, then the impairment-free transmission distance is 100 km for which we need eight amplifiers. In this case, the noise-penalty factor is

$$10 \log F_{path}(G) = 10 \log \left[\frac{1}{100} \left(\frac{100-1}{\ln 100} \right)^2 \right] = 10 \log 4.62 = 6.6 \, \text{dB}$$

11.5.3 Preamplifiers

An optical amplifier can be used as a preamplifier to improve the sensitivity of direct-detection receivers that are limited by thermal noise.[28] First, assume the receiver noise is represented by the electrical power level N. Let S_{min} be the minimum value of the electrical signal power S that is required for the receiver to perform with a specific acceptable bit-error rate. The acceptable signal-to-noise ratio then is S_{min}/N. If we now use an optical preamplifier with gain G, the electrical received signal power is $G^2 S'$ and the signal-to-noise ratio is

$$\left(\frac{S}{N}\right)_{preamp} = \frac{G^2 S'}{N + N'} \tag{11-37}$$

where the noise term N' is the spontaneous emission from the optical preamplifier that gets converted by the photodiode in the receiver to an additional background noise. If S'_{min} is the new minimum detectable electrical signal level needed to maintain the same signal-to-noise ratio, then we need to have

$$\frac{G^2 S'_{min}}{N + N'} = \frac{S_{min}}{N} \tag{11-38}$$

For an optical preamplifier to enhance the received signal level, we must have $S'_{min} < S_{min}$, so that

$$\frac{S_{min}}{S'_{min}} = G^2 \frac{N}{N + N'} > 1 \tag{11-39}$$

This ratio of S_{min} to S'_{min} represents the *improvement of minimum detectable signal* or *detector sensitivity*.

Example 11-9. Consider an EDFA used as an optical preamplifier. Assume that N is due to thermal noise and that the noise N' introduced by the preamplifier is dominated by signal–ASE beat noise. We want to see under what conditions Eq. (11-39) holds. For sufficiently high gain G, Eq. (11-39) becomes

$$G^2 - 1 \approx G^2 > \frac{N'}{N} \approx \frac{\sigma^2_{s\text{-ASE}}}{\sigma^2_T}$$

Substituting Eqs. (6-17) and (11-27) into this expression, using Eq. (11-24) for S_{ASE}, and solving for $P_{s,in}$ yields

$$P_{s,in} < \frac{k_B T \, h\nu}{R n_{sp} \eta^2 q^2}$$

If $T = 300$ K, $R = 50\,\Omega$, $\lambda = 1550$ nm, $n_{sp} = 2$, and $\eta = 0.65$, then $P_{s,in} < 490\,\mu$W. This level is much higher than any expected received signal, so the condition in Eq. (11-39) is always satisfied. However, note that this only specifies the upper bound on $P_{s,in}$. It does not mean that by making G sufficiently high, the improvement in sensitivity can be made arbitrarily large, since there is a minimum received optical power level that is needed to achieve a specific BER (see Prob. 11-15).

11.5.4 Multichannel Operation

An advantage of both semiconductor optical amplifiers and EDFAs is their ability to amplify multiple optical channels, provided the bandwidth of the multichannel signal is smaller than the amplifier bandwidth.[29] For both SOAs and

EDFAs, this bandwidth ranges from 1 to 5 THz. A disadvantage of SOAs is their sensitivity to interchannel crosstalk arising from carrier-density modulation due to beating of signals from adjacent optical channels.[30] For SOAs, this beating occurs whenever the channel spacing is less than 10 GHz.

This crosstalk does not occur in EDFAs, as long as the channel spacing is greater than 10 kHz, which holds in practice. Thus, EDFAs are ideally suited for multichannel amplification. For multichannel operation in an EDFA, the signal power for N channels is given by

$$P_s = \sum_{i=1}^{N} P_{s,i} \qquad (11\text{-}40)$$

where $P_{s,i}$ is the signal power in channel i; that is, at the optical carrier frequency ν_i.

Another characteristic of an EDFA is that its gain is wavelength-dependent in its normal operating window of 1530–1560 nm.[31] If it is not equalized over the spectral range of operation in a multichannel system, this gain variation will create a large signal-to-noise ratio differential among the channels after passing through a cascade of EDFAs. Numerous techniques, such as the use of gain-compensating fiber gratings, have been tried for this equalization.[32–34] In addition, when combined with a Raman fiber amplifier that boosts the gain at higher wavelengths, a gain that is flat to within 3 dB can be extended to 1616 nm. Figure 11-12 shows typical results for two commercially available EDFAs that have been gain-compensated in the 1528-to-1563-nm and the 1568-to-1603-nm spectral bands, respectively.

In the preceding discussions and analyses, we have looked only at transmission in one direction along a fiber. However, bidirectional propagation is also possible in a fiber link carrying multiple wavelengths in each direction through a cascaded chain of EDFAs.[35]

11.5.5 In-Line Amplifier Gain Control

In a long-distance fiber transmission system using optical amplifiers, it is desirable to keep the output power of the in-line amplifiers constant when there are fluctuations in the input power level.[16] For example, such fluctuations could occur from loss variations in the optical cable or from degradation in a preceding optical amplifier. Changes in the number of channels caused by network reconfiguration will also produce a power transition in the optical amplifier output.

A practical way to keep the output power constant is to operate the optical amplifier in the gain-compression (saturation) region, as shown in Fig. 11-13. In this *signal-controlled automatic gain control* (AGC) method, when the input power to the amplifier decreases, the gain becomes higher to yield a higher output power. Conversely, if the input power increases, the gain drops to compensate for this variation. The exact amount of compensation depends on the relationship between the gain and the input power (as given by the slope of the curve in the

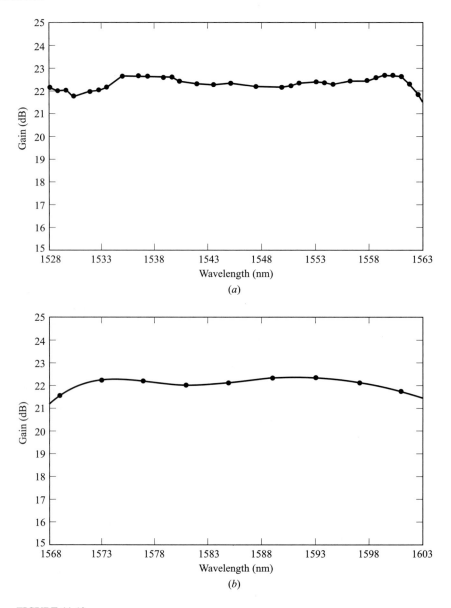

FIGURE 11-12
(*a*) The gain of a commercially available EDFA that has been flattened over the 1528-to-1563-nm spectral band. (*b*) The gain of a commercially available EDFA that has been flattened over the 1568-to-1603-nm spectral band. (Curves are provided courtesy of AFC Technologies, Inc.; www.afctek.com).

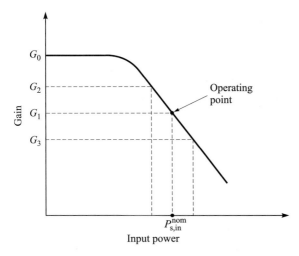

FIGURE 11-13
A passive signal-controlled gain control method in which the amplifier operates in the saturation region with a nominal input $P_{s,in}^{nom}$. A decrease in the input power will raise the gain toward G_2, whereas an input-power increase will lower the gain toward G_3.

saturation region).[19,36-38] In a cascaded chain of optical amplifiers, if the span loss between amplifiers happens to increase somewhere so that the input power to an optical amplifier is reduced, the signal power is largely restored in the following several amplifiers. This is known as a *self-healing effect* in optically amplified communication systems.

The limitation of the signal-controlled AGC scheme is that the gain is low, since the amplifier operates in the saturation region. More versatile *dynamic gain control* methods that keep the per-wavelength output power constant when the number of channels changes have also been demonstrated.[39-41] For example, in one dynamic AGC implementation, when a channel is suddenly dropped, the amplifier output power is restored to its original level within 1 ms with the transient level change being less than 0.5 dB.

11.6 WAVELENGTH CONVERTERS

An optical wavelength converter is a device that can directly translate information contained on an incoming wavelength to a new wavelength without entering the electrical domain. This is an important component in all-optical networks, since the wavelength of the incoming signal may already be in use by another information channel residing on the destined outgoing path. Converting the incoming signal to a new wavelength will allow both information channels to traverse the same fiber simultaneously. Here, we will describe two classes of wavelength converters, with one example from each class.

11.6.1 Optical-Gating Wavelength Converters

A wide variety of optical-gating techniques using devices such as semiconductor optical amplifiers, semiconductor lasers, or nonlinear optical-loop mirrors have been investigated to achieve wavelength conversion. The use of a SOA in a cross-phase modulation (XPM) mode has been one of the most successful techniques for implementing single-wavelength conversion.[42-46] The configurations for implementing this scheme include the Mach-Zehnder or the Michelson interferometer setups shown in Fig. 11-14.

The XPM scheme relies on the dependency of the refractive index on the carrier density in the active region of the SOA. As depicted in Fig. 11-14, the basic concept is that an incoming information-carrying signal at wavelength λ_s and a continuous-wave (CW) signal at the desired new wavelength λ_c (called the *probe beam*) are simultaneously coupled into the device. The two waves can be either copropagating or counterpropagating. However, the noise in the latter case is higher.[47,48] The signal beam modulates the gain of the SOA by depleting the carriers, which produces a modulation of the refractive index. When the CW beam encounters the modulated gain and refractive index, its amplitude and phase are changed, so that it now carries the same information as the input signal. As shown in Fig. 11-14, the SOAs are placed in an asymmetric configuration so that the phase change in the two amplifiers is different. Consequently, the CW light is modulated according to the phase difference. A typical splitting ratio is 69/31 percent. This type of converters readily handle data rates of at least 10 Gb/s.

A limitation of the XPM architecture is that it only converts one wavelength at a time. In addition, it has limited transparency in terms of the data format. Any information that is in the form of phase, frequency, or analog amplitude is lost during the wavelength conversion process. Consequently, this scheme is restricted to converting digital signal streams.

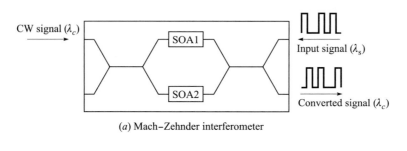

(a) Mach–Zehnder interferometer

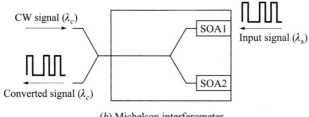

(b) Michelson interferometer

FIGURE 11-14

(a) Mach-Zehnder interferometer and (b) Michelson interferometer setups using a pair of SOAs for implementing the cross-phase modulation wavelength-conversion scheme.

11.6.2 Wave-Mixing Wavelength Converters

Wavelength conversion based on nonlinear optical wave mixing offers important advantages compared with other methods.[49-55] These include a multiwavelength conversion capability and transparency to the modulation format. The mixing results from nonlinear interactions among optical waves traversing a nonlinear material. The outcome is the generation of another wave whose intensity is proportional to the product of the intensities of the interacting waves. The phase and frequency of the generated wave are a linear combination of those of the interacting waves. Therefore, the wave mixing preserves both amplitude and phase information, and, consequently, is the only wavelength-conversion category that offers strict transparency to the modulation format.

Two successful schemes are four-wave mixing (FWM) in either passive waveguides or SOAs, and difference-frequency generation in waveguides. For wavelength conversion, the FWM scheme employs the mixing of three distinct input waves to generate a fourth distinct output wave. In this method, an intensity pattern resulting from two input waves interacting in a nonlinear material forms a grating. For example, in SOAs there are three physical mechanisms that can form a grating. These are carrier-density modulation, dynamic carrier heating, and spectral hole burning. The third input wave in the material gets scattered by this grating, thereby generating an output wave. The frequency of the generated output wave is offset from that of the third wave by the frequency difference between the first two waves. If one of the three incident waves contains amplitude, phase, or frequency information, and the other two waves are constant, then the generated wave will contain the same information.

Difference-frequency generation in waveguides is based on the mixing of two input waves. Here, the nonlinear interaction of the material is with a pump and a signal wave. Figure 11-15 gives an example of the simultaneous conversion of eight input wavelengths in the 1546-to-1560-nm region to a set of eight output wavelengths in the 1524-to-1538-nm region.[53]

PROBLEMS

11-1. Consider an InGaAsP semiconductor optical amplifier that has the following parameter values:

Symbol	Parameter	Value
w	Active area width	$5\,\mu$m
d	Active area thickness	$0.5\,\mu$m
L	Amplifier length	$200\,\mu$m
Γ	Confinement factor	0.3
τ_r	Time constant	1 ns
a	Gain coefficient	$1 \times 10^{-20}\,\text{m}^2$
v_g	Group velocity	$2.0 \times 10^8\,\text{m/s}$
n_{th}	Threshold density	$1.0 \times 10^{24}\,\text{m}^{-3}$

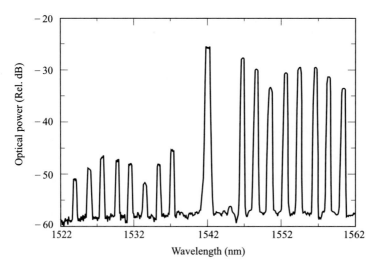

FIGURE 11-15
Simultaneous conversion of eight input wavelengths (1546, 1548, 1550, 1552, 1554, 1556, 1558, 1560 nm) to a set of eight output wavelengths (1538, 1536, 1534, 1532, 1530, 1528, 1526, 1524 nm) using difference-frequency generation. The 1542-nm spike is a second-order spectrometer response to the pump wave at 771 nm. (Reproduced with permission from Yoo et al.,[53] © 1997, OSA.)

If a 100-mA bias current is applied, find (a) the pumping rate R_p, (b) the maximum (zero-signal) gain, (c) the saturation photon density, (d) the photon density if a 1-μW signal at 1310 nm enters the amplifier. Compare the results of (c) and (d).

11-2. Verify that Eq. (11-15) follows from Eq. (11-14).

11-3. Solving Eq. (11-15) numerically, make plots of the normalized amplifier gain (G/G_0) versus the normalized output power $(P_{s,out}/P_{amp,sat})$ for unsaturated amplifier gain values of $G_0 = 30$ dB, 15 dB, and 10 dB.

11-4. The *output saturation power* $P_{out,sat}$ is defined as the amplifier output power for which the amplifier gain G is reduced by 3 dB (a factor of 2) from its unsaturated value G_0. Assuming $G_0 \gg 1$, show that in terms of the amplifier saturation power $P_{amp,sat}$, the output saturation power is

$$P_{out,sat} = \frac{G_0 \ln 2}{(G_0 - 2)} P_{amp,sat}$$

11-5. Since the gain constant a depends on the frequency, the amplifier gain is also frequency-dependent. The 3-dB bandwidth (full-width at half-maximum, FWHM) is defined as the frequency for which the power gain $G(v)$ is reduced by a factor of 2. Assume the gain parameter g has a gaussian profile

$$g(v) = \frac{g_0}{1 + 4(v - v_0)^2/(\Delta v)^2}$$

where Δv is the optical bandwidth (the spectral width of the gain profile) and v_0 is the maximum-gain frequency. Show that the ratio of the 3-dB bandwidth $2(v - v_0)$ to the optical bandwidth Δv is

$$\frac{2(v - v_0)}{\Delta v} = \left[\log_2(G_0/2)\right]^{-1/2}$$

where $\log_2 X$ is the base-2 logarithm of X. What does this equation show concerning the relationship between the amplifier gain and the optical bandwidth?

11-6. Assume the gain profile of an optical amplifier is

$$g(\lambda) = g_0 e^{-(\lambda - \lambda_0)^2/2(\Delta\lambda)^2}$$

where λ_0 is the peak-gain wavelength and $\Delta\lambda$ is the spectral width of the amplifier gain. If $\Delta\lambda = 25$ nm, find the FWHM (the 3-dB gain) of the amplifier gain if the peak gain at λ_0 is 30 dB.

11-7. (*a*) Compare the maximum theoretical PCE for 980-nm and 1475-nm pumping in an EDFA for a 1545-nm signal. Contrast this with actual measured results of PCE = 50.0 percent and 75.6 percent or 980-nm and 1475-nm pumping, respectively.

(*b*) Using the actual results for PCE given in (*a*), plot the maximum signal output power as a function of pump power for $0 \le P_{p,\text{in}} \le 200$ mW for pump wavelengths of 980 nm and 1475 nm.

11-8. Assume we have an EDFA power amplifier that produces $P_{s,\text{out}} = 27$ dBm for an input level of 2 dBm at 1542 nm.

(*a*) Find the amplifier gain.

(*b*) What is the minimum pump power required?

11-9. (*a*) To see the relative contributions of the various noise mechanisms in an optical amplifier, calculate the values of the five noise terms in Eq. (11-29) for operational gains of $G = 20$ dB and 30 dB. Assume the optical bandwidth is equal to the spontaneous emission bandwidth (30-nm spectral width) and use the following parameter values:

Symbol	Parameter	Value
η	Photodiode quantum efficiency	0.6
$\mathcal{R}$	Responsivity	0.73 A/W
P_{in}	Input optical power	$1\,\mu\text{W}$
λ	Wavelength	1550 nm
Δv_{opt}	Optical bandwidth	3.77×10^{12} Hz
B	Receiver bandwidth	1×10^9 Hz
n_{sp}	Spontaneous emission factor	2
R_L	Receiver load resistor	$1000\,\Omega$

(*b*) To see the effect of using a narrowband optical filter at the receiver, let $\Delta v_{\text{opt}} = 1.25 \times 10^{11}$ Hz (125 GHz at 1550 nm) and find the same five noise terms for $G = 20$ dB and 30 dB.

11-10. Plot the penalty factor $F(G)$ given by Eq. (11-36) as a function of amplifier gain for gains ranging from 0 to 30 dB. Assuming the fiber attenuation is 0.2 dB/km, draw a distance axis parallel to the gain axis to show the transmission distances corresponding to the gain values.

11-11. Consider a cascaded chain of k fiber-plus-EDFA combinations, as shown in Fig. P11-11.

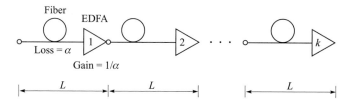

FIGURE P11-11

(a) Show that the path-averaged signal power is

$$\langle P \rangle_{\text{path}} = P_{s,\text{in}} \frac{G - 1}{G \ln G}$$

(b) Derive the path-averaged ASE power given by Eq. (11-35).

11-12. Consider a long-distance transmission system containing a cascaded chain of EDFAs. Assume each EDFA is operated in the saturation region and that the slope of the gain-versus-input power curve in this region is -0.5; that is, the gain changes by ± 3 dB for a ∓ 6-dB variation in input power. Let the link have the following operational parameters:

Symbol	Parameter	Value
G	Nominal gain	7.1 dB
$P_{s,\text{out}}$	Nominal output optical power	3.0 dBm
$P_{s,\text{in}}$	Nominal input optical power	-4.1 dBm

(a) Suppose there is a sudden 6-dB drop in signal level at some point in the link. Find the power output levels after the degraded signal has passed through 1, 2, 3, and 4 succeeding amplifier stages.

(b) Repeat part (a) for a signal-level drop of 12 dB.

11-13. Let the electric field of an optical signal at a carrier frequency ν_i be

$$E_i(t) = \sqrt{2P_i} \cos(2\pi \nu_i t + \phi_i)$$

where P_i is the signal power at the carrier frequency ν_i and ϕ_i is the carrier phase. If N optical signals each at a different frequency ν_i are traveling along a fiber, show that the signal power is

$$P_s = \sum_{i=1}^{N} P_{s,i} + \sum_{j}^{N} \sum_{k \neq j}^{N} 2\sqrt{P_j P_k} \cos(\Omega_{jk} t + \phi_j - \phi_k)$$

where $\Omega_{jk} = 2\pi(\nu_j - \nu_k)$ represents the beat frequency at which the carrier population oscillates.

11-14. Consider an EDFA with a gain of 26 dB and a maximum output power of 0 dBm.

(a) Compare the output signal levels per channel for 1, 2, 4, and 8 wavelength channels, where the input power is 1 μW for each signal.

(b) What are the output signal levels per channel in each case if the pump power is doubled?

11-15. Recall from Eq. (7-51) that the bit-error rate (BER) can be given in terms of a Q factor, where, from Eq. (7-53),

$$Q = \frac{b_{\text{on}} - b_{\text{off}}}{\sqrt{\sigma_{\text{on}}^2} + \sqrt{\sigma_{\text{off}}^2}}$$

When thermal noise is dominant, we have $\sigma_{\text{on}}^2 = \sigma_{\text{off}}^2$, as we saw in Chap. 7. However, for an EDFA the existence of signal–ASE beat noise produces the condition $\sigma_{\text{on}}^2 > \sigma_{\text{off}}^2$. In this case, $\sigma_{\text{on}}^2 = \sigma_{\text{total}}^2$ from Eq. (11-29),

$$\sigma_{\text{off}}^2 = \sigma_T^2 + \sigma_{\text{shot-ASE}}^2 + \sigma_{\text{ASE-ASE}}^2$$

and

$$b_{\text{on}} - b_{\text{off}} = (GI_s + I_N) - I_N = G\mathcal{R}P_{s,\text{in}}$$

Noting that the receiver sensitivity P_R is half the signal power of a transmitted 1 bit for a uniform distribution of ones and zeros (i.e., $P_s = 2P_R$), show that in terms of the Q factor the receiver sensitivity is

$$P_R = h\nu\, B\Bigg\{ FQ^2 + Q\bigg[\frac{2n_{\text{sp}}}{\eta}\bigg(\frac{G-1}{G^2}\bigg)\frac{\Delta\nu_{\text{opt}}}{B} + n_{\text{sp}}^2\bigg(\frac{G-1}{G}\bigg)^2\bigg(2\frac{\Delta\nu_{\text{opt}}}{B} - 1\bigg)$$

$$+ \frac{4k_B T}{R_L(\eta q)^2 G^2 B}\bigg]^{1/2}\Bigg\}$$

where F is the noise figure given by Eq. (11-34).

11-16. Using the expression for the receiver sensitivity P_R given in Prob. 11-15, plot P_R as a function of G for gain values ranging from 10 to 40 dB. Use the following values of $\Delta\nu_{\text{opt}}$: 2.5, 12.5, 125, and 675 GHz (corresponding to spectral bandpasses of 0.02, 0.1, 1, and 5 nm, respectively at 1550 nm). Let $Q = 6$ and assume the following values for the parameters in the expression for P_R:

Symbol	Parameter	Value
η	Photodiode quantum efficiency	0.6
T	Temperature	300 K
λ	Wavelength	1550 nm
B	Receiver bandwidth	1.25 GHz
n_{sp}	Spontaneous emission factor	2
R_L	Receiver load resistor	50 Ω

11-17. Consider the expression for the receiver sensitivity P_R given in Prob. 11-15. For sufficiently large values of G, the thermal noise term is negligible and the first term in the square-root expression is small compared with the second term. In this case, and for $2\Delta\nu_{\text{opt}} \gg 1$, the expression for P_R becomes

$$P_R \approx h\nu\, BF\Bigg\{ Q^2 + \frac{Q}{2\eta}\bigg(\frac{\Delta\nu_{\text{opt}}}{B}\bigg)^{1/2}\Bigg\}$$

Letting $Q = 6$ (for a 10^{-9} BER), plot P_R as a function of the relative optical filter bandwidth $\Delta\nu_{\text{opt}}/B$ for the range $10 \le \Delta\nu_{\text{opt}}/B \le 1000$ for the following three values of F: 3 dB, 5 dB, and 7 dB. Assume a wavelength of 1550 nm, a photodiode quantum efficiency of 0.5, and an electrical bandwidth of 1.25 GHz

11-18. If the input power and the optical bandwidth are relatively large, the relative intensity noise inherent in the laser transmitter may affect the noise figure of an optical amplifier. In this case, the noise figure F relative to $2n_{sp}$ (the noise figure when signal–ASE beating is the dominant noise) is given by[56]

$$\frac{F}{2n_{sp}} = 1 + \frac{P_{ASE}}{P_s} + \frac{(RIN)\Delta\nu_{opt}}{8}\frac{2B}{\Delta\nu_{opt}} + \frac{(RIN)\Delta\nu_{opt}}{4}\frac{P_s}{P_{ASE}}$$

where $P_{ASE} = n_{sp}\,h\nu\,\Delta\nu_{opt}$ is the ASE noise power. Here, the third term comes from the interaction between the RIN and the ASE, and the fourth term arises from the interaction between the RIN and the signal. Plot $F/2n_{sp}$ as a function of P_s/P_{ASE} for the range $-10\,\text{dB} \le P_s/P_{ASE} \le 20\,\text{dB}$. Use the following values of $(RIN)\,\Delta\nu_{opt}$: 0.01, 0.1, 1, and 10.

REFERENCES

1. S. Shimata and H. Ishio, *Optical Amplifiers and Their Applications*, Wiley, Chichester, UK, 1994.
2. C. R. Giles, M. Newhouse, J. Wright, and K. Hagimoto, eds., Special Issue on "System and Network Applications of Optical Amplifiers," *J. Lightwave Tech.*, vol. 13, May 1995.
3. R. Giles and T. T. Li, "Optical amplifiers transform long-distance lightwave telecommunications," *Proc. IEEE*, vol. 84, pp. 870–883, June 1996.
4. M. Potenza, "Optical fiber amplifiers for telecommunication systems," *IEEE Commun. Mag.*, vol. 34, pp. 96–102, Oct. 1996.
5. M. J. O'Mahoney, "Semiconductor laser optical amplifiers for use in future fiber systems," *J. Lightwave Tech.*, vol. 6, pp. 531–544, Apr. 1988.
6. T. Saitoh and T. Mukai, "Recent progress in semiconductor laser amplifiers," *J. Lightwave Tech.*, vol. 6, pp. 1656–1664, Nov. 1988.
7. P. C. Becker, N. A. Olsson, and J. R. Simpson, *Erbium-Doped Fiber Amplifiers*, Academic, New York, 1999.
8. H. Ghafouri-Shiraz, *Fundamentals of Laser Diode Amplifiers*, Wiley, New York, 1996.
9. A. Yariv, *Optical Electronics*, Saunders College Publ., Orlando, FL, 4th ed., 1991.
10. P. E. Green, Jr., *Fiber Optic Networks*, Prentice Hall, New York, 1993.
11. R. Ramaswami and P. Humblet, "Amplifier induced crosstalk in multichannel optical networks," *J. Lightwave Tech.*, vol. 8, pp. 1882–1896, Apr. 1990.
12. B. Clesca, D. Bayart, and J. L. Beylat, "1.5-μm fluoride-based fiber amplifiers for wideband multichannel transport networks," *Opt. Fiber Technol.*, vol. 1, pp. 135–157, Mar. 1995.
13. T. J. Whitley, "A review of recent system demonstrations incorporating 1.3-μm praseodymium-doped fluoride fiber amplifiers," *J. Lightwave Tech.*, vol. 13, pp. 744–760, May 1995.
14. J. Lucas, "Review: Fluoride glasses," *J. Mater. Sci.*, vol. 24, pp. 1–13, Jan. 1989.
15. (a) W. Miniscalco, "Erbium-doped glasses for fiber amplifiers at 1500 nm," *J. Lightwave Tech.*, vol. 9, pp. 234–250, Feb. 1991.
    (b) W. Miniscalco, "Optical and electronic properties of rare earth ions in glasses," in M. J. F. Digonnet, ed., *Rare Earth Doped Fiber Lasers and Amplifiers*, Dekker, New York, 1993, chap. 2, pp. 19–133.
16. E. Desurvire, *Erbium-Doped Fiber Amplifiers*, Wiley, New York, 1994.
17. (a) C.-C. Fan, J.-D. Peng, J.-H. Li, X. Jiang, G.-S. Wu, and B.-K. Zhou, "Theoretical and experimental investigations on erbium-doped fiber amplifiers," *Fiber & Integrated Optics*, vol. 13, no. 3, pp. 247–260, 1994.
    (b) F. Fontana and G. Grasso, "The erbium-doped fiber amplifier: Technology and applications," *Fiber & Integrated Optics*, vol. 13, no. 3, pp. 135–145, 1994.

18. B. Pedersen, K. Dybdal, C. D. Hansen, A. Bjarklev, J. H. Povlsen, H. Vendeltorp-Pommer, and C. C. Larsen, "Detailed theoretical and experimental investigation of high-gain erbium-doped fiber amplifier," *IEEE Photonics Tech. Lett.*, vol. 2, pp. 863–865, Dec. 1990.

19. D. A. Chapman, "Erbium-doped fiber amplifiers," *Electron. & Commun. Eng. J.*, vol. 6, pp. 59–67, Apr. 1994.

20. ITU-T Recommendation G.662, *Generic Characteristics of Optical Fibre Amplifier Devices and Sub-systems*, July 1995.

21. C. Lester, A. Bjarklev, T. Rasmussen, and P. G. Dinesen, "Modeling of Yb-sensitized Er-doped silica waveguide amplifiers," *J. Lightwave Tech.*, vol. 13, pp. 740–743, May 1995.

22. H. Masuda, S. Kawai, K.-I. Suzuki, and K. Aida, "Ultrawide 75-nm gain-band optical amplification with erbium-doped fluoride fiber amplifiers and distributed Raman amplifiers," *IEEE Photonics Tech. Lett.*, vol. 10, pp. 516–518, Apr. 1998.

23. C. R. Giles and E. Desurvire, "Modeling erbium-doped fiber amplifiers," *J. Lightwave Tech.*, vol. 9, pp. 271–283, Feb. 1991.

24. T. Li, "The impact of optical amplifiers on long-distance telecommunications," *Proc. IEEE*, vol. 81, pp. 1568–1579, June 1993.

25. G. R. Walker, N. G. Walker, R. C. Steele, M. J. Creaner, and M. C. Brain, "Erbium-doped fiber amplifier cascade for multichannel coherent optical transmission," *J. Lightwave Tech.*, vol. 9, pp. 182–193, Feb. 1991.

26. A. Hardy and R. Oron, "Signal amplification in strongly pumped fiber amplifiers," *IEEE J. Quantum Electron.*, vol. 33, pp. 307–313, Mar. 1997.

27. J. P. Gordon and L. F. Mollenauer, "Effects of fiber nonlinearities and amplifier spacing on ultra-long distance transmission," *J. Lightwave Tech.*, vol. 9, pp. 170–173, Feb. 1991.

28. (*a*) T. T. Ha, G. E. Keiser, and R. L. Borchart, "Bit error probabilities of OOK lightwave systems with optical amplifiers," *J. Opt. Commun.*, Vol. 18, pp. 151–155, Aug. 1997.
    (*b*) Y. K. Park and S. W. Granlund, "Optical preamplifier receivers: Applications to digital long-haul transmission," *Opt. Fiber Technol.*, vol. 1, pp. 59–71, Oct. 1994.

29. F. Matera and M. Settembre, "Performance of optical links with optical amplifiers," *Fiber & Integrated Optics*, vol. 15, no. 2, pp. 89–107, 1996.

30. K. Inoue, "Crosstalk and its power penalty in multichannel transmission due to gain saturation in a semiconductor laser amplifier," *J. Lightwave Tech.*, vol. 7, pp. 1118–1124, July 1989.

31. A. E. Willner and S.-M. Hwang "Transmission of many WDM channels through a cascade of EDFAs in long distance links and ring networks," *J. Lightwave Tech.*, vol. 13, pp. 802–816, May 1995.

32. J.-C. Dung, S. Chi, and S. Wen, "Gain flattening of erbium-doped fiber amplifier using fiber Bragg gratings," *Electron. Lett.*, vol. 34, pp. 555–556, Mar. 1998.

33. H. S. Kim, S. H. Yun, N. Park, and B. Y. Kim, "Actively gain-flattened erbium-doped fiber amplifier over 35 nm using all-fiber acousto-optic tunable filters," *IEEE Photonics Tech. Lett.*, vol. 10, pp. 790–792, June 1998.

34. H. Ono, M. Yamada, T. Kanamori, S. Sudo, and Y. Ohishi, "1.58-μm band gain-flattened erbium-doped fiber amplifiers for WDM transmission systems." *J. Lightwave Tech.*, vol. 17, pp. 490–496, Mar. 1999.

35. C. Delisle and J. Conradi, "Model for bidirectional transmission in an open cascade of optical amplifiers," *J. Lightwave Tech.*, vol. 15, pp. 749–757, May 1997.

36. R. L. Mortenson, B. S. Jackson, S. Shapiro, and W. F. Sirocky, "Undersea optically amplified repeatered technology, products, and challenges," *AT&T Tech. J.*, vol. 74, pp. 33–46, Jan./Feb. 1995.

37. T. Otani, K. Goto, T. Kawazawa, H. Abe, and M. Tanaka, "Effect of span loss increase on the optically amplified communication system," *J. Lightwave Tech.*, vol. 15, pp. 737–742, May 1997.

38. A. K. Srivastava, Y. Sun, J. L. Zyskind, and J. W. Sulhoff, "EDFA transient response to channel loss in WDM transmission system," *IEEE Photonics Tech. Lett.*, vol. 9, pp. 386–388, Mar. 1997.

39. D. H. Richards, J. L. Jackel, and M. A. Ali, "Multichannel EDFA chain control: A comparison of two approaches," *IEEE Photonics Tech. Lett.*, vol. 10, pp. 156–158, Jan. 1998.

40. H. Suzuki, N. Takachio, O. Ishida, and M. Koga, "Dynamic gain control by maximum signal power channel in optical linear repeaters for WDM photonic transport networks," *IEEE Photonics Tech. Lett.*, vol. 10, pp. 734–736, May 1998.

41. S. Y. Park, H. Y. Kim, G. Y. Lyu, S. M. Kang, and S. Y. Shin, "Dynamic gain and output power control in a gain-flattened erbium-doped fiber amplifier," *IEEE Photonics Tech. Lett.*, vol. 10, pp. 787–789, June 1998.

42. M. J. Offside, J. E. Carroll, M. E. Bray, and A. Hadjifotiou, "Optical wavelength converters," *Electron. & Commun. Eng. J.*, vol. 7, pp. 59–71, Apr. 1995.

43. M. E. Bray and M. J. O'Mahoney, "Cascading gain-saturation semiconductor laser-amplifier wavelength translators," *IEE Proc.—Optoelectron.*, vol. 143, pp. 1–6, Feb. 1996.

44. T. Durhuus, B. Mikkelsen, C. Joergensen, S. L. Danielsen, and K. E. Stubkjaer, "All-optical wavelength conversion by semiconductor optical amplifiers," *J. Lightwave Tech.*, vol. 14, pp. 942–954, June 1996.

45. S. J. B. Yoo, "Wavelength conversion technologies for WDM network applications," *J. Lightwave Tech.*, vol. 14, pp. 955–966, June 1996.

46. M. Asghari, I. H. White, and R. V. Penty, "Wavelength conversion using semiconductor optical amplifiers," *J. Lightwave Tech.*, vol. 15, pp. 1181–1190, July 1997.

47. K. Obermann, S. Kindt, D. Breuer, K. Petermann, C. Schmidt, S. Diez, and H. G. Weber, "Noise characteristics of semiconductor optical amplifiers used for wavelength conversion via cross-gain and cross-phase modulation," *IEEE Photonics Tech. Lett.*, vol. 9, pp. 312–314, Mar. 1997.

48. S. L. Danielsen, P. B. Hansen, K. E. Stubkjaer, M. Schilling, K. Wünstel, W. Idler, P. Doussiere, and F. Pommerau, "All optical wavelength conversion for increased input power dynamic range," *IEEE Photonics Tech. Lett.*, vol. 10, pp. 60–62, Jan. 1998.

49. A. D'Ottavi, E. Iannone, A. Mecozzi, S. Scotti, P. Spano, R. Dall'Ara, J. Eckner, and G. Guekos, "Efficiency and noise performance of wavelength converters based on FWM in semiconductor optical amplifiers," *IEEE Photonics Tech. Lett.*, vol. 7, pp. 357–359, Apr. 1995.

50. G. P. Bava, P. Debernardi, and G. Osella, "Frequency conversion in traveling wave semiconductor laser amplifiers with bulk and quantum well structures," *IEE Proc.—Optoelectron.*, vol. 143, pp. 119–125, Apr. 1996.

51. K. Obermann, I. Koltchanov, K. Petermann, S. Diez, R. Ludwig, and H. G. Weber, "Noise analysis of frequency converters utilizing semiconductor-laser amplifiers," *IEEE J. Quantum Electron.*, vol. 33, pp. 81–88, Jan. 1997.

52. D. F. Geraghty, R. B. Lee, K. J. Vahala, M. Verdiell, M. Ziari, and A. Mathur, "Wavelength conversion up to 18 nm at 10 Gb/s by four-wave mixing in a semiconductor optical amplifier," *IEEE Photonics Tech. Lett.*, vol. 9, pp. 452–454, Apr. 1997.

53. S. J. B. Yoo, A. Rajhel, C. Caneau, R. Bhat, and M. A. Koza, "Multichannel polarization-independent wavelength conversion by difference-frequency-generation in AlGaAs waveguides," *OSA/IEEE OFC '97 Technical Digest*, pp. 78–80, Feb. 1997.

54. T. F. Morgen, J. P. R. Lacey and R. C. Tucker, "Widely tunable FWM in semiconductor optical amplifiers with constant conversion frequency," *IEEE Photonics Tech. Lett.*, vol. 10, pp. 1401–1403, Oct. 1998.

55. J. M. Yates, J. P. R. Lacey, M. P. Rumsewicz, and M. A. Summerfield, "Performance of networks using wavelength converters based on FWM in semiconductor optical amplifiers," *J. Lightwave Tech.*, vol. 17, pp. 782–791, May 1999.

56. I. Jacobs, "Dependence of optical amplifier noise figure on relative intensity noise," *J. Lightwave Tech.*, vol. 13, pp. 1461–1465, July 1995.

CHAPTER
12

OPTICAL NETWORKS

Chapter 9 covered the performance features of point-to-point links, where the optical fiber system serves as a simple connection between two sets of electrical signal-processing equipment. This chapter treats more complex networks that can be utilized in local-, metropolitan-, or wide-area networks to connect hundreds or thousands of users with a wide range of transmission capacities and speeds.[1-8] A major motivation for developing these sophisticated networks has been the rapid proliferation of information exchange desired by institutions such as commerce, finance, education, health, government, security, and entertainment. The potential for this information exchange arose from the ever-increasing power of computers and data-storage devices, which now need to be interconnected by high-speed, high-capacity networks.[9-11]

In Sec. 12.1 we describe some of the basic topologies that are possible for fiber optic networks and examine the design tradeoffs among them. Next, Sec. 12.2 looks at the closely coupled SONET and SDH standards, which specify formats for optical signals so they can be shared between networks. The basic SONET/SDH characteristics covered include the standard data-frame structure, the optical interface specifications, and the fundamental ring architectures. Also shown are possible network architectures that can be constructed from SONET/SDH rings.

Methods for simultaneously accessing a number of nodes in an optical network can be implemented either in the spectral domain or in the time domain. We first look at WDMA (WDM access), which is an example of a multiple-access architecture operating in the spectral domain. In this context, Sec. 12.3 addresses broadcast-and-select networks, in which the optical signal from one station is sent to a large number of other stations by using a passive all-optical distribution architecture. As an example, each station could transmit at a different wavelength, with the receiver using an optical filter to extract the desired wavelength destined for it from the others.

To overcome a number of problems associated with the broadcast-and-select technique, the next level of sophistication uses either a passive or an active method to route individual incoming wavelengths to a particular destination. Section 12.4 discusses these wavelength-routing networks, which utilize wavelength-conversion technologies.

The design of a WDM network requires careful planning of fiber selection, component tuning, and network layout to combat system-degradation processes and to create a network that is easy to operate and maintain. Section 12.5 addresses these topics. Included here are descriptions of various performance-degradation effects (such as inelastic scattering of photons and nonlinear processes in fibers), and dispersion-compensation schemes. In addition to these effects, designing optically amplified WDM links and networks requires careful consideration of factors such as bandwidth, BER requirements, and crosstalk between optical channels. Section 12.6 describes these issues.

Advanced communication techniques for optical systems include soliton transmission, optical code-division multiple access (optical CDMA), and ultrafast time-division multiplexing (TDM). These topics are addressed in Sec. 12.7 through 12.9, respectively. Further topics in this rapidly changing field can be found on the web site for this book (see Sec. 1.5.4).

12.1 BASIC NETWORKS

Before delving into network details, let us briefly define some terms. Suppose there is a collection of devices with which users wish to communicate. These devices are called *stations*, and may be computers, terminals, telephones, or other equipment for communicating. Stations are also referred to as data terminal equipment (DTE) in the networking world. To establish connections between these stations, one interconnects them by transmission paths to form a *network*. Within this network, a *node* is a point where one or more communication lines terminate and/or where stations are connected. Stations also can connect directly to a transmission line. The *topology* is the logical manner in which nodes are linked together by information-transmission channels to form a network. The transfer of information from source to destination through a series of intermediate nodes is called *switching*, and the selection of a suitable path through a network is referred to as *routing*. Thus, a *switched communication network* consists of an interconnected collection of nodes, in which information streams that enter the network from a station are routed to the destination by being switched from node to node. When two networks that use different information-exchange rules (protocols) are interconnected, a device called a *router* is used at the interconnection point to translate the control information from one protocol to another.

Networks are traditionally divided into the following three broad categories:

1. *Local-area networks* (LANs) interconnect users in a localized area such as a department, a building, an office or factory complex, or a university campus.
2. *Metropolitan-area networks* (MANs) provide user interconnection within a city or in the metropolitan area surrounding a city.
3. *Wide-area networks* (WANs) cover a large geographical area ranging from connections between nearby cities to connections of users across a country.

In this section we will consider star, linear bus, and ring topologies used for fiber optic networks, and we will compare the performance of linear bus and star configurations in terms of a detailed power-budget analysis in a LAN environment. A popular protocol used in optical LANs is the Fiber Distributed Data Interface (FDDI).[12] Section 12.2 describes the SONET and SDH protocols, which are widely used on a ring network with active nodes in MANs and WANs.

12.1.1 Network Topologies

Figure 12-1 shows the three common topologies used for fiber optic networks. These are the *linear-bus*, *ring*, and *star configurations*. Each has its own particular advantages and limitations in terms of reliability, expandability, and performance characteristics.

Nonoptical bus networks, such as standard Ethernet, employ coaxial cable as the transmission medium. The primary advantages of such a network are the totally passive nature of the transmission medium and the ability to easily install low-perturbation (high-impedance) taps on the coaxial line without disrupting the operating network. In contrast to the coaxial bus, a fiber-optic-based bus network is more difficult to implement. The impediment is that there are no low-perturbation optical-tap equivalents to coax taps for efficiently coupling optical signals into and out of the main optical fiber trunk line. Access to an optical data bus is achieved by means of a coupling element, which can be either active or passive. An *active coupler* converts the optical signal on the data bus to its electric baseband counterpart before any data processing (such as injecting additional data into the signal stream or merely passing on the received data) is carried out. A *passive coupler* employs no electronic elements. It is used passively to tap off a portion of the optical power from the bus. Examples of these are the 2×2 couplers described in Chap. 10.

In a ring topology, consecutive nodes are connected by point-to-point links that are arranged to form a single closed path. Information in the form of data packets (a group of information bits plus overhead bits) is transmitted from node to node around the ring. The interface at each node is an active device that has the ability to recognize its own address in a data packet in order to accept messages. The active node forwards those messages that are not addressed to itself on to its next neighbor.

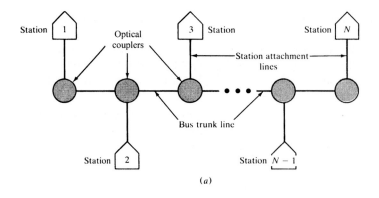

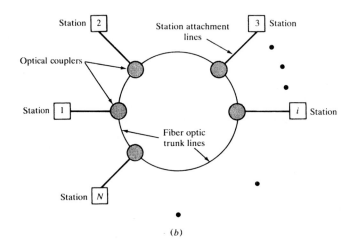

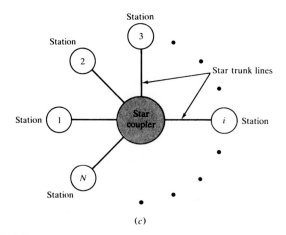

FIGURE 12-1

Three common topologies used for fiber optic networks: (*a*) bus, (*b*) ring, and (*c*) star.

In a star architecture, all nodes are joined at a single point called the *central node* or *hub*. The central node can be an active or a passive device. Using an active hub, one can control all routing of messages in the network from the central node. This is useful when most of the communications are between the central and the outlying nodes, as opposed to information exchange between the attached stations. If there is a great deal of message traffic between the outlying nodes, then a heavy switching burden is placed on an active central node. In a star network with a passive central node, a power splitter is used at the hub to divide the incoming optical signals among all the outgoing lines to the attached stations.

12.1.2   Performance of Passive Linear Buses

To evaluate the performance of a passive linear bus, let us examine the various locations of power loss along the transmission path. We consider this in terms of the fraction of power lost at a particular interface or within a particular component. First, as described in Sec. 3-1, over an optical fiber of length x (in kilometers), the ratio A_0 of received power $P(x)$ to transmitted power $P(0)$ is given by

$$A_0 = \frac{P(x)}{P(0)} = 10^{-\alpha x/10} \qquad (12\text{-}1)$$

where α is the fiber attenuation in units of dB/km.

The losses encountered in a passive coupler used for a linear bus are shown schematically in Fig. 12-2. This is nominally a cascaded combination of two directional couplers where two ports (one in each directional coupler) are not used. For simplicity, we do not show these unused ports here. The coupler thus has four functioning ports: two for connecting the coupler onto the fiber bus, one for receiving tapped-off light, and one for inserting an optical signal onto the line after the tap-off to keep the signal out of the local receiver. If a fraction F_c of optical power is lost at each port of the coupler, then the *connecting loss* L_c is

$$L_c = -10 \log(1 - F_c) \qquad (12\text{-}2)$$

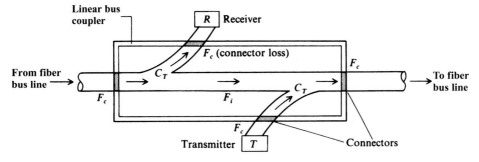

FIGURE 12-2
Losses encountered in a passive linear-bus coupler consisting of a cascade of two directional couplers.

For example, if we take this fraction to be 20 percent, then L_c is about 1 dB; that is, the optical power gets reduced by 1 dB at any coupling junction.

Let C_T represent the fraction of power that is removed from the bus and delivered to the detector port. The power extracted from the bus is called a *tap loss* and is given by

$$L_{\text{tap}} = -10 \log C_T \tag{12-3}$$

For a symmetric coupler, C_T also is the fraction of power that is coupled from the transmitting input port to the bus. Thus, if P_0 is the optical power launched from a source flylead, the power coupled to the bus is $C_T P_0$. Note that, in general, when calculating the throughput power for intermediate stations, the transmission path through the bus coupler passes two tap points, since optical power is extracted at both the receiving and the transmitting taps of the device. The power removed at the transmitting tap goes out of the unused port and thus is lost from the system. The *throughput coupling loss* L_{thru} in decibels is then given by

$$L_{\text{thru}} = -10 \log(1 - C_T)^2 = -20 \log(1 - C_T) \tag{12-4}$$

In addition to connection and tapping losses, there is an intrinsic transmission loss L_i associated with each bus coupler. If the fraction of power lost in the coupler is F_i, then the *intrinsic transmission loss* L_i in decibels is

$$L_i = -10 \log(1 - F_i) \tag{12-5}$$

Generally, a linear bus will consist of a number of stations separated by various lengths of bus line. However, for analytical simplicity, here we consider a simplex linear bus of N stations uniformly separated by a distance L, as shown in Fig. 12-3. Thus, from Eq. (12-1) the fiber attenuation between any two adjacent stations in decibels is

$$L_{\text{fiber}} = -10 \log A_0 = \alpha L \tag{12-6}$$

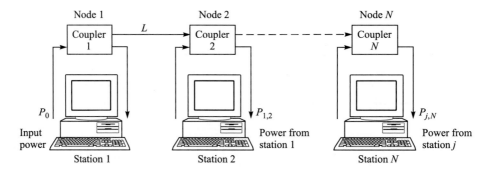

FIGURE 12-3
Topology of a simplex linear bus consisting of N uniformly spaced stations.

The term *simplex* means that, in this configuration, information flows only from left to right. To have *full-duplex* communication, in which stations can communicate in either direction, one needs an analogous configuration with a separate parallel line and another set of N couplers. The information flow in the second line would then be from right to left.

To determine the power budget, we shall first examine the link in terms of fractional power losses at each link element. The examples will then show the power-budget calculation (in decibels) using a tabular form. For the fractional-loss method, we use the notation $P_{j,k}$ to denote the optical power received at the detector of the kth station from the transmitter of the jth station. For simplicity, we assume that a bus coupler exists at every terminal of the bus, including the two end stations.

NEAREST-NEIGHBOR POWER BUDGET. The smallest distance in transmitted and received power occurs for adjacent stations, such as between stations 1 and 2 in Fig. 12-3. If P_0 is the optical power launched from a source flylead at station 1, then the power detected at station 2 is

$$P_{1,2} = A_0 C_T^2 (1 - F_c)^4 (1 - F_i)^2 P_0 \tag{12-7}$$

since the optical power flow encounters the following loss-inducing mechanisms:

- One fiber path with attenuation A_0.
- Tap points at both the transmitter and the receiver, each with coupling efficiencies C_T.
- Four connecting points, each of which passes a fraction $(1 - F_c)$ of the power entering them.
- Two couplers which pass only the fraction $(1 - F_i)$ of the incident power owing to intrinsic losses.

Using Eqs. (12-2) through (12-4) and Eq. (12-6), the expression for the losses between stations 1 and 2 can be expressed in logarithmic form as

$$10 \, \log\left(\frac{P_0}{P_{1,2}}\right) = \alpha L + 2L_{\text{tap}} + 4L_c + 2L_i \tag{12-8}$$

LARGEST-DISTANCE POWER BUDGET. The largest distance for transmitted and received power occurs between stations 1 and N. At the transmitting end the fractional power level coupled into the first length of cable from the transmitter flylead through the bus coupler at station 1 is

$$F_1 = (1 - F_c)^2 C_T (1 - F_i) \tag{12-9a}$$

Similarly, at station N the fraction of power from the bus-coupler input port that emerges from the detector port is

$$F_N = (1 - F_c)^2 C_T (1 - F_i) \tag{12-9b}$$

For each of the $(N - 2)$ intermediate stations, the fraction of power passing through each coupling module (shown in Fig. 12-2) is

$$F_{\text{coup}} = (1 - F_c)^2(1 - C_T)^2(1 - F_i) \tag{12-10}$$

since, from the input to the output of each coupler, the power flow encounters two connector losses, two tap losses, and one intrinsic loss. Combining the expressions from Eqs. (12-9a), (12-9b), and (12-10), and the transmission losses of the $N - 1$ intervening fibers, we find that the power received at station N from station 1 is

$$
\begin{aligned}
P_{1,N} &= A_0^{N-1} F_1 F_{\text{coup}}^{N-2} F_N P_0 \\
&= A_0^{N-1}(1 - F_c)^{2N}(1 - C_T)^{2(N-2)} C_T^2 (1 - F_i)^N P_0 \tag{12-11}
\end{aligned}
$$

Using Eqs. (12-2) through (12-6), the power budget for this link then is

$$
\begin{aligned}
10 \log\left(\frac{P_0}{P_{1,N}}\right) &= (N - 1)\alpha L + 2NL_c + (N - 2)L_{\text{thru}} + 2L_{\text{tap}} + NL_i \\
&= [\text{fiber} + \text{connector} + \text{coupler throughput} \\
&\quad + \text{ingress/egress} + \text{coupler intrinsic}]\ \text{losses} \\
&= N(\alpha L + 2L_c + L_{\text{thru}} + L_i) - \alpha L - 2L_{\text{thru}} + 2L_{\text{tap}} \tag{12-12}
\end{aligned}
$$

The last expression shows that the losses (in decibels) of the linear bus increase linearly with the number of stations N.

Example 12-1. Let us compare the power budgets of three linear buses, having 5, 10, and 50 stations, respectively. Assume that $C_T = 10$ percent, so that $L_{\text{tap}} = 10$ dB and $L_{\text{thru}} = 0.9$ dB. Let $L_i = 0.5$ dB and $L_c = 1.0$ dB. If the stations are relatively close together, say 500 m, as they might be in a LAN, then for an attenuation of 0.4 dB/km at 1300 nm the fiber loss is 0.2 dB. Using Eq. (12-12), the power budgets for these three cases can be calculated using the tabular method shown in Table 12-1. For actual calculations, it is best to use a spreadsheet on a computer.

TABLE 12-1
Comparison of the power budgets of three linear buses that have 5, 10, and 50 stations, respectively

Coupling/loss factor	Loss expression	Loss (dB)	Losses for 5 stations	Losses for 10 stations	Losses for 50 stations
Source connector	Eq. (12-2)	1.0	1.0	1.0	1.0
Coupling (tap) loss	Eq. (12-3)	2×10.0	20.0	20.0	20.0
Coupler-to-fiber loss	Eq. (12-2)	$2(N - 1) \times 1.0$	8.0	18.0	98.0
Fiber loss (500 m)	Eq. (12-6)	$(N - 1) \times 0.2$	0.8	1.8	9.8
Coupler throughput	Eq. (12-4)	$(N - 2) \times 0.9$	2.7	7.2	43.2
Intrinsic coupler loss	Eq. (12-5)	$N \times 0.5$	2.5	5.0	25.0
Receiver connector	Eq. (12-2)	1.0	1.0	1.0	1.0
Total loss (dB)	Eq. (12-12)	—	36.0	54.0	198.0

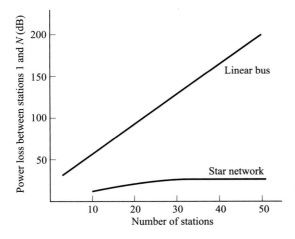

FIGURE 12-4
Total loss as a function of the number of attached stations for linear-bus and star architectures.

The total loss values given in Table 12-1 are plotted in Fig. 12-4, which shows that the loss (in dB) increases linearly with the number of stations.

▓▓▓ **Example 12-2.** For the applications given in Example 12-1, suppose that for implementing a 10-Mb/s bus we have a choice of an LED that emits -10 dBm or a laser diode capable of emitting $+3$ dBm of optical power from a fiber flylead. At the destination, assume we have an APD receiver with a sensitivity of -48 dBm at 10 Mb/s in the 1300-nm window. In the LED case, the power loss allowed between the source and the receiver is 38 dB. As shown in Fig. 12-4, this allows up to 5 stations on the bus. For the laser diode source, we have an additional 13 dB of margin, so for this example we can have a maximum of 8 stations connected to the bus.

DYNAMIC RANGE. Owing to the serial nature of the linear bus, the optical power available at a particular node decreases with increasing distance from the source. Thus, a performance quantity of interest is the system dynamic range. This is the maximum optical power range to which any detector must be able to respond. The worst-case dynamic range (DR) is found from the ratio of Eq. (12-7) to Eq. (12-11):

$$DR = 10 \log\left(\frac{P_{1,2}}{P_{1,N}}\right) = 10 \log\left\{\frac{1}{[A_0(1 - F_c)^2(1 - C_T)^2(1 - F_i)]^{N-2}}\right\}$$
$$= (N - 2)(\alpha L + 2L_c + L_{\text{thru}} + L_i) \tag{12-13}$$

For example, this could be the difference in power levels received at station N from station $(N - 1)$ and from station 1 (i.e., $P_{1,2} = P_{N-1,N}$).

▓▓▓ **Example 12-3.** Consider the linear buses described in Example 12-1. For $N = 5$, from Eq. (12-13) the dynamic range is

$$DR = 3[0.2 + 2(1.0) + 0.9 + 0.5]\, dB = 10.8\, dB$$

For $N = 10$ stations,

$$\text{DR} = 8[0.2 + 2(1.0) + 0.9 + 0.5]\,\text{dB} = 28.8\,\text{dB}$$

12.1.3   Performance of Star Architectures

To see how a star coupler can be applied to a given network, let us examine the various optical power losses associated with the coupler. Section 10.2.4 gives the details of how an individual star coupler works. As a quick review, the excess loss is defined as the ratio of the input power to the total output power. That is, it is the fraction of power lost in the process of coupling light from the input port to all the output ports. From Eq. (10-25), for a single input power P_{in} and N output powers, the excess loss in decibels is given by

$$\text{Fiber star excess loss} = L_{\text{excess}} = 10\log\left(\frac{P_{\text{in}}}{\sum\limits_{i=1}^{N} P_{\text{out,i}}}\right) \tag{12-14}$$

In an ideal star coupler the optical power from any input is evenly divided among the output ports. The total loss of the device consists of its splitting loss plus the excess loss in each path through the star. The *splitting loss* is given in decibels by

$$\text{Splitting loss} = L_{\text{split}} = -10\log\left(\frac{1}{N}\right) = 10\log N \tag{12-15}$$

To find the power-balance equation, we use the following parameters:

- P_S is the fiber-coupled output power from a source in dBm.
- P_R is the minimum optical power in dBm required at the receiver to achieve a specific bit-error rate.
- α is the fiber attenuation.
- All stations are located at the same distance L from the star coupler.
- L_c is the connector loss in decibels.

Then, the power-balance equation for a particular link between two stations in a star network is

$$\begin{aligned}
P_S - P_R &= L_{\text{excess}} + \alpha(2L) + 2L_c + L_{\text{split}} \\
&= L_{\text{excess}} + \alpha(2L) + 2L_c + 10\log N \tag{12-16}
\end{aligned}$$

Here, we have assumed connector losses at the transmitter and the receiver. This equation shows that, in contrast to a passive linear bus, for a star network the loss increases much slower as $\log N$. Figure 12-4 compares the performance of the two architectures.

Example 12-4. Consider two star networks that have 10 and 50 stations, respectively. Assume each station is located 500 m from the star coupler and that the fiber attenuation is 0.4 dB/km. Assume the excess loss is 0.75 dB for the 10-station network and 1.25 dB for the 50-station network. Let the connector loss be 1.0 dB. For $N = 10$, from Eq. (12-16) the power margin between the transmitter and the receiver is

$$P_S - P_R = [0.75 + 0.4(1.0) + 2(1.0) + 10 \log 10] \, dB = 13.2 \, dB$$

For $N = 50$,

$$P_S - P_R = [1.25 + 0.4(1.0) + 2(1.0) + 10 \log 50] \, dB = 20.6 \, dB$$

Using the transmitter output and receiver sensitivity values given in Example 12-2, we see that an LED transmitter can easily accommodate the losses in this 50-station star network. In comparison, a laser transmitter could not even meet the 10-station design in a passive linear bus.

12.2 SONET/SDH

With the advent of fiber optic transmission lines, the next step in the evolution of the digital time-division-multiplexing (TDM) scheme was a standard signal format called *synchronous optical network (SONET)* in North America and *synchronous digital hierarchy (SDH)* in other parts of the world. This section addresses the basic concepts of SONET/SDH, its optical interfaces, and fundamental network implementations. The aim here is to discuss only the physical-layer aspects of SONET/SDH as they relate to optical transmission lines and optical networks. Topics such as the detailed data format structure, SONET/SDH operating specifications, and the relationships of switching methodologies such as asynchronous transfer mode (ATM) with SONET/SDH are beyond the scope of this text. These can be found in numerous sources, such as those listed in Refs. 13–15.

12.2.1 Transmission Formats and Speeds

In the mid-1980s, several service providers in the USA started efforts on developing a standard that would allow network engineers to interconnect fiber optic transmission equipment from various vendors through multiple-owner trunking networks. This soon grew into an international activity, which, after many differences of opinion of implementation philosophy were resolved, resulted in a series of ANSI T1.105 standards[16] for SONET and a series of ITU-T recommendations for SDH.[17] Of particular interest here are the ANSI T1.105.06 standard and the ITU-T G.957 recommendation. Although there are some implementation differences between SONET and SDH, all SONET specifications conform to the SDH recommendations.

Figure 12-5 shows the basic structure of a SONET frame. This is a two-dimensional structure consisting of 90 columns by 9 rows of bytes, where one byte

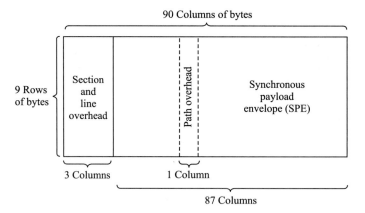

FIGURE 12-5
Basic structure of an STS-1 SONET frame.

is eight bits. Here, in standard SONET terminology, a *section* connects adjacent pieces of equipment, a *line* is a longer link that connects two SONET devices, and a *path* is a complete end-to-end connection. The fundamental SONET frame has a 125-μs duration. Thus, the transmission bit rate of the basic SONET signal is

$$\text{STS-1} = (90 \text{ bytes/row})(9 \text{ rows/frame})(8 \text{ bits/byte})/(125\,\mu s/\text{frame}) = 51.84 \text{ Mb/s}$$

This is called an STS-1 signal, where STS stands for *synchronous transport signal*. All other SONET signals are integer multiples of this rate, so that an STS-N signal has a bit rate equal to N times 51.84 Mb/s. When an STS-N signal is used to modulate an optical source, the *logical STS-N signal* is first scrambled to avoid long strings of ones and zeros and to allow easier clock recovery at the receiver. After undergoing electrical-to-optical conversion, the resultant *physical-layer optical signal* is called OC-N, where OC stands for *optical carrier*. In practice, it has become common to refer to SONET links as OC-N links. Algorithms have been developed for values of N ranging between 1 and 255. However, in the range from 1 to 192, the ANSI T1.105 standard recognizes only the values $N = 1, 3, 12, 24, 48,$ and 192.

In SDH the basic rate is equivalent to STS-3, or 155.52 Mb/s. This is called the *synchronous transport module—level 1* (STM-1). Higher rates are designated by STM-M. (Note: Although the SDH standard uses the notation "STM-N," here we use the designation "STM-M" to avoid confusion when comparing SDH and SONET rates.) Values of M supported by the ITU-T recommendations are $M = 1, 4, 16,$ and 64. These are equivalent to SONET OC-N signals, where $N = 3M$ (i.e., $N = 3, 12, 48,$ and 192). This shows that, in practice, to maintain compatibility betwen SONET and SDH, N is a multiple of three. Analogous to SONET, SDH first scrambles the logical signal. In contrast to SONET, SDH does not distinguish between a logical electrical signal (e.g., STS-N in SONET) and a physical optical signal (e.g., OC-N), so that both signal types are designated by STM-M. Table 12-2 lists commonly used values of OC-N and STM-M.

TABLE 12-2
Commonly used SONET and SDH transmission rates

SONET level	Electrical level	Line rate (Mb/s)	SDH equivalent
OC-1	STS-1	51.84	—
OC-3	STS-3	155.52	STM-1
OC-12	STS-12	622.08	STM-4
OC-24	STS-24	1244.16	STM-8
OC-48	STS-48	2488.32	STM-16
OC-96	STS-96	4976.64	STM-32
OC-192	STS-192	9953.28	STM-64

Referring to Fig. 12-5, the first three columns comprise transport overhead bytes that carry network management information. The remaining field of 87 columns is called the *synchronous payload envelope* (SPE) and carries user data plus nine bytes of path overhead (POH). The POH supports performance monitoring by the end equipment, status, signal labeling, a tracing function, and a user channel. The nine path-overhead bytes are always in a column and can be located anywhere in the SPE. An important point to note is that the synchronous byte-interleaved multiplexing in SONET/SDH (unlike the asynchronous bit interleaving used in earlier TDM standards) facilitates add/drop multiplexing of information channels in optical networks (see Sec. 12.2.4).

For values of N greater than 1, the columns of the frame become N times wider, with the number of rows remaining at nine, as shown in Fig. 12-6. Thus, an STS-3 (or STM-1) frame is 270 columns wide with the first nine columns containing overhead information and the next 261 columns being payload data. The line and section overhead bytes differ somewhat between SONET and SDH, so that a translation mechanism is needed to interconnect them. To obtain further details on the contents of the frame structure and the population schemes for the payload field, the reader is referred to the SONET and SDH specifications.[16,17]

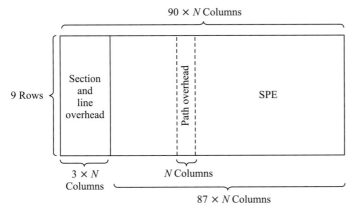

FIGURE 12-6
Basic format of an STS-N SONET frame.

12.2.2 Optical Interfaces

To ensure interconnection compatibility between equipment from different manufacturers, the SONET and SDH specifications provide details for the optical source characteristics, the receiver sensitivity, and transmission distances for various types of fibers. Six transmission ranges are defined, with different terminology for SONET and SDH, as Table 12-3 indicates.

The optical fibers specified in ANSI T1.105.06 and ITU-T G.957 fall into the following three categories and operational windows:

1. Graded-index multimode in the 1310-nm window.
2. Conventional non-dispersion-shifted single-mode in the 1310-nm and 1550-nm windows.
3. Dispersion-shifted single-mode in the 1550-nm window.

Table 12-4 shows the wavelength ranges and attenuations specified in these fibers for transmission distances up to 80 km.[17d]

Depending on the attenuation and dispersion characteristics for each hierarchical level shown in Table 12-4, feasible optical sources include light-emitting diodes (LEDs), multilongitudinal mode (MLM) lasers, and single-longitudinal mode (SLM) lasers. Of the various spectral characteristics of the optical sources detailed in ANSI T1.105.06 and ITU-T G.957, only the launched-power range is given here. These powers are listed in Table 12-5 for transmissions rates up to OC-48 or STM-16 (2.5 Gb/s). The system objective is to achieve a bit-error rate (BER) of less than 10^{-10} for rates less than 1 Gb/s and 10^{-12} for higher rates and/or higher-performance systems. This performance is with a link power penalty of less than 1 dB.

The specified receiver sensitivities are the worst-case, end-of-life values. They are defined as the minimum-acceptable, average, received power needed to achieve a 10^{-10} BER. The values take into account extinction ratio, pulse rise and fall times, optical return loss at the source, receiver connector degrada-

TABLE 12-3
Transmission distances and their SONET and SDH designations for long-haul, very long haul, and ultralong-haul distances; the fiber types are specified by ITU-T G.652, G.653, and G.655, respectively (see Ref. 17)

Transmission distance	SONET terminology	SDH terminology
$\leq$ 2 km	Short-reach	Intraoffice
15 km	Intermediate-reach	Short-haul
40 km at 1310 nm 80 km at 1550 nm	Long-reach	Long-haul
120 km at 1550 nm 160 km at 1550 nm		Very long-haul Ultralong-haul

TABLE 12-4
Wavelength ranges and fiber attenuation for transmission distances up to 80 km

Transmission distance	1310-nm window	1550-nm window	Attenuation at 1310 nm	Attenuation at 1550 nm
≤ 15 km	1260–1360 nm	1430–1580 nm	3.5 dB/km	Not specified
≤ 40 km	1260–1360 nm	1430–1580 nm	0.8 dB/km	0.5 dB/km
≤ 80 km	1280–1335 nm	1480–1580 nm	0.5 dB/km	0.3 dB/km

tions, and measurement tolerances. The receiver sensitivity does not include power penalties associated with dispersion, jitter, or reflections from the optical path, since these are included in the maximum optical path penalty. Table 12-5 lists the receiver sensitivities for various link configurations up through long-haul distances (80 km).

Longer transmission distances are possible using higher-power lasers. To comply with eye-safety standards, an upper limit is imposed on fiber-coupled powers. If the maximum total output power (including ASE) is set at the Class-3A laser limit of + 17 dBm, then for ITU-T G.655 fiber this allows transmission distances of 160 km for a single-channel link.

In a WDM link, as the number of channels increases, the maximum transmission distance decreases, since the allowed power per channel becomes lower. The ITU-T Recommendation G.692 describes these conditions.[17] This recommendation defines interface parameters for systems of 4, 8, and 16 channels operating at bit rates of up to OC-48/STM-16. The specified fibers are described by ITU-T Recommendations G.652, G.653, and G.655 with nominal span lengths of 80, 120, and 160 km, respectively. (Note: This is an evolving recommendation, with 32 and higher channel numbers, OC-192/STM-64 rates, and bidirectional transmission aspects being under consideration.) Using the + 17-dBm *total* power limit, the maximum nominal optical powers per channel will be as shown in Table 12-6 (see ITU-T Recommendation G.692).

TABLE 12-5
Source output range and receiver sensitivity for various rates and distances up to 80 km (see Ref. 17d)

Distance	SONET rate	SDH rate	Source output range (dBm)	Receiver sensitivity (dBm)
≤ 15 km	OC-3	STM-1	−15 to −8	−23
	OC-12	STM-4	−15 to −8	−23
	OC-48	STM-16	−10 to −3	−18
≤ 40 km	OC-3	STM-1	−15 to −8	−28
	OC-12	STM-4	−15 to −8	−28
	OC-48	STM-16	−5 to 0	−18
≤ 80 km	OC-3	STM-1	−5 to 0	−34
	OC-12	STM-4	−3 to +2	−28
	OC-48	STM-16	−2 to +3	−27

TABLE 12-6
Maximum nominal optical power per wavelength channel based on the total optical power being + 17 dBm (see ITU-T Recommendation G.692, Oct. 1998, Ref. 17*f*)

Number of wavelengths (channels)	Nominal power per channel (dBm)
1	17.0
2	14.0
3	12.2
4	11.0
5	10.0
6	9.2
7	8.5
8	8.0

12.2.3   SONET/SDH Rings

A key characteristic of SONET and SDH is that they are usually configured as a ring architecture. This is done to create *loop diversity* for uninterrupted service protection purposes in case of link or equipment failures. The SONET/SDH rings are commonly called *self-healing rings*, since the traffic flowing along a certain path can automatically be switched to an alternate or standby path following failure or degradation of the link segment.

Three main features, each with two alternatives, classify all SONET/SDH rings, thus yielding eight possible combinations of ring types. First, there can be either two or four fibers running between the nodes on a ring. Second, the operating signals can travel either clockwise only (which is termed a *unidirectional ring*) or in both directions around the ring (which is called a *bidirectional ring*). Third, protection switching can be performed either via a line-switching or a path-switching scheme.[18–21] Upon link failure or degradation, *line switching* moves all signal channels of an entire OC-*N* channel to a protection fiber. Conversely, *path switching* can move individual payload channels within an OC-*N* channel (e.g., an STS-1 subchannel in an OC-12 channel) to another path.

Of the eight possible combinations of ring types, the following two architectures have become popular for SONET and SDH networks:

● two-fiber, unidirectional, path-switched ring (two-fiber UPSR).

● two-fiber or four-fiber, bidirectional, line-switched ring (two-fiber or four-fiber BLSR).

The common abbreviations of these configurations are given in parentheses. They are also referred to as unidirectional or bidirectional self-healing rings (USHRs or BSHRs), respectively.

Figure 12-7 shows a two-fiber unidirectional path-switched ring network. By convention, in a unidirectional ring the normal working traffic travels clockwise around the ring, on the *primary path*. For example, the connection from node 1 to

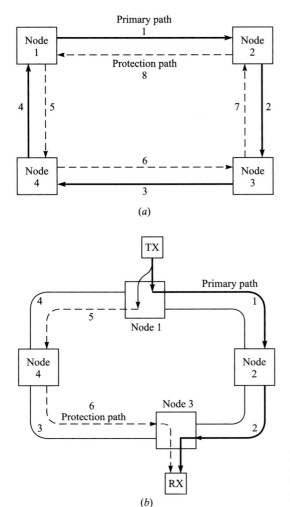

FIGURE 12-7
(*a*) Generic two-fiber unidirectional network with a counter-rotating protection path. (*b*) Flow of primary and protection traffic from node 1 to node 3.

node 3 uses links 1 and 2, whereas the traffic from node 3 to node 1 traverses links 3 and 4. Thus, two communicating nodes use a specific bandwidth capacity around the entire perimeter of the ring. If nodes 1 and 3 exchange information at an OC-3 rate in an OC-12 ring, then they use one-quarter of the capacity around the ring on all the primary links. In a unidirectional ring the counter-clockwise path is used as an alternate route for protection against link or node failures. This *protection path* (links 5–8) is indicated by dashed lines. To achieve protection, the signal from a transmitting node is dual-fed into both the primary and protection fibers. This establishes a designated protection path on which traffic flows counterclockwise; namely, from node 1 to node 3 via links 5 and 6, as shown in Fig. 12-7.

Consequently, two identical signals from a particular node arrive at their destination from opposite directions, usually with different delays, as denoted in Fig. 12-7*b*. The receiver normally selects the signal from the primary path. However, it continuously compares the fidelity of each signal and chooses the alternate signal in case of severe degradation or loss of the primary signal. Thus, each path is individually switched based on the quality of the received signal. For example, if path 2 breaks or equipment in node 2 fails, then node 3 will switch to the protection channel to receive signals from node 1.

Figure 12-8 illustrates the architecture of a four-fiber bidirectional line-switched ring. Here, two primary fiber loops (with fiber segments labeled 1p through 8p) are used for normal bidirectional communication, and the other two secondary fiber loops are standby links for protection purposes (with fiber segments labeled 1s through 8s). In contrast to the two-fiber UPSR, the four-fiber BLSR has a capacity advantage since it uses twice as much fiber cabling and because traffic between two nodes is sent only partially around the ring. To see this, consider the connection between nodes 1 and 3. The traffic from node 1 to node 3 flows in a clockwise direction along links 1p and 2p. Now, however, in the return path the traffic flows counterclockwise from node 3 to node 1 along links 7p and 8p. Thus, the information exchange between nodes 1 and 3 does not tie up any of the primary channel bandwidth in the other half of the ring.

To see the function and versatility of the standby links in the four-fiber BLSR, consider first the case where a transmitter or receiver circuit card used on the primary ring fails in either node 3 or 4. In this situation, the affected nodes detect a loss-of-signal condition and switch both primary fibers connecting them to the secondary protection pair, as shown in Figure 12-9. The protection segment between nodes 3 and 4 now becomes part of the primary bidirectional loop. The exact same reconfiguration scenario will occur when the primary fiber connecting nodes 3 and 4 breaks. Note that in either case the other links remain unaffected.

Now suppose an entire node fails, or both the primary and the protection fibers in a given span are severed, which could happen if they are in the same cable

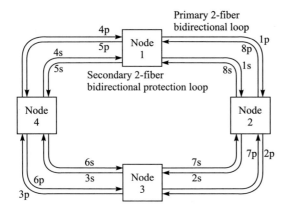

FIGURE 12-8
Architecture of a four-fiber bidirectional line-switched ring (BLSR).

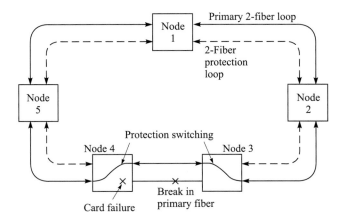

FIGURE 12-9
Reconfiguration of a four-fiber BLSR under transceiver or line failure.

duct between two nodes. In this case, the nodes on either side of the failed inter-nodal span internally switch the primary-path connections from their receivers and transmitters to the protection fibers, in order to loop traffic back to the previous node. This process again forms a closed ring, but now with all of the primary and protection fibers in use around the entire ring, as shown in Fig. 12-10.

12.2.4 SONET/SDH Networks

Commercially available SONET/SDH equipment allows the configuration of a variety of network architectures, as shown in Fig. 12-11. For example, one can

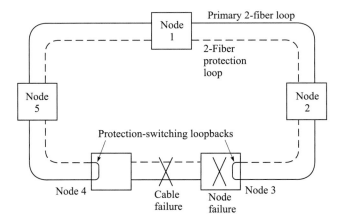

FIGURE 12-10
Reconfiguration of a four-fiber BLSR under node or fiber-cable failure.

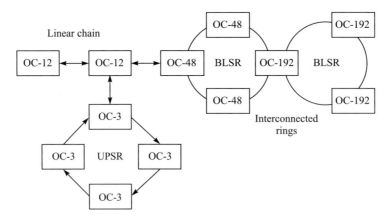

FIGURE 12-11

Generic configuration of a large SONET network consisting of linear chains and various types of interconnected rings.

build point-to-point links, linear chains, unidirectional path-switched rings (UPSR), bidirectional line-switched rings (BLSR), and interconnected rings. The OC-192 four-fiber BLSR could be a large national backbone network with a number of OC-48 rings attached in different cities. The OC-48 rings can have lower-capacity localized OC-12 or OC-3 rings or chains attached to them, thereby providing the possibility of attaching equipment that has an extremely wide range of rates and sizes. Each of the individual rings has its own failure–recovery mechanism and SONET/SDH network management procedures.

An important SONET/SDH network element is the *add/drop multiplexer* (ADM).[22] This piece of equipment is a fully synchronous, byte-oriented multiplexer that is used to add and drop subchannels within an OC-*N* signal. Figure 12-12 shows the functional concept of an ADM. Here, various OC-12s and OC-3s are multiplexed into an OC-48 stream. Upon entering an ADM, these subchannels can be individually dropped by the ADM and others can be added. For example,

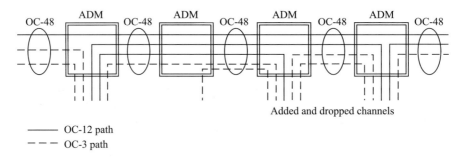

FIGURE 12-12

Functional concept of an add/drop multiplexer for SONET applications.

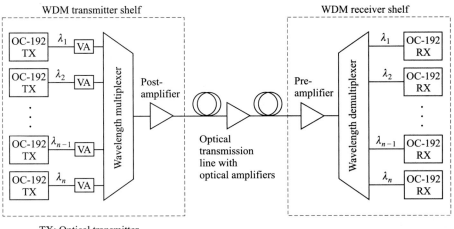

TX: Optical transmitter
RX: Optical receiver
VA: Variable attenuator

FIGURE 12-13
Dense WDM deployment of n wavelengths in an OC-192 trunk ring.

in Fig.12-12, one OC-12 and two OC-3 channels enter the left-most ADM as part of an OC-48 channel. The OC-12 is passed through and the two OC-3s are dropped by the first ADM. Then, two more OC-12s and one OC-3 are multiplexed together with the OC-12 channel that is passing through and the aggregate (partially filled) OC-48 is sent to another ADM node downstream.

The SONET/SDH architectures can also be implemented with multiple wavelengths. For example, Fig. 12-13 shows a dense WDM deployment on an OC-192 trunk ring for n wavelengths (e.g., one could have $n = 16$). The different wavelength outputs from each OC-192 transmitter are passed first through a variable attenuator to equalize the output powers. These are then fed into a wavelength multiplexer, possibly amplified by a post-transmitter optical amplifier, and sent out over the transmission fiber. Additional optical amplifiers might be located at intermediate points and/or at the receiving end.

12.3 BROADCAST-AND-SELECT WDM NETWORKS

As another step toward realizing the full potential of optical fiber transmission capacity, researchers have looked at all-optical WDM networks to extend the versatility of communication networks beyond architectures such as those provided by SONET. These networks can be classified as either *broadcast-and-select* or *wavelength-routing* networks.[8,23] In general, broadcast-and-select techniques employing passive optical stars, buses, or wavelength routers are used for local network applications, whereas active optical components form the basis for constructing wide-area wavelength-routing networks. This section addresses broadcast-and-select networks; section 12.4 discusses wavelength-routing architectures. Broadcast-and-select networks can be categorized as *single-hop* or *multi-hop networks*. Single-hop refers to networks where information transmitted in the form of light reaches its destination without being converted to an electrical form at any intermediate point. On the other hand, intermediate electro-optical conversion can occur in a multihop network.

12.3.1 Broadcast-and-Select Single-Hop Networks

Figure 12-14 shows two alternate physical architectures for a WDM-based local network. Here, N sets of transmitters and receivers are attached to either a star coupler or a passive bus. Each transmitter sends its information at a different fixed wavelength. All the transmissions from the various nodes are combined in a passive star coupler or coupled onto a bus and the result is sent out to all receivers.

Each receiver sees all wavelengths and uses a tunable filter to select the one wavelength addressed to it. In addition to point-to-point links, this configuration can also support multicast or broadcast services, where one transmitter sends the same information to several nodes. Figure 12-15 illustrates this concept for a star network. Workstations at nodes 4 and 2 communicate using λ_2, whereas a user at node 1 broadcasts information to workstations at nodes 3 and 5 using λ_1. The same concepts are applicable to bus structures, although the losses encountered in the star and bus architectures are different.

An interesting point to note is that the WDM setup in Fig. 12-15 is *protocol transparent*. This means that different sets of communicating nodes can use different information-exchange rules (protocols) without affecting the other nodes in the network. This is analogous to standard time-division-multiplexed telephone lines in which voice, data, or facsimile services are sent in different time slots without interfering with each other. However, here the rates differ widely.

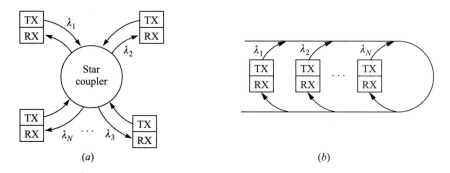

(a)   (b)

FIGURE 12-14
Two alternate physical architectures for a WDM-based local network: (*a*) star, (*b*) bus.

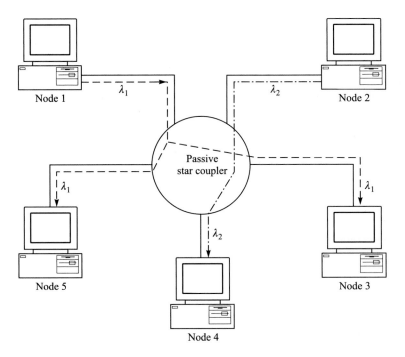

FIGURE 12-15
Architecture of a single-hop broadcast-and-select network.

Although the architectures of single-hop broadcast-and-select networks are fairly simple, there needs to be careful dynamic coordination between the nodes. For example, if the network is set up so that a transmitter sends its information at a unique fixed wavelength, then the destined receiver needs to be informed when a message is being sent to it, so that it can tune its selective filter to that wavelength. Also, conflicts need to be resolved in cases where two stations transmitting at different wavelengths want to send information to the same recipient simultaneously.

Alternatively, if a node received at a fixed wavelength and the transmitter is tunable to different wavelengths, then two sending stations need to coordinate their transmissions so that collisions of information streams at the same wavelength do not occur. More details of this are beyond the scope of the discussion in this text, but can be found in the literature.[5–8,23]

12.3.2 Broadcast-and-Select Multihop Networks

A drawback of single-hop networks is the need for rapidly tunable lasers or receiver optical filters. The designs of multihop networks avoid this need. Multihop networks generally do not have direct paths between each node pair. Each node has a small number of fixed-tuned optical transmitters and receivers.

Figure 12-16 shows an example of a four-node broadcast-and-select multihop network where each node transmits on one set of two fixed wavelengths and receives on another set of two fixed wavelengths. Stations can send information directly only to those nodes that have a receiver tuned to one of the two transmit wavelengths. Information destined for other nodes will have to be routed through intermediate stations.

To visualize this operation, consider a very simplified transmission scheme in which messages are sent as packets with a data field and an address header containing source and destination identifiers (i.e., routing information) together with other control bits, as shown in Fig. 12-17. At each intermediate node, the optical signal is converted to an electrical format, and the address header is decoded to examine the routing information field, which will indicate where the packet should go. Using this routing information, the packet is then switched electronically to the specific optical transmitter that will appropriately direct the packet to the next node in the logical path toward its final destination.

The flow of traffic can be seen from Fig. 12-16. If node 1 wants to send a message to node 2, it first transmits the message to node 3 using λ_1. Then node 3 forwards the message to node 2 using λ_6. In contrast to single-hop networks, with this scheme there are no destination conflicts or packet collisions in the network, since each wavelength channel is dedicated to a particular source–destination link. However, for H hops between nodes, there is a network throughput penalty of at least $1/H$.

12.3.3   The ShuffleNet Multihop Network

Various topologies have been suggested for multihop lightwave networks. Among these are the ShuffleNet graph,[24–26] the de Bruijn graph,[27] and the toroidal Manhattan Street Network.[28] Here, we will consider the ShuffleNet as an example.

A scheme called the *perfect shuffle* is widely used to form processor-interconnect patterns in multiprocessors.[29] For optical networks, the extension of this to the logical configuration consists of a cylindrical arrangement of k columns, each having p^k nodes, where p is the number of fixed transceiver pairs per node. The total number of nodes is then

$$N = kp^k \tag{12-17}$$

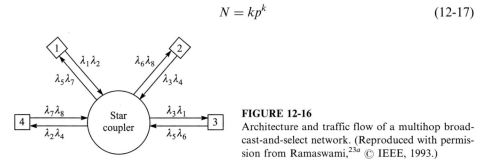

FIGURE 12-16
Architecture and traffic flow of a multihop broadcast-and-select network. (Reproduced with permission from Ramaswami,[23a] © IEEE, 1993.)

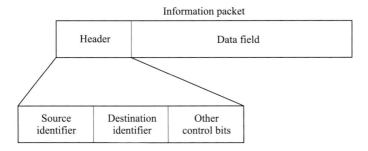

FIGURE 12-17
Simple representation of the fields contained in a data packet.

with $k = 1, 2, 3, \ldots$ and $p = 1, 2, 3, \ldots$ Given that each node requires p wavelengths to transmit information, the total number of wavelengths N_λ needed in the network is

$$N_\lambda = pN = kp^{k+1} \tag{12-18}$$

Figure 12-18 illustrates a $(p, k) = (2, 2)$ ShuffleNet, where the $(k + 1)$th column represents the completion of a trip around the cylinder back to the first column, as indicated by the return arrow. In this example, there are eight nodes and sixteen wavelengths.

An important performance parameter for the ShuffleNet is the average number of hops between any two randomly chosen nodes. Since all nodes have p output wavelengths, p nodes can be reached from any node in one hop, p^2 additional nodes can be reached in two hops, and so on, until all the $(p^k - 1)$ other nodes are visited. The maximum number of hops is

$$H_{\max} = 2k - 1 \tag{12-19}$$

As an example, in Fig. 12-18, consider the connections between nodes 1 and 5 and between nodes 1 and 7. In the first case, the hop number is one. In the second case, three hops are needed with the routes being either 1–6–4–7 or 1–5–2–7. In general, the average number of hops $\bar{H}$ of a ShuffleNet is[24,25]

$$
\begin{aligned}
\bar{H} &= \frac{1}{N-1}\left[\sum_{j=1}^{k-1} jp^j + \sum_{j=0}^{k-1}(k+j)(p^k - p^j)\right] \\
&= \frac{kp^k(p-1)(3k-1) - 2k(p^k-1)}{2(p-1)(kp^k-1)}
\end{aligned} \tag{12-20}
$$

As a result of multihopping, only part of the capacity of a particular link directly connecting two nodes is actually utilized for carrying traffic between them. The rest of the link capacity is used to forward messages from other nodes. Since the system has $Np = kp^{k+1}$ links, the total network capacity C is

$$C = \frac{kp^{k+1}}{\bar{H}} \tag{12-21}$$

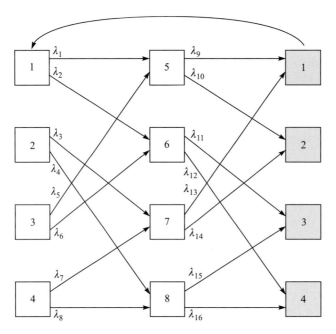

FIGURE 12-18
Logical interconnection pattern and wavelength assignment of a $(p, k) = (2, 2)$ ShuffleNet.

and the per-user throughput S is

$$S = \frac{C}{N} = \frac{p}{\bar{H}} \tag{12-22}$$

Different (p, k) combinations result in different throughputs, so one can make some tradeoffs among the variables to get a better network performance. For example, given that the number of nodes is fixed at N, one can reduce the average number of hops by increasing p (which decreases k) to boost the capacity and the throughput.

12.4 WAVELENGTH-ROUTED NETWORKS

Two problems arise in broadcast-and-select networks when trying to extend them to wide-area networks. First, more wavelengths are needed as the number of nodes in the network grows. Typically, there are at least as many wavelengths as there are nodes, unless several nodes time-share a wavelength (which has an adverse impact on delay and efficiency). Second, without the widespread use of optical booster amplifiers, a large number of users spread over a wide area cannot readily be interconnected with a broadcast-and-select network. This is because the network employs passive star couplers in which the splitting losses could be prohibitively high for many attached stations.

Wavelength-routed networks overcome these limitations through wavelength reuse, wavelength conversion, and optical switching. The physical topology of a wavelength-routed network consists of optical wavelength routers interconnected by pairs of point-to-point fiber links in an arbitrary mesh configuration, as illustrated in Fig. 12-19. Each link can carry a certain number of wavelengths, which can be directed independently to different output paths at a node. Each node may have logical connections with several other nodes in the network, where each connection uses a particular wavelength. Provided the paths taken by any two connections do not overlap, they can use the same wavelength. Thereby, the number of wavelengths is greatly reduced. For example, in Fig. 12-19 the connection from node 1 to node 2 and from node 2 to node 3 can both be on λ_1, whereas the connection between nodes 4 and 5 requires a different wavelength (λ_2).

12.4.1 Optical Cross-Connects

A high degree of path modularity, capacity scaling, and flexibility in adding or dropping channels at a user site can be achieved by introducing the concept of an optical cross-connect architecture in the physical path structure (the so-called *path layer*) of an optical network. These optical cross-connects (OXCs) operate directly in the optical domain and can route very high-capacity WDM data streams over a network of interconnected optical paths.[30-33]

To visualize the operations of an OXC, consider first the OXC architecture shown in Fig. 12-20 that uses space switching without wavelength conversion. The space switches can be constructed of a cascade of electronically controlled optical directional-coupler elements or semiconductor–optical-amplifier switching

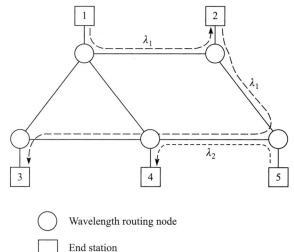

Wavelength routing node

End station

FIGURE 12-19
Wavelength reuse on a mesh network.

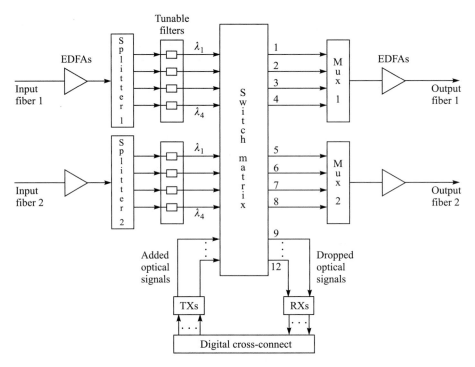

FIGURE 12-20
Optical cross-connect architecture using optical space switches and no wavelength converters.

gates.[34–39] Each of the input fibers carries M wavelengths (in this case, four), any or all of which could be added or dropped at a node. At the input, the arriving aggregate of signal wavelengths is amplified and passively divided into N streams by a power splitter. Tunable filters then select individual wavelengths, which are directed to an optical space-switching matrix. Alternatively, a waveguide-grating demultiplexer could be used to divide the incoming aggregate stream into individual wavelength channels. The switch matrix directs the channels either to one of the eight output lines if it is a through-traveling signal, or to a particular receiver attached to the OXC at output ports 9 through 12 if it has to be dropped to a user at that node. Signals that are generated locally by a user get connected electrically via the digital cross-connect matrix (DXC) to an optical transmitter. From here, they enter the switch matrix, which directs them to the appropriate output line. The M output lines, each carrying separate wavelengths, are fed into a wavelength multiplexer (popularly known as a *mux*, and a demultiplexer is a *demux*) to form a single aggregate output stream. An optical amplifier to boost the signal level for transmission over the trunk fiber normally follows this.

Contentions could arise in the architecture shown in Fig. 12-20 when channels having the same wavelength but traveling on different input fibers enter the OXC and need to be switched simultaneously to the same output fiber. This could be resolved by assigning a fixed wavelength to each optical path throughout the network, or by dropping one of the channels and retransmitting it at another wavelength. However, in the first case, wavelength reuse and network scalability (expandability) are reduced; in the second case, the add/drop flexibility of the OXC is lost. These blocking characteristics can be eliminated by using wavelength conversion at any output of the OXC, as shown in the following example.

Example 12-5. Consider the 4×4 OXC shown in Fig. 12-21. Here, two input fibers are each carrying two wavelengths. Either wavelength can be switched to any of the four output ports. The OXC consists of three 2×2 switch elements. Suppose that λ_2 on input fiber 1 needs to be switched to output fiber 2 and that λ_1 on input fiber 2 needs to be switched to output fiber 1. This is achieved by having the first two switch elements set in the bar state (the straight-through configuration) and the third element set in the cross state, as indicated in Fig. 12-21. Obviously, without wavelength conversion there would be wavelength contention at both mux output ports. By using wavelength converters ahead of the multiplexer, the cross-connected wavelengths can be converted to noncontending wavelengths.

12.4.2 Performance Evaluation of Wavelength Conversion

Numerous studies have been carried out to quantify the benefits of wavelength conversion.[40-45] These efforts employed either probabilistic models or they used deterministic algorithms on specific network topologies. The studies indicate that the benefits are greater in a mesh network than in a ring or fully connected network.

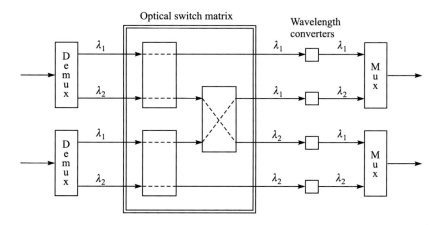

FIGURE 12-21
Example of a simple 4×4 optical cross-connect architecture using optical space switches and wavelength converters.

To illustrate the effect of wavelength conversion, we show a simple model that is based on standard series independent-link assumptions commonly used in circuit-switched networks.[42,46] In this simplified example, during a request for establishing a light-path connection between two stations, the usage of a wavelength on a fiber is statistically independent of other fiber links and other wavelengths. Although this model tends to overestimate the probability that a wavelength is blocked along a path, it provides insight into the network performance improvement when using wavelength conversion.

Assume that there are H links (or hops) between two nodes that need to be connected, which we will call nodes A and B. Take the number of available wavelengths per fiber link to be F, and let ρ be the probability that a wavelength is used on any fiber link. Then, since ρF is the expected number of busy wavelengths on any link, ρ is a measure of the *fiber utilization* along the path.

First, consider a network with wavelength conversion. In this case, a connection request between nodes A and B is blocked if one of the H intervening fibers is full; that is, the fiber is already supporting F independent sessions on different wavelengths. Thus, the probability P'_b that the connection request from A to B is blocked is the probability that there is a fiber link in this path with all F wavelengths in use, so that

$$P'_b = 1 - (1 - \rho^F)^H \qquad (12\text{-}23)$$

If q is the *achievable utilization* for a given blocking probability in a network *with* wavelength conversion, then[42]

$$q = [1 - (1 - P'_b)^{1/H}]^{1/F} \approx \left(\frac{P'_b}{H}\right)^{1/F} \qquad (12\text{-}24)$$

where the approximation holds for small values of P'_b/H. Figure 12-22 shows the achievable utilization q for $P'_b = 10^{-3}$ as a function of the number of wavelengths for $H = 5, 10,$ and 20 hops. The effect of path length is small, and q rapidly approaches 1 as F becomes large.

Now consider a network without wavelength conversion. Here, a connection request between A and B can be honored only if there is a free wavelength; that is, if there is a wavelength that is unused on each of the H intervening fibers. Thus, the probability P_b that the connection request from A to B is blocked is the probability that each wavelength is used on at least one of the H links, so that

$$P_b = [1 - (1 - \rho)^H]^F \qquad (12\text{-}25)$$

Letting p be the *achievable utilization* for a given blocking probability in a network *without* wavelength conversion, then[42]

$$p = 1 - (1 - P_b^{1/F})^{1/H} \approx -\frac{1}{H}\ln(1 - P_b^{1/F}) \qquad (12\text{-}26)$$

where the approximation holds for large values of H and for $P_b^{1/F}$ not too close to unity. In this case, the achievable utilization is inversely proportional to the length of the path H between A and B, as one would expect. Figure 12-23 shows this

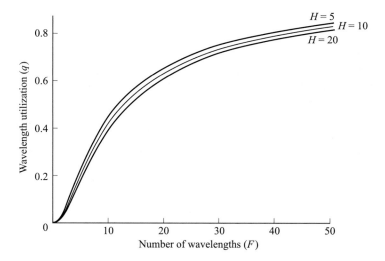

FIGURE 12-22
Achievable wavelength utilization as a function of the number of wavelengths for a 10^{-3} blocking probability in a network using wavelength conversion. (Reproduced with permission from Barry and Humblet,[42] © IEEE, 1996.)

effect. Analogous to Fig. 12-22, this depicts the achievable utilization p for $P_b = 10^{-3}$ as a function of the number of wavelengths for $H = 5$, 10, and 20 hops. In contrast to the previous case, here the effect of path length (i.e., the number of links) is dramatic.

To measure the benefit of wavelength conversion, define the gain $G = q/p$ to be the increase in fiber or wavelength utilization for the same blocking probability. Setting $P_b' = P_b$ in Eqs. (12-23) and (12-25), we have

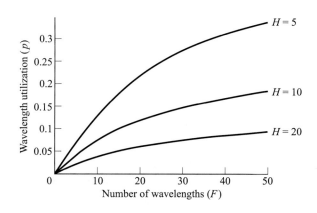

FIGURE 12-23
Achievable wavelength utilization as a function of the number of wavelengths for a 10^{-3} blocking probability in a network not using wavelength conversion. (Reproduced with permission from Barry and Humblet,[42] © IEEE, 1996.)

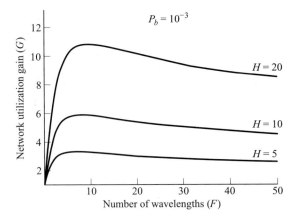

FIGURE 12-24
Increase in network utilization as a function of the number of wavelengths for a 10^{-3} blocking probability when wavelength conversion is used. (Reproduced with permission from Barry and Humblet,[42] © IEEE, 1996.)

$$G \equiv \frac{q}{p} = \frac{[1 - (1 - P_b)^{1/H}]^{1/F}}{1 - (1 - P_b^{1/F})^{1/H}}$$

$$\approx H^{1 - 1/F} \frac{P_b^{1/F}}{-\ln(1 - P_b^{1/F})} \tag{12-27}$$

As an example, Fig. 12-24 shows G as a function of $H = 5$, 10, and 20 links for a blocking probability of $P_b = 10^{-3}$. This figure shows that as F increases, the gain increases, and peaks at about $H/2$. The gain then slowly decreases, since large trunking networks are more efficient than small ones.

12.5 NONLINEAR EFFECTS ON NETWORK PERFORMANCE

Optical fiber transmission technology has found widespread use in networks of diversified speeds and sizes. A laudable accomplishment has been to create networks that span continents or oceans to interconnect communication assets in major urban areas. Important challenges in designing such networks include the following:

- Transmission of the different wavelength channels at the highest possible bit rate.
- Transmission over the longest possible distance with the smallest number of optical amplifiers.
- Network architectures that allow simple and efficient network operation, control, and management.

To meet these challenges, careful design practices must be followed, since various signal-impairment effects that are inherent in optical fiber transmission links can seriously degrade network performance. The following are among these effects:[47–53]

1. Group velocity dispersion (GVD), which limits the bit rate by temporally spreading a transmitted optical pulse. Whether one implements high-speed single-wavelength or WDM networks, dispersion-induced pulse spreading can be minimized by operation in a low-dispersion window. This is the 1310-nm band for standard fiber and the 1550-nm window for dispersion-shifted fiber.

2. Nonuniform gain across the desired wavelength range of EDFAs in WDM links. This characteristic of an EDFA can be equalized over the desired wavelength range by techniques such as the use of grating filters, as described in Chap. 11.

3. Polarization-mode dispersion (PMD), which arises from orthogonal polarization modes traveling at slightly different speeds owing to fiber birefringence. This effect cannot be easily mitigated and is a very serious impediment for links operating at 10 Gb/s and higher (see Secs. 3.2.6 and 13.4.5).

4. Reflections from splices and connectors that can cause instabilities in laser sources. These can be eliminated by the use of optical isolators.[54]

5. Nonlinear inelastic scattering processes, which are interactions between optical signals and molecular or acoustic vibrations in a fiber.

6. Nonlinear variations of the refractive index in a silica fiber that occur because the refractive index is dependent on intensity changes in the signal.

This section addresses the origins of the two nonlinear categories and shows the limitations they place on system performance. The first category encompasses the nonlinear inelastic scattering processes. These are stimulated Raman scattering (SRS) and stimulated Brillouin scattering (SBS). The second category of nonlinear effects arises from intensity-dependent variations in the refractive index in a silica fiber. This produces effects such as self-phase modulation (SPM), cross-phase modulation (XPM), and four-wave mixing (FWM). In the literature, FWM is also referred to as four-photon mixing (FPM), and XPM is sometimes designated by CPM.

The processes SBS, SRS, and FWM result in gains or losses in a wavelength channel that are dependent on the optical signal intensity. These nonlinear processes provide gains to some channels while depleting power from others, thereby producing crosstalk between the wavelength channels. In analog video systems, SBS significantly degrades the carrier-to-noise ratio when the scattered power is equivalent to the signal power in the fiber. Both SPM and XPM affect only the phase of signals, which causes chirping in digital pulses. This can worsen pulse broadening due to dispersion, particularly in very high-rate systems (> 10 Gb/s).

When any of these nonlinear effects contribute to signal impairment, an additional amount of power will be needed at the receiver to maintain the same BER as in their absence. This additional power (in decibels) is known as the *power penalty* for that effect.

Viewing these nonlinear processes in a little more detail, Sec. 12.5.1 first shows how to define the distances over which the processes are important.

Sections 12.5.2 and 12.5.3 then qualitatively describe the different ways in which the stimulated scattering mechanisms physically affect a light-wave system. Section 12.5.4 presents the origins of SPM and XPM and shows their degradation effect on system performance. The mechanisms giving rise to FWM and the resultant physical effects on link operation are outlined in Sec. 12.5.5. The FWM can be largely suppressed through clever arrangements of fibers that have different dispersion characteristics. Section 12.5.6 describes these dispersion-compensation techniques in which one tailors the various fiber segments in the link to have high local dispersion, but an overall low dispersion. The low average dispersion minimizes pulse spreading, whereas the high local dispersion destroys the carrier-frequency phase relationships that give rise to FWM inter-modulation products.

12.5.1 Effective Length and Area

Modeling the nonlinear processes can be quite complicated, since they depend on the transmission length, the cross-sectional area of the fiber, and the optical power level in the fiber. The difficulty arises from the fact that the impact of the nonlinearity on signal fidelity increases with distance. However, this is offset by the continuous decrease in signal power along the fiber due to attenuation. In practice, one can use a simple but sufficiently accurate model that assumes the power is constant over a certain fiber length, which is less than or equal to the actual fiber length. This *effective length* L_{eff}, which takes into account power absorption along the length of the fiber (i.e., the optical power decays exponentially with length), is given by[8]

$$L_{eff} = \frac{1}{P_0} \int_0^L P(z)\ dz = \int_0^L e^{-\alpha z}\ dz = \frac{1 - e^{-\alpha L}}{\alpha} \qquad (12\text{-}28)$$

Given a typical attenuation of 0.22 dB/km (or, equivalently, 5.07×10^{-2} km^{-1}) at 1550 nm yields an effective length of about 20 km when $L \gg 1/\alpha$. When there are optical amplifiers in a link, the signal impairments owing to the nonlinearities do not change as the signal passes through the amplifier. In this case, the effective length is the sum of the effective lengths of the individual spans between optical amplifiers. If the total amplified link length is L_A and the span length between amplifiers is L, the effective length is approximately[8]

$$L_{eff} = \frac{1 - e^{-\alpha L}}{\alpha} \frac{L_A}{L} \qquad (12\text{-}29)$$

Figure 12-25 illustrates the effective length as a function of the actual system length. The two curves shown are for a nonamplified link with $\alpha = 0.22$ dB/km and an amplified link with a 75-km amplifier spacing. As indicated by Eq. (12-29), the total effective length decreases as the amplifier span increases.

The effects of nonlinearities increase with the light intensity in a fiber. For a given optical power, this intensity is inversely proportional to the cross-sectional area of the fiber core. Since the power is not distributed uniformly across the

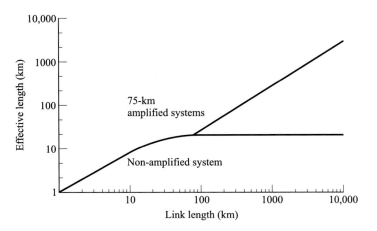

FIGURE 12-25
Effective length as a function of the actual link length for a nonamplified link [from Eq. (12-28)] and a link with optical amplifiers spaced 75 km apart [from Eq. (12-29)].

fiber-core cross section, for convenience one can use an *effective cross-sectional area* A_{eff}. In general, although it can be calculated from mode-overlap integrals, the effective area is close to the actual core area.[51] As a rule of thumb, standard non-dispersion-shifted single-mode fibers have effective areas of $80\,\mu\text{m}^2$, dispersion-shifted fibers have effective areas of $55\,\mu\text{m}^2$, and dispersion-compensating fibers have effective areas on the order of $20\,\mu\text{m}^2$.

12.5.2 Stimulated Raman Scattering

Stimulated Raman scattering is an interaction between light waves and the vibrational modes of silica molecules.[48–53] If a photon with energy $h\nu_1$ is incident on a molecule having a vibrational frequency ν_m, the molecule can absorb some energy from the photon. In this interaction, the photon is scattered, thereby attaining a lower frequency ν_2 and a corresponding lower energy $h\nu_2$. The modified photon is called a *Stokes photon*. Because the optical signal wave that is injected into a fiber is the source of the interacting photons, it is often called the *pump wave*, since it supplies power for the generated wave.

This process generates scattered light at a wavelength longer than that of the incident light. If another signal is present at this longer wavelength, the SRS light will amplify it and the pump-wavelength signal will decrease in power; Fig. 12-26 illustrates this effect. Consequently, SRS can severely limit the performance of a multichannel optical communication system by transferring energy from short-wavelength channels to neighboring higher-wavelength channels. This is a broadband effect that can occur in both directions. Powers in WDM channels separated by up to 16 THz (125 nm) can be coupled through the SRS effect, as Fig. 12-27 illustrates in terms of the Raman gain coefficient g_R as a function of the channel separation $\Delta\nu_s$. This shows that, owing to SRS, the power transferred from a lower-wavelength channel to a higher-wavelength channel increases approxi-

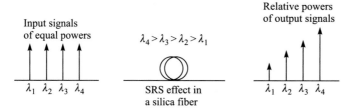

FIGURE 12-26
SRS transfers optical power from shorter wavelengths to longer wavelengths.

mately linearly with channel spacing up to a maximum of about $\Delta\nu_c = 16$ THz (or $\Delta\lambda_c = 125$ nm in the 1550-nm window), and then drops off sharply for larger spacings.

To see the effects of SRS, consider a WDM system that has N channels equally spaced in a 30-nm band centered at 1545 nm. Channel 0, which is at the lowest wavelength, is affected the worst since power gets transferred from this channel to all longer-wavelength channels. For simplicity, assume that the transmitted power P is the same on all channels, that the Raman gain increases linearly as shown by the dashed line in Fig. 12-27, and that there is no interaction between the other channels. If $F_{out}(j)$ is the fraction of power coupled from channel 0 to channel j, then the total fraction of power coupled out of channel 0 to all the other channels is (see Buck[55] for details)

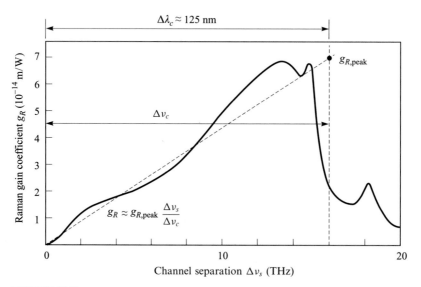

FIGURE 12-27
Raman gain coefficient g_R as a function of the wavelength channel separation. The dashed line is a linear approximation for g_R, as given by the inset equation, used for computational simplicity for channel separations up to 8 THz (or 60 nm in the 1550-nm window).

$$F_{\text{out}} = \sum_{j=1}^{N-1} F_{\text{out}}(j) = \sum_{j=1}^{N-1} g_{R,\text{peak}} \frac{j\Delta\nu_s}{\Delta\nu_c} \frac{PL_{\text{eff}}}{2A_{\text{eff}}} = \frac{g_{R,\text{peak}}\Delta\nu_s PL_{\text{eff}}}{2\Delta\nu_c A_{\text{eff}}} \frac{N(N-1)}{2} \quad (12\text{-}30)$$

The power penalty for this channel then is $-10 \log (1 - F_{\text{out}})$. To keep the penalty below 0.5 dB, we need to have $F_{\text{out}} < 0.1$. Using Eq. (12-30), and with $A_{\text{eff}} = 55\,\mu\text{m}^2$, this gives the criterion[48,55]

$$[NP][N - 1]\Delta\nu_s L_{\text{eff}} < 5 \times 10^3\,\text{mW} \cdot \text{THz} \cdot \text{km} \quad (12\text{-}31)$$

Here, NP is the total power coupled into the fiber, $(N - 1)\Delta\nu_s$ is the total occupied optical bandwidth, and L_{eff} is the effective length, which takes into account power absorption along the length of the fiber.

Example 12-6. The limits indicated by Eqs. (12-31) are illustrated in Fig. 12-28 for systems with four and eight wavelength channels. The curves show the maximum power per channel as a function of the number of wavelengths for three different channel spacings (recall that a 125-GHz frequency spacing is equivalent to a 1-nm wavelength spacing at 1550 nm), a fiber attenuation of 0.2 dB/km (or, equivalently, $4.61 \times 10^{-2}\,\text{km}^{-1}$), and an amplifier spacing of 75 km (which yields an effective length of $L_{\text{eff}} = 22$ km).

The results in Fig. 12-28 were calculated for the worst-case scenario caused by SRS. In general, if the optical power per channel is not excessively high (e.g.,

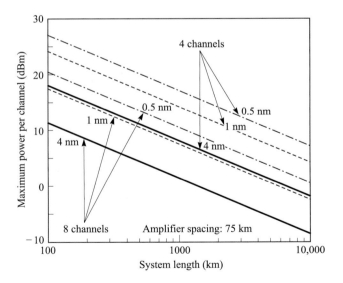

FIGURE 12-28
Maximum allowable power per wavelength channel versus transmission length for three different channel spacings. The curves are for the power levels that ensure an SRS degradation of less than 1 dB for all channels. (Reproduced with permission from O'Mahoney, Simeinidou, Yu, and Zhou, *J. Lightwave Tech.*, vol. 13, pp. 817–828, © IEEE, May 1995.)

less than 1 mW each), then the effects of SRS do not contribute significantly to the eye-closure penalty as a function of transmission distance.[30,50]

12.5.3 Stimulated Brillouin Scattering

Stimulated Brillouin scattering arises when lightwaves scatter from acoustic waves.[52,53,56–60] The resultant scattered wave propagates principally in the backward direction in single-mode fibers. This backscattered light experiences gain from the forward-propagating signals, which leads to depletion of the signal power. The frequency of the scattered light experiences a Doppler shift given by

$$\nu_B = 2nV_s/\lambda \qquad (12\text{-}32)$$

where n is the index of refraction and V_s is the velocity of sound in the material. In silica, this interaction occurs over a very narrow *Brillouin linewidth* of $\Delta\nu_B = 20$ MHz at 1550 nm. For $V_s = 5760$ m/s in fused silica, the frequency of the backward-propagating light at 1550 nm is downshifted by 11 GHz (0.09 nm) from the original signal. This shows that the SBS effect is confined within a single wavelength channel in a WDM system. Thus, the effects of SBS accumulate individually for each channel, and, consequently, occur at the same power level in each channel as occurs in a single-channel system.

System impairment starts when the amplitude of the scattered wave is comparable to the signal power. For typical fibers, the threshold power for this process is around 10 mW for single-fiber spans. In a long fiber chain containing optical amplifiers, there are normally optical isolators to prevent backscattered signals from entering the amplifier. Consequently, the impairment due to SBS is limited to the degradation occurring in a single amplifier-to-amplifier span.

One criterion for determining at what point SBS becomes a problem is to consider the SBS threshold power P_{th}. This is defined to be the signal power at which the backscattered light equals the fiber-input power. The calculation of this expression is rather complicated, but an approximation is given by[53]

$$P_{th} \approx 21 \frac{A_{eff}b}{g_B L_{eff}} \left(1 + \frac{\Delta\nu_{source}}{\Delta\nu_B}\right) \qquad (12\text{-}33)$$

Here, A_{eff} is the effective cross-sectional area of the propagating wave, and the polarization factor b lies between 1 and 2 depending on the relative polarizations of the pump and Stokes waves. The effective length L_{eff} is given in Eq. (12-28) and g_B is the *Brillouin gain coefficient*, which is approximately 4×10^{-11} m/W, independent of the wavelength. Equation (12-33) shows that the SBS threshold power increases as the source linewidth becomes larger.

Example 12-7. Consider an optical source with a 40-MHz linewidth. Using the values $\Delta\nu_B = 20$ MHz at 1550 nm, $A_{eff} = 55 \times 10^{-12}$ m^2 (for a typical dispersion-shifted single-mode fiber), $L_{eff} = 20$ km, and assuming a value of $b = 2$, then from Eq. (12-33) we have $P_{th} = 8.6$ mW $= 9.3$ dBm.

Figure 12-29 illustrates the effect of SBS on the signal power once the threshold is reached.[59] The plots give the Brillouin-scattered power and the signal power transmitted through a 13-km dispersion-shifted fiber as a function of the signal-input power. Below the SBS threshold, the transmitted power increases linearly with the input level. Once the scattered power reaches the SBS threshold, the transmitted power remains constant for higher inputs. This occurs since power is extracted from the signal to feed the scattered wave.

Figure 12-30 illustrates the SBS-induced impairment on the carrier-to-noise ratio (CNR) of an amplitude-modulated vestigial-sideband (AM-VSB) video signal for the same fiber as in Fig. 12-29. Here, the CNR grows with increasing fiber-injected power up to the SBS threshold. Beyond this point the CNR starts to decrease.[60]

Several schemes are available for reducing the power-penalty effects of SBS. These include[8,53,57]

1. Keeping the optical power per WDM channel below the SBS thresholds. For long-haul systems, this may require a reduction in the amplifier spacing.

2. Increasing the linewidth of the source, since the gain bandwidth of SBS is very small. This can be achieved through direct modulation of the source (as opposed to external modulation), since this causes the linewidth to broaden because of chirping effects. However, a large dispersion penalty may result from this.

3. Slightly dithering the laser output in frequency, at roughly tens of kilohertz. This is effective since SBS is a narrowband process. The dither frequency should scale as the ratio of the injected power to the SBS threshold. For

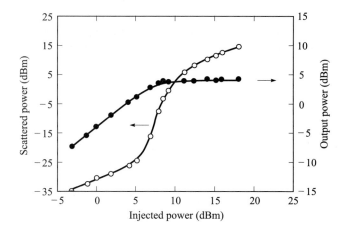

FIGURE 12-29
The effect of SBS on signal power. The signal power reaches a plateau once the SBS threshold is reached. (Adapted with permission from Mao, Tkach, Chraplyvy, Jopson, and Dorosier,[59] © IEEE, 1992.)

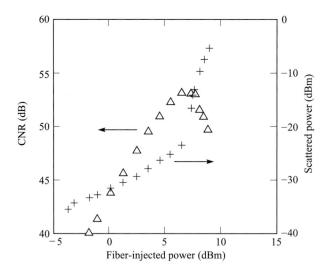

FIGURE 12-30

The SBS impairment on the CNR of an AM-VSB signal. The triangles are the CNR and the crosses represent the backscattered power. (Adapted with permission from Mao, Bodeep, Tkach, Chraplyvy, Darcie, and Dorosier,[60] © IEEE, 1992.)

high injected powers of 1 W, the dither frequency should be around 10 MHz. It has been shown that dithering can achieve near total suppression of SBS.[57]

12.5.4 Self-Phase Modulation and Cross-Phase Modulation

The refractive index n of many optical materials has a weak dependence on optical intensity I (equal to the optical power per effective area in the fiber) given by

$$n = n_0 + n_2 I = n_0 + n_2 \frac{P}{A_{\text{eff}}} \quad (12\text{-}34)$$

where n_0 is the ordinary refractive index of the material and n_2 is the *nonlinear index coefficient*. In silica, the factor n_2 varies from 2.2 to 3.4 $\times 10^{-8}$ $\mu\text{m}^2/\text{W}$. The nonlinearity in the refractive index is known as the *Kerr nonlinearity*. This nonlinearity produces a carrier-induced phase modulation of the propagating signal, which is called the *Kerr effect*. In single-wavelength links, this gives rise to *self-phase modulation* (SPM), which converts optical power fluctuations in a propagating light wave to spurious phase fluctuations in the same wave.[52,61]

To see the effect of SPM, consider what happens to the optical pulse shown in Fig. 12-31 as it propagates in a fiber. Here, the time axis is normalized to the parameter t_0, which is the pulse half-width at the $1/e$-intensity point. The edges of the pulse represent a time-varying intensity, which rises rapidly from zero to a maximum value, and then returns to zero. In a medium that has an intensity-dependent refractive index, a time-varying signal intensity will produce a time-varying refractive index. Thus, the index at the peak of the pulse will be slightly different than the value in the wings of the pulse. The leading edge will see a positive dn/dt, whereas the trailing edge will see a negative dn/dt.

This temporally varying index change results in a temporally varying phase change, shown by $d\phi/dt$ in Fig. 12-31. The consequence is that the instantaneous optical frequency differs from its initial value ν_0 across the pulse. That is, since the phase fluctuations are intensity-dependent, different parts of the pulse undergo different phase shifts. This leads to what is known as *frequency chirping*, in that the rising edge of the pulse experiences a red shift in frequency (toward higher frequencies), whereas the trailing edge of the pulse experiences a blue shift in frequency (toward lower frequencies). Since the degree of chirping depends on the transmitted power, SPM effects are more pronounced for higher-intensity pulses.

For some types of fibers, the time-varying phase may result in a power penalty owing to a GVD-induced spectral broadening of the pulse as it travels along the fiber. In the wavelength region where chromatic dispersion is negative [or, from Eq. (3-17), where $\beta_2 > 0$], the red-shifted leading edge of the pulse travels faster and thus moves away from the center of the pulse. The blue-shifted trailing edge travels slower, and also moves away from the center of the pulse. Therefore, in this case, chirping worsens the effects of GVD-induced pulse broadening. On the other hand, in the wavelength region where chromatic dispersion is positive, the red-shifted leading edge of the pulse travels slower and moves toward the center of the pulse. Similarly, the blue-shifted trailing edge travels faster, and

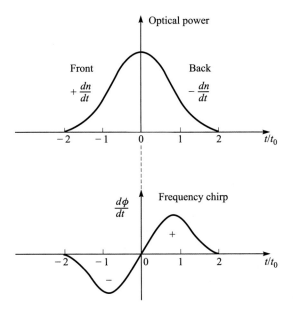

FIGURE 12-31
Phenomenological description of spectral broadening of a pulse due to self-phase modulation.

also moves toward the center of the pulse. In this case, SPM causes the pulse to narrow, thereby partly compensating for chromatic dispersion.

In WDM systems, the refractive index nonlinearity gives rise to *cross-phase modulation* (XPM), which converts power fluctuations in a particular wavelength channel to phase fluctuations in other copropagating channels.[48,52] This can be greatly mitigated in WDM systems operating over standard non-dispersion-shifted single-mode fiber, but can be a significant problem in WDM links operating at 10 Gb/s and higher over dispersion-shifted fiber. When combined with fiber dispersion, the spectral broadening from SPM and XPM can be a significant limitation in very long transmission links, such as cross-country or undersea systems.[8,62–64]

12.5.5 Four-Wave Mixing

Dense WDM transmission in which individual wavelength channels are modulated at rates of 10 Gb/s offers capacities of $N \times 10$ Gb/s, where N is the number of wavelengths. To transmit such high capacities over long distances requires operation in the 1550-nm window of dispersion-shifted fiber. In addition, to preserve an adequate signal-to-noise ratio, a 10-Gb/s system operating over long distances and having nominal optical repeater spacings of 100 km needs optical launch powers of around 1 mW per channel. For such WDM systems, the simultaneous requirements of high launch power and low dispersion give rise to the generation of new frequencies due to four-wave mixing.[8,53,65–67]

Four-wave mixing (FWM) is a third-order nonlinearity in silica fibers that is analogous to intermodulation distortion in electrical systems. When wavelength channels are located near the zero-dispersion point, three optical frequencies (ν_i, ν_j, ν_k) will mix to produce a fourth intermodulation product ν_{ijk} given by

$$\nu_{ijk} = \nu_i + \nu_j - \nu_k \qquad \text{with} \qquad i, j \neq k \qquad (12\text{-}35)$$

When this new frequency falls in the transmission window of the original frequencies, it can cause severe crosstalk.

Figure 12-32 shows a simple example for two waves at frequencies ν_1 and ν_2. As these wave copropagate along a fiber, they mix and generate sidebands at $2\nu_1 - \nu_2$ and $2\nu_2 - \nu_1$. Similarly, three copropagating waves will create nine new optical sideband waves at frequencies given by Eq. (12-35). These sidebands will travel along with the original waves and will grow at the expense of signal-

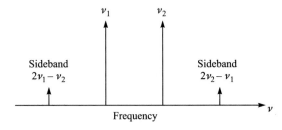

FIGURE 12-32

Two optical waves at frequencies ν_1 and ν_2 mix to generate two third-order sidebands.

strength depletion. In general, for N wavelengths launched into a fiber, the number of generated mixing products M is

$$M = \frac{N^2}{2}(N - 1) \tag{12-36}$$

If the channels are equally spaced, a number of the new waves will have the same frequencies as the injected signals. Thus, the resultant crosstalk interference plus the depletion of the original signal waves can severely degrade multichannel system performance unless steps are taken to diminish it.

The efficiency of four-wave mixing depends on fiber dispersion and the channel spacings. Since the dispersion varies with wavelength, the signal waves and the generated waves have different group velocities. This destroys the phase matching of the interacting waves and lowers the efficiency at which power is transferred to newly generated frequencies. The higher the group velocity mismatches and the wider the channel spacings, the lower the four-wave mixing.

At the exit of a fiber of length L and attenuation α, the power P_{ijk} that is generated at frequency ν_{ijk} due to the interaction of signals at frequencies ν_i, ν_j, and ν_k that have fiber-input powers P_i, P_j, and P_k, respectively, is[53]

$$P_{ijk}(L) = \eta(\mathscr{D}\kappa)^2 P_i(0)P_j(0)P_k(0)\exp(-\alpha L) \tag{12-37}$$

where the *nonlinear interaction constant* κ is

$$\kappa = \frac{32\pi^3 \chi_{1111}}{n^2 \lambda c}\left(\frac{L_{\text{eff}}}{A_{\text{eff}}}\right) \tag{12-38}$$

Here, χ_{1111} is the third-order nonlinear susceptibility; η is the efficiency of the four-wave mixing; n is the fiber refractive index; and $\mathscr{D}$ is the degeneracy factor, which has the value of 3 or 6 for two waves mixing or three waves mixing, respectively. The effective length L_{eff} is given by Eq. (12-29) and A_{eff} is the effective cross-sectional area of the fiber. Figure 12-33 gives examples of η as a function of channel spacing for three equally spaced frequencies for dispersion values of a conventional single-mode fiber [16 ps/(nm · km) average in the 1550-nm window] and a dispersion-shifted fiber [1 ps/(nm · km) average in the 1550-nm window]. These curves show the frequency-spacing range over which the FWM process is efficient for these two dispersion values (see Prob. 12-19 for a detailed expression for η, which leads to an oscillatory behavior of P_{ijk} as a function of channel spacing). For example, in the conventional single-mode fiber, only frequencies with separations less than 20 GHz will mix efficiently. In contrast, the FWM mixing efficiencies are greater than 20 percent for channel separations up to 50 GHz for dispersion-shifted fibers.

Example 12-8. Consider a 75-km link of dispersion-shifted single-mode fiber carrying two wavelengths at 1540.0 and 1540.5 nm. The new frequencies generated due to FWM are at

$$\nu_{112} = 2\nu_1 - \nu_2 = 2(1540.0 \text{ nm}) - 1540.5 \text{ nm} = 1539.5 \text{ nm}$$

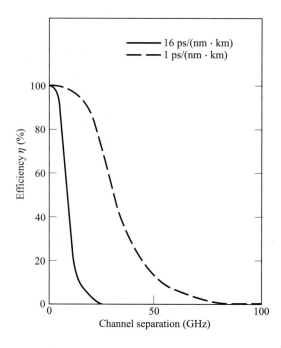

FIGURE 12-33
Efficiencies of four-wave mixing as a function of wavelength channel spacing. The solid curve is for standard single-mode fiber with 16-ps/ (nm · km) dispersion. The dashed curve is for dispersion-shifted fiber with 1-ps/(nm · km) dispersion. (Reproduced with permission from Chraplyvy,[48] © IEEE, 1990.)

and

$$\nu_{221} = 2\nu_2 - \nu_1 = 2(1540.5 \text{ nm}) - 1540.0 \text{ nm} = 1541.0 \text{ nm}$$

Assume the fiber has an attenuation of $\alpha = 0.20$ dB/km $= 0.0461$ km^{-1}, a refractive index of 1.48, and a 9.0-μm core diameter, so that $L_{\text{eff}} = 22$ km and $A_{\text{eff}} = 6.4 \times 10^{-11}$ m^2. From Fig. 12-33 we find $\eta \approx 5$ percent for a 62-GHz (0.5-nm) channel spacing. If each channel has an input power of 1 mW, then, using the values $\chi_{1111} = 6 \times 10^{-16}$ cm^3/erg $= 6 \times 10^{-15}$ m^3/(W · s) and $\mathscr{D} = 3$, we find

$$P_{112} = 0.05(3)^2 \left[\frac{32\pi^3 6 \times 10^{-15} \dfrac{\text{m}^3}{\text{W} \cdot \text{s}}}{(1.48)^2(1.54 \times 10^{-6} \text{ m})3 \times 10^8 \text{ m/s}} \right]^2$$

$$\times \left(\frac{22 \times 10^3 \text{ m}}{6.4 \times 10^{-11} \text{ m}^2} \right)^2 (1.0 \times 10^{-3} \text{ W})^3 \exp[-(0.0461/\text{km})75 \text{ km}]$$

$$= 5.80 \times 10^{-8} \text{ mW}$$

12.5.6 Dispersion Management

A large base of dispersion-shifted fiber has been installed throughout the world for use in single-wavelength transmission systems. Four-wave mixing can be a significant problem for these links when one attempts to upgrade them with high-speed dense WDM technology in which the channel spacings are less than 100 GHz and the bit rates are in excess of 2.5 Gb/s.

One approach to reducing the effect of FWM is to use passive dispersion compensation.[68–77] This consists of inserting into the link a loop of fiber having a dispersion characteristic that negates the accumulated dispersion of the transmission fiber. This process is called *dispersion compensation* and the fiber loop is referred to as a *dispersion-compensating fiber* (DCF). If the transmission fiber has a low positive dispersion [say, 2.3 ps/(nm · km)], then the DCF will have a large negative dispersion [say, −16 ps/(nm · km)]. With this technique, the total accumulated dispersion is zero after some distance, but the absolute dispersion per length is nonzero at all points along the fiber. The nonzero absolute value causes a phase mismatch between wavelength channels, thereby destroying the possibility of effective FWM production.

Figure 12-34 shows that the DCF can be inserted at either the beginning or the end of an installed fiber span between two optical amplifiers. A third option is to have a DCF at both ends. In *precompensation* schemes, the DCF is located right after the optical amplifier and thus just before the transmission fiber. Conversely, in *postcompensation* schemes, the DCF is placed right after the transmission fiber and just before the optical amplifier. Figure 12-34 also shows plots of the accumulated dispersion and the power level as functions of distance along the fiber. These plots are called *dispersion maps* and *power maps*, respectively. As Fig. 12-34a illustrates, in precompensation the DCF causes the dispersion to drop quickly to a low negative level from which it slowly rises toward zero (at the next

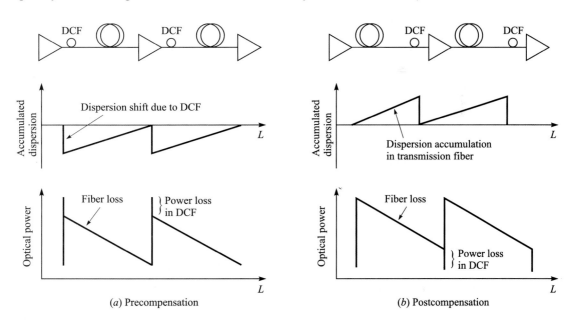

(a) Precompensation   (b) Postcompensation

FIGURE 12-34
Dispersion maps and power maps for (a) precompensation and (b) postcompensation methods.

optical amplifier) with increasing distance along the trunk fiber. This process repeats itself following amplification. The power map shows that the optical amplifier first boosts the power level to a high value. Since the DCF is a loop of fiber, there is a small drop in power level before the signal enters the actual transmission path, in which it decays exponentially before being amplified once more.

Similar processes occur in postcompensation, as shown in Fig. 12-34b. In either case, the accumulated dispersion is near zero after some distance to minimize the effects of pulse spreading, but the absolute dispersion per length is nonzero at all points, thereby causing a phase mismatch between different wavelengths, which effectively destroys FWM.

12.6   PERFORMANCE OF WDM + EDFA SYSTEMS

In addition to the performance limitations due to nonlinear effects discussed in Sec. 12.5, designing optically amplified WDM links and networks requires careful consideration of the system operating conditions. Among these are the link bandwidth, optical power requirements for a specific bit-error rate, and crosstalk between optical channels.

12.6.1   Link Bandwidth

If the N transmitters in a WDM link, such as shown in Fig. 12-13, operate at bit rates of B_1 through B_N, respectively, then the total bandwidth is

$$B = \sum_{i=1}^{N} B_i \tag{12-39}$$

When all the bit rates are equal, then the system capacity is enhanced by a factor N as compared with a single-channel link. For example, if the bandwidth of each channel is 2.5 Gb/s, then the total bandwidth of the WDM link for eight channels is 20 Gb/s and for 40 channels it is 100 Gb/s.

The total capacity of a WDM link depends on the bandwidth of the optical amplifier and on how closely the channels can be spaced in the available transmission window. The standard wavelength spacing is suggested to be 100 GHz by the ITU-T Recommendation G.692. As noted in that document, the central frequency is 193.100 THz (1552.524 nm), with the wavelength grid for a standard erbium-doped fiber amplifier ranging from 1537 nm to 1563 nm. This window can be extended to 1616 nm using a Raman amplifier or a silicate-based fiber instead of a silica-based fiber. Through the use of closer channel spacings and an extended EDFA range, vendors are making commercially available dense WDM links with 128 wavelengths.

12.6.2 Optical Power Requirements for a Specific BER

At the outputs of the demultiplexer, system parameters that need to be considered include the signal level, noise level, and crosstalk. The bit-error rate (BER) of a WDM channel is determined by the optical signal-to-noise ratio (SNR) delivered to the photodetector. For an acceptably low BER in an ideal link, this should be approximately 14 dB measured in a 0.01-nm optical bandwidth. For commercial systems, taking into account likely variations in realistic components and including reasonable amounts of system margin (usually between 3 and 6 dB), one typically needs SNRs of 18–20 dB. These values then determine the amount of optical power that must be launched into each wavelength channel, the number of EDFAs needed over the desired link length, and the fiber attenuation that can be tolerated in the spans between optical amplifiers.

An important factor to keep in mind is the difference in noise effects between an optically amplified WDM link and a conventional link without amplifiers, where the transmission performance is dominated by receiver noise. In an optically amplified link, the main noise factor for a digital "1" arises from the signal mixing with the ASE noise from the EDFA, whereas for a digital "0" signal the probability of error is determined by the ASE noise alone.

For a given channel transmitted over a link containing several optical amplifiers, the SNR starts out at a high level. It then decreases at each amplifier as the ASE noise accumulates through the length of the link, as shown in Fig. 12-35. The higher the gain in the amplifier, the faster the ASE noise builds up. However, although the SNR decreases quickly in the first few amplifications, the

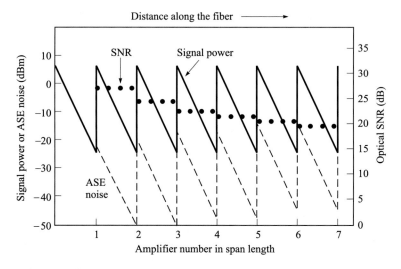

FIGURE 12-35
SNR degradation as a function of link distance over which the ASE noise increases with the number of amplifiers.

incremental effect of adding another EDFA diminishes rapidly with an increasing number of amplifiers. As a consequence, although the SNR drops by 3 dB when the number of EDFAs increases from three to six, it also drops by 3 dB when the number of amplifiers is further increased from six to twelve.

12.6.3   Crosstalk

The narrow channel spacings in dense WDM links give rise to *crosstalk*, which is defined as the feedthrough of one channel's signal into another channel.[78–81] Crosstalk can be introduced by almost any component in a WDM system, including optical filters, wavelength multiplexers and demultiplexers, optical switches, optical amplifiers, and the fiber itself.

    The two types of crosstalk that can occur in WDM systems are intrachannel and interchannel crosstalk. Both of these cause power penalties in the system performance. *Interchannel crosstalk* arises when an interfering signal comes from a neighboring channel that operates at a different wavelength. This nominally occurs when a wavelength-selecting device imperfectly rejects or isolates the signals from other nearby wavelength channels. Crosstalk then arises since these spurious neighboring signals could fall partially within the receiver passband. Figure 12-36 shows an example of crosstalk in a demultiplexer.

    For *intrachannel crosstalk*, the interfering signal is at the same wavelength as the desired signal. This effect is more severe than interchannel crosstalk, since the interference falls completely within the receiver bandwidth. Figure 12-37 gives an example of the origin of intrachannel crosstalk. Here, two independent signals, each at a wavelength λ_1, enter an optical switch. This switch routes the signal entering port 1 to output port 4, and routes the signal entering port 2 to output port 3. Within the switch, a spurious fraction of the optical power entering port 1 gets coupled to port 3, where it interferes with the signal from port 2.

    If the average received intrachannel crosstalk power is a fraction ε of the average received signal power P, then in an amplified system, where the dominant noise component is signal-dependent, the intrachannel power penalty is[8]

$$\text{Penalty}_{\text{intra}} = -5\ \log(1 - 2\sqrt{\varepsilon}) \qquad (12\text{-}40)$$

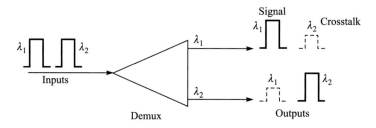

FIGURE 12-36
Example of the origin of interchannel crosstalk in a WDM system.

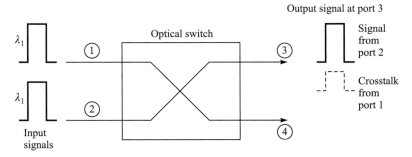

FIGURE 12-37
Example of the origin of intrachannel crosstalk in a WDM system.

If there are N interfering channels in a WDM system, each contributing an average crosstalk power $\varepsilon_i P$, then the factor ε in Eq. (12-40) is given by

$$\sqrt{\varepsilon} = \sum_{i=1}^{N} \sqrt{\varepsilon_i} \qquad (12\text{-}41)$$

For interchannel crosstalk, again let the received crosstalk power be a fraction ε of the average received signal power P. Again, considering an amplified system, the power penalty then is[8]

$$\text{Penalty}_{\text{inter}} = -5 \, \log(1 - \varepsilon) \qquad (12\text{-}42)$$

In this case, for N interfering channels, each with an average crosstalk power $\varepsilon_i P$, the factor ε in Eq. (12-42) is given by

$$\varepsilon = \sum_{i=1}^{N} \varepsilon_i \qquad (12\text{-}43)$$

Figure 12-38 illustrates the power penalties from intrachannel and interchannel crosstalk for 8 and 16 WDM channels as a function of the individual crosstalk level. Here, each channel contributes an equal amount of crosstalk power. This shows that the intrachannel effect is more severe, since it falls completely within the receiver passband. As an example, a 1-dB penalty arises when the intrachannel level is 38.7 dB below the signal level. For interchannel crosstalk this 1-dB-penalty level can be much higher (about 16 dB below the signal).

12.7 SOLITONS

As detailed in Chap. 3, group velocity dispersion (GVD) causes most pulses to broaden in time as they propagate through an optical fiber. However, a particular pulse shape known as a *soliton* takes advantage of nonlinear effects in silica, particularly self-phase modulation (SPM) resulting from the Kerr nonlinearity, to overcome the pulse-broadening effects of GVD.[82–90]

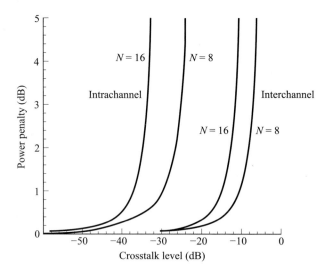

FIGURE 12-38
Power penalties from intrachannel and interchannel crosstalk for 8 and 16 WDM channels as a function of the individual crosstalk level.

The term "soliton" refers to special kinds of waves that can propagate undistorted over long distances and remain unaffected after collisions with each other. John Scott Russell made the first recorded observation of a soliton in 1838, when he saw a peculiar type of wave generated by boats in narrow Scottish canals.[91] The resulting water wave was of great height and traveled rapidly and unattenuated over a long distance. After passing through slower waves of lesser height, the waves emerged from the interaction undistorted, with their identities unchanged.

In an optical communication system, solitons are very narrow, high-intensity optical pulses that retain their shape through the interaction of balancing pulse dispersion with the nonlinear properties of an optical fiber. If the relative effects of SPM and GVD are controlled just right, and the appropriate pulse shape is chosen, the pulse compression resulting from SPM can exactly offset the pulse broadening effect of GVD. Depending on the particular shape chosen, the pulse either does not change its shape as it propagates, or it undergoes periodically repeating changes in shape. The family of pulses that do not change in shape are called *fundamental solitons*, and those that undergo periodic shape changes are called *higher-order solitons*. In either case, attenuation in the fiber will eventually decrease the soliton energy. Since this weakens the nonlinear interaction needed to counteract GVD, periodically spaced optical amplifiers are required in a soliton link to restore the pulse energy.

12.7.1 Soliton Pulses

Let us look at the soliton pulse features in more detail. No optical pulse is monochromatic, since it excites a spectrum of frequencies. For example, as Eq. (10-1) shows, if an optical source emits power in a wavelength band $\Delta\lambda$, its spectral spread is Δv. This is important, because in an actual fiber a pulse is affected by both the GVD and the Kerr nonlinearity. This is particularly significant for high-intensity optical excitations. Since the medium is dispersive, the pulse width will spread in time with increasing distance along the fiber owing to GVD. In addition, when a high-intensity optical pulse is coupled to a fiber, the optical power modulates the refractive index seen by the optical excitation. This induces phase fluctuations in the propagating wave, thereby producing a chirping effect in the pulse, as shown in Fig. 12-31 in Sec. 12.5.4. The result is that the front of the pulse (at smaller times) has lower frequencies and the back of the pulse (at later times) has higher frequencies than the carrier frequency.

When such a pulse traverses a medium with a positive GVD parameter β_2 for the constituent frequencies, the leading part of the pulse is shifted toward a longer wavelength (lower frequencies), so that the speed in that portion increases. Conversely, in the trailing half, the frequency rises so the speed decreases. This causes the trailing edge to be further delayed. Consequently, in addition to a spectral change with distance, the energy in the center of the pulse is dispersed to either side, and the pulse eventually takes on a rectangular-wave shape. Figure 12-39 illustrates these intensity changes as the pulse travels along such a fiber. The plot is in terms of the normalized time. These effects will severely limit high-speed long-distance transmission if the system is operated in this condition.

On the other hand, when a narrow high-intensity pulse traverses a medium with a negative GVD parameter for the constituent frequencies, GVD counteracts the chirp produced by SPM. Now, GVD retards the low frequencies in the front end of the pulse and advances the high frequencies at the back. The result is that the high-intensity sharply peaked soliton pulse changes neither its shape nor its

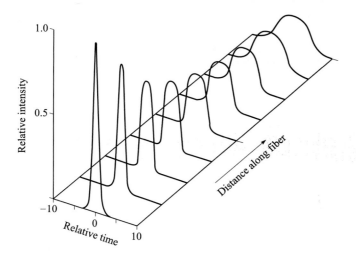

FIGURE 12-39
Temporal changes in a narrow high-intensity pulse that is subjected to the Kerr effect as it travels through a nonlinear dispersive fiber that has a positive GVD parameter.

spectrum as it travels along the fiber. Figure 12-40 illustrates this for a fundamental soliton. Provided the pulse energy is sufficiently strong, this pulse shape is
maintained as it travels along the fiber. In a standard optical fiber, there is a zero-
dispersion point around 1320 nm (see Fig. 3-24). For wavelengths shorter than
1320 nm β_2 is positive, and for longer wavelengths it is negative. Thus, soliton
operation is limited to the region greater than 1320 nm.

To derive the evolution of the pulse shape required for soliton transmission,
one needs to consider the *nonlinear Schrödinger (NLS) equation*[87,88]

$$-j\frac{\partial u}{\partial z} = \frac{1}{2}\frac{\partial^2 u}{\partial t^2} + N^2|u|^2u - j(\alpha/2)u \qquad (12\text{-}44)$$

Here, $u(z, t)$ is the pulse envelope function, z is the propagation distance along the
fiber, N is an integer designating the *order* of the soliton, and α is the coefficient of
energy gain per unit length, with negative values of α representing energy loss.
Following conventional notation, the parameters in Eq. (12-44) have been
expressed in special soliton units to eliminate scaling constants in the equation.

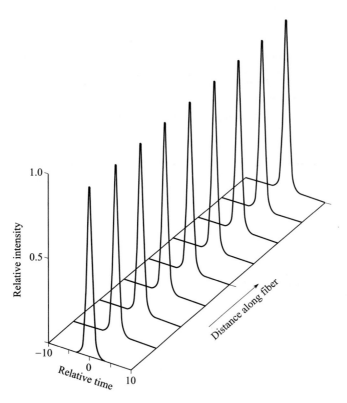

FIGURE 12-40
Characteristics of a high-intensity sharply peaked soliton pulsed that is subject to the Kerr effect as it
travels through a nonlinear dispersive fiber that has a negative GVD parameter.

These parameters (defined in Sec. 12.7.2) are the *normalized time* T_0, the *dispersion length* L_{disp}, and the *soliton peak power* P_{peak}.

For the three right-hand terms in Eq. (12-44):

1. The first term represents GVD effects of the fiber. Acting by itself, dispersion tends to broaden pulses in time.
2. The second nonlinear term denotes the fact that the refractive index of the fiber depends on the light intensity. Through the self-modulation process, this physical phenomenon broadens the frequency spectrum of a pulse.
3. The third term represents the effects of energy loss or gain; for example, due to fiber attenuation or optical amplification, respectively.

Solving the NLS equation analytically yields a pulse envelop that is either independent of z (for the fundamental soliton with $N = 1$) or that is periodic in z (for higher-order solitons with $N \geq 2$). The general theory of solitons is mathematically complex and can be found in the literature.[84,85] Several excellent and comprehensive overviews of solitons, which are beyond the scope of this book, are recommended to readers who want more details.[86–90] Here, we present the basic concepts for fundamental solitons. The solution to Eq. (12-44) for the fundamental soliton is given by

$$u(z, t) = \text{sech}(t)\exp(jz/2) \tag{12-45}$$

where sech(t) is the hyperbolic secant function. This is a bell-shaped pulse, as Fig. 12-41 illustrates. The time scale is given in units normalized to the $1/e$ width of the pulse. Since the phase term $\exp(jz/2)$ in Eq. (12-45) has no influence on the shape of the pulse, the soliton is independent of z and hence is nondispersive in the time domain.

When examining the NLS equation, one finds that the first-order effects of the dispersive and nonlinear terms are just complementary phase shifts. For a pulse given by Eq. (12-45), these phase shifts are

$$d\phi_{\text{nonlin}} = |u(t)|^2\,dz = \text{sech}^2(t)\,dz \tag{12-46}$$

for the nonlinear process, and

$$d\phi_{\text{disp}} = \left(\frac{1}{2u}\frac{\partial^2 u}{\partial t^2}\right)dz = [\tfrac{1}{2} - \text{sech}^2(t)]dz \tag{12-47}$$

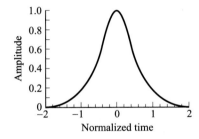

FIGURE 12-41
The hyperbolic secant function used for soliton pulses. The time scale is given in units normalized to the $1/e$ width of the pulse.

for the dispersion effect. Figure 12-42 shows plots of these terms and their sum, which is a constant. Upon integration, the sum simply yields a phase shift of $z/2$, which is common to the entire pulse. Since such a phase shift changes neither the temporal nor the spectral shape of a pulse, the soliton remains completely non-dispersive in both the temporal and frequency domains.

12.7.2   Soliton Parameters

Recall that the full-width half-maximum (FWHM) of a pulse is defined as the full width of the pulse at its *half-maximum power level* (see Fig. 12-43). For the solutions to Eq. (12-44), the power is given by the square of the envelope function in Eq. (12-40). Thus, the FWHM T_s of the fundamental soliton pulse in normalized time is found from the relationship $\text{sech}^2(\tau) = \frac{1}{2}$ with $\tau = T_s/(2T_0)$, where T_0 is the basic normalized time unit. This yields

$$T_0 = \frac{T_s}{2 \cosh^{-1} \sqrt{2}} = \frac{T_s}{1.7627} \approx 0.567 T_s \qquad (12\text{-}48)$$

▓▓ **Example 12-9.** Typical soliton FWHM pulse widths T_s range from 15 to 50 ps, so that the normalized time T_0 is on the order of 9–30 ps.

The *normalized distance* parameter (also called *dispersion length*) L_{disp} is a characteristic length for the effects of the dispersion term. As described later, L_{disp} is a measure of the period of a soliton. This parameter is given by

$$L_{\text{disp}} = \frac{2\pi c}{\lambda^2} \frac{T_0^2}{D} = \frac{1}{[2 \cosh^{-1} \sqrt{2}]^2} \frac{2\pi c}{\lambda^2} \frac{T_s^2}{D} \approx 0.322 \frac{2\pi c}{\lambda^2} \frac{T_s^2}{D} \qquad (12\text{-}49)$$

where c is the speed of light, λ is the wavelength in vacuum, and D is the dispersion of the fiber.

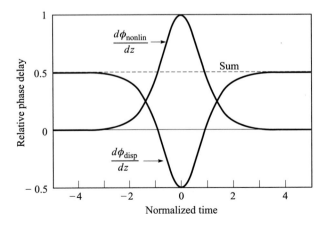

FIGURE 12-42
Dispersion and nonlinear phase shifts of a soliton pulse. Their sum is a constant, which yields a common phase shift for the entire pulse.

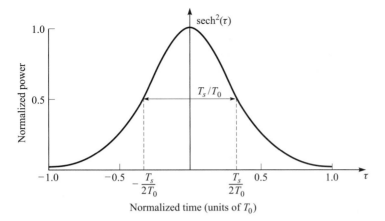

FIGURE 12-43
Definition of the half-maximum soliton width in terms of normalized time units.

■ **Example 12-10.** Consider a dispersion-shifted fiber having $D = 0.5$ ps/(nm · km) at 1550 nm. If $T_s = 20$ ps, we have

$$L_{disp} = \frac{1}{(1.7627)^2} \frac{2\pi(3 \times 10^8 \text{ m/s})}{(1550 \text{ nm})^2} \frac{(20 \text{ ps})^2}{0.5 \text{ ps/(nm · km)}} = 202 \text{ km}$$

which shows that L_{disp} is on the order of hundreds of kilometers.

The parameter P_{peak} is the *soliton peak power* and is given by

$$P_{peak} = \frac{A_{eff}}{2\pi n_2} \frac{\lambda}{L_{disp}} = \left(\frac{1.7627}{2\pi}\right)^2 \frac{A_{eff}\lambda^3}{n_2 c} \frac{D}{T_s^2} \tag{12-50}$$

where A_{eff} is the effective area of the fiber core, n_2 is the nonlinear intensity-dependent refractive-index coefficient [see Eq. (12-34)], and L_{disp} is measured in km.

■ **Example 12-11.** For $\lambda = 1550$ nm, $A_{eff} = 50 \, \mu m^2$, $n_2 = 2.6 \times 10^{-16} \text{ cm}^2/\text{W}$, and with the value of $L_{disp} = 202$ km from Example 12-10, we have that the soliton peak power P_{peak} is

$$P_{peak} = \frac{A_{eff}}{2\pi n_2} \frac{\lambda}{L_{disp}} = \frac{(50 \, \mu m^2)}{2\pi(2.6 \times 10^{-16} \text{ cm}^2/\text{W})} \frac{1550 \text{ nm}}{202 \text{ km}} = 2.35 \text{ mW}$$

This shows that when L_{disp} is on the order of hundreds of kilometers, P_{peak} is on the order of a few milliwatts.

For $N > 1$, the soliton pulse experiences periodic changes in its shape and spectrum as it propagates through the fiber. It resumes its initial shape at multiple distances of the *soliton period*, which is given by

$$L_{period} = \frac{\pi}{2} L_{disp} \tag{12-51}$$

As an example, Fig. 12-44 shows the evolution of a second-order soliton ($N = 2$).

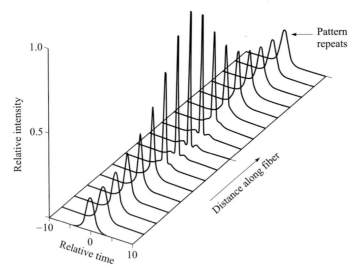

FIGURE 12-44
Propagation characteristics of a second-order soliton ($N = 2$).

12.7.3   Soliton Width and Spacing

The soliton solution to the NLS equation holds to a reasonable approximation only when individual pulses are well separated. To ensure this, the soliton width must be a small fraction of the bit slot. This eliminates use of the non-return-to-zero (NRZ) format that is commonly implemented in standard digital systems. Consequently, the return-to-zero (RZ) format is used. This condition thus constrains the achievable bit rate, since there is a limit on how narrow a soliton pulse can be generated.

If T_B is the width of the bit slot, then we can relate the bit rate B to the soliton half-maximum width T_s by

$$B = \frac{1}{T_B} = \frac{1}{2s_0 T_0} = \frac{1.7627}{2s_0 T_s} \tag{12-52}$$

where the factor $2s_0 = T_B/T_0$ is the *normalized separation* between neighboring solitons.

The physical explanation of the separation requirement is that the overlapping tails of closely spaced solitons create nonlinear interactive forces between them. These forces can be either attractive or repulsive, depending on the initial relative phase of the solitons. For solitons that are initially in phase and separated by $2s_0 \gg 1$, the soliton separation is periodic with an *oscillation period*:[92,93]

$$\Omega = \frac{\pi}{2} \exp(s_0) \tag{12-53}$$

The mutual interactive force between in-phase solitons thus results in periodic attraction, collapse, and repulsion. The *interaction distance* is[90]

$$L_I = \Omega L_{disp} = L_{period} \exp(s_0) \tag{12-54}$$

This interaction distance, and particularly the ratio L_I/L_{disp}, determine the maximum bit rate allowable in soliton systems.

These types of interactions are not desirable in a soliton system, since they lead to jitter in the soliton arrival times. One method for avoiding this situation is to increase s_0, since the interaction between solitons depends on their spacing. Since Eq. (12-52) is accurate for $s_0 > 3$, this equation together with the criterion that $\Omega L_{disp} \gg L_T$, where L_T is the total transmission distance, is suitable for system designs in which soliton interaction can be ignored.

Using Eq. (12-49) for L_{disp}, Eq. (12-52) for T_0, and Eq. (3-17) for D, the design condition $\Omega L_{disp} \gg L_T$ becomes

$$B^2 L_T \ll \left(\frac{2\pi}{s_0\lambda}\right)^2 \frac{c}{16D} \exp(s_0) = \frac{\pi}{8s_0^2|\beta_2|} \exp(s_0) \tag{12-55}$$

When written in this form, Eq. (12-55) shows the effects on the bandwidth B or the total transmission distance L_T for selected values of s_0.

Example 12-12. Suppose we want to transmit information at a rate of 10 Gb/s over an 8600-km trans-Pacific soliton link.

(a) Since this is a high data rate over a long distance, we start by selecting a value of $s_0 = 8$. Then, from Eq. (12-53) we have $\Omega = 4682$. Given that the dispersion length is at least 100 km, then $\Omega L_{disp} > 4.7 \times 10^5$ km, which for all practical purposes satisfies the condition $\Omega L_{disp} \gg L_T = 8600$ km.

(b) If $D = 0.5$ ps/(nm · km) at 1550 nm, then Eq. (12-55) yields

$$B^2 L_T \ll 2.87 \times 10^7 \text{ km(Gb/s)}^2$$

For a 10-Gb/s data rate, the transmission distance then must satisfy the condition

$$L_T \ll 2.87 \times 10^5 \text{ km}$$

This is satisfied, since the right-hand side is 33 times greater than the desired length.

(c) Using Eq. (12-52), we find the FWHM soliton pulse width to be

$$T_s = \frac{0.881}{s_0 B} = \frac{0.881}{8(10 \times 10^9 \text{ b/s})} = 11 \text{ ps}$$

(d) The fraction of the bit slot occupied by a soliton when $s_0 = 8$ is

$$\frac{T_s}{T_B} = \frac{0.881}{s_0} = \frac{0.881}{8} = 11\%$$

Note that for a given value of s_0, this is independent of the bit rate. For example, if the data rate is 20 Gb/s, then the FWHM pulse width is 5.5 ps, which also occupies 11 percent of the bit slot.

12.8 OPTICAL CDMA

In long-haul optical fiber transmission links and networks, the information consists of a multiplexed aggregate data stream originating from many individual subscribers and normally is sent in a well-timed synchronous format. The design goal of this TDM process is to make maximum use of the available optical fiber bandwidth for information transmission, since the multiplexed information stream requires very high-capacity links. To increase the capacity even further, WDM techniques that make use of the wide spectral transmission window in optical fibers are employed. As an alternative to these techniques in a local-area network (LAN), optical code-division multiple access (CDMA) has been examined.[94–105] This scheme can provide multiple access to a network without using wavelength-sensitive components as in WDM, and without employing very high-speed electronic data-processing devices as are needed in TDM networks. In the simplest configuration, CDMA achieves multiple access by assigning a unique code to each user. To communicate with another node, users imprint their agreed-upon code onto the data. The receiver can then decode the bit stream by locking onto the same code sequence.

The principle of optical CDMA is based on spread-spectrum techniques, which have been widely used in mobile-satellite and digital-cellular communication systems.[106] The concept is to spread the energy of the optical signal over a frequency band that is much wider than the minimum bandwidth required to send the information. For example, a signal that conveys 10^3 b/s may be spread over a 1 MHz bandwidth. This spreading is done by a code that is independent of the signal itself. Thus, an optical encoder is used to map each bit of information into a high-rate (longer code-length) *optical sequence*.

The symbols in the spreading code are called *chips*, and the energy density of the transmitted waveform is distributed more or less uniformly over the entire spread-spectrum bandwidth. The set of optical sequences becomes a set of unique *address codes* or *signature sequences* for the individual network users. In this addressing scheme, each 1 data bit is encoded into a waveform or signature sequence $s(n)$ consisting of N chips, which represents the destination address of that bit. The 0 data bits are not encoded. Ideally, all of the signature sequences would be mutually orthogonal, and each receiver would process only the address signals intended for it. However, in practice, "nearly orthogonal" is the best that has been accomplished. Consequently, there is some amount of cross correlation between the various addresses. Figure 12-45 illustrates the encoding scheme. Here, the signature sequence contains six chips. When the data signal contains a 1 data bit, the six-chip sequence is transmitted; no chips are sent for a 0 data bit.

From a simplistic point of view, this unique address-encoding scheme can be considered analogous to having numerous pairs of people, in the same room, talking simultaneously using different languages. Ideally, each communicating pair will understand only their own language, so that interference generated by other speakers is minimal. Thus, time-domain optical CDMA allows a number of users to access a network simultaneously through the use of a common wave-

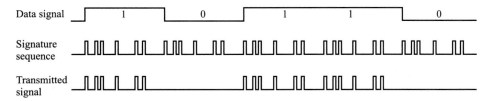

FIGURE 12-45
Example of a six-chip optical CDMA encoding scheme.

length. This is particularly useful in ultrahigh-speed LANs where bit rates of more than 100 Gb/s will be utilized (see Sec. 12.9). A basic limitation of optical CDMA using a coded sequence of pulses is that as the number of users increases, the code length has to be increased in order to maintain the same performance. Since this leads to shorter and shorter pulses, various ideas for mitigating this effect have been proposed. Alternatively, frequency-domain methods based on spectral encoding of broadband incoherent sources (e.g., LEDs or Fabry-Perot lasers) have been proposed.[103,107]

Both asynchronous and synchronous optical CDMA techniques have been examined. Each of these has its strengths and limitations. In general, since synchronous accessing schemes follow rigorous transmission schedules, they produce more successful transmissions (higher throughputs) than asynchronous methods where network access is random and collisions between users can occur. In applications that require real-time transmission, such as voice or interactive video, synchronous accessing techniques are most efficient. When the traffic tends to be bursty in nature or when real-time communication requirements are relaxed, such as in data transmission or file transfers, asynchronous multiplexing schemes are more efficient than synchronous multiplexing.

Figure 12-46 shows an optical CDMA network that is based on the use of a coded sequence of pulses. The setup consists of N transmitter and receiver pairs interconnected in a star network. To send information from node j to node k, the address code for node k is impressed upon the data by the encoder at node j. At the destination, the receiver differentiates between codes by means of correlation detection. That is, each receiver correlates its own address $f(n)$ with the received signal $s(n)$. The receiver output $r(n)$ is

$$r(n) = \sum_{k=1}^{N} s(k)f(k - n) \qquad (12\text{-}56)$$

If the received signal arrives at the correct destination, then $s(n) = f(n)$, and Eq. (12-56) represents an autocorrelation function. At an incorrect destination, $s(n) \neq f(n)$, and Eq. (12-56) represents a cross-correlation function. For a receiver to be able to distinguish the proper address correctly, it is necessary to maximize the autocorrelation function and minimize the cross-correlation function. This is accomplished by selecting the appropriate code sequence.

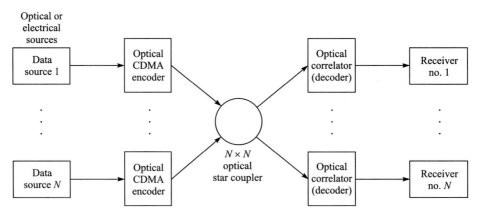

FIGURE 12-46

Example of an optical CDMA network based on using a coded sequence of pulses.

Prime-sequence codes and optical orthogonal codes (OOCs) are the commonly used spreading sequences in optical CDMA systems.[108,109] In addition, bipolar codes have been investigated.[101,110–112] These are particularly suitable for hybrid WDMA/CDMA systems. As noted earlier, some amount of cross correlation exists among the various optical sequences, since the codes devised so far are not perfectly orthogonal. In optical orthogonal codes, for example, there are many more zeros than ones in order to minimize the overlap of different code sequences. If the code words overlap in too many positions, then interference among users will be high. In an OOC system the number of simultaneous users N is bounded by[94]

$$N \le \left\lfloor \frac{F-1}{K(K-1)} \right\rfloor \qquad (12\text{-}57)$$

where F is the length of the code sequence and K is the *weight* or the number of ones in the sequence. Here, the symbol $\lfloor x \rfloor$ denotes the integer portion of the real value of x.

Example 12-13. Consider an optical orthogonal code of length $F = 32$ and weight $K = 4$. Then, from Eq. (12-57) the maximum number of simultaneous users is $N = 2$. If the code length is increased to 100 and the same weight is used, then eight simultaneous users can be supported. A code with $F = 341$ and $K = 5$ can have $N = 17$ unique addresses.

12.9 ULTRAHIGH CAPACITY NETWORKS

A major challenge in devising new optical communication systems has been the desire to fully exploit the enormous bandwidth of at least 25 THz that optical fiber channels can provide. Advances in very dense WDM technology, ultrafast optical TDM, and the creation of clever techniques for mitigating signal impairments have already allowed transfer rates in excess of 1 Tb/s to be achieved on a single fiber.[113-120] In addition to using dense WDM techniques to increase the capacity of long-haul transmission links, ultrafast optical TDM schemes have been devised.[121-127] These are particularly attractive in local-area networks (LANs) or metropolitan-area networks (MANs). In such networks, the system performance over the relatively short transmission distances is not affected as adversely by nonlinear dispersion effects as in long-haul links. In particular, researchers have examined the application of 100-Gb/s optical TDM schemes to shared-media local networks. Two candidate methods are bit-interleaved TDM and time-slotted TDM. Although these two techniques and WDM are identical at some level of mathematical abstraction, they are very different in practical applications.

12.9.1 Ultrahigh Capacity WDM Systems

Higher and higher capacities are continually being demonstrated in dense WDM systems. Two approaches are popular for achieving these increased capacities. The first is to widen the spectral bandwidth of EDFAs from 30 to 80 nm. In conventional optical amplifiers, this bandwidth ranges from 1530 to 1560 nm, whereas by using broadening techniques, such as boosting the gain at higher wavelengths with a Raman fiber amplifier, the useable EDFA bandwidth can cover the 1530-to-1610-nm range. Thus, the number of wavelengths that can be sent through the system increases greatly. The second method for increasing the capacity of a WDM link is to improve the spectral efficiency of the WDM signals. This will increase the total transmission capacity independent of any expansion of the EDFA bandwidth.[119]

Most of the demonstrations use a rate of 20 Gb/s for each individual wavelength in order to avoid nonlinear effects. Two examples are as follows:

- A 50-channel WDM system operating at an aggregated 1-Tb/s rate over a 600-km link.[118]
- A 132-channel WDM system operating at an aggregated 2.6-Tb/s rate over a 120-km link.[119]

12.9.2 Bit-Interleaved Optical TDM

Bit-interleaved TDM is similar to WDM in that the access nodes share many small channels operating at a peak rate that is a fraction of the media rate.[114] For example, the channel rates could vary from 100 Mb/s to 1 Gb/s, whereas the time-multiplexed media rate is around 100 Gb/s. Figure 12-47 illustrates the basic concept of point-to-point transmission using bit-interleaved optical TDM. A laser source produces a regular stream of very narrow return-to-zero optical pulses at a repetition rate B. This rate typically ranges from 2.5 to 10 Gb/s, which

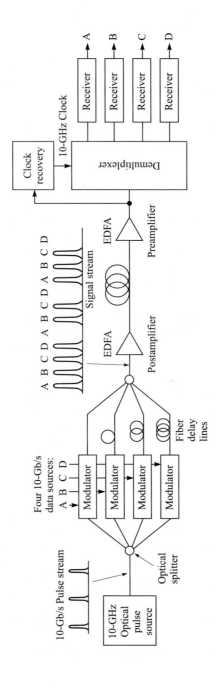

FIGURE 12-47
Example of an ultrafast point-to-point transmission system using optical TDM. (Adapted with permission from Cotter, Lucek, and Marcenac,[114] © IEEE, 1997.)

corresponds to the bit rate of the electronic data tributaries feeding the system. An optical splitter divides the pulse train into N separate streams. In the example in Fig. 12-47, the pulse stream is 10 Gb/s and $N = 4$. Each of these channels is then individually modulated by an electrical tributary data source at a bit rate B. The modulated outputs are delayed individually by different fractions of the clock period, and are then interleaved through an optical combiner to produce an aggregate bit rate of $N \times B$.

Optical postamplifiers and preamplifiers are generally included in the link to compensate for splitting and attenuation losses. At the receiving end, the aggregate pulse stream is demultiplexed into the original N independent data channels for further signal processing. In this technique, a clock-recovery mechanism operating at the base bit rate B is required at the receiver to drive and synchronize the demultiplexer.

12.9.3   Time-Slotted Optical TDM

In *time-slotted TDM*, the access nodes share one fast channel, which is capable of sending burst rates at 100 Gb/s. In such systems, the generation of a high-speed signal with uniform pulse separations is important for suppressing crosstalk effects from adjacent pulses and to minimize the jitter during timing extraction.[117] The most important features of time-slotted optical TDM networks are to provide a backbone to interconnect high-speed networks, to transfer quickly very large data blocks, to switch large aggregations of traffic, and to provide both high-rate and low-rate access to users. These types of networks can provide truly flexible bandwidth-on-demand services in a local environment for bursty users that may operate at speeds ranging from 10 to 100 Gb/s, as well as accommodating aggregates of lower-speed users. High-end customers include high-speed video servers, terabyte media banks, and supercomputers. The advantages of using high-speed time-slotted optical TDM are that, depending on the user rates and traffic statistics, it can provide improvements in terms of shorter user-access time, lower delay, and higher throughput. In addition, end-node equipment is conceptually simpler for single-channel versus multiple-channel approaches.

PROBLEMS

12-1. An engineer plans to construct an in-line optical fiber data bus operating at 10 Mb/s. The stations are to be separated by 100 m, for which optical fibers with a 3-dB/km attenuation are used. The optical sources are laser diodes having a 500-μW (-3 dBm) output from a fiber flylead, and the detectors are avalanche photodiodes with a 1.6-nW (-58-dBm) sensitivity. The couplers have a power tap-off factor of $C_T = 5$ percent and a 10-percent fractional intrinsic loss F_i. The power loss at the connectors is 20 percent (1 dB).

(a) Make a plot of P_{1N} in dBm as a function of the number of stations N for $2 \le N \le 12$.

(b) What is the operating margin of the system for eight stations?

(c) What is the worst-case dynamic range for the maximum allowable number of stations if a 6-dB power margin is required?

12-2. Consider an N-node star network in which 0 dBm of optical power is coupled from any given transmitter into the star. Let the fiber loss be 0.3 dB/km. Assume the stations are located 2 km from the star, the receiver sensitivity is −38 dBm, each connector has a 1-dB loss, the excess loss in the star coupler is 3 dB, and the link margin is 3 dB.

(a) Determine the maximum number of stations N that can be incorporated on this network.

(b) How many stations can be attached if the receiver sensitivity is −32 dBm?

12-3. A two-story office building has two 10-feet-wide hallways per floor that connect four rows of offices with eight offices per row as is shown in Fig. P12-3. Each office is a 15 feet × 15 feet square. The office ceiling height is 9 feet with a false ceiling hung 1 foot below the actual ceiling. Also, as shown in Fig. P12-3, there is a wiring room for LAN interconnection and control equipment in one corner of each floor. Every office has a local area-network socket on each of the two walls that are perpendicular to the hallway wall. If we assume that cables can be run only in the walls and in the ceilings, estimate the length of cable (in feet) that is required for the following configurations:

(a) A coaxial cable bus with a twisted-pair wire drop from the ceiling to each outlet.

(b) A fiber-optic star that connects each outlet to the wiring room on the corresponding floor and a vertical fiber-optic riser that connects the stars in each wiring room.

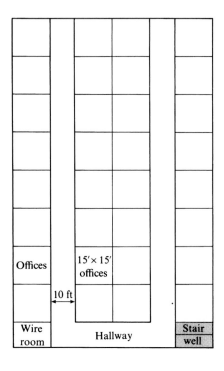

FIGURE P12-3

12-4. Consider the $M \times N$ grid of stations shown in Fig. P12-4 that are to be connected by a local-area network. Let the stations be spaced a distance d apart and assume that interconnection cables will be run in ducts that connect nearest-neighbor stations (i.e., ducts are not run diagonally in Fig. P12-4). Show that for the following configurations, the cable length for interconnecting the stations is as stated:

(a) $(MN - 1)d$ for a bus configuration.

(b) MNd for a ring topology.

(c) $MN(M + N - 2)d/2$ for a star topology where each subscriber is connected *individually* to the network hub which is located *in one corner* of the grid.

12-5. Consider the $M \times N$ rectangular grid of computer stations shown in Fig. P12-4, where the spacing between stations is d. Assume these stations are to be connected by a star-configured LAN using the duct network shown in the figure. Furthermore, assume that each station is connected to the central star by means of its own dedicated cable.

(a) If m and n denote the relative position of the star, show that the total cable length L needed to connect the stations is given by

$$L = [MN(M + N + 2)/2 - Nm(M - m + 1) - Mn(N - n + 1)]d$$

(b) Show that if the star is located in one corner of the grid, then this expression becomes

$$L = MN(M + N - 2)d/2$$

(c) Show that the shortest cable length is obtained when the star is at the center of the grid.

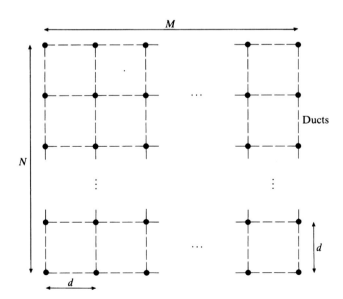

FIGURE P12-4

12-6. Here, do not resort to wavelength conversion but assume that wavelengths can be reused in different parts of a network.

(*a*) Show that the minimum number of wavelengths required to connect N nodes in a WDM network is as follows:

- $N - 1$ for a star network.
- $(N/2)^2$ for N even or $(N - 1)(N + 1)/4$ for N odd in a bus network.
- $N(N - 1)/2$ for a ring network.

(*b*) Draw example networks and their wavelength assignments for $N = 3$ and 4.

(*c*) Plot the number of wavelengths versus the number of nodes for star, bus, and ring networks for $2 \leq N \leq 20$.

12-7. Compare the system margins for 40-km and 80-km long-haul OC-48 (STM-16) links at 1550 nm for the minimum and maximum source output ranges. Assume there is a 1.5-dB coupling loss at each end of the link. Use Tables 12-4 and 12-5.

12-8. Verify that the maximum optical powers per wavelength channel given in Table 12-6 yield a total power level of $+17$ dBm in an optical fiber.

12-9. Two SONET rings need to be interconnected at two mutual nodes in order to ensure redundant paths under failure conditions. Draw the interconnection between two bidirectional line-switched (BLSR) SONET rings showing the primary and secondary path setups that designate the signal flows under normal and failure conditions. In designing the interface, consider the following possible failure conditions of the two mutual nodes:

(*a*) Failure of a transmitter or receiver in one of the nodes.

(*b*) Failure of an entire node.

(*c*) A fiber break in the link between the two nodes.

12-10. Similar to Prob. 12-9, draw the interconnection between a UPSR and a BLSR.

12-11. Consider the four-node network shown in Fig. P12-11. Each node uses a different combination of three wavelengths to communicate with the other nodes, so that there are six different wavelengths in the network. Given that node 1 uses λ_2, λ_4, and λ_6 for information exchange with the other nodes (i.e., these wavelengths are added and dropped at node 1, and the remaining wavelengths from other nodes pass through), establish wavelength assignments for the other nodes. Show this by means of the following table, where we assume that a transmitter at any given node does not communicate to the receiver at that node:

	R_1	R_2	R_3	R_4	R_5	R_6
T_1	—					
T_2		—				
T_3			—			
T_4				—		
T_5					—	
T_6						—

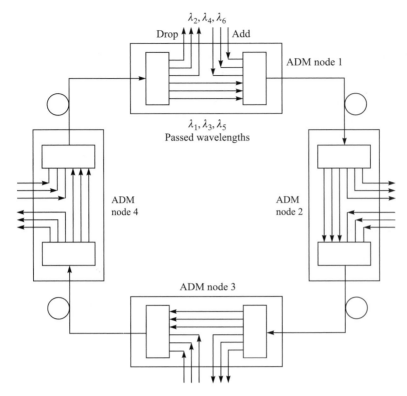

λ₂, λ₄, λ₆

FIGURE P12-11

12-12. Consider a $(p, k) = (3, 2)$ ShuffleNet.
    (*a*) Draw the interconnections between the nodes as in Fig. 12-18.
    (*b*) How many wavelengths are needed in the network?
    (*c*) Find the average number of hops.
    (*d*) If each wavelength carries a bit rate of 1 Gb/s, find the total network capacity.

12-13. If the routing algorithm in ShuffleNet balances the traffic load on all channels, then the channel efficiency is given by $\eta = 1/\bar{H}$, where $\bar{H}$ is the average number of hops given by Eq. (12-20). Plot the channel efficiency η as a function of the degree of connectivity p (i.e., the number of transceiver pairs per node) for values of $k = 2, 3,$ and 4 over the range $1 \leq p \leq 10$.

12-14. Derive Eq. (12-20) by counting the number of nodes that are h hops away from any given source node for a (p, k) ShuffleNet, where $1 \leq h \leq 2k - 1$, and then summing these numbers over h.

12-15. Figure P12-15 shows two architectures in which switches share wavelength converters for cost-saving purposes.[40] Find a set of wavelength connections that can be set up with the share-per-node architecture in Fig. P12-15*a* but not with the share-per-link architecture in Fig. P12-15*b*, and vice versa.

12-16. Consider the network consisting of three interconnected rings, shown in Fig. P12-16. Here, the circles represent nodes that contain optical switches and wavelength converters. These nodes can receive two wavelengths from any direction and can

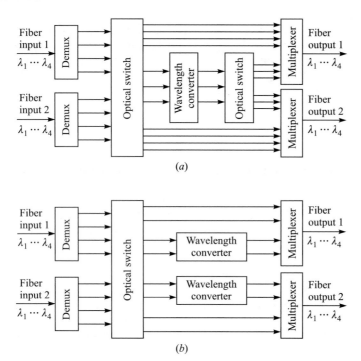

(a)

(b)

FIGURE P12-15

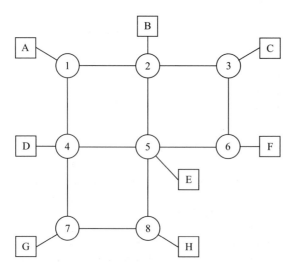

FIGURE P12-16

transmit them out over any line. The squares are access stations that have tunable optical transmitters and receivers (both wavelengths can be transmitted and received at any access station). Suppose the network has two wavelengths available to set up the following paths:

(a) A-1-2-5-6-F

(b) B-2-3-C

(c) B-2-5-8-H

(d) G-7-8-5-6-F

(e) A-1-4-7-G

Assign the two wavelengths to these paths and show at which nodes wavelength conversion is needed.

12-17. Analogous to Fig. 12-24, use Eq. (12-27) to plot the gain $G = q/p$ as a function of the number of wavelengths F for blocking probabilities $P_b = 10^{-4}$ and 10^{-5}. Use the same hopping number values $H = 5$, 10, and 20. For $H = 20$, plot G versus F on the same graph for blocking probabilities $P_b = 10^{-3}$, 10^{-4}, and 10^{-5}. What is the effect on the gain as the blocking probability increases?

12-18. Consider three copropagating optical signals at frequencies v_1, v_2, and v_3.

(a) If these frequencies are evenly spaced so that $v_1 = v_2 - \Delta v$ and $v_3 = v_2 + \Delta v$, where Δv is an incremental frequency change, list the third-order waves that are generated due to FWM and plot them in relation to the original three waves. Note that several of these FWM-generated waves coincide with the original frequencies.

(b) Now examine the case when $v_1 = v_2 - \Delta v$ and $v_3 = v_2 + (3/2)\Delta v$.

Find the FWM-generated frequencies and plot them in relation to the original three waves.

12-19. A detailed expression for the FWM efficiency η is given by[53]

$$\eta = \frac{\alpha^2}{\alpha^2 + \Delta\beta^2}\left[1 + \frac{4\exp(-\alpha L)\sin^2(\Delta\beta\, L/2)}{[1 - \exp(-\alpha L)]^2}\right]$$

where the factor $\Delta\beta$ is the difference of the propagation constants of the various waves due to dispersion, and is given by

$$\Delta\beta = \frac{2\pi\lambda^2}{c}|v_i - v_k| \times |v_j - v_k|\left[D(v_0) + \frac{\lambda^2}{2c}\frac{dD}{d\lambda}(|v_i - v_0| + |v_j - v_0|)\right]$$

Here, the value of the dispersion $D(v_0)$ and its slope $dD/d\lambda$ are taken at the optical frequency v_0. Using these expressions in Eq. (12-37), plot the ratio of the generated power P_{112} to the transmitted channel power P_1 as a function of the channel spacing for two $+7$-dBm channels. Find this ratio for the following dispersion and wavelength values:

(a) $D = 0$ ps/(nm · km) and $\lambda = 1556.6$ nm.

(b) $D = 0.13$ ps/(nm · km) and $\lambda = 1556.1$ nm.

(c) $D = 1.64$ ps/(nm · km) and $\lambda = 1537.2$ nm.

Let the frequency spacing of the two channels range from 0 to 250 GHz. In each case, take $dD/d\lambda = 0.08$ ps/(nm^2 · km), $\alpha = 0.0461$ km^{-1}, $L = 11$ km, and $A_{\text{eff}} = 55$ μm^2. For χ_{1111} and $\mathscr{D}$, use the values given in Example 12-8.

12-20. A soliton transmission system operates at 1550 nm with fibers that have a dispersion of 1.5 ps/(nm · km) and an effective core area of 50 μm^2. Find the peak power required for fundamental solitons that have a 16-ps FWHM width. Use the value

$n_2 = 2.6 \times 10^{-16}$ cm²/W. What are the dispersion length and the soliton period? What is the required peak power for 30-ps pulses?

12-21. A telecommunications service provider wants a single-wavelength soliton transmission system that is to operate at 40 Gb/s over a 2000-km distance. How would you design such a system? You are free to choose whatever components and design parameters are needed.

12-22. Create a cost model for the soliton system designed in Prob. 12-21, so that the service provider can determine the funding required for the project.

12-23. Consider a WDM system that utilizes two soliton channels at wavelengths λ_1 and λ_2. Since different wavelengths travel at slightly different velocities in a fiber, the solitons of the faster channel will gradually overtake and pass through the slower-channel solitons. If the collision length L_{coll} is defined as the distance between the beginning and end of the pulse overlap at the half-power points, then

$$L_{coll} = \frac{2T_s}{D \Delta \lambda}$$

where $\Delta \lambda = \lambda_1 - \lambda_2$, T_s is the pulse FWHM, and D is the dispersion parameter.
(a) What is the collision length for $T_s = 16$ ps, $D = 0.5$ ps/(nm·km), and $\Delta \lambda = 0.8$ nm?
(b) Four-wave-mixing effects arise betwen the soliton pulses during their collision, but then collapse to zero afterwards. To avoid amplifying these effects, the condition $L_{coll} \geq 2L_{amp}$ should be satisfied, where L_{amp} is the amplifier spacing. What is the upper bound for L_{amp} for the above case?

12-24. Based on the conditions described in Prob. 12-23, what is the maximum number of allowed wavelength channels spaced 0.4 nm apart in a WDM soliton system when $L_{amp} = 25$ km, $T_s = 20$ ps, and $D = 0.4$ ps/(nm·km)?

12-25. The bit-error probability of an ideal (0, 1)-sequence code, such as an optical orthogonal code, can be described by its upper bound[103]

$$P_B(L) \approx \frac{1}{2} \sum_{i=k}^{L-1} \frac{(L-1)!}{(L-1-i)!i!} \left(\frac{K^2}{2F} \right)^i \left(1 - \frac{K^2}{2F} \right)^{L-1-i}$$

where, as described in Eq. (12-57), F is the length of the code sequence, K is the weight of the code, and L denotes the number of active users on the network. The maximum value of L is given by Eq. (12-57). Plot $P_B(L)$ as a function of the number of active users L (over the range $K \leq L \leq N$) for the following three values of K and F:
(a) $K = 3$ and $F = 121$
(b) $K = 4$ and $F = 241$
(c) $K = 5$ and $F = 401$
Note which code weight and sequence length gives the lowest error probability.

REFERENCES

1. (a) F. Callegati, M. Casoni, C. Raffaelli, and B. Bostica, "Packet optical networks for high-speed TCP-IP backbones," *IEEE Commun. Mag.*, vol. 37, pp. 124–129, Jan. 1999.
(b) H. Yoshimura, K.-I. Sato, and N. Takachio, "Future photonic transport networks based on WDM technologies," *IEEE Commun. Mag.*, vol. 37, pp. 74–81, Feb. 1999.

(*c*) A. G. Malis, "Reconstructing transmission networks using ATM and DWDM," *IEEE Commun. Mag.*, vol. 37, pp. 140–145, June 1999.

2. R. D. Feldman, E. E. Harstead, S. Jiang, T. H. Wood, and M. Zirngibl, "An evaluation of architectures incorporating wavelength-division multiplexing for broadband fiber access," *J. Lightwave Tech.*, vol. 16, pp. 1546–1559, Sept. 1998.

3. (*a*) P. E. Green, Jr., "Optical networking update," *IEEE J. Sel. Areas Commun.*, vol. 14, pp. 764–779, June 1996.

    (*b*) P. E. Green, Jr., *Fiber Optic Networks*, Prentice Hall, New York, 1993.

4. B. Fabianek, K. Fitchew, S. Myken, and A. Houghton, "Optical network research and development in European Community programs: From RACE to ACTS," *IEEE Commun. Mag.*, vol. 35, pp. 50–56, Apr. 1997.

5. I. P. Kaminow, "Advanced multiaccess lightwave networks," in I. P. Kaminow and T. L. Koch, eds., *Optical Fiber Telecommunications—III*, vol. A, Academic, New York, 1997, chap. 15, pp. 560–593.

6. D. J. G. Mestdagh, *Fundamentals of Multiaccess Optical Fiber Networks*, Artech House, Boston, 1995.

7. B. Mukherjee, *Optical Communication Networks*, McGraw-Hill, New York, 1997.

8. R. Ramaswami and K. N. Sivarajan, *Optical Networks*, Morgan Kaufmann, San Francisco, 1998.

9. (*a*) H. Akimaru and M. R. Finley, Jr., "Elements of the emerging broadband information highway," *IEEE Commun. Mag.*, vol. 35, pp. 84–94, June 1997.

    (*b*) C.-J. L. van Driel, P. A. M. van Grinsven, V. Pronk, and W. A. M. Snijders, "The (r)evolution of access networks for the Information Superhighway," *IEEE Commun. Mag.*, vol. 35, pp. 104–112, June 1997.

10. D. K. Hunter, M. H. M. Nizam, M. C. Chia, I. Andonovic, K. M. Guild, A. Tzanakaki, M. J. O'Mahony, J. D. Bainbridge, M. F. C. Stephens, R. V. Penty, and I. H. White, "WASPNET: A wavelength switched packet network," *IEEE Commun. Mag.*, vol. 37, pp. 120–129, Mar. 1999.

11. G. Hill, L. Fernandez, and R. Cadeddu, "Building the road to optical networks," *Brit. Telecomm. Eng.*, vol. 16, pp. 2–12, Apr. 1997.

12. S. Saunders, *The McGraw-Hill High-Speed LANs Handbook*, McGraw-Hill, New York, 1996.

13. (*a*) W. J. Goralski, *SONET: A Guide to Synchronous Optical Networks*, McGraw-Hill, New York, 1997.

    (*b*) C. A. Siller, Jr. and M. Shafi (eds.), *SONET/SDH*, IEEE Press, New York, 1996.

14. T. M. Chen and S. S. Liu, *ATM Switching Systems*, Artech House, Boston, 1995.

15. (*a*) D. E. McDysan and D. L. Spohn, *ATM: Theory and Application*, McGraw-Hill, New York, 1995.

    (*b*) Z. Dziong, *ATM Network Resource Management*, McGraw-Hill, New York, 1997.

16. American National Standards Institute (ANSI), New York.

    (*a*) ANSI T1.105—1995, Telecommunications, *Synchronous Optical Network (SONET)—Basic Description Including Multiplex Structures, Rates, and Formats*, Oct. 1995.

    (*b*) ANSI T1.105.01—1995, *Synchronous Optical Network (SONET)—Automatic Protection Switching*, Nov. 1995.

    (*c*) ANSTI T1.105.06—1996 *Synchronous Optical Network (SONET)—Physical Layer Specifications*, Mar. 1996.

17. International Telecommunication Union—Telecommunication Standardization Sector (ITU-T), http://www.itu.int.

    (*a*) ITU-T Recommendation G.652, *Characteristics of Single-Mode Optical Fiber Cable*, 1993.

    (*b*) ITU-T Recommendation G.653, *Characteristics of Dispersion-Shifted Single-Mode Optical Fiber Cable*, 1993.

    (*c*) ITU-T Recommendation G.655, *Characteristics of Non-zero Dispersion-Shifted Single-Mode Optical Fiber Cable*, 1996.

    (*d*) ITU-T Recommendation G.957, *Optical Interfaces for Equipment and Systems Relating to the Synchronous Digital Hierarchy*, July 1995.

    (*e*) ITU-T Recommendation G.691, *Optical Interfaces for Single-Channel Systems with Optical*

Amplifiers, 1998.

(*f*) ITU-T Recommendation G.692, *Optical Interfaces for Multichannel Systems with Optical Amplifiers*, Oct. 1998.

18. F. Tong, C. K. Chan, L. K. Chen and D. Lam, "Fault surveillance schemes for optical components and systems using fiber Bragg gratings and optical amplifiers as monitoring sources," in R. A. Barry, ed., *Optical Networks and Their Applications*, TOPS vol. 20, Optical Society of America, Washington, DC, 1998, pp. 84–90.

19. I. Haque, W. Kremer, and K. Raychaudhuri, "Self-healing rings in a synchronous environment," in C. A. Siller, Jr. and M. Shafi, eds., *SONET/SDH*, IEEE Press, New York, 1996, pp. 131–138.

20. T.-H. Wu, "Emerging technologies for fiber network survivability," *IEEE Commun. Mag.*, vol. 33, pp. 58–74, Feb. 1995.

21. E. Ayanoglu and R. D. Gitlin, "Broadband network restoration," *IEEE Commun. Mag.*, vol. 34, pp. 110–119, July 1996.

22. J. M. Simmons, E. L. Goldstein, and A. A. M. Saleh, "Quantifying the benefit of wavelength add-drop in WDM rings with distance-independent and dependent traffic," *J. Lightwave Tech.*, vol. 17, pp. 48–57, Jan. 1999.

23. (*a*) R. Ramaswami, "Multiwavelength lightwave networks for computer communications," *IEEE Commun. Mag.*, vol. 31, pp. 78–88, Feb. 1993.

(*b*) E. Hall, J. Kravitz, R. Ramaswami, M. Halvorson, S. Tenbrink, and R. Tomsen, "The Rainbow II gigabit optical network," *IEEE J. Sel. Areas Commun.*, vol. 14, pp. 814–823, June 1996.

24. A. S. Acampora, "The scalable lightwave network," *IEEE Commun. Mag.*, vol. 32, pp. 36–42, Dec. 1994.

25. M. G. Hluchyj and M. J. Karol, "ShuffleNet: An application of generalized perfect shuffles to multihop lightwave networks," *J. Lightwave Tech.*, vol. 9, pp. 1386–1397, Oct. 1991.

26. M. I. Irshid and M. Kavehrad, "A WDM cross-connected star topology for multihop lightwave networks," *J. Lightwave Tech.*, vol. 10, pp. 828–835, June 1992.

27. K. Sivarajan and R. Ramaswami, "Lightwave network based on de Bruijn graphs," *IEEE/ACM Trans. on Networking*, vol. 2, pp. 70–79, Feb. 1994.

28. J. Brassil, A. K. Choudhury, and N. F. Maxemchuk, "The Manhattan Street Network: A high-performance, highly reliable metropolitan area network," *Computer Networks & ISDN Sys.*, vol. 26, pp. 841–858, Mar. 1994.

29. (*a*) H. J. Siegel, *Interconnection Networks for Large-Scale Parallel Processing*, McGraw-Hill, New York, 1990.

(*b*) K. Hwang, *Advanced Computer Architecture*, McGraw-Hill, New York, 1993.

30. E. Iannone and R. Sabella, "Optical path technologies: A comparison among different cross-connect architectures," *J. Lightwave Tech.*, vol. 14, pp. 2184–2196, Oct. 1996.

31. S. Okamoto, K. Oguchi, and K.-I. Sato, "Network architecture for optical path transport networks," *IEEE Trans. Commun.*, vol. 45, pp. 968–977, Aug. 1997.

32. Y.-K. Chen and C.-C. Lee, "Fiber Bragg grating-based large nonblocking multiwavelength cross-connects," *J. Lightwave Tech.*, vol. 16, pp. 1746–1756, Oct. 1998.

33. A Carena, M. D. Vaughn, R. Gaudino, M. Shell, and D. J. Blumenthal, "OPERA: An optical packet experimental routing architecture with label-swapping capability," *J. Lightwave Tech.*, vol. 16, pp. 2135–2145, Dec. 1998.

34. H. S. Hinton, J. R. Erickson, T. J. Cloonan, and G. W. Richards, "Space-division switching," in J. E. Midwinter, ed., *Photonics in Switching*, vol. 2, Academic, New York, 1993.

35. M. Fujiwara and T. Sawano, "Photonic space-division switching technologies for broadband networks," *IEICE Trans. Commun.*, vol. E77-B, pp. 110–118, Feb. 1994.

36. D. J. Blumenthal, P. R. Prucnal, and J. R. Sauer, "Photonic packet switches: Architectures and experimental implementations," *Proc. IEEE*, vol. 82, pp. 1650–1667, Nov. 1994.

37. T. A. Tumolillo, Jr., M. Donckers, and W. H. G. Horsthuis, "Solid state optical space switches for network cross-connect and protection applications," *IEEE Commun. Mag.*, vol. 35, pp. 124–130, Feb. 1997.

38. H. Okayama, T. Kamijoh, and M. Kawahara, "Multiwavelength highway photonic switches using wavelength sorting elements—Design," *J. Lightwave Tech.*, vol. 15, pp. 607–615, Apr. 1997.

39. Special Issue on "Photonic Packet Switching Technologies, Techniques, and Systems," *J. Lightwave Tech.*, vol. 16, Dec. 1998.

40. K.-C. Lee and V. O. K. Li, "A wavelength-convertible optical network," *J. Lightwave Tech.*, vol. 11, pp. 962–970, May/June 1993.

41. R. Ramaswami and K. Sivarajan, "Design of logical topologies for wavelength-routed optical networks," *IEEE J. Sel. Areas Commun.*, vol. 14, pp. 840–851, June 1996.

42. R. A. Barry and P. Humblet, "Models of blocking probability in all-optical networks with and without wavelength conversion," *IEEE J. Sel. Areas Commun.*, vol. 14, pp. 858–867, June 1996.

43. M. Kovacevic and A. S. Acampora, "Benefits of wavelength translation in all-optical clear-channel networks," *IEEE J. Sel. Areas Commun.*, vol. 14, pp. 868–880, June 1996.

44. S. Baroni and P. Bayvel, "Wavelength requirements in arbitrarily connected wavelength-routed optical networks," *J. Lightwave Tech.*, vol. 15, pp. 242–251, Feb. 1997.

45. D. Nesset, T. Kelly, and D. Marcenac, "All-optical wavelength conversion using SOA non-linearities," *IEEE Commun. Mag.*, vol. 36, pp. 56–61, Dec. 1998.

46. J. Y. Hui, *Switching and Traffic Theory for Integrated Broadband Networks*, Kluwer, Norwell, MA, 1990.

47. R. G. Waarts, A. A. Friesem, E. Lichtman, H. H. Yaffe, and R.-P. Braun, "Nonlinear effects in coherent multichannel transmission through optical fibers," *Proc. IEEE*, vol. 78, pp. 1344–1368, Aug. 1990.

48. A. R. Chraplyvy, "Limitations on lightwave communications imposed by optical-fiber non-linearities," *J. Lightwave Tech.*, vol. 8, pp. 1548–1557, Oct. 1990.

49. A. R. Chraplyvy and R. W. Tkach, "What is the actual capacity of single-mode fibers in amplified lightwave systems?" *IEEE Photonics Tech. Lett.*, vol. 5, pp. 666–668, June 1993.

50. X. Y. Zou, M. I. Hayee, S.-M. Hwang, and A. E. Willner, "Limitations in 10-Gb/s WDM optical fiber transmission when using a variety of fiber types to manage dispersion and nonlinearities," *J. Lightwave Tech.*, vol. 14, pp. 1144–1152, June 1996.

51. R. H. Stolen, "Nonlinear properties of optical fibers," in S. E. Miller and A. G. Chynoweth, eds., *Optical Fiber Telecommunications*, Academic, New York, 1979.

52. G. P. Agrawal, *Nonlinear Fiber Optics*, Academic, New York, 2nd ed., 1995.

53. F. Forghieri, R. W. Tkach, and A. R. Chraplyvy, "Fiber Nonlinearities and their Impact on Transmission Systems," in I. P. Kaminow and T. L. Koch, eds., *Optical Fiber Telecommunications—III*, vol. A, Academic, New York, 1997, chap. 8, pp. 196–264.

54. B. Zhang and L. Lu, "Isolators protect fiber optic systems and optical amplifiers," *Laser Focus World (http://www.lfw.com)*, vol. 34, pp. 147–152, Nov. 1998.

55. J. A. Buck, *Fundamentals of Optical Fibers*, Wiley, New York, 1995.

56. C. McIntosh, A. Yeniay, J. Toulouse, and J.-M. P. Delavaux, "Stimulated Brillouin scattering in dispersion-compensating fiber," *Opt. Fiber Technol.*, vol. 3, pp. 173–176, Apr. 1997.

57. D. A. Fishman and J. A. Nagel, "Degradations due to stimulated Brillouin scattering in multi-gigabit intensity-modulated fiber-optic systems," *J. Lightwave Tech.*, vol. 11, pp. 1721–1728, Nov. 1993.

58. N. Ohkawa and Y. Hayashi, "Bit rate degradation caused by SBS in CPFSK coherent optical repeaterless systems with booster amplifiers," *J. Lightwave Tech.*, vol. 13, pp. 914–922, May 1995.

59. X. P. Mao, R. W. Tkach, A. R. Chraplyvy, R. M. Jopson, and R. M. Dorosier, "Stimulated Brillouin threshold dependence on fiber type and uniformity," *IEEE Photonics Tech. Lett.*, vol. 4, pp. 66–69, Jan. 1992.

60. X. P. Mao, G. E. Bodeep, R. W. Tkach, A. R. Chraplyvy, T. E. Darcie, and R. M. Dorosier, "Brillouin scattering in externally modulated lightwave AM-VSB CATV transmission systems," *IEEE Photonics Tech. Lett.*, vol. 4, pp. 287–289, Mar. 1992.

61. N. Kikuchi and S. Sasaki, "Analytical evaluation technique of self-phase modulation effect on the performance of cascaded optical amplifier systems," *J. Lightwave Tech.*, vol. 13, pp. 868–878, May 1995.

62. N. Kikuchi, K. Sekine, and S. Sasaki, "Analysis of XPM effect on WDM transmission performance," *Electron. Lett.*, vol. 33, pp. 653–654, Apr. 1997.

63. L. Rapp, "Experimental investigation of signal distortions induced by CPM combined with dispersion," *IEEE Photonics Tech. Lett.*, vol. 9, pp. 1592–1594, Dec. 1997.

64. M. Shtaif, M. Eiselt, R. W. Tkach, R. H. Stolen, and A. H. Gnauck, "Crosstalk in WDM systems caused by CPM in EDFAs," *IEEE Photonics Tech. Lett.*, vol. 10, pp. 1796–1798, Dec. 1998.

65. R. W. Tkach, A. R. Chraplyvy, F. Forghieri, A. H. Gnauck, and R. M. Dorosier, "Four-photon mixing and high-speed WDM systems," *J. Lightwave Tech.*, vol. 13, pp. 841–849, May 1995.

66. W. Zeiler, F. Di Pasquale, P. Bayvel, and J. E. Midwinter, "Modeling of four-wave mixing and gain peaking in amplified WDM optical communication systems and networks," *J. Lightwave Tech.*, vol. 14, pp. 1933–1941, Sept. 1996.

67. N. Shibata, R. P. Braun, and R. G. Waarts, "Phase-match dependence of efficiency of wave generation through four-wave mixing in a single-mode optical fiber," *IEEE J. Quantum Electron.*, vol. 23, pp. 1205–1210, July 1987.

68. C. Kurtzke, "Suppression of fiber nonlinearities by appropriate dispersion management," *IEEE Photonics Tech. Lett.*, vol. 5, pp. 1250–1253, Oct. 1993.

69. E. Lichtman, "Performance limitations imposed on all-optical ultralong communication systems," *J. Lightwave Tech.*, vol. 13, pp. 898–905, May 1995.

70. B. Jopson and A. H. Gnauck, "Dispersion compensation for optical fiber systems," *IEEE Commun. Mag.*, vol. 33, pp. 96–102, June 1995.

71. H. Taga, "Long distance transmission experiments using the WDM technology," *J. Lightwave Tech.*, vol. 14, pp. 1287–1298, June 1996.

72. K. Hinton, "Long-haul system issues with Bragg fiber grating-based dispersion compensation," *Opt. Fiber Technol.*, vol. 5, pp. 145–164, Apr. 1999.

73. D. Garthe, G. Milner, and Y. Cai, "System performance of broadband dispersion-compensating gratings," *Electron. Lett.*, vol. 34, pp. 582–583, Mar. 1998.

74. A. N. Pilipetskii, V. J. Mazurczyk, and C. J. Chen, "The effect of dispersion compensation on system performance when nonlinearities are important," *IEEE Photonics Tech. Lett.*, vol. 11, pp. 284–286, Feb. 1999.

75. D. Breuer, H. J. Ehrke, F. Küppers. R. Ludwig, K. Petermann, H. G. Weber, and K. Weich, "Unrepeated 40-Gb/s RZ single-channel transmission at 1.55 μm using various fiber types," *IEEE Photonics Tech. Lett.*, vol. 10, pp. 822–824, June 1998.

76. C. Casper, H.-M. Foisel, R. Freund, and B. Strebel, "Four-channel 10-Gb/s transmission over 15-wavelength selective cross-connect paths and 1175-km dispersion-compensated single-mode fiber," *IEEE Photonics Tech. Lett.*, vol. 10, pp. 1479–1480, Oct. 1998.

77. D. Breuer, K. Obermann, and K. Petermann, "Comparison of N × 40 Gb/s and 4N × 10 Gb/s WDM transmission over standard single-mode fiber at 1.55 μm," *IEEE Photonics Tech. Lett.*, vol. 10, pp. 1793–1795, Dec. 1998.

78. C.-S. Li and F. Tong, "Crosstalk and interference penalty in all-optical networks using static wavelength routers," *J. Lightwave Tech.*, vol. 14, pp. 1120–1126, June 1996.

79. P. A. Humblet and W. M. Hamdy, "Crosstalk analysis and filter optimization of single- and double-cavity Fabry-Perot filters," *IEEE J. Sel. Areas Commun.*, vol. 8, pp. 1095–1107, Aug. 1990.

80. K.-P. Ho and J. M. Kahn, "Methods for crosstalk measurement and reduction in dense WDM systems," *J. Lightwave Tech.*, vol. 14, pp. 1127–1135, June 1996.

81. L. W. Couch II, *Digital and Analog Communication Systems*, Prentice Hall, Upper Saddle River, NJ, 5th ed., 1997.

82. L. F. Mollenauer, J. P. Gordon, and S. G. Evangelides, "Multigigabit soliton transmissions traverse ultra-long distance," *Laser Focus World*, vol. 27, pp. 165–170, Nov. 1991.

83. H. A. Haus, "Optical fiber solitons: Their properties and uses," *Proc. IEEE*, vol. 81, pp. 970–983, July 1993.

84. A. Hasegawa and Y. Kodama, *Solitons in Optical Communication*, Clarendon Press, Oxford, 1995. This book presents detailed mathematical aspects of the theory of solitons.

85. Y. Lai and H. A. Haus, "Quantum theory of solitons in optical fibers. II. Exact solution," *Phys. Rev.*, vol. 40, pp. 854–866, July 1989.

86. H. Haus and W. S. Wong, "Solitons in optical communications," *Rev. Mod. Phys.*, vol. 68, pp. 432–444, 1996.

87. L. F. Mollenauer, J. P. Gordon, and P. V. Mamyshev, "Solitons in high bit-rate, long-distance transmission," in I. P. Kaminow and T. L. Koch, eds., *Optical Fiber Telecommunications—III*, vol. A, Academic, New York, 1997, chap. 12, pp. 373–460.

88. E. Iannone, F. Matera, A. Mecozzi, and M. Settembre, *Nonlinear Optical Communication Networks*, Wiley, New York, 1998, chap. 5, pp. 138–251.

89. E. Desurvire, *Erbium-Doped Fiber Amplifiers*, Wiley, New York, 1994.

90. G. P. Agrawal, *Fiber Optic Communication Systems*, Wiley, New York, 2nd ed., 1997, chap. 10.

91. J. S. Russell, *Reports of the Meetings of the British Association for the Advancement of Science*, 1844.

92. J. P. Gordon, "Interaction forces among solitons in optical fibers," *Opt. Lett.*, vol. 8, no. 11, pp. 596–598, 1983.

93. F. M. Mitsche and L. F. Mollenauer, "Experimental observation of interaction forces between solitons in optical fibers," *Opt. Lett.*, vol. 12, no. 5, pp. 355–357, 1987.

94. J. A. Salehi, "Code-division multiple-access techniques in optical fiber networks—Part I: Fundamental principles," *IEEE Trans. Commun.*, vol. 37, pp. 824–833, Aug. 1989.

95. J. A. Salehi and C. A. Brackett, "Code-division multiple-access techniques in optical fiber networks—Part II: System performance analysis," *IEEE Trans. Commun.*, vol. 37, pp. 834–842, Aug. 1989.

96. W. C. Kwong, P. A. Perrier, and P. R. Prucnal, "Performance comparison of asynchronous and synchronous CDMA techniques for fiber-optic local area networks," *IEEE Trans. Commun.*, vol. 39, pp. 1625–1634, Nov. 1991.

97. E. L. Walker, "A theoretical analysis of the performance of CDMA communications over multimode fiber channels—Part I: Transmission and detection," *IEEE J. Sel. Areas Commun.*, vol. 12, pp. 751–761, May 1994.

98. E. L. Walker, "A theoretical analysis of the performance of CDMA communications over multimode fiber channels—Part II: System performance evaluation," *IEEE J. Sel. Areas Commun.*, vol. 12, pp. 976–983, June 1994.

99. M. Kavehrad and D. Zaccarin, "Optical code-division-multiplexed systems based on spectral encoding of noncoherent sources," *J. Lightwave Tech.*, vol. 13, pp. 534–545, Mar. 1995.

100. S. V. Maric, O. Moreno, and C. J. Corrada, "Multimedia transmission in fiber-optic LANs using optical CDMA," *J. Lightwave Tech.*, vol. 14, pp. 2149–2153, Oct. 1996.

101. C.-L. Lin and J. Wu, "A synchronous fiber-optic CDMA system using adaptive optical hard limiter," *J. Lightwave Tech.*, vol. 16, pp. 1393–1403, Aug. 1998.

102. N. Karafolas and D. Uttamchandani, "Local area network ocmmunications using optical spread spectrum and serial correlation of bipolar codes," *Opt. Fiber Technol.*, vol. 3, no. 3, pp. 253–266, July 1997.

103. J. Mückenheim, K. Iversen, and D. Hampicke, "Construction of high-efficient optical CDMA computer networks: Statistical design," *Proc. IEEE Int. Conf. Commun. (ICC) 98*, pp. 1289–1293, June 1998.

104. S. W. Lee and D. H. Green, "Performance analysis of optical orthogonal codes in CDMA LANs," *IEE Proc.—Commun.*, vol. 145, pp. 265–271, Aug. 1998.

105. E. D. J. Smith, R. J. Blaikie, and D. P. Taylor, "Performance enhancement of spectral-amplitude-coding optical CDMA using pulse-position modulation," *IEEE Trans. Commun.*, vol. 46, pp. 1176–1186, Sept. 1998.

106. R. C. Dixon, *Spread Spectrum Systems with Commercial Applications*, Wiley, New York, 3rd ed., 1994.

107. G.-C. Yang and W. C. Kwong, "Performance comparison of multiwavelength CDMA and WDMA + CDMA for fiber-optic networks," *IEEE Trans. Commun.*, vol. 45, pp. 1426–1434, Nov. 1997.

108. F. R. K. Chung, J. A. Salehi, and V. K. Wei, "Optical orthogonal codes: Design, analysis and applications," *IEEE Trans. Inform. Theory*, vol. 35, pp. 595–604, May 1989.

109. S. V. Maric, M. Hahm, and E. L. Titlebaum, "Construction and performance analysis of a new family of optical orthogonal codes for CDMA fiber-optic networks," *IEEE Trans. Commun.*, vol. 43, pp. 485–489, Feb./Mar./Apr. 1995.

110. D. Zaccarin and M. Kavehrad, "New architecture for incoherent optical CDMA to achieve bipolar capacity," *Electron. Lett.*, vol. 30, pp. 258–259, Feb. 1994.

111. D. Zaccarin and M. Kavehrad, "Performance evaluation of optical CDMA systems using non-coherent detection and bipolar codes," *J. Lightwave Tech.*, vol. 12, pp. 96–105, Jan. 1994.

112. L. Nguyen, T. Dennis, B. Aazhang, and J. F. Young, "Experimental demonstration of bipolar codes for optical spectral amplitude CDMA communication," *J. Lightwave Tech.*, vol. 15, pp. 1647–1653, Sept. 1997.

113. Special Issue on "Ultrafast Electronics, Photonics, and Optoelectronics," *IEEE J. Sel. Topics Quantum Electron.*, vol. 2, no. 3, Sept. 1996.

114. D. Cotter, J. K. Lucek, and D. D. Marcenac, "Ultra-high-bit-rate networking: From the trans-continental backbone to the desktop," *IEEE Commun. Mag.*, vol. 35, pp. 90–95, Apr. 1997.

115. A. R. Chraplyvy and R. W. Tkach, "Terabit/second transmission experiments," *IEEE J. Quantum Electron.*, vol. 34, pp. 2103–2108, Nov. 1998.

116. D. Cotter and A. D. Ellis, "Asynchronous digital optical regeneration and networks," *J. Lightwave Tech.*, vol. 16, pp. 2068–2080, Dec. 1998.

117. V. W. S. Chan, K. L. Hall, E. Modiano, and K. A. Rauschenbach, "Architectures and technologies for high-speed optical data networks," *J. Lightwave Tech.*, vol. 16, pp. 2146–2168, Dec. 1998.

118. S. Aisawa, T. Sakamoto, M. Fukui, J. Kani, M. Jinno, and K. Oguchi, "Ultra-wideband, long-distance WDM demonstration of 1 Tb/s 600-km transmission using 1550 and 1580 nm wavelength bands," *Electron. Lett.*, vol. 34, pp. 1127–1129, May 1998.

119. T. Ono and Y. Yano, "Key technologies for terabit/second WDM systems with high spectral efficiency of over 1 bit/Hz," *IEEE J. Quantum Electron.*, vol. 34, pp. 2080–2088, Nov. 1998.

120. H. Taga, M. Suzuki, N. Edagawa, S. Yamamoto, and S. Akiba, "Long-distance WDM transmission experiments using the dispersion slope compensator," *IEEE J. Quantum Electron.*, vol. 34, pp. 2055–2063, Nov. 1998.

121. S. Kawanishi, "Ultrahigh-speed optical TDM transmission technology based on optical signal processing," *IEEE J. Quantum Electron.*, vol. 34, pp. 2064–2079, Nov. 1998.

122. L. F. Mollenauer and P. V. Mamyshev, "Massive WDM with solitons," *IEEE J. Quantum Electron.*, vol. 34, pp. 2089–2102, Nov. 1998.

123. A. D. Ellis, D. M. Patrick, D. Flannery, R. J. Manning, D. A. O. Davies, and D. M. Spirit, "Ultra-high speed OTDM networks using semiconductor amplifier-based processing nodes," *J. Lightwave Tech.*, vol. 13, pp. 761–770, May 1995.

124. S.-W. Seo, K. Bergman, and P. R. Prucnal, "Transparent optical networks with time-division multiplexing," *IEEE J. Sel. Areas Commun.*, vol. 14, pp. 1039–1051, June 1996.

125. C.-K. Chan, L.-K Chen, and K.-W. Cheung, "A fast channel-tunable optical transmitter for ultra-high-speed all-optical time-division multiaccess networks," *IEEE J. Sel. Areas Commun.*, vol. 14, pp. 1052–1056, June 1996.

126. Ö. Boyraz, J. W. Lou, K. H. Ahn, Y. Liang, T. J. Xia, Y.-H. Kao, and M. N. Islam, "Demonstration and performance analysis for the off-ramp portion of an all-optical access network," *J. Lightwave Tech.*, vol. 17, pp. 998–1010, June 1999.

127. B. Y. Yu, I. Glesk, and P. R. Prucnal, "Analysis of a dual-receiver node with high fault tolerance for ultrafast OTDM packet-switched shuffle networks," *J. Lightwave Tech.*, vol. 16, pp. 736–744, May 1998.

CHAPTER
13

MEASUREMENTS

The design and installation of an optical fiber communication system require measurement techniques for verifying the operational characteristics of the constituent components. Of particular importance are accurate and precise measurements of the optical fiber, since this component cannot be readily replaced once it has been installed. Two basic groups of people are interested in fiber characterization. These are the manufacturers, who are concerned with the materials composition and fabrication effects on fiber properties, and the system engineers, who must have sufficient data on the fiber to perform meaningful design calculations and to evaluate systems during installation and operation.

During the design phase, the fiber parameters of interest for multimode fibers are the core and cladding diameters, numerical aperture, refractive-index profile, fiber attenuation, and dispersion. For single-mode fibers, the important characteristics are the effective cutoff wavelength, the mode-field diameter, fiber attenuation, and dispersion. Generally, the fiber manufacturer can supply the values for these parameters. The fiber geometry, refractive-index profile, numerical aperture, cutoff wavelength, and mode-field diameter are not expected to change during cable manufacture, installation, and operation. Thus, once these parameters are known, there is no need to remeasure them.

However, the attenuation and dispersion of a fiber can change during fiber cabling and cable installation. In single-mode fibers, chromatic and polarization-mode dispersions are important factors that limit the bandwidth–distance product. Furthermore, modal dispersion effects also need to be examined in multimode fibers. For example, microbends can cause additional loss in the fiber, and modal redistribution at fiber joints can significantly affect the modal dispersion when several sections of fiber cables are connected in series. Chromatic dispersion effects are of particular importance in high-speed WDM links, and polarization-mode dispersion can ultimately limit the highest achievable data rate in single-mode links. Measurement procedures for these parameters are thus of interest to the user, as are methods for locating breaks and faults in optical fiber cables.

In addition to optical fiber parameters, system engineers are interested in knowing the characteristics of passive splitters, connectors, and couplers, and those of electro-optic components, such as sources, photodetectors, and optical amplifiers. Furthermore, when a link is being installed and tested, the operational parameters of interest include bit-error rate, timing jitter, and signal-to-noise ratio as indicated by the eye pattern. During actual operation, measurements are needed for maintenance and monitoring functions to determine factors such as fault locations in fibers and the status of remotely located optical amplifiers.

This chapter discusses measurements and performance tests of interest to designers and users of fiber optic links and networks. Measurements of the invariant fiber characteristics (geometry, MFD, refractive indices, etc.) are not considered here, owing to space limitation. The interested reader is referred to several comprehensive works in this area.[1-8] Of particular interest here are measurements for WDM links. Figure 13-1 shows some of the relevant test parameters and at what points in a WDM link they are of importance.

In Secs. 13.1 and 13.2, respectively, this chapter first addresses measurement standards and basic test equipment for optical fiber communication links. Section 13.3 discusses attenuation measurements, and Sec. 13.4 explains methods for measuring modal, chromatic, and polarization-mode dispersion. The operation of an instrument for field evaluations of fibers is explained in Sec. 13.5. The final two sections, respectively, cover eye patterns and their evaluations, and the use of an optical spectrum analyzer for characterizing the spectral response of optical sources and for testing EDFAs.

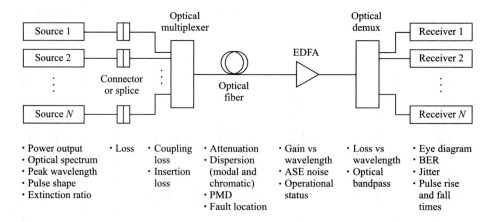

FIGURE 13-1
Components of a typical WDM link and some performance-measurement parameters of user interest.

13.1 MEASUREMENT STANDARDS AND TEST PROCEDURES

Before delving into measurement techniques, let us first look at what standards exist for fiber optics. There are three basic classes of standards: primary standards, component testing standards, and system standards.

Primary standards refer to measuring and characterizing fundamental physical parameters, such as attenuation, bandwidth, mode-field diameter for single-mode fibers, and optical power. In the United States, the main organization involved in primary standards is the National Institute of Standards and Technology (NIST). This organization carries out fiber optic and laser standardization work at Boulder, Colorado,[9] and it sponsors an annual conference on optical fiber measurements.[10] Other national organizations include the National Physical Laboratory (NPL) in the United Kingdom[11] and the Physikalisch-Technische Bundesanstalt (PTB) in Germany.[12]

Component testing standards define relevant tests for fiber-optic component performance, and they establish equipment-calibration procedures. Several different organizations are involved in formulating testing standards, some very active ones being the Telecommunication Industries Association (TIA)[13] in association with the Electronic Industries Association (EIA),[14] the Telecommunication Standardization Sector of the International Telecommunication Union (ITU-T),[15] and the International Electrotechnical Commission (IEC).[16] The TIA has a list of over 120 fiber optic test standards and specifications under the general designation TIA/EIA-455-*XX*-*YY*, where *XX* refers to a specific measurement technique and *YY* refers to the publication year. These standards are also called Fiber Optic Test Procedures (FOTP), so that TIA/EIA-455-*XX* becomes FOTP-*XX*. These include a wide variety of recommended methods for testing the response of fibers, cables, passive devices, and electro-optic components to environmental factors and operational conditions. For example, TIA/EIA-455-60-1997, or FOTP-60, is a method published in 1997 for measuring fiber or cable length.

System standards refer to measurement methods for links and networks. The major organizations involved here are the American National Standards Institute (ANSI),[17] the Institute of Electrical and Electronic Engineers (IEEE),[18] and the ITU-T. Of particular interest for fiber optic systems are test standards and recommendations from the ITU-T. Existing and emerging ITU-T recommendations aimed at all aspects of optical networking include the following:[19]

1. *Recommendation G.ons*, "Network Node Interface for the Optical Transport Network," includes definitions of optical-layer overheads for functions such as supervision of the transport wavelengths.

2. *Recommendation G.872*, "Architectures of Optical Transport Networks," Feb. 1999.

3. *Recommendation G.798* gives the functional characteristics of optical network elements.

4. *Recommendation G.onc*, "Optical Network Components and Subsystems," addresses transmission aspects of components and subsystems, such as add/drop multiplexers and optical cross-connects.

5. *Recommendation* G.983, "High-Speed Optical Access Systems Based on Passive Optical Network," Oct. 1998.

6. *Recommendation G.959.1*, "Physical Layer Aspects of Optical Networks," addresses point-to-point WDM systems optimized for long-haul transport.

7. *Recommendation G.onm* deals with the items indicated in its title, "Management of Optical Network Elements."

8. *Recommendation G.871*, "Frame of Optical Networking Elements," outlines the links between the various recommendations and the rationale followed in preparing them.

13.2 TEST EQUIPMENT

As light signals pass through the various parts of an optical link, they need to be measured and characterized in terms of the three fundamental areas of optical power, polarization, and spectral content. The basic pieces of test equipment for carrying out such measurements on optical fiber components and systems include optical power meters, attenuators, tunable laser sources, spectrum analyzers, and time-domain reflectometers. These come in a variety of capabilities, with sizes ranging from portable, hand-held units for field use to sophisticated briefcase-sized instruments for laboratory applications.[20] In general, the field units do not need to have the very high precision of laboratory instruments, but they need to be more rugged to maintain reliable and accurate measurements under extreme environmental conditions of temperature, humidity, dust, and mechanical stress. However, even the hand-held equipment for field use has reached a high degree of sophistication with automated microprocessor-controlled test features and computer-interface capabilities.

More sophisticated instruments, such as polarization analyzers and optical communication analyzers, are available for measuring and analyzing polarization-mode dispersion (PMD), eye diagrams, and pulse waveforms. These instruments enable a variety of statistical measurements to be made at the push of a button, after the user has keyed in the parameters to be tested and the desired measurement range.

13.2.1 Optical Power Meters

Optical power measurement is the most basic function in fiber optic metrology. Some form of optical power detection is involved in almost every piece of light-wave test equipment. Hand-held instruments come in a wide variety of types with different levels of capabilities. Multiwavelength optical power meters that use photodetectors are the most common instrument for measuring optical signal power levels. Usually, the meter outputs are given in dBm (where 0 dBm = 1 mW) or dBμ (where 0 dBμ = 1μW).

Figure 13-2 shows a compact hand-held Model FOT-90A fiber optic power meter from EXFO. In this versatile instrument, various photodetector heads that have different performance characteristics are available. For example, using a Ge detector allows a measuring range of +18 to −60 dBm in the 780-to-1600-nm wavelength band, whereas an InGaAs detector allows a measuring range of +3 to −73 dBm in the 840-to-1650-nm wavelength band. In each case, the power measurements can be made at 20 calibrated wavelengths with a ±20-dB accuracy. An RS-232 interface together with application softwave allows a user to download measurements and view, export, or print them in either tabular or graphic form. The permanent memory registers can store 512 readings manually or 400 readings automatically at a programmable time interval.

Figure 13-3 shows another hand-held tester that also contains optical sources so it can do more sophisticated optical power measuring. For example, this instrument can function as a power meter, an optical-loss tester for automatically measuring loss in a fiber in two directions at two wavelengths, an optical return-loss tester for measuring the quality of optical patch cords, a visual fault indicator for locating breaks and failures in a fiber cable, and a talk set for full-duplex communications between field personnel.

13.2.2 Optical Attenuators

In many laboratory or production tests, the characteristics of a high optical signal level may need to be measured. If the level is very high, such as a strong output from an optical amplifier, the signal may need to be precisely attenuated before being measured. This is done to prevent instrument damage or to avoid overload

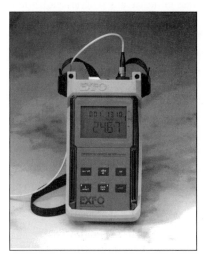

FIGURE 13-2
Example of a versatile hand-held optical fiber power meter. (Model FOT-90A, provided courtesy of EXFO E.O. Engineering, Inc.)

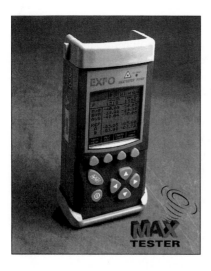

FIGURE 13-3
Compact, portable multipurpose test instrument for use in field environments. (Model FOT-920, provided courtesy of EXFO E.O. Engineering, Inc.)

distortion in the measurements. Optical attenuators allow a user to reduce an optical signal level—for example, up to 60 dB (a factor of 10^6)—in precise steps at a specified wavelength, which is usually 1310 or 1550 nm. The capabilities of attenuators range from simple tape-cassette-sized devices for quick field measurements that may only need to be accurate to 0.5 dB, to laboratory instruments that have an attenuation precision of 0.001 dB.

13.2.3 Tunable Laser Sources

Tunable laser sources are important instruments for tests that measure the wavelength-dependent response of an optical component or link. Figure 13-4 shows an example from Hewlett-Packard (Model 8168B) that generates a true single-mode laser line for every selected wavelength point. The source is an external-cavity semiconductor laser. A movable diffraction grating is used as a tunable filter for wavelength selection. Depending on the source and grating combination, a typical instrument is tunable over either the 1280-to-1330-nm band or the 1450-to-1565-

FIGURE 13-4
Example of a tunable laser source. (Model HP-8168B, provided courtesy of Hewlett-Packard Co.)

nm band. Wavelength scans, with an output power that is flat across the scanned spectral band, can be done automatically. The minimum output power of these instruments is −10 dBm and the absolute wavelength accuracy is typically ±0.1 nm

13.2.4 Optical Spectrum Analyzers

Optical spectrum analyzers measure optical power as a function of wavelength.[21] The most common implementation uses a diffraction-grating-based optical filter, which yields wavelength resolutions to less than 0.1 nm. Higher wavelength accuracy (±0.001 nm) is achieved with wavelength meters based on Michelson interferometry. To measure very narrow linewidths—for example, the 10-MHz linewidth of a typical single-frequency semiconductor laser—optical spectrum analyzers employing homodyne and heterodyne techniques are used. Figure 13-5 shows a general-purpose optical spectrum analyzer with a typical measurement trace on the display screen.

13.2.5 Optical Time-Domain Reflectometer (OTDR)

The long-term workhorse instrument in fiber optic systems is the OTDR. In addition to locating faults within an optical link, this instrument measures parameters such as attenuation, length, connector and splice losses, and reflectance levels. A typical OTDR consists of an optical source and receiver, a data-acquisition module, a central processing unit (CPU), an information-storage unit for retaining data either in the internal memory or on an external disk, and a display. Figure 13-6 shows an example of a versatile portable unit for making measurements in the field.

Figure 13-7 shows the basis of the OTDR technique. An OTDR is fundamentally an optical radar. It operates by periodically launching narrow laser pulses into one end of a fiber under test by using either a directional coupler or a beam splitter. The properties of the optical fiber link are then determined by analyzing the amplitude and temporal characteristics of the waveform of the

FIGURE 13-5

A general-purpose optical spectrum analyzer (OSA) with a typical measurement trace on the display screen. (Model HP-70951A, provided courtesy of Hewlett-Packard Co.)

FIGURE 13-6
Example of a portable universal test set used here as an optical time-domain reflectometer (OTDR). (Model FTB-300, provided courtesy of EXPO E.O. Engineering, Inc.)

backscattered light. The various applications of an OTDR and the interpretation of the backscattered waveform are described in Sec. 13.5.

13.2.6 Multifunction Optical Test Systems

For laboratory, manufacturing, and quality-control environments, there are instruments with exchangeable modules for performing a variety of measurements. Figure 13-8 shows an example from EXFO, which includes a basic modular mainframe and an expansion unit. The mainframe is a microprocessor-based unit that coordinates data compilation and analyses from variety of test instruments. This test system can control external instruments that have RS-232 communication capability, and it has a networking capability for remote access from a

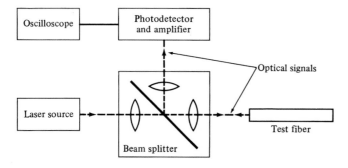

FIGURE 13-7
Operating principle of an optical time-domain reflectometer (OTDR).

FIGURE 13-8
Multifunction PC-based optical test system. (Model IQ-203 mainframe and IQ-206 expansion unit, provided courtesy of EXFO E.O. Engineering, Inc.)

computer. The plug-in modules cover a wide range of test capabilities. Example functions include single-channel or multichannel power meter, ASE broadband source, tunable-laser source, variable attenuator, optical spectrum analyzer, and PMD analyzer.

13.3 ATTENUATION MEASUREMENTS

Attenuation of optical power in a fiber waveguide is a result of absorption processes, scattering mechanisms, and waveguide effects. The manufacturer is generally interested in the magnitude of the individual contributions to attenuation, whereas the system engineer who uses the fiber is more concerned with the total transmission loss of the fiber. Here, we shall treat only measurement techniques for total transmission loss as described in various FOTPs and in ITU-T Recommendations G.650 for single-mode fibers and G.651 for graded-index multimode fibers.

Three basic methods are available for determining attenuation in fibers. The earliest devised and most common approach involves measuring the optical power transmitted through a long and a short length of the same fiber using identical input couplings. This method is known as the cutback technique. A less accurate but nondestructive method is the insertion-loss method, which is useful for cables with connectors on them. These two methods are described in this section. Section 13.5 describes the third technique, which involves the use of an OTDR.

13.3.1 The Cutback Technique

The *cutback technique*,[22,23] which is a destructive method requiring access to both ends of the fiber, is illustrated in Fig. 13-9. Measurements may be made at one or more specific wavelengths, or, alternatively, a spectral response may be required over a range of wavelengths. To find the transmission loss, the optical power is first measured at the output (or far end) of the fiber. Then, without disturbing the

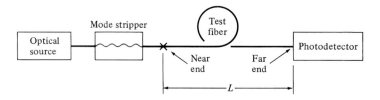

FIGURE 13-9
Schematic experimental setup for determining fiber attenuation by the cutback technique. The optical power is first measured at the far end, then the fiber is cut at the near end, and the power output there is measured.

input condition, the fiber is cut off a few meters from the source, and the output power at this near end is measured. If P_F and P_N represent the output powers of the far and near ends of the fiber, respectively, the average loss α in decibels per kilometer is given by

$$\alpha = \frac{10}{L} \log \frac{P_N}{P_F} \tag{13-1}$$

where L (in kilometers) is the separation of the two measurement points. The reason for following these steps is that it is extremely difficult to calculate the exact amount of optical power launched into a fiber. By using the cutback method, the optical power emerging from the short fiber length is the input power to the fiber of length L

In carrying out this measurement technique, special attention must be paid to how optical power is launched into the fiber. This is because, in a multimode fiber, different launch conditions can yield different loss values. The effects of modal distributions in the multimode fiber that result from different numerical apertures and spot sizes on the launch end of the fiber are shown in Fig. 13-10. If the spot size is small and its NA is less than that of the fiber core, the optical power is concentrated in the center of the core, as Fig. 13-10a shows. In this case, the attenuation contribution arising from higher-order-mode power loss is negligible. In Fig. 13-10b the spot size is larger than the fiber core and the spot NA is larger than that of the fiber. For this overfilled condition, those parts of the incident light beam that fall outside of the fiber core and outside of the fiber NA are lost. In addition, there is a large contribution to the attenuation arising from higher-mode power loss (see Secs. 5.1 and 5.3).

Steady-state equilibrium-mode distributions are typically achieved by the *mandrel-wrap* method.[24] In this procedure, excess higher-order cladding modes that are launched by initially overexciting the fiber are filtered out by wrapping several turns of fiber around a mandrel, which is about 1.0–1.5 cm in diameter. In single-mode fibers, this type of mode filter is used to eliminate cladding modes from the fiber.

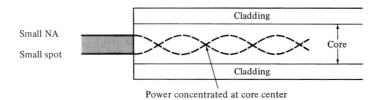

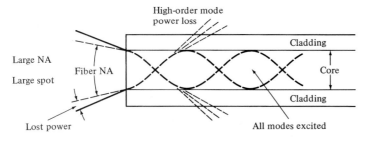

FIGURE 13-10

The effects of launch numerical aperture and spot size on the modal distribution. (*a*) Underfilling the fiber excites only lower-order modes; (*b*) an overfilled fiber has excess attenuation from higher-order mode loss.

13.3.2   Insertion-Loss Method

For cables with connectors, one cannot use the cutback method. In this case, one commonly uses an *insertion-loss technique*.[25] This is less accurate than the cutback method, but is intended for field measurements to give the total attenuation of a cable assembly in decibels.

The basic setup is shown in Fig. 13-11, where the launch and detector couplings are made through connectors. The wavelength-tunable light source is coupled to a short length of fiber that has the same basic characteristics as the fiber to be tested. For multimode fibers, a mode scrambler is used to ensure that the fiber core contains an equilibrium-mode distribution. In single-mode fibers, a cladding-mode stripper is employed so that only the fundamental mode is allowed to propagate along the fiber. A wavelength-selective device, such as an optical filter, is generally included to find the attenuation as a function of wavelength.

To carry out the attenuation tests, the connector of the short-length launching fiber is attached to the connector of the receiving system and the launch-power level $P_1(\lambda)$ is recorded. Next, the cable assembly to be tested is connected between the launching and receiving systems, and the received-power level $P_2(\lambda)$ is recorded. The attenuation of the cable in decibels is then

$$A = 10 \log \frac{P_1(\lambda)}{P_2(\lambda)} \tag{13-2}$$

This attenuation is the sum of the loss of the cabled fiber and the connector between the launch connector and the cable.

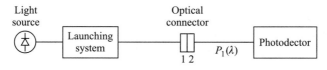

(a) Reference measurement

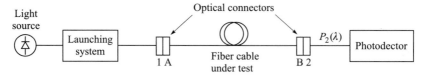

(b) Cable attenuation measurement

FIGURE 13-11
Test setup for measuring attenuation by the insertion-loss technique of cables where the launch and detector couplings are made through connectors.

13.4   DISPERSION MEASUREMENTS

Three basic forms of dispersion produce pulse broadening of light-wave signals in optical fibers, thereby limiting the information-carrying capacity. In multimode fibers, intermodal dispersion arises from the fact that each mode in an optical pulse travels a slightly different distance and thus arrives at the fiber end at slightly offset times. Chromatic dispersion stems from the variation in the propagation speed of the individual wavelength components of an optical signal. Polarization-mode dispersion arises from the splitting of a polarized signal into orthogonal polarization modes, each of which has a different propagation speed.

There are many ways to measure the various dispersion effects. Here, we look at some common methods.

13.4.1   Intermodal Dispersion

For practical purposes in evaluating intermodal dispersion, the fiber can be considered as a filter characterized by an impulse response $h(t)$ or by a power transfer function $H(f)$, which is the Fourier transform of the impulse response.[26,27] Either of these can be measured to determine the pulse dispersion. The impulse-response measurements are made in the time domain, whereas the power transfer function is measured in the frequency domain.

Both the time-domain and frequency-domain dispersion measurements assume that the fiber behaves quasi-linearly in power; that is, the individual overlapping output pulses from an optical waveguide can be treated as adding linearly. The behavior of such a system in the time domain is described simply as

$$p_{\text{out}}(t) = h(t) * p_{\text{in}}(t) = \int_{-T/2}^{T/2} p_{\text{in}}(t - \tau)h(\tau)\, d\tau \qquad (13\text{-}3)$$

That is, the output pulse response $p_{out}(t)$ of the fiber can be calculated through the convolution (denoted by $*$) of the input pulse $p_{in}(t)$ and the power impulse function $h(t)$ of the fiber. The period T between the input pulses should be taken to be wider than the expected time spread of the output pulses.

In the frequency domain, Eq. (13-3) can be expressed as the product (see App. E)

$$P_{out}(f) = H(f)P_{in}(f) \qquad (13\text{-}4)$$

Here, $H(f)$, the power transfer function of the fiber at the baseband frequency f, is the Fourier transform of $h(t)$:

$$H(f) = \int_{-\infty}^{\infty} h(t)\, e^{-j2\pi ft} dt \qquad (13\text{-}5)$$

and $P_{out}(f)$ and $P_{in}(f)$ are the Fourier transforms of the output and input pulse responses $p_{out}(t)$ and $p_{in}(t)$, respectively:

$$P(f) = \int_{-\infty}^{\infty} p(t)\, e^{-j2\pi ft} dt \qquad (13\text{-}6)$$

The transfer function of a fiber optic cable is of importance because it contains the bandwidth information of the system. For pulse dispersion to be negligible in digital systems, one of the following approximately equivalent conditions should be satisfied: (1) the fiber transfer function must not roll off to less than 0.5 of its low-frequency value for frequencies up to half the desired bit rate, or (2) the rms width of the fiber impulse response must be less than one-quarter of the pulse spacing.

13.4.2 Time-Domain Intermodal Dispersion Measurements

The simplest approach for making pulse-dispersion measurements in the time domain is to inject a narrow pulse of optical energy into one end of an optical fiber and detect the broadened output pulse at the other end.[28,29] Figure 13-12 illustrates a setup for this. Here, output pulses from a laser source are coupled through a mode scrambler into a test fiber. The output of the fiber is measured with a sampling oscilloscope that has a built-in optical receiver, or the signal can be detected with an external photodetector and then measured with a regular sampling oscilloscope. Next, the shape of the input pulse is measured the same way by replacing the test fiber with a short reference fiber that has a length less than 1 percent of the test fiber length. This reference fiber can be a short length cut from the test fiber or it can be a fiber segment that has similar properties. The variable delay in the trigger line is used to offset the difference in delay between the test fiber and the shorter reference fiber.

From the output pulse shape, an rms pulse width σ, as defined in Fig. 13-13, can be calculated by

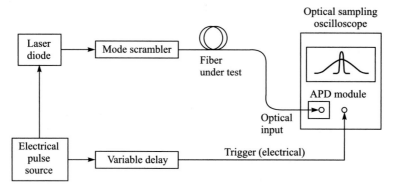

FIGURE 13-12
Test setup for making pulse-dispersion measurements in the time domain.

$$\sigma^2 = \frac{\displaystyle\int_{-\infty}^{\infty} (t - \bar{t})^2 p_{\text{out}}(t)\, dt}{\displaystyle\int_{-\infty}^{\infty} p_{\text{out}}(t)\, dt} \tag{13-7}$$

where the center time $\bar{t}$ of the pulse is determined from

$$\bar{t} = \frac{\displaystyle\int_{-\infty}^{\infty} t p_{\text{out}}(t)\, dt}{\displaystyle\int_{-\infty}^{\infty} p_{\text{out}}(t)\, dt} \tag{13-8}$$

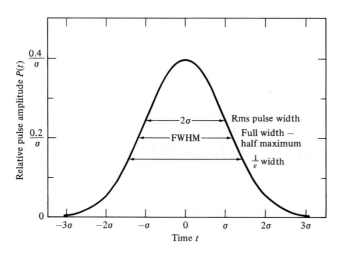

FIGURE 13-13
Definitions of pulse-shape parameters.

The evaluation of Eq. (13-8) requires a numerical integration. An easier method is to assume that the output response of a fiber can be approximated by a gaussian described by

$$p_{\text{out}}(t) = \frac{1}{\sigma\sqrt{2\pi}} = \exp\left(-\frac{t^2}{2\sigma^2}\right) \tag{13-9}$$

where the parameter σ determines the pulse width, as shown in Fig. 13-13. This figure also shows the parameter t_{FWHM}, which is the full width of the pulse at its half-maximum value. As denoted in Eq. (8-13), this is equal to $2\sigma(2\ln 2)^{1/2} = 2.355\sigma$. As described in Sec. 8.1, the optical bandwidth of the fiber can be defined through a Fourier transform. Normally, this is done in terms of the 3-dB bandwidth, which is the modulation frequency at which the optical power has fallen to one-half the value of the zero frequency modulation (dc value). From Eq. (8-14), this is

$$f_{\text{3-dB optical}} = \frac{0.440}{t_{\text{FWHM}}} = \frac{0.187}{\sigma}\,\text{Hz} \tag{13-10}$$

where "3-dB optical" means a 50-percent optical power reduction. Electrical bandwidths are related to optical bandwidths by $1/\sqrt{2}$, so that

$$f_{\text{3-dB electrical}} = \frac{1}{\sqrt{2}} f_{\text{3-dB optical}} = \frac{0.311}{t_{\text{FWHM}}} = \frac{0.133}{\sigma}\,\text{Hz} \tag{13-11}$$

13.4.3 Frequency-Domain Intermodal Dispersion Measurements

Frequency-domain intermodal dispersion measurements yield information on amplitude-versus-frequency response and phase-versus-frequency response.[28,30] These data are often more useful for system designers than time-domain pulse-dispersion measurements, especially if equalization techniques are to be performed on the detected signal at the receiver. The dispersion measurements can be made by sinusoidally modulating a narrowband continuous-wave (CW) light signal about a fixed level. The baseband frequency response is then found from the ratio of the sine wave amplitudes at the beginning and end of the fiber.

Figure 13-14 shows an experimental arrangement for finding fiber baseband frequency response. A swept-frequency RF source or a microwave signal source is used to modulate an optical carrier sinusoidally. The optical signal is coupled through a mode scrambler to the test fiber. At the exit end of the fiber, a photodetector measures $P_{\text{out}}(f)$, the output power as a function of the modulation frequency. The input signal is then measured by substituting a short reference fiber for the test fiber, thereby yielding $P_{\text{in}}(f)$.

Comparison of the spectrum at the fiber output with the spectrum at the fiber input provides the baseband frequency response $H(f)$ of the fiber under test:

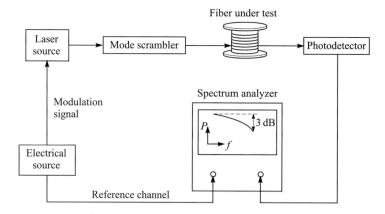

FIGURE 13-14
Test setup for finding baseband fiber frequency response.

$$H(f) = \frac{P_{\text{out}}(f)}{P_{\text{in}}(f)} \tag{13-12}$$

As the modulation frequency is increased, the optical power level at the fiber output will eventually start to decrease. The fiber bandwidth is defined as the lowest frequency at which $H(f)$ has been reduced to 0.5.

13.4.4 Chromatic Dispersion

Chromatic dispersion is the primary dispersive mechanism in single-mode fibers. Here, we present one method for its measurement.[31] Other techniques and measurements for multimode fibers are found in Refs. 32 through 36.

Figure 13-15 shows a setup for measuring chromatic dispersion by the *modulation phase-shift method.* An electric signal generator intensity modulates the output of a narrowband, tunable optical source by means of an external modulator. After detecting the transmitted signal with a photodiode receiver, a vector voltmeter is used to measure the phase of the modulation of the received signal relative to the electrical modulation source. The phase measurement is repeated at wavelength intervals $\Delta\lambda$ across the spectral band of interest. Using the measurements at any two adjacent wavelengths, the change in group delay (in ps) over the wavelength interval between them is[28]

$$\Delta\tau_\lambda = \frac{\phi_{\lambda+\Delta\lambda/2} - \phi_{\lambda-\Delta\lambda/2}}{360 f_m} \times 10^6 \tag{13-13}$$

where λ is the wavelength at the center of the interval, f_m is the modulation frequency in MHz, and ϕ is the phase of the measured modulation in degrees.

These data points are then plotted to yield the typical curve shown in Fig. 13-15. The dispersion can be calculated by applying the curve-fitting equations described in Chap. 3 to the pulse-delay data.

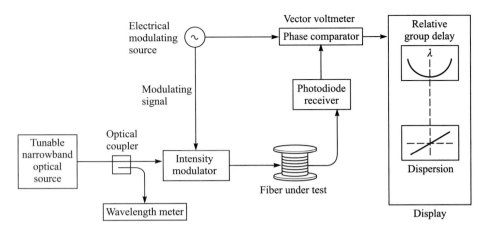

FIGURE 13-15
Test setup and display output for measuring chromatic dispersion by the phase-shift method.

13.4.5 Polarization-Mode Dispersion

As Sec. 3.2.6 describes (see Fig. 3-17), signal energy at a given wavelength occupies two orthogonal polarization modes. Since no fiber is perfectly round and materially symmetrical along its length, it has a varying birefringence along its length so that each polarization mode will travel at a slightly different group velocity and the polarization orientation will rotate with distance. The resulting difference in propagation times $\Delta\tau_{pol}$ between the two orthogonal polarization modes at a given wavelength will result in pulse spreading. This is the so-called *polarization-mode dispersion* (PMD).[37,38] Ultimately, PMD can limit the highest achievable data rates for single-mode systems and its measurement is thus of high importance (see Prob. 13-8).

An important point to note is that, in contrast to chromatic dispersion which is a relatively stable phenomenon along a fiber, PMD varies randomly along a fiber, because of the randomness of the underlying geometric and stress irregularities. Thus, statistical predictions are needed to account for its effects. A useful means of characterizing PMD is in terms of the mean or expected value of the differential group delay $\langle\Delta\tau_{pol}\rangle$ averaged over time. In contrast to the instantaneous value $\Delta\tau_{pol}$, which varies over time and type of source, the expected value does not change from day to day or from source to source. As noted in Sec. 3.2.6, mean values of the polarization-mode dispersion parameter D_{PMD} vary from 0.03 to 1.3 ps/$\sqrt{km}$, depending on the cable environment.

At least seven different methods have been developed for measuring PMD.[28] Here, we present only the *fixed-analyzer method*.[39,40] In this technique, the mean differential group delay is evaluated statistically from the number of peaks and valleys appearing in the optical power as it is transmitted through a polarizer and scanned as a function of wavelength. Figure 13-16 shows a simple setup using a spectrum analyzer. A typical spectrum analyzer trace showing the transmitted

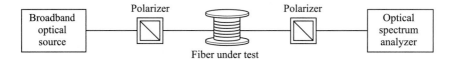

FIGURE 13-16
Setup for measuring polarization-mode dispersion using an optical spectrum analyzer.

power level as a function of wavelength is given in Fig. 13-17. Automatic methods using extrema counting and Fourier analysis are generally used to extract the PMD information from the measurement data. Using extrema counting, the expected value of the differential group delay of the fiber (or of any other device) under test can be calculated from the relationship

$$\langle \Delta \tau_{\text{pol}} \rangle_\lambda = \frac{k N_e \lambda_{\text{start}} \lambda_{\text{stop}}}{2(\lambda_{\text{start}} - \lambda_{\text{stop}})c} \tag{13-14}$$

where λ_{start} and λ_{stop} are the beginning and end, respectively, of the wavelength measurement sweep, N_e represents the number of extrema occurring in the scan, and c is the speed of light. The dimensionless mode-coupling factor k statistically accounts for the wavelength dependence of the polarization states. Its value is 0.84 for randomly mode-coupled fibers and 1.0 for non-mode-coupled fibers and devices.[39] The subscript λ on the $\langle \Delta \tau_{\text{pol}} \rangle$ terms means that the expected value of the differential group delay is determined over a wavelength span.

13.5 OTDR FIELD APPLICATIONS

Evolution of the OTDR followed from a publication in 1976 by Barnoski and Jensen that described backscattering in optical fibers and illustrated its potential for characterizing fibers.[41] The OTDR is now one of the fundamental instruments for making single-ended measurements of optical link characteristics such as fiber attenuation, connector and splice losses, reflectance levels from link components,

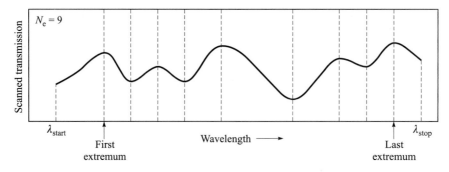

FIGURE 13-17
Typical OSA trace for PMD showing transmitted power level as a function of wavelength.

and chromatic dispersion.[42-45] The latter parameter requires a four-wavelength OTDR, whereas the other measurements can use a single-wavelength device.[46] In addition to measurement of these parameters, OTDRs are also used for link maintenance to locate fiber breaks quickly and accurately.

13.5.1 OTDR Trace

Figure 13-18 shows a typical trace as would be seen on the display screen of an OTDR. The scale of the vertical axis is logarithmic and measures the returning (back-reflected) signal in decibels. The horizontal axis denotes the distance between the instrument and the measurement point in the fiber. The backscattered waveform has four distinct features:

1. A large initial pulse resulting from Fresnel reflection at the input end of the fiber.
2. A long decaying tail resulting from Rayleigh scattering in the reverse direction as the input pulse travels along the fiber.
3. Abrupt shifts in the curve caused by optical loss at joints or connectors in the fiber line.
4. Positive spikes arising from Fresnel reflection at the far end of the fiber, at fiber joints, and at fiber imperfections.

Fresnel reflection and Rayleigh scattering principally produce the back-scattered light. *Fresnel reflection* occurs when light enters a medium that has a different index of refraction. For a glass-air interface, when light of power P_0 is incident perpendicular to the interface, the reflected power P_{ref} is

$$P_{ref} = P_0 \left(\frac{n_{fiber} - n_{air}}{n_{fiber} + n_{air}} \right)^2 \qquad (13\text{-}15)$$

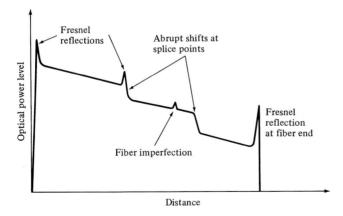

FIGURE 13-18
Representative trace of backscattered optical power as displayed on an OTDR screen.

where n_{fiber} and n_{air} are the refractive indices of the fiber core and air, respectively. A perfect fiber end reflects about 4 percent of the power incident on it. However, since fiber ends are generally not perfectly polished and perpendicular to the fiber axis, the reflected power tends to be much lower than the maximum possible value.

Two important performance parameters of an OTDR are dynamic range and measurement range.[47] *Dynamic range* is defined as the difference between the initial backscatter power level and the noise level after 3 minutes of measurement time. It is expressed in decibels of one-way fiber loss. Dynamic range provides information on the maximum fiber loss that can be measured and denotes the time required to measure a given fiber loss. Thus, it is often used to rank the capabilities of an OTDR. A basic limitation of an OTDR is the tradeoff between dynamic range and resolution. For high spatial resolution, the pulse width has to be as small as possible. However, this reduces the signal-to-noise ratio and thus lowers the dynamic range. For example, using an HP-8147 OTDR, a 100-ns pulse width allows a 24-dB dynamic range, whereas a 20-μs pulse width increases the dynamic range to 40 dB.[48]

Measurement range deals with the capability of identifying events in the link, such as splice points, connection points, or fiber breaks. It is defined as the maximum allowable attenuation between an OTDR and an event that still enables the OTDR to measure the event accurately. Normally, for definition purposes, a 0.5-dB splice is selected as the event to be measured.

13.5.2   Attenuation Measurements

Rayleigh scattering reflects light in all directions throughout the length of the fiber. This factor is the dominant loss mechanism in most high-quality fibers. The optical power that is Rayleigh-scattered in the reverse direction inside the fiber can be used to determine attenuation.

The optical power at a distance x from the input coupler can be written as

$$P(x) = P(0)\exp\left[-\int_0^x \beta(y)\,dy\right] \qquad (13\text{-}16)$$

Here, $P(0)$ is the fiber input power and $\beta(y)$ is the fiber loss coefficient in km^{-1}, which may be position-dependent; that is, the loss may not be uniform along the fiber. The parameter 2β can be measured in natural units called *nepers*, which are related to the loss $\alpha(y)$ in decibels per kilometer through the relationship (see App. D)

$$\beta(\text{km}^{-1}) = 2\beta\,(\text{nepers}) = \frac{\alpha(\text{dB})}{10\log e} = \frac{\alpha(\text{dB})}{4.343} \qquad (13\text{-}17)$$

Under the assumption that the scattering is the same at all points along the optical waveguide and is independent of the modal distribution, the power $P_R(x)$ scattered in the reverse direction at the point x is

$$P_R(x) = SP(x) \qquad (13\text{-}18)$$

Here, S is the fraction of the total power that is scattered in the backward direction and trapped in the fiber. Thus, the backscattered power from point x that is seen by the photodetector is

$$P_D(x) = P_R(x) \exp\left[-\int_0^x \beta_R(y) \, dy\right]$$ (13-19)

where $\beta_R(y)$ is the loss coefficient for the reverse-scattered light. Since the modes in the fiber excited by the backscattered light can be different from those launched in the forward direction, the parameter $\beta_R(y)$ may be different from $\beta(y)$.

Substituting Eqs. (13-16), (13-17), and (13-18) into Eq. (13-19) yields

$$P_D(x) = SP(0) \exp\left[-\frac{2\bar{\alpha}(x)x}{10 \log e}\right]$$ (13-20)

where the average attenuation coefficient $\bar{\alpha}(x)$ is defined as

$$\bar{\alpha}(x) = \frac{1}{2x} \int_0^x [\alpha(y) + \alpha_R(y)] \, dy$$ (13-21)

Using this equation, the average attenuation coefficient can be found from experimental semilog data plots such as the one shown in Fig. 13-18. For example, the average attenuation between two points x_1 and x_2, where $x_1 > x_2$, is

$$\bar{\alpha} = -\frac{10[\log P_D(x_2) - \log P_D(x_1)]}{2(x_2 - x_1)}$$ (13-22)

13.5.3 Fiber Fault Location

In addition to measuring attenuation and component losses, an OTDR can also be used to locate breaks and imperfections in an optical fiber. The fiber length L (and, hence, the position of the break or fault) can be calculated from the time difference between the pulses reflected from the front and far ends of the fiber. If this time difference is t, then the length L is given by

$$L = \frac{ct}{2n_1}$$ (13-23)

where n_1 is the core refractive index of the fiber. The factor "2" accounts for the fact that light travels a length L from the source to the break point and then another length L on the return trip.

Since the OTDR is based on using a pulsed probe signal, the *spatial resolution* or *sampling spacing* of where some event occurs in a fiber is limited by the pulse width of the source. Finer resolution is achievable with shorter pulse widths. The relationship between spatial resolution Δx and pulse width is given by

$$\Delta x = \frac{c}{2n} \Delta t_s$$ (13-24)

where Δt_s is the system response time, which is equal to the pulse width if the receiver has a sufficiently fast response. Normally, since it is not practical to increase the resolution by using a higher data sampling rate, an OTDR will use an interleaving scheme, which can improve spatial resolution down to the centimeter range. This is done through a composition of individual measurement shots that are delayed by a fraction of the sampling time. For example, launching a series of pulses delayed by a quarter sampling time yields a fourfold improvement in the sampling spacing.

As an example of the distance resolution possible with commercial test equipment, consider the HP-E6000A mini-OTDR from Hewlett-Packard.[48] The total distance accuracy D_{acc} is specified as

$$D_{acc} = \text{offset error} \pm (\text{scale error}) \times (\text{distance}) \pm \text{sampling error}$$
$$= \text{offset error} \pm (\text{scale error}) \times (\text{distance}) \pm 0.5 \times (\text{sampling spacing})$$

Thus, given that the offset error is ± 1 m, the scale error is $\pm 10^{-4}$ (i.e., the instrument has an 0.01 percent time-base accuracy) and the sampling spacing is 1 m (a 100-MHz sampling rate with no interleaving), then, for a break occurring after a 10-km fiber segment, the distance error is ± 2.5 m.

13.6   EYE PATTERNS

The eye-pattern technique is a simple but powerful measurement method for assessing the data-handling ability of a digital transmission system.[49,50] This method has been used extensively for evaluating the performance of wire systems and can also be applied to optical fiber data links. The eye-pattern measurements are made in the time domain and allow the effects of waveform distortion to be shown immediately on an oscilloscope.

Figure 13-19 shows a basic equipment setup for making eye-pattern measurements. The output from a pseudorandom data pattern generator is applied to the vertical input of an oscilloscope and the data rate is used to trigger the horizontal sweep. This results in the type of display shown in Fig. 13-20, which is called the *eye pattern*. To see how the display pattern is formed, consider the eight possible 3-bit-long NRZ combinations shown in Fig. 13-21. When these eight patterns are superimposed simultaneously, an eye pattern as shown in Fig. 13-20 is formed. The basic upper and lower bounds are determined by the logic one and zero levels, shown by b_{on} and b_{off}, respectively. Some key features of this pattern include the following:

- The opening (height) and width of the eye
- The 20-to-80-percent rise and fall times
- Overshoot on logic ones and zeros
- Undershoot on a logic zero
- Jitter in the eye pattern

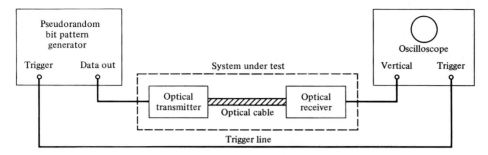

FIGURE 13-19
Basic equipment setup for making eye-diagram measurements.

   To measure system performance with the eye-pattern technique, a variety of word patterns should be provided. A convenient approach is to generate a random data signal, because this is the characteristic of data streams found in practice. This type of signal generates ones and zeros at a uniform rate but in a random manner. A variety of pseudorandom pattern generators are available for this purpose. The word *pseudorandom* means that the generated combination or sequence of ones and zeros will eventually repeat but that it is sufficiently

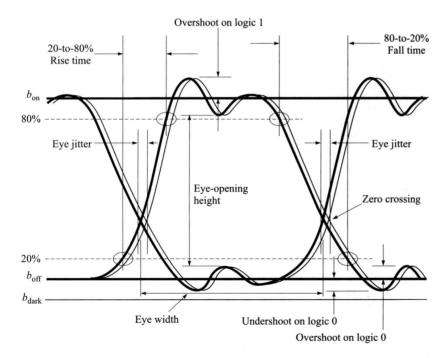

FIGURE 13-20
General configuration of an eye diagram showing definitions of fundamental measurement parameters.

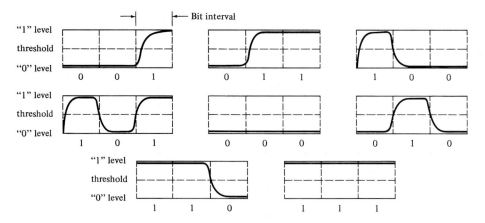

FIGURE 13-21
Eight possible 3-bit-long NRZ combinations of pulses that have moderate rise and fall times.

random for test purposes. A *pseudorandom binary sequence* (PRBS) comprises four different 2-bit-long combinations, eight different 3-bit-long combinations, sixteen different 4-bit-long combinations, and so on (i.e., sequences of 2^N different N-bit-long combinations), up to a limit set by the instrument. These combinations are randomly selected. The PRBS pattern length is of the form $2^N - 1$, where N is an integer. This choice assures that the pattern-repetition rate is not harmonically related to the data rate. Typical values of N are 7, 10, 15, 20, 23, and 31. After this limit has been reached, the data sequence will repeat.

A great deal of system-performance information can be deduced from the eye-pattern display. To interpret the eye pattern, consider Fig. 13-20 and the simplified drawing shown in Fig. 13-22. The following information regarding the signal amplitude distortion, timing jitter, and system rise time can be derived:

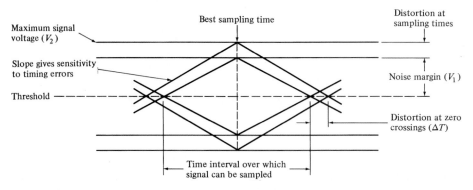

FIGURE 13-22
Simplified eye diagram showing key performance parameters.

- The width of the eye opening defines the time interval over which the received signal can be sampled without error from intersymbol interference.
- The best time to sample the received waveform is when the height of the eye opening is largest. This height is reduced as a result of amplitude distortion in the data signal. The vertical distance between the top of the eye opening and the maximum signal level gives the maximum distortion. The more the eye closes, the more difficult it is to distinguish between ones and zeros in the signal.
- The height of the eye opening at the specified sampling time shows the noise margin or immunity to noise. *Noise margin* is the percentage ratio of the peak signal voltage V_1 for an alternating bit sequence (defined by the height of the eye opening) to the maximum signal voltage V_2 as measured from the threshold level, as shown in Fig. 13-22. That is,

$$\text{Noise margin (percent)} = \frac{V_1}{V_2} \times 100 \text{ percent} \qquad (13\text{-}25)$$

- The rate at which the eye closes as the sampling time is varied (i.e., the slope of the eye-pattern sides) determines the sensitivity of the system to timing errors. The possibility of timing errors increases as the slope becomes more horizontal.
- *Timing jitter* (also referred to as *edge jitter* or *phase distortion*) in an optical fiber system arises from noise in the receiver and pulse distortion in the optical fiber. If the signal is sampled in the middle of the time interval (i.e., midway between the times when the signal crosses the threshold level), then the amount of distortion ΔT at the threshold level indicates the amount of jitter. Timing jitter is thus given by

$$\text{Timing jitter (percent)} = \frac{\Delta T}{T_b} \times 100 \text{ percent} \qquad (13\text{-}26)$$

where T_b is a bit interval.
- Traditionally, the *rise time* is defined as the time interval between the point where the rising edge of the signal reaches 10 percent of its final amplitude and the time it reaches 90 percent of its final amplitude. However, when measuring optical signals, these points are often obscured by noise and jitter effects. Thus, the more distinct values at the 20-percent and 80-percent threshold points are normally measured. To convert from the 20-to-80-percent rise time to a 10-to-90-percent rise time, one can use the approximate relationship

$$T_{10\text{-}90} = 1.25 \times T_{20\text{-}80} \qquad (13\text{-}27)$$

A similar approach is used to determine the fall time.
- Any nonlinearities of the channel transfer characteristics will create an asymmetry in the eye pattern. If a purely random data stream is passed through a purely linear system, all the eye openings will be identical and symmetrical.

13.7   OPTICAL SPECTRUM ANALYZER APPLICATIONS

The widespread implementation of WDM systems calls for making optical spectrum analyses to characterize the spectral behavior of various telecommunication network elements.[20,51] A variety of *optical spectrum analyzers* (OSAs) with different degrees of capabilities, such as wavelength resolution, are available to measure the optical output or transfer characteristics of a device as a function of wavelength. The *wavelength resolution* is determined by the bandwidth of the optical filter in the OSA. The term *resolution bandwidth* describes the width of this optical filter. Typical OSAs have selectable filters ranging from 10 nm down to 0.1 nm. The OSA normally sweeps across a spectral band making measurements at discretely spaced wavelength points. This spacing depends on the bandwidth-resolution capability of the instrument, and is known as the *trace-point spacing*. Here, we will look at spectral measurements of optical sources and EDFAs.

13.7.1   Characterization of Optical Sources

The three basic light sources used for optical fiber systems are the light-emitting diode (LED), the Fabry-Perot (FP) laser, and the distributed-feedback (DFB) laser. Each of these has a completely different output-versus-wavelength behavior. An OSA is a highly versatile instrument for quickly and accurately measuring the spectral output characteristics of these devices.

The LEDs emit light as a continuous spectrum over a broad wavelength band, with FWHM spectral patterns of 30–150 nm. Figure 13-23 shows a typical OSA trace for an LED spectrum centered at 1300 nm. Some parameters of interest that an OSA will automatically measure and display include the following:

- The *total power output*, which is the summation of the power P_i at each trace point i normalized by the ratio of the trace-point spacing to the resolution bandwidth. That is, if the instrument makes N measurements during the spectral sweep, then

$$P_{\text{total}} = \sum_{i=1}^{N} \left(P_i \frac{\text{trace-point spacing}}{\text{resolution bandwidth}} \right) \qquad (13\text{-}28)$$

- The *mean wavelength*, which represents the center of mass of the trace points. This parameter is given by

$$\lambda_{\text{mean}} = \sum_{i=1}^{N} \left(\frac{\lambda_i P_i}{P_{\text{total}}} \frac{\text{trace-point spacing}}{\text{resolution bandwidth}} \right) \qquad (13\text{-}29)$$

- The *peak wavelength*, which is where the peak of the LED spectrum occurs.
- The *full-width at half-maximum* (FWHM), which designates the half-power points; that is, the points where the power-spectral density is one-half that of

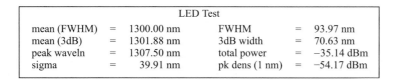

LED Test					
mean (FWHM)	=	1300.00 nm	FWHM	=	93.97 nm
mean (3dB)	=	1301.88 nm	3dB width	=	70.63 nm
peak waveln	=	1307.50 nm	total power	=	−35.14 dBm
sigma	=	39.91 nm	pk dens (1 nm)	=	−54.17 dBm

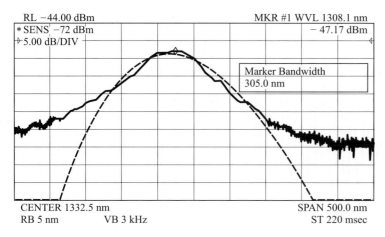

FIGURE 13-23

Spectrum of a light-emitting diode as recorded by an HP-71450 optical spectrum analyzer. The display shows measurement parameters such as spectral FWHM, mean-wavelength position, and peak-power density. A gaussian or lorentzian curve fit may also be displayed. (Courtesy of Hewlett-Packard Co.)

the peak amplitude. As denoted in Eq. (8-13), assuming a continuous, gaussian power distribution, this is given by

$$\text{FWHM} = 2.355\sigma \tag{13-30}$$

where σ is the rms spectral width of the LED, which is measured by the OSA as

$$\sigma^2 = \sum_{i=1}^{N} \left[(\lambda_i - \lambda_{\text{mean}})^2 \left(\frac{P_i}{P_{\text{total}}} \frac{\text{trace-point spacing}}{\text{resolution bandwidth}} \right) \right] \tag{13-31}$$

- The *3-dB spectral width* of the LED, which is defined as the separation of the two wavelengths on either side of the peak of the LED spectrum where the spectral density is one-half the peak-power spectral density.

Parameters for a Fabry-Perot laser that an OSA will automatically measure include the spectral FWHM or envelope bandwidth, center wavelength, mode spacing, and total power of the laser. Figure 13-24 shows a trace of a typical Fabry-Perot laser. The total power and the mean wavelength are calculated analogously to Eqs. (13-28) and (13-29), respectively, but without the normalization factor, since the FP laser does not have a continuous spectrum like the LED.

Distributed-feedback lasers are similar to FP lasers, except that here there is only one narrow spectral component. The automatic DFB-laser measurement

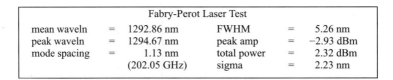

Fabry-Perot Laser Test			
mean waveln	= 1292.86 nm	FWHM	= 5.26 nm
peak waveln	= 1294.67 nm	peak amp	= −2.93 dBm
mode spacing	= 1.13 nm	total power	= 2.32 dBm
	(202.05 GHz)	sigma	= 2.23 nm

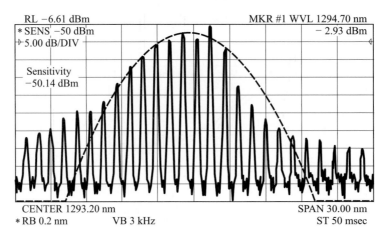

FIGURE 13-24

Spectrum of a Fabry-Perot laser as recorded by an HP-71450 optical spectrum analyzer. The display shows measurement parameters such as spectral FWHM, center wavelength, mode spacing, and total power of the laser. (Courtesy of Hewlett-Packard Co.)

function on an OSA provides values such as the center wavelength, side-mode suppression ratios, peak power, and stop-band characterization. The *side-mode suppression ratio* is the amplitude difference between the main spectral component and the largest side mode. The *stop band* is the wavelength spacing between the upper and lower side modes adjacent to the main mode. Figure 13-25 shows a trace of a typical DFB laser.

13.7.2   EDFA Gain and Noise-Figure Testing

The amplifier gain and noise figure are the two most important parameters of optical amplifiers when applying them in an optical communications link.[52–55] Amplifier gain can be measured with either an optical power meter, an electrical spectrum analyzer, or an optical spectrum analyzer. Noise-figure measurements can be carried out with either an electrical or an optical spectrum analyzer. Each has its strengths, limitations, and levels of measurement difficulty. Here, we shall look at using an OSA for measuring the amplifier gain and noise figure of an EDFA. Since these operational parameters are dependent on input power level and wavelength, it is necessary to measure the response of amplifier gain and noise figure in relation to these two factors.

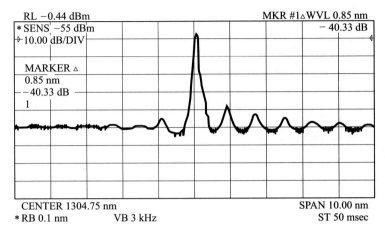

DFB Laser Test					
peak waveln	=	1304.80 nm	SMSR	=	40.33 dBc
mode offset	=	0.85 nm	peak amp	=	−0.32 dBm
stop band	=	1.83 nm	bandwidth	=	0.212 nm
cntr offset	=	0.06 nm			(al −20.70 dB)

RL −0.44 dBm   MKR #1△WVL 0.85 nm

*SENS −55 dBm   −40.33 dB

↦10.00 dB/DIV

MARKER △

0.85 nm

−−40.33 dB

1

CENTER 1304.75 nm   SPAN 10.00 nm
*RB 0.1 nm   VB 3 kHz   ST 50 msec

FIGURE 13-25
Spectrum of a DFB laser diode as recorded by an HP-71450 optical spectrum analyzer. The DFM laser measurements shown on the display include center wavelength, side-mode suppression ratios, peak power, and stop-band characterization. (Courtesy of Hewlett-Packard Co.)

GAIN MEASUREMENTS. Figure 13-26 illustrates the basic setup and the resultant OSA display for measuring optical amplifier gain. The setup contains a tunable laser, which has an adjustable output power level, and an OSA. First, the source is connected to the OSA without the EDFA inserted in the line in order to measure the output level from the source without amplification. This yields the bottom trace of the spectrum-versus-wavelength display shown in Fig. 13-26. Next, the EDFA is inserted to obtain the gain measurement. This gives the top trace of the display in Fig. 13-26. The difference in amplitudes is the amplifier gain G.

This technique can also be extended to the WDM case when the EDFA amplifies the signals from several different optical sources. However, since the cost and complexity of these WDM tests increase with the number of channels, a reduced-source approximation can be used for EDFA gain testing. In this method, a single source represents the ensemble of WDM signals in the spectral band of interest. The details of this are given in the literature.[56]

NOISE-FIGURE MEASUREMENTS. *Noise figure* is defined as the ratio of the signal-to-noise ratios at the input and output of the amplifier under the following conditions: shot-noise-limited photodetection, shot-noise-limited input signal, and an optical bandwidth approaching zero. A basic problem in making noise-figure measurements is that the light sources needed for generating a variable

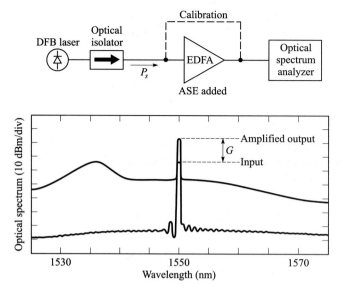

FIGURE 13-26
Basic setup for measuring optical amplifier gain and a typical measurement-display on an OSA.

input power and a variable wavelength range also emit a broad LED-like spectrum called *source spontaneous emission* (SSE). This SSE gets amplified along with the signal and is added to the EDFA output power. Thus, to measure only the signal and the ASE contribution from the EDFA, the SSE has to be eliminated from the measurement data.

The three basic methods for measuring optical amplifier noise are (1) optical-source-subtraction technique, (2) polarization nulling, and (3) time-domain extinction or pulse method.[54,57,58] For illustration purposes, we shall look only at the first technique. The test setup is illustrated in Fig. 13-27. In this *optical-source-subtraction method*, the SSE spectral density P_{SSE} of the laser source is determined during calibration of the test setup (with no optical amplifier in the link) and is stored in a calibration file in the OSA. Next, the optical amplifier is inserted in the test setup and the total noise spectral density P_{ASE} from the EDFA, which includes the SSE, is measured. In addition, the signal power P_{sig} entering the EDFA and the total amplifier output power P_{out}, which includes ASE and the

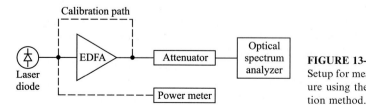

FIGURE 13-27
Setup for measuring EDFA noise figure using the optical-source-subtraction method.

amplified SSE, are measured. Knowing these values, the gain G and the quantum-limited noise figure NF can be calculated from the expressions

$$G = (P_{\text{out}} - P_{\text{ASE}})/P_{\text{sig}} \qquad (13\text{-}32)$$

and

$$NF = \frac{P_{\text{ASE}}}{G\ h\nu\ B_0} + \frac{1}{G} - \frac{P_{\text{SSE}}}{h\nu\ B_0} \qquad (13\text{-}33)$$

where ν is the frequency of the light at which the measurement was made and B_0 is the optical filter bandwidth at the receiver. The last term in Eq. (13-33) represents the subtraction of the amplified SSE.

PROBLEMS

13-1. Optical spectrum analyzers directly measure the wavelength of light in an air environment. However, most wavelength measurements are quoted in terms of wavelengths or optical frequencies in a vacuum. This can lead to errors, particularly in DWDM systems, since the index of refraction of air is a function of wavelength, temperature, pressure, and gas composition. The wavelength dependence of the index of refraction n_{air} of standard dry air at 760 torr and 15°C is[51]

$$n_{\text{air}} = 1 + 10^{-8}\left(8342.13 + \frac{2406030}{130 - \frac{1}{\lambda^2}} + \frac{15997}{38.9 - \frac{1}{\lambda^2}}\right)$$

where λ is measured in micrometers.

(a) Given that $\lambda_{\text{vacuum}} = \lambda_{\text{air}} n_{\text{air}}$, what is the error in wavelength measurement at 1550 nm if the effect of the index of refraction of air is ignored? What impact would this have on 0.8-nm-spaced WDM channels in the 1550-nm window?

(b) To compensate for temperature and pressure effects on the value of n_{air}, one can use the relationship

$$n(T, P) = 1 + \frac{(n_{\text{air}} - 1)(0.00138823P)}{1 + 0.003671T}$$

where P is measured in torr and T is in Celsius. How much does $n(T, P)$ vary from n_{air} when the pressure is 640 torr and the temperature is 0°C (which would be at a higher elevation and a lower temperature)?

13-2. An engineer wants to find the attenuation at 1310 nm of an 1895-m long fiber. The only available instrument is a photodetector, which gives an output reading in volts. Using this device in a cutback-attenuation setup, the engineer measures an output of 3.31 V from the photodiode at the far end of the fiber. After cutting the fiber 2 m from the source, the output voltage from the photodetector now reads 3.78 V. What is the attenuation of the fiber in dB/km?

13-3. Consider the cutback attenuation measurement technique described by Eq. (13-1). Using a photodetector, the power measurements are proportional to the detector output voltage. If the uncertainty in the voltage readings for the two power measurements are ±0.1 percent each, what is the uncertainty in the attenuation accuracy? How long must the fiber be to get a sensitivity better than ±0.05 dB/km?

13-4. (*a*) Verify that the full-width at half-maximum of a gaussian pulse is given by $t_{FWHM} = 2\sigma(2 \ln 2)^{1/2}$.

(*b*) Derive Eq. (13-10), which describes the 3-dB bandwidth of a fiber based on a gaussian output response.

13-5. A gaussian approximation of $|H(f)|$ in the form

$$|H(f)| \approx \exp[-(2\pi f \sigma)^2/2]$$

has been found to be accurate to at least the 0.75-amplitude point in the frequency domain. Using this relationship, plot $P(f)/P(0)$ as a function of frequency from 0 to 1000 MHz for fibers that have impulse responses of full rms pulse widths 2σ equal to 2.0, 1.0, and 0.5 ns. What are the 3-dB bandwidths of these fibers?

13-6. Consider the data shown in Fig. P13-6 of the group delay versus wavelength for a 10-km long fiber. From this data, plot the chromatic dispersion D as a function of wavelength. What is the value of the zero-dispersion slope S_0 in the relationship $D(\lambda) = S_0(\lambda - \lambda_0)$?

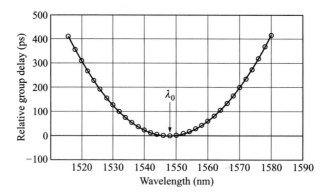

FIGURE P13-6
Chromatic-dispersion measurement of a 10-km fiber.

13-7. Determine the value of the expected differential group delay from the polarization-mode-dispersion measurement response shown in Fig. P13-7 for a non-mode-coupled fiber.

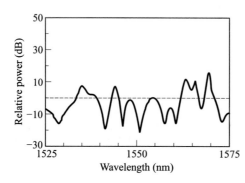

FIGURE P13-7
PMD-induced power excursions about the mean power level (dashed line).

13-8. The influence of polarization-mode dispersion can be neglected for data rates up to a few Gb/s for intensity-modulated direct-detection systems. However, for higher rates, PMD can cause intersymbol interference (ISI) in long-span links. The ISI power penalty in decibels for PMD is approximately[58]

$$P_{ISI} \approx 26 \frac{\langle \Delta\tau \rangle^2 \gamma(1-\gamma)}{T^2}$$

where T is a bit period (1/bit rate) and γ is the power-splitting ratio between principal polarization states. The maximum power penalty occurs when $\gamma = \frac{1}{2}$. If typical values of the expected value of the differential group delay $\langle \Delta\tau \rangle$ are 1 ps over a 100-km link and 10 ps over a 1000-km link, find the maximum PMD power penalty for data rates of 10 and 100 Gb/s over these two link distances.

13-9. The optical power in a fiber at a distance x from the input end is given by Eq. (13-16). By assuming that the loss coefficient is uniform along the fiber, use this equation to derive Eq. (13-1).

13-10. Assuming that Rayleight scattering is approximately isotropic (uniform in all directions), show that the fraction S of scattered light trapped in a multimode fiber in the backward direction is given by

$$S \approx \frac{\pi(NA)^2}{4\pi n^2} = \frac{1}{4} \left(\frac{NA}{n} \right)^2$$

where NA is the fiber numerical aperture, n is the core refractive index, and NA/n represents the half-angle of the cone of captured rays. If NA = 0.20 and $n = 1.50$, what fraction of the scattered light is recaptured by the fiber in the reverse direction?

13-11. Three 500-m-long fibers have been spliced together in series and an OTDR is used to measure the attenuation of the resultant fiber. The reduced data of the OTDR display is shown in Fig. P13-11. What are the attenuations in decibels per kilometer of the three individual fibers? What are the splice losses in decibels? What are some

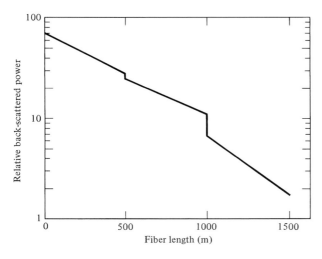

FIGURE P13-11
An OTDR trace of three 500-km spliced fibers.

possible reasons for the large splice loss occurring between the second and third fibers?

13-12. Let α be the attenuation of the forward-propagating light, α_s the attenuation of the backscattered light, and S the fraction of the total output power scattered in the backward direction, as described in Eq. (13-18). Show that the backscatter response of a rectangular pulse of width W from a point a distance L down the fiber is

$$P_s(L) = S\frac{\alpha_s}{\alpha} P_0\, e^{-2\alpha L}(1 - e^{-\alpha W})$$

when $L \geq W/2$, and

$$P_s(L) = S\frac{\alpha_s}{\alpha} P_0\, e^{-\alpha W}(1 - e^{-2\alpha L})$$

for $0 \leq L \leq W/2$.

13-13. Using the expression given in Prob. 13-12 for the backscattered power $P_s(L)$ from a rectangular pulse of width W, show that for very short pulse widths the backscattered power is proportional to the pulse duration. Note: This is the basis of operation of an OTDR.

13-14. The uncertainty U of OTDR loss measurements as a function of the signal-to-noise ratio SNR can be approximated by[42]

$$\log|U| = -0.2\text{SNR} + 0.6$$

Here, U and SNR are given in decibels. If a 0.5-dB splice is located near the far end of a 50-km fiber, what dynamic range must the OTDR have to measure the insertion loss of this splice event with a ± 0.5-dB accuracy. Assume the fiber attenuation is 0.33 dB/km.

13-15. Show that, when using an OTDR, an optical pulse width of 5 ns or less is required to locate a fiber fault to within ± 0.05 m of its true position.

REFERENCES

1. D. Marcuse, *Principles of Optical Fiber Measurements*, Academic, New York, 1981.
2. G. Cancellieri and U. Ravaioli, *Measurements of Optical Fibers and Devices*, Artech House, Dedham, MA, 1984.
3. M. L. Dakss, "Optical fiber measurements," in E. E. Basch, ed., *Optical Fiber Transmission*, H. W. Sams, Indianapolis, IN, 1987.
4. L. G. Cohen, P. Kaiser, P. L. Lazay, and H. M. Presby, "Fiber characterization," in S. E. Miller and A. G. Chynoweth, eds., *Optical Fiber Telecommunications*, Academic, New York, 1979.
5. D. L. Philen and W. T. Anderson, "Optical fiber transmission evaluation," in S. E. Miller and I. P. Kaminow, eds., *Optical Fiber Telecommunicationss—II*, Academic, New York, 1988.
6. Special Issue on "Fiber Measurements," *J. Lightwave Tech.*, vol. 7, Aug. 1989.
7. F. P. Kapron, "Fiber optic test methods," in F. C. Allard, ed., *Fiber Optics Handbook for Engineers and Scientists*, McGraw-Hill, New York, 1990.
8. Various TIA/EIA and ITU-T test standards and recommendations (see Refs. 10, 13–15).
9. National Institute of Standards and Technology (NIST), 325 Broadway, Boulder, CO 80303 (http://www.nist.gov).
10. *Technical Digest—Symposium on Optical Fiber Measurements*, annual conference, starting in 1980, sponsored by National Institute of Standards and Technology (NIST); Note: called National Bureau of Standards (NBS) until 1988.

11. National Physical Laboratory (NPL), Queens Road, Teddington, Middlesex TW11 0LW, UK (http://www.npl.co.uk).

12. Physikalisch-Technische Bundesanstalt (PTB), Braunschweig, Germany (http://www.ptb.de).

13. Telecommunication Industries Association (TIA) (http://www.tiaonline.org).

14. Electronic Industries Association (EIA), 2001 Eye Street, Washington, DC 20006.

15. Telecommunication Standardization Sector of the International Telecommunication Union (ITU-T), Place des Nations, CH-1211 Geneva 20, Switzerland (http://www.itu.ch).

16. International Electrotechnical Commission (IEC), 3 rue de Varambé, CH-1211 Geneva 20, Switzerland (http://www.iec.ch).

17. American National Standards Institute (ANSI), 1430 Broadway, New York, NY 10018 (http://www.ansi.org).

18. Institute of Electrical and Electronic Engineers (IEEE), 345 E. 45th Street, New York, NY 10017 (http://www.ieee.org).

19. A McGuire and P. A. Bonenfant, "Standards: The blueprints for optical networks," *IEEE Commun. Mag.*, vol. 36, pp. 68–78, Feb. 1998.

20. See vendor catalogs, such as Ando, Anritsu Wiltron, EXFO, Fotec Hewlett-Packard, ILX Lightwave, Photonetics, Tektronix, etc.

21. D. Derickson, ed., contributors from Hewlett-Packard Co., *Fiber Optic Test and Measurement*, Prentice Hall, Upper Saddle River, NJ, 1998.

22. TIA/EIA FOTP-46, *Spectral Attenuation Measurement for Long-Length, Graded-Index Optical Fibers*, 1990.

23. TIA/EIA FOTP-78, *Spectral-Attenuation Cutback Measurement for Single-Mode Optical Fibers*, May 1990.

24 (a) TIA/EIA FOTP-50, *Light-Launch Conditions of Long-Length Graded-Index Fiber Spectral Attenuation Measurements*, Sept. 1998.
 (b) TIA/EIA FOTP-77, *Procedures to Qualify Higher-Order Mode Filter for Measurements on Single-Mode Fiber*, 1991.

25. (a) ITU-T Recommendation G.650, *Definition and Test Methods for the Relevant Parameters of Single-Mode Fibers*, Mar. 1993.
 (b) ITU-T Recommendation G.651, *Characteristics of a 50/125-μm Multimode Graded-Index Optical Fiber Cable*, Mar. 1993.

26. S. D. Personick, "Baseband linearity and equalization in digital fiber optical communication systems," *Bell Sys. Tech. J.*, vol. 52, pp. 1175–1194, Sept. 1973.

27. S. D. Personick, "Receiver design for digital fiber optic communication systems," *Bell Sys. Tech. J.*, vol. 52, pp. 843–874, July/Aug. 1973.

28. P. Hernday, "Dispersion measurements," in D. Derickson, ed., *Fiber Optic Test and Measurement*, Prentice Hall, Upper Saddle River, NJ, 1998.

29. TIA/EIA FOTP-51, *Pulse Distortion Measurement of Multimode Glass Optical Fiber Information Transmission Capacity*, May 1991.

30. TIA/EIA FOTP-30, *Frequency Domain Measurement of Multimode Optical Fiber Information Transmission Capacity*, 1991.

31. TIA/EIA FOTP-169, *Chromatic Dispersion Measurement of Single-Mode Optical Fibers by the Phase-Shift Method*, Oct. 1992.

32. TIA/EIA FOTP-175, *Chromatic Dispersion Measurement of Single-Mode Optical Fibers by the Differential Phase-Shift Method*, 1992.

33. B. Christensen, J. Mark, G. Jacobsen, and E. Bodtker, "Simple dispersion measurement technique with high resolution," *Electron. Lett.*, vol. 29, pp. 132–134, 1993.

34. L. G. Cohen, "Comparison of single-mode fiber dispersion techniques," *J. Lightwave Tech.*, vol. LT-3, pp. 958–966, Oct. 1985.

35. TIA/EIA FOTP-168, *Chromatic Dispersion Measurement of Multimode Graded-Index and Single-Mode Optical Fibers by Spectral Group Delay Measurement in the Time Domain*, Mar. 1992.

36. M. J. Hackert, "Development of chromatic dispersion measurement on multimode fiber using the relative time of flight technique," *IEEE Photonics Tech. Lett.*, vol. 4, pp. 198–200, 1992.

37. A Galtarossa, G. Gianello, C. G. Someda, and M. Schiano, "In-field comparison among polarization-mode-dispersion measurement techniques," *J. Lightwave Tech.*, vol. 14, pp. 42–49, Jan. 1996.

38. (*a*) G. J. Foshini and C. D. Poole, "Statistical theory of polarization dispersion in single-mode fibers," *J. Lightwave Tech.*, vol. 9, pp. 1439–1456, Nov. 1991.
    (*b*) F. Bruyère, "Impact of first- and second-order PMD in optical digital transmission systems, *Opt. Fiber Technol.*, vol. 2, pp. 269–280, July 1996.

39. C. D. Poole and D. L. Favin, "Polarization-mode dispersion measurements based on transmission spectra through a polarizer," *J. Lightwave Tech.*, vol. 12, pp. 917–929, June 1994.

40. TIA/EIA FOTP-113, *Polarization Mode Dispersion Measurement for Single-Mode Optical Fibers by the Fixed-Analyzer Method*, 1997.

41. M. K. Barnoski and S. M. Jensen, "Fiber waveguides: A novel technique for investigating attenuation characteristics," *Appl. Opt.*, vol. 15, pp. 2112–2115, Sept. 1976.

42. J. Beller, "OTDRs and backscatter measurements," in D. Derickson, ed., *Fiber Optic Test and Measurement*, Prentice Hall, Upper Saddle River, NJ, 1998.

43. TIA/EIA FOTP-59, *Measurement of Fiber Point Defects Using an OTDR*, Nov. 1989.

44. TIA/EIA FOTP-60, *Measurement of Fiber or Cable Length Using an OTDR*, Nov. 1989.

45. TIA/EIA FOTP-61, *Measurement of Fiber or Cable Attenuation Using an OTDR*, Nov. 1989.

46. K. Nakajima, M. Ohashi, and M. Tateda, "Chromatic dispersion distribution measurement along a single-mode optical fiber," *J. Lightwave Tech.*, vol. 15, pp. 1095–1101, July 1997.

47. *Generic Requirements for OTDR Type Equipment*, Bellcore Document GR-196-CORE, Issue 1, 1995.

48. *Lightwave Test and Measurement Catalog*, Hewlett-Packard Co., 1997, pp. 146–149.

49. TIA/EIA-526-4, *Optical Eye Pattern Measurement Procedure*, 1995.

50. S. W. Hinch and C. M. Miller, "Analysis of digital modulation on optical carriers," in D. Derickson, ed., *Fiber Optic Test and Measurement*, Prentice Hall, Upper Saddle River, NJ, 1998.

51. J. Vobis and D. Derickson, "Optical spectrum analysis," in D. Derickson, ed., *Fiber Optic Test and Measurement*, Prentice Hall, Upper Saddle River, NJ, 1998.

52. ITU-T Recommendation G.611, *Definition and Test Methods for the Relevant Generic Parameters of Optical Fiber Amplifiers*, Oct. 1998.

53. E. Leckel, J. Sang, R. Müller, C. Rück, and C. Hentschel, "Erbium-doped fiber amplifier test system," *Hewlett-Packard J.*, vol. 46, pp. 13–19, Feb. 1995.

54. D. M. Baney, "Characterization of erbium-doped fiber amplifiers," in D. Derickson, ed., *Fiber Optic Test and Measurement*, Prentice Hall, Upper Saddle River, NJ, 1998.

55. D. Bonnedal, "Single-setup characterization of optical fiber amplifiers," *IEEE Photonics Tech. Lett.*, vol. 5, pp. 1193–1195, Oct. 1993.

56. D. M. Baney and J. Stimple, "WDM EDFA gain characterization with a reduced set of saturating channels," *IEEE Photonics Tech. Lett.*, vol. 8, pp. 1615–1617, Dec. 1996.

57. S. Poole, "Noise figure measurement in optical fiber amplifiers," *NIST Tech. Digest—Symp. on Optical Fiber Measurements*, Boulder, CO., NIST Special Publ. 864, pp. 1–6, 1994.

58. M. Movassaghi, M. K. Jackson, V. M. Smith, and W. J. Hallam, "Noise figure of EDFAs in saturated operation," *J. Lightwave Tech.*, vol. 16, pp. 812–817, May 1998.

APPENDIX
A

INTERNATIONAL SYSTEM OF UNITS

Quantity	Unit	Symbol	Dimensions
Length	meter	m	
Mass	kilogram	kg	
Time	second	s	
Temperature	kelvin	K	
Current	ampere	A	
Frequency	hertz	Hz	$1/s$
Force	newton	N	$(kg \cdot m)/s^2$
Pressure	pascal	Pa	N/m^2
Energy	joule	J	$N \cdot m$
Power	watt	W	J/s
Electric charge	coulomb	C	$A \cdot s$
Potential	volt	V	J/C
Conductance	siemens	S	A/V
Resistance	ohm	Ω	V/A
Capacitance	farad	F	C/V
Magnetic flux	weber	Wb	$V \cdot s$
Magnetic induction	tesla	T	Wb/m^2
Inductance	henry	H	Wb/A

APPENDIX
B

USEFUL MATHEMATICAL RELATIONS

Some of the mathematical relations encountered in this text are listed for convenient reference. More comprehensive listings are available in various handbooks.[1-4]

B.1 TRIGONOMETRIC IDENTITIES

$$e^{\pm j\theta} = \cos\theta \pm j\sin\theta$$

$$\sin^2\theta + \cos^2\theta = 1$$

$$\cos^2\theta - \sin^2\theta = \cos 2\theta$$

$$4\sin^3\theta = 3\sin\theta - \sin 3\theta$$

$$4\cos^3\theta = 3\cos\theta + \cos 3\theta$$

$$8\sin^4\theta = 3 - 4\cos 2\theta + \cos 4\theta$$

$$8\cos^4\theta = 3 + 4\cos 2\theta + \cos 4\theta$$

$$\sin(\alpha \pm \beta) = \sin\alpha\cos\beta \pm \cos\alpha\sin\beta$$

$$\cos(\alpha \pm \beta) = \cos\alpha\cos\beta \mp \sin\alpha\sin\beta$$

$$\tan(\alpha \pm \beta) = \frac{\tan\alpha \pm \tan\beta}{1 \mp \tan\alpha\tan\beta}$$

B.2 VECTOR ANALYSIS

The symbols $\mathbf{e}_x$, $\mathbf{e}_y$, and $\mathbf{e}_z$, denote unit vectors lying parallel to the x, y, and z axes, respectively, of the rectangular coordinate system. Similarly, $\mathbf{e}_r$, $\mathbf{e}_\phi$, and $\mathbf{e}_z$, are unit vectors for cylindrical coordinates. The unit vectors $\mathbf{e}_r$ and $\mathbf{e}_\phi$ vary in direction as the angle ϕ changes. The conversion from cylindrical to rectangular coordinates is made through the relationships

$$x = r\cos\phi \qquad y = r\sin\phi \qquad z = z$$

B.2.1 Rectangular Coordinates

$$\text{Gradient } \nabla f = \frac{\partial f}{\partial x}\mathbf{e}_x + \frac{\partial f}{\partial y}\mathbf{e}_y + \frac{\partial f}{\partial z}\mathbf{e}_z$$

$$\text{Divergence } \nabla\cdot\mathbf{A} = \frac{\partial A_x}{\partial x} + \frac{\partial A_y}{\partial y} + \frac{\partial A_z}{\partial z}$$

$$\text{Curl } \nabla \times \mathbf{A} = \begin{vmatrix} \mathbf{e}_x & \mathbf{e}_y & \mathbf{e}_z \\ \dfrac{\partial}{\partial x} & \dfrac{\partial}{\partial y} & \dfrac{\partial}{\partial z} \\ A_x & A_y & A_z \end{vmatrix}$$

$$\text{Laplacian } \nabla^2 f = \frac{\partial^2 f}{\partial x^2} + \frac{\partial^2 f}{\partial y^2} + \frac{\partial^2 f}{\partial z^2}$$

B.2.2 Cylindrical Coordinates

$$\text{Gradient } \nabla f = \frac{\partial f}{\partial r}\mathbf{e}_r + \frac{1}{r}\frac{\partial f}{\partial \phi}\mathbf{e}_\phi + \frac{\partial f}{\partial z}\mathbf{e}_z$$

$$\text{Divergence } \nabla\cdot\mathbf{A} = \frac{1}{r}\frac{\partial(rA_r)}{\partial r} + \frac{1}{r}\frac{\partial A_\phi}{\partial \phi} + \frac{\partial A_z}{\partial z}$$

$$\text{Curl } \nabla \times \mathbf{A} = \begin{vmatrix} \dfrac{1}{r}\mathbf{e}_r & \mathbf{e}_\phi & \dfrac{1}{r}\mathbf{e}_z \\ \dfrac{\partial}{\partial r} & \dfrac{\partial}{\partial \phi} & \dfrac{\partial}{\partial z} \\ A_r & rA_\phi & A_z \end{vmatrix}$$

$$\text{Laplacian } \nabla^2 f = \frac{1}{r}\frac{\partial}{\partial r} + \left(r\frac{\partial f}{\partial r}\right) + \frac{1}{r^2}\frac{\partial^2 f}{\partial \phi^2} + \frac{\partial^2 f}{\partial z^2}$$

B.2.3 Vector Identities

$$\nabla \times (\nabla \times \mathbf{A}) = \nabla(\nabla\cdot\mathbf{A}) - \nabla^2\mathbf{A}$$

$$\nabla^2\mathbf{A} = \nabla^2 A_x\mathbf{e}_x + \nabla^2 A_y\mathbf{e}_y + \nabla^2 A_z\mathbf{e}_z$$

B.3 INTEGRALS

$$\int \sin x \; dx = -\cos x$$

$$\int \cos x \; dx = \sin x$$

$$\int \sqrt{a^2 - x^2} \; dx = \tfrac{1}{2}\left(x\sqrt{a^2 - x^2} + a^2 \arcsin\frac{x}{a}\right)$$

$$\int x\sqrt{a^2 - x^2} \; dx = -\tfrac{1}{3}(a^2 - x^2)^{3/2}$$

$$\int x^2 \sin^2 x \; dx = \frac{x^3}{6} - \left(\frac{x^2}{6} - \frac{1}{8}\right)\sin 2x - \frac{x \cos 2x}{4}$$

$$\int \frac{dx}{\cos^n x} = \frac{1}{n-1}\frac{\sin x}{\cos^{n-1} x} + \frac{n-2}{n-1}\int \frac{dx}{\cos^{n-2} x}$$

$$\int u \; dv = uv - \int v \; du$$

$$\int e^{ax} dx = \frac{1}{a}e^{ax}$$

$$\int \sin^2 x \; dx = \frac{x}{2} - \frac{1}{4}\sin 2x$$

$$\int \sin^n x \; dx = -\frac{\sin^{n-1} x \cos x}{n} + \frac{n-1}{n}\int \sin^{n-2} x \; dx$$

$$\int \cos^2 x \; dx = \frac{x}{2} + \frac{1}{4}\sin 2x$$

$$\int \cos^n x \; dx = \frac{1}{n}\cos^{n-1} x \sin x + \frac{n-1}{n}\cos^{n-2} x \; dx$$

$$\int_{-\infty}^{\infty} \frac{e^{jpx}}{(\beta + jx)^n} dx = \begin{cases} 0 & \text{if } p < 0 \\ \dfrac{2\pi(p)^{n-1}e^{-\beta p}}{\Gamma(n)} & \text{if } p \geq 0 \end{cases} \qquad \text{where } \Gamma(n) = (n-1)!$$

$$\int_{-\infty}^{\infty} e^{-p^2 x^2 + qx} dx = e^{q^2/4p^2}\frac{\sqrt{\pi}}{p}$$

$$\int_{-\infty}^{\infty} \frac{1}{1 + (x/a)^2} dx = \frac{\pi a}{2}$$

$$\frac{2}{\sqrt{\pi}}\int_0^t e^{-x^2} dx = \text{erf}(t)$$

B.4 SERIES EXPANSIONS

$$(1+x)^n = 1 + nx + \frac{n(n-1)}{2!}x^2 + \frac{n(n-1)(n-2)}{3!}x^3 + \cdots \quad \text{for} \quad |nx| < 1$$

$$e^x = 1 + x + \frac{x^2}{2!} + \frac{x^3}{3!} + \cdots$$

$$\sin x = x - \frac{x^3}{3!} + \frac{x^5}{5!} - \cdots$$

$$\cos x = 1 - \frac{x^2}{2!} + \frac{x^4}{4!} - \cdots$$

REFERENCES

1. M. Kurtz, *Handbook of Applied Mathematics for Engineers and Scientists*, McGraw-Hill, New York, 1991.
2. D. Zwillinger, ed., *Standard Mathematical Tables and Formulae*, CRC Press, Boca Raton, FL, 30th ed., 1995.
3. M. Abramowitz and I. A. Stegun, *Handbook of Mathematical Functions*, Dover, New York, 10th ed., 1972.
4. I. S. Gradshteyn and I. M. Ryzhik, *Table of Integrals, Series, and Products*, Academic, New York, 5th ed., 1994.

APPENDIX
C

BESSEL
FUNCTIONS

This appendix lists the definitions and some recurrence relations for integer-order Bessel functions of the first kind $J_v(z)$ and modified Bessel functions $K_v(z)$. Detailed mathematical properties of these and other Bessel functions can be found in Refs. 24 through 26 of Chap. 2. Here, the parameter v is any integer and n is a positive integer or zero. The parameter $z = x + jy$.

C.1 BESSEL FUNCTIONS OF THE FIRST KIND

C.1.1 Various Definitions

A Bessel function of the first kind of order n and argument z, commonly denoted by $J_n(z)$, is defined by

$$J_n(z) = \frac{1}{2\pi} \int_{-\pi}^{\pi} e^{jz \, \sin\theta - jn\theta} d\theta$$

or, equivalently,

$$J_n(z) = \frac{1}{\pi} \int_0^{\pi} \cos(z \sin\theta - n\theta) d\theta$$

Just as the trigonometric functions can be expanded in power series, so can the Bessel function $J_v(z)$:

$$J_v(z) = \sum_{k=0}^{\infty} \frac{(-1)^k (\frac{1}{2}z)^{v+2k}}{k!(v+k)!}$$

In particular, for $v = 0$,

$$J_0(z) = 1 - \frac{\frac{1}{4}z^2}{(1!)^2} + \frac{(\frac{1}{4}z^2)^2}{(2!)^2} - \frac{(\frac{1}{4}z^2)^3}{(3!)^2} + \cdots$$

For $v = 1$,

$$J_1(z) = \frac{1}{2}z - \frac{(\frac{1}{2}z)^3}{2!} + \frac{(\frac{1}{2}z)^5}{2!3!} - \cdots$$

and so on for higher values of v.

C.1.2 Recurrence Relations

$$J_{v-1}(z) + J_{v+1}(z) = \frac{2v}{z} J_v(z)$$

$$J_{v-1}(z) - J_{v+1}(z) = 2J_v'(z)$$

$$J_v'(z) = J_{v-1}(z) - \frac{v}{z} J_v(z)$$

$$J_v'(z) = -J_{v+1}(z) + \frac{v}{z} J_v(z)$$

$$J_0'(z) = -J_1(z)$$

C.2 MODIFIED BESSEL FUNCTIONS

C.2.1 Integral Representations

$$K_0(z) = \frac{-1}{\pi} \int_0^{\pi} e^{\pm z \cos \theta} [\gamma + \ln(2z \sin^2 \theta)] d\theta$$

where Euler's constant $\gamma = 0.57722$.

$$K_v(z) = \frac{\pi^{1/2}(\frac{1}{2}z)^v}{\Gamma(v + \frac{1}{2})} \int_0^{\infty} e^{-z \cosh t} \sinh^{2v} t \, dt$$

$$K_0(x) = \int_0^{\infty} \cos(x \sinh t) \, dt = \int_0^{\infty} \frac{\cos(xt)}{\sqrt{t^2 + 1}} dt \qquad (x > 0)$$

$$K_v(x) = \sec(\tfrac{1}{2}v\pi) \int_0^{\infty} \cos(x \sinh t) \cosh(vt) \, dt \qquad (x > 0)$$

C.2.2   Recurrence Relations

If $L_\nu = e^{j\pi\nu} K_\nu$, then

$$L_{\nu-1}(z) - L_{\nu+1}(z) = \frac{2\nu}{z} L_\nu(z)$$

$$L_\nu'(z) = L_{\nu-1}(z) - \frac{\nu}{z} L_\nu(z)$$

$$L_{\nu-1}(z) + L_{\nu+1}(z) = 2L_\nu'(z)$$

$$L_\nu'(z) = L_{\nu+1}(z) + \frac{\nu}{z} L_\nu(z)$$

C.3   ASYMPTOTIC EXPANSIONS

For fixed ν $(\neq -1, -2, -3, \ldots)$ and $z \to 0$,

$$J_\nu(z) \simeq \frac{(\frac{1}{2}z)^\nu}{\Gamma(\nu+1)}$$

For fixed ν and $|z| \to \infty$,

$$J_\nu(z) \simeq \left(\frac{2}{\pi z}\right)^{1/2} \cos\left(z - \frac{\nu\pi}{2} - \frac{\pi}{4}\right)$$

For fixed ν and large $|z|$,

$$K_\nu(z) \simeq \left(\frac{\pi}{2z}\right)^{1/2} e^{-z}\left[1 - \frac{\mu-1}{8z} + \frac{(\mu-1)(\mu-9)}{2!(8z)^2} + \cdots\right]$$

where $\mu = 4\nu^2$.

C.4   GAMMA FUNCTION

$$\Gamma(z) = \int_0^\infty t^{z-1} e^{-t}\, dt$$

For integer n,

$$\Gamma(n+1) = n!$$

For fractional values,

$$\Gamma(\tfrac{1}{2}) = \pi^{1/2} = (-\tfrac{1}{2})! \simeq 1.77245$$
$$\Gamma(\tfrac{3}{2}) = \tfrac{1}{2}\pi^{1/2} = (\tfrac{1}{2})! \simeq 0.88623$$

D.1 DEFINITION

In designing and implementing an optical fiber link, it is of interest to establish, measure, and/or interrelate the signal levels at the transmitter, at the receiver, at the cable connection and splice points, and in the cable. A convenient method for this is to reference the signal level either to some absolute value or to a noise level. This is normally done in terms of a power ratio measured in *decibels* (dB) defined as

$$\text{Power ratio in dB} = 10\log\frac{P_2}{P_1} \qquad \text{(D-1)}$$

where P_1 and P_2 are electric or optical powers.

The logarithmic nature of the decibel allows a large ratio to be expressed in a fairly simple manner. Power levels differing by many orders of magnitude can be simply compared when they are in decibel form. Some very helpful figures to remember are given in Table D-1. For example, doubling the power means a 3-dB gain (the power level increases by 3 dB), halving the power means a 3-dB loss (the power level decreases by 3 dB), and power levels differing by factors of 10^N or 10^{-N} have decibel differences of $+10N$ dB and $-10N$ dB, respectively.

TABLE D-1
Examples of decibel measures of power ratios

Power ratio	10^N	10	2	1	0.5	0.1	10^{-N}
dB	$+10N$	$+10$	$+3$	0	-3	-10	$-10N$

D.2   THE dBm

The decibel is used to refer to ratios or relative units. For example, we can say that a certain optical fiber has a 6-dB loss (the power level gets reduced by 75 percent in going through the fiber) or that a particular connector has a 1-dB loss (the power level gets reduced by 20 percent at the connector). However, the decibel gives no indication of the absolute power level. One of the most common derived units for doing this in optical fiber communications is the *dBm*. This is the decibel power level referred to 1 mW. In this case, the power in dBm is an absolute value defined by

$$\text{Power level} = 10 \ \log \frac{P}{1 \ \text{mW}} \tag{D-2}$$

An important relationship to remember is that 0 dBm = 1 mW. Other examples are shown in Table D-2.

TABLE D-2
Examples of dBm units (decibel measure of power relative to 1 mW)

Power (mW)	100	10	2	1	0.5	0.1	0.01	0.001
Value (dBm)	+20	+10	+3	0	−3	−10	−20	−30

D.3   THE NEPER

The *neper* (N) is an alternative unit that is sometimes used instead of the decibel. If P_1 and P_2 are two power levels, with $P_2 > P_1$, then the power ratio in nepers is given as the natural (or naperian) logarithm of the power ratio:

$$\text{Power ratio in nepers} = \frac{1}{2} \ln \frac{P_2}{P_1} \tag{D-3}$$

where

$$\ln e = \ln 2.71828 = 1$$

To convert nepers to decibels, multiply the number of nepers by

$$20 \ \log e = 8.686$$

TOPICS
FROM
COMMUNICATION
THEORY

E.1 CORRELATION FUNCTIONS

A spectral density function is often used in communication theory to describe signals in the frequency domain. To define this, we first introduce the *autocorrelation function* $R_v(\tau)$, defined as

$$R_v(\tau) = \frac{1}{T_0} \int_{-T_0/2}^{T_0/2} v(t)v(t-\tau)dt \qquad \text{(E-1)}$$

where $v(t)$ is a periodic signal of period T_0. The autocorrelation function measures the dependence on the time τ of $v(t)$ and $v(t-\tau)$. An important property of $R_v(\tau)$ is that, if $v(t)$ is an energy or power signal, then

$$R_v(0) = \|v\|^2 = \frac{1}{T_0} \int_{-T_0/2}^{T_0/2} v^2(t)\, dt \qquad \text{(E-2)}$$

where $\|v\|^2$ represents the signal energy or power, respectively.

E.2 SPECTRAL DENSITY

Since $R_v(\tau)$ gives information about the time-domain behavior of $v(t)$, the spectral behavior of $v(t)$ in the frequency domain can be found from the Fourier transform of $R_v(\tau)$. We thus define the *spectral density $G_v(f)$* by

$$G_v(f) = F[R_v(\tau)] = \int_{-\infty}^{\infty} R_v(\tau)e^{-j2\pi f\tau} d\tau \qquad \text{(E-3)}$$

The fundamental property of $G_v(f)$ is that by integrating over all frequencies we obtain $\|v\|^2$:

$$\int_{-\infty}^{\infty} G_v(f)df = R_v(0) = \|v\|^2 \qquad \text{(E-4)}$$

The spectral density thus tells how energy or power is distributed in the frequency domain. When $v(t)$ is a current or voltage waveform feeding a $1 - \Omega$ load resistor, then $G_v(f)$ is measured in units of watts/hertz. For loads other than $1\,\Omega$, $G_v(f)$ is expressed either in volts^2/hertz or amperes^2/hertz. The symbol S is often used in electronics books to denote the spectral density.

Consider a signal $x(t)$ that is passed through a linear system having a transfer function $h(t)$. The output signal is given by

$$y(t) = h(t)x(t)$$

If $x(t)$ is modeled in the frequency domain by a spectral density $G_x(f)$ and the linear system has a transfer function $H(f)$, then the spectral density $G_y(f)$ of the output signal $y(t)$ is

$$G_y(f) = |H(f)|^2 G_x(f) \qquad \text{(E-5)}$$

E.3   NOISE-EQUIVALENT BANDWIDTH

A *white-noise* signal $n(t)$ is characterized by a spectral density that is flat or constant over all frequencies; that is,

$$G_n(f) = \frac{\eta}{2} = \text{constant} \qquad \text{(E-6)}$$

The factor $\frac{1}{2}$ indicates that half the power is associated with positive frequencies and half with negative frequencies.

If white noise is applied at the input of a linear system that has a transfer function $H(f)$, the spectral density $G_0(f)$ of the output noise is

$$G_0(f) = |H(f)|^2 G_n(f) = \frac{\eta}{2}|H(f)|^2 \qquad \text{(E-7)}$$

Thus, the output noise power N_0 is given by

$$N_0 = \int_{-\infty}^{\infty} G_0(f)\, df = \eta \int_{0}^{\infty} |H(f)|^2\, df \qquad \text{(E-8)}$$

If the same noise comes from an ideal low-pass filter of bandwidth B and amplitude $H(0)$—that is, the magnitude of an arbitrary filter transfer function at zero frequency—then,

$$N_0 = \eta|H(0)|^2 B \qquad \text{(E-9)}$$

By combining Eqs. (E-8) and (E-9), we define a *noise-equivalent bandwidth B* as

$$B = \frac{1}{|H(0)|^2} \int_0^\infty |H(f)|^2 \, df \tag{E-10}$$

E.4 CONVOLUTION

Convolution is an important mathematical operation used by communication engineers. The convolution of two real-valued functions $p(t)$ and $q(t)$ of the same variable is defined as

$$\begin{aligned} p(t) * q(t) &= \int_{-\infty}^\infty p(x)q(t-x) \, dx \\ &= \int_{-\infty}^\infty q(x)p(t-x) \, dx \\ &= q(t) * p(t) \end{aligned} \tag{E-11}$$

where the symbol $*$ denotes convolution. Note that convolution is commutative.

Two important properties of convolutions are

$$F[p(t) * q(t)] = P(f)Q(f) \tag{E-12}$$

that is, the convolution of two signals in the time domain corresponds to the multiplication of their Fourier transforms in the frequency domain, and

$$F[p(t)q(t)] = P(f) * Q(f) \tag{E-13}$$

that is, the multiplication of two functions in the time domain corresponds to their convolution in the frequency domain.

APPENDIX
F

FACTORS CONTRIBUTING TO DISPERSION

As noted in Sec. 2.4, the wave propagation constant β is a function of the wavelength, or, equivalently, the angular frequency ω. Since β is a slowly varying function of this angular frequency, one can see where various dispersion effects arise by expanding β in a Taylor series about a central frequency ω_0. Doing so to third order yields

$$\beta(\omega) \approx \beta_0(\omega_0) + \beta_1(\omega_0)(\omega - \omega_0) + \frac{1}{2}\beta_2(\omega_0)(\omega - \omega_0)^2 + \frac{1}{6}\beta_3(\omega_0)(\omega - \omega_0)^3 \quad \text{(F-1)}$$

where $\beta_m(\omega_0)$ denotes the mth derivative of β with respect to ω evaluated at $\omega = \omega_0$; that is,

$$\beta_m = \left(\frac{\partial^m \beta}{\partial \omega^m}\right)_{\omega=\omega_0} \quad \text{(F-2)}$$

The first term causes a phase shift, and the product $[\beta_{0x}(\omega_0) - \beta_{0y}(\omega_0)] \cdot z$ (i.e., z times the difference in the x and y components of β_0) describes the polarization evolution of the optical wave.

In the second term of Eq. (F-1), the factor $\beta_1(\omega_0)z$ produces a group delay $\tau_g = z/V_g$, where z is the distance traveled by the pulse and $V_g = 1/\beta_1$ is the group velocity [see Eqs. (3-13) and (3-14)]. Hence, the expression

$$\Delta\tau_{\text{pol}} = z|\beta_{1x} - \beta_{1y}| \quad \text{(F-3)}$$

is called the *polarization mode dispersion* (PMD) of the ideal uniform fiber [see Eq. (3-28)]. Note that in a real fiber the PMD varies statistically and is calculated according to Eq. (3-29).

In the third term of Eq. (F-1), the factor β_2 shows that the group velocity of a monochromatic wave depends on the wave frequency. This means that the different group velocities of the frequency components of a pulse cause it to broaden as it travels along a fiber. This spreading of the group velocities is known as *chromatic dispersion* or *group velocity dispersion* (GVD). The factor β_2 is thus known as the *GVD parameter* (see Sec. 3.2.2). As noted in Eq. (3-17), the dispersion D is related to β_2 through the expression

$$D = -\frac{2\pi c}{\lambda^2}\beta_2 \tag{F-4}$$

In the fourth term of Eq. (F-1), the factor β_3 is known as the *third-order dispersion*. This term is important around the wavelength at which β_2 equals zero. The third-order dispersion can be related to the dispersion D and the *dispersion slope* $S_0 = \partial D/\partial\lambda$ (the variation in the dispersion D with wavelength) by transforming the derivative with respect to ω into a derivative with respect to λ. Thus we have

$$\begin{aligned}
\beta_3 &= \frac{\partial\beta_2}{\partial\omega} = -\frac{\lambda^2}{2\pi c}\frac{\partial\beta_2}{\partial\lambda} = -\frac{\lambda^2}{2\pi c}\frac{\partial}{\partial\lambda}\left[-\frac{\lambda^2}{2\pi c}D\right]\\
&= \frac{\lambda^2}{(2\pi c)^2}(\lambda^2 S_0 + 2\lambda D)
\end{aligned} \tag{F-5}$$

INDEX

M

Y